Modern Birkhäuser Classics

Many of the original research and survey monographs in pure and applied mathematics published by Birkhäuser in recent decades have been groundbreaking and have come to be regarded as foundational to the subject. Through the MBC Series, a select number of these modern classics, entirely uncorrected, are being re-released in paperback (and as eBooks) to ensure that these treasures remain accessible to new generations of students, scholars, and researchers.

J. Donald Monk

Cardinal Invariants
on Boolean Algebras

Reprint of the 1996 Edition

Birkhäuser Verlag
Basel · Boston · Berlin

Author:

J. Donald Monk
Department of Mathematics, CB0395
University of Colorado
Boulder, CO 80309-0395
USA
e-mail: don.monk@colorado.edu

Originally published under the same title as volume 142 in the Progress in
Mathematics series by Birkhäuser Verlag, Switzerland, ISBN 978-3-7643-5402-2
© 1996 Birkhäuser Verlag, P.O. Box 133, CH-4010 Basel, Switzerland

1991 Mathematics Subject Classification 03E10, 03G05, 04A10, 06E05, 54A25

Library of Congress Control Number: 2009937810

Bibliographic information published by Die Deutsche Bibliothek
Die Deutsche Bibliothek lists this publication in the Deutsche Nationalbibliografie;
detailed bibliographic data is available in the Internet at <http://dnb.ddb.de>.

ISBN 978-3-0346-0333-1 Birkhäuser Verlag AG, Basel · Boston · Berlin

© 2010 Birkhäuser Verlag AG
Basel · Boston · Berlin
P.O. Box 133, CH-4010 Basel, Switzerland
Part of Springer Science+Business Media
Printed on acid-free paper produced of chlorine-free pulp. TCF ∞

ISBN 978-3-0346-0333-1   e-ISBN 978-3-0346-0334-8

9 8 7 6 5 4 3 2 1   www.birkhauser.ch

To Dorothy, An, and Steve

Foreword

This is a greatly revised and expanded version of the book **Cardinal functions on Boolean algebras**, Birkhäuser 1990. Known mistakes in that book have been corrected, and many of the problems stated there have solutions in the present treatment. At the same time, many new problems are formulated here; some as development of the solved problems from the earlier work, but most as a result of more careful study of the notions. The book is supposed to be self-contained, and for that reason many classical results are included.

For help on this book I wish to thank E. K. van Douwen, K. Grant, L. Heindorf, I. Juhász, S. Koppelberg, P. Koszmider, P. Nyikos, D. Peterson, M. Rubin, S. Shelah, and S. Todorčević. Unpublished results of some of these people are contained here, sometimes with proofs, with their permission. As the reader will see, my greatest debt is to Saharon Shelah, who has worked on, and solved, many of the problems stated in the 1990 book as well as in preliminary versions of this book.

Of course I am always eager to hear about solutions of problems, mistakes, etc. Electronic lists of errata and the status of the open problems are maintained, initially on the anonymous ftp server of euclid.colorado.edu, directory pub/babib; on www, go to ftp://euclid.colorado.edu/pub/babib.

J. Donald Monk
Boulder, Colorado
monkd@euclid.colorado.edu
July, 1995

Contents

Appendices

0. Introduction

This book is concerned with the theory of the most common functions k which assign to each infinite Boolean algebra A a cardinal number kA. Examples of such functions are the cardinality of the algebra A, and $\sup\{|X| : X$ is a family of pairwise disjoint elements of $A\}$. We have selected 21 such functions as the most important ones, and others are briefly treated. In Chapter 24 we list most of the additional functions mentioned in the book, as well as some new ones. For each function one can consider two very general questions: (1) How does the function behave with respect to algebraic operations, e.g., what is the value of k on a subalgebra of A in terms of its value on A? (2) What can one say about other cardinal functions naturally derived from a given one, e.g., what is $\sup\{kB : B$ is a homomorphic image of $A\}$? Another very general kind of question concerns the relationships between the various cardinal functions: some of them are always less or equal certain others. We shall shortly be more specific about what these three general questions amount to. The purpose of this book is to survey this area of the theory of BAs, giving proofs for a large number of results, some of which are new, mentioning most of the known results, and formulating open problems. Some of the open problems are somewhat vague ("Characterize..." or something like that), but frequently these are even more important than the specific problems we state; so we have opted to enumerate problems of both sorts in order to focus attention on them. But there are some natural questions which are not given formally as problems, since we have not thought much about them.

The framework that we shall set forth and then follow in investigating cardinal functions seems to us to be important for several reasons. First of all, the functions themselves seem intrinsically interesting. Many of the questions which naturally arise can be easily answered on the basis of our current knowledge of the structure of Boolean algebras, but some of these answers require rather deep arguments of set theory, algebra, or topology. This provides another interest in their study: as a natural source of applications of set-theoretical, algebraic, or topological methods. Some of the unresolved questions are rather obscure and uninteresting, but some of them have a general interest. Altogether, the study of cardinal functions seems to bring a unity and depth to many isolated investigations in the theory of BAs.

There are several surveys of cardinal functions on Boolean algebras, or, more generally, on topological spaces: See Arhangelskiĭ [78], Comfort [71], van Douwen [89], Hodel [84], Juhász [71], Juhász [80], Juhász [84], Monk [84], and Monk [90] (upon which this book is based). We shall not assume any acquaintance with any of these. On the other hand, we shall frequently refer to results proved in Part I of the Handbook of Boolean Algebras, Koppelberg [89a].

Definition of the cardinal functions considered.

Cellularity. A subset X of a BA A is called *disjoint* if its members are pairwise disjoint. The *cellularity* of A, denoted by cA, is

$\sup\{|X| : X$ is a disjoint subset of $A\}$.

Depth. Depth$A$ is

$\sup\{|X| : X$ is a subset of A well-ordered by the Boolean ordering$\}$.

Topological density. The *density* of a topological space X, denoted by dX, is the smallest cardinal κ such that X has a dense subspace of cardinality κ. The *topological density* of a BA A, also denoted by dA, is the density of its Stone space Ult$A$.

π-weight. A subset X of a BA A is *dense* in A if for all $a \in A^+$ there is an $x \in X^+$ such that $x \leq a$. The *π-weight* of a BA A, denoted by πA, is the smallest cardinal κ such that A has a dense subset of cardinality κ. This could also be called the *algebraic density* of A. (Recall that for any subset X of a BA, X^+ is the collection of nonzero elements of X.)

Length. Length$A$ is

$\sup\{|X| : X$ is a subset of A totally ordered by the Boolean ordering$\}$.

Irredundance. A subset X of a BA A is *irredundant* if for all $x \in X$, $x \notin \langle X\backslash\{x\}\rangle$. (Recall that $\langle Y \rangle$ is the subalgebra generated by Y.) The *irredundance* of A, denoted by Irr$A$, is

$\sup\{|X| : X$ is an irredundant subset of $A\}$.

Cardinality. This is just $|A|$.

Independence. A subset X of A is called *independent* if X is a set of free generators for $\langle X \rangle$. Then the *independence* of A, denoted by Ind$A$, is

$\sup\{|X| : X$ is an independent subset of $A\}$.

π-character. For any ultrafilter F on A, let $\pi\chi F = \min\{|X| : X$ is dense in $F\}$. Note here that it is not required that $X \subseteq F$. Then the *π-character* of A, denoted by $\pi\chi A$, is

$\sup\{\pi\chi F : F$ an ultrafilter of $A\}$.

Tightness. For any ultrafilter F on A, let t$F = \min\{\kappa :$ if Y is contained in Ult$A$ and F is contained in $\bigcup Y$, then there is a subset Z of Y of power at most κ such that F is contained in $\bigcup Z\}$. Then the *tightness* of A, denoted by tA, is

$\sup\{$t$F : F$ is an ultrafilter on $A\}$.

Spread. The *spread* of A, denoted by sA, is

$\sup\{|D| : D \subseteq$ Ult$A$, and D is discrete in the relative topology$\}$.

Character. The *character* of A, denoted by χA, is

$\min\{\kappa :$ every ultrafilter on A can be generated by at most κ elements$\}$.

Hereditary Lindelöf degree. For any topological space X, the *Lindelöf degree* of X is the smallest cardinal LX such that every open cover of X has a subcover with at most LX elements. Then the *hereditary Lindelöf degree* of A, denoted by hLA, is

> $\sup\{LX : X$ is a subspace of Ult$A\}$.

Hereditary density. The *hereditary density* of A, hd$A$, is

> $\sup\{dS : S$ is a subspace of Ult$A\}$.

Incomparability. A subset X of A is *incomparable* if for any two distinct elements $x, y \in X$ we have $x \nleq y$ and $y \nleq x$. The *incomparability* of A, denoted by Inc$A$, is

> $\sup\{|X| : X$ is an incomparable subset of $A\}$.

Hereditary cofinality. This cardinal function, h-cof$A$, is

> $\min\{\kappa :$ for all $X \subseteq A$ there is a $C \subseteq X$ with $|C| \leq \kappa$ and C cofinal in $X\}$.

Number of ultrafilters. Of course, this is the same as the cardinality of the Stone space of A, and is denoted by $|\mathrm{Ult}A|$.

Number of automorphisms. We denote by Aut$A$ the set of all automorphisms of A. So this cardinal function is $|\mathrm{Aut}A|$.

Number of endomorphisms. We denote by End$A$ the set of all endomorphisms of A, and hence this cardinal function is $|\mathrm{End}A|$.

Number of ideals of A. We denote by IdA the set of all ideals of A, so here we have the cardinal function $|\mathrm{Id}A|$.

Number of subalgebras of A. We denote by Sub$A$ the set of all subalgebras of A; $|\mathrm{Sub}A|$ is this cardinal function.

Some classifications of cardinal functions

Some theorems which we shall present, especially some involving unions or ultra-products, are true for several of our functions, with essentially the same proof. For this reason we introduce some rather ad hoc classifications of the functions. Some of the statements below are proved later in the book.

A cardinal function k is an *ordinary sup-function* with respect to P if P is a function assigning to every infinite BA A a subset PA of $\mathscr{P}A$ so that the following conditions hold for any infinite BA A:

(1) $kA = \sup\{|X| : X \in PA\}$;
(2) If B is a subalgebra of A, then $PB \subseteq PA$ and $X \cap B \in PB$ for any $X \in PA$.
(3) For each infinite cardinal κ there is a BA C of size κ such that there is an $X \in PC$ with $|X| = \kappa$.

Table 0.1 lists some ordinary sup-functions.

Given any ordinary sup-function k with respect to a function P and any infinite cardinal κ, we say that A satisfies the $\kappa - k - chain\ condition$ provided that $|X| < \kappa$ for all $X \in PA$.

Table 0.1	
Function	The subset PA
cA	$\{X : X$ is disjoint$\}$
Depth$A$	$\{X : X$ is well-ordered by the Boolean ordering of $A\}$
Length$A$	$\{X : X$ is totally ordered by the Boolean ordering of $A\}$
Irr$A$	$\{X : X$ is irredundant$\}$
Ind$A$	$\{X : X$ is independent$\}$
sA	$\{X : X$ is ideal-independent$\}$
Inc$A$	$\{X : X$ is incomparable$\}$

A cardinal function k is an *ultra-sup function with respect to P* if P is a function assigning to each infinite BA a subset PA of $\mathscr{P}A$ such that the following conditions hold:

(1) $kA = \sup\{|X| : X \in PA\}$.
(2) If $\langle A_i : i \in I \rangle$ is a sequence of BAs, F is an ultrafilter on I, and $X_i \in PA_i$ for all $i \in I$, then $\{f/F : fi \in X_i$ for all $i \in I\} \in P\left(\prod_{i \in I} A_i/F\right)$.

All of the above ordinary sup-functions except Depth are also ultra-sup functions.

    For the next classification, extend the first-order language for BAs by adding two unary relation symbols $\mathbf{F}$ and $\mathbf{P}$. Then we say that k is a *sup-min function* if there are sentences $\varphi(\mathbf{F}, \mathbf{P})$ and $\psi(\mathbf{F})$ in this extended language such that:

(1) For any BA A we have $kA = \sup\{\min\{|P| : (A, F, P) \models \varphi\} : A$ is infinite and $(A, F) \models \psi\}$.
(2) φ has the form $\forall x \in \mathbf{P}(x \neq 0 \wedge \varphi'(\mathbf{F})) \wedge \forall x_0 \ldots x_{n-1} \in \mathbf{F}\, \exists y \in \mathbf{P}\varphi''(\mathbf{F})$.
(3) $(A, F) \models \psi(\mathbf{F}) \to \exists x(x \neq 0 \wedge \varphi'(\mathbf{F}))$.

Some sup-min functions are listed in Table 0.2, where $\mu(\mathbf{F})$ is the formula saying that $\mathbf{F}$ is an ultrafilter.

Table 0.2		
Function	$\psi(\mathbf{F})$	$\varphi(\mathbf{F}, \mathbf{P})$
π	$\forall x \mathbf{F}x$	$\forall x \in \mathbf{P}(x \neq 0) \wedge \forall x \in \mathbf{F}\, \exists y \in \mathbf{P}(x \neq 0 \to y \leq x)$
$\pi\chi$	$\mu(\mathbf{F})$	$\forall x \in \mathbf{P}(x \neq 0) \wedge \forall x \in \mathbf{F}\, \exists y \in \mathbf{P}(y \leq x)$
χ	$\mu(\mathbf{F})$	$\forall x \in \mathbf{P}(x \neq 0 \wedge x \in \mathbf{F}) \wedge \forall x \in \mathbf{F}\, \exists y \in \mathbf{P}(y \leq x)$
h-cof	$\forall x \mathbf{F}x$	$\forall x \in \mathbf{P}(x \neq 0 \wedge x \in \mathbf{F}) \wedge \forall x \in \mathbf{F}\, \exists y \in \mathbf{P}(y \geq x)$

A cardinal function k is an *order-independence* function if there exists a sentence φ in the language of $(\omega, <, \omega, \omega)$ such that the following two conditions hold:

(1) For any infinite BA A we have $kA = \sup\{\lambda :$ there exists a sequence $\langle a_\alpha : \alpha < \lambda\rangle$ of elements of A such that for all finite $G, H \subseteq \lambda$ such that $(\lambda, <, G, H) \models \varphi$ we have $\prod_{\alpha \in G} a_\alpha \cdot \prod_{\alpha \in H} -a_\alpha \neq 0\}$.

(2) If λ is an infinite cardinal, $(\lambda, <, G, H) \models \varphi$, $G', H' \subseteq \lambda$, and f is a one-to-one function from $G \cup H$ onto $G' \cup H'$ such that for all $\alpha, \beta \in G \cup H$, if $\alpha < \beta$ then $f\alpha < f\beta$, then $(\lambda, <, G', H') \models \varphi$.

Some order-independence functions are listed in Table 0.3.

Table 0.3	
Function	φ
t	$\forall x \in \mathbf{G}\, \forall y \in \mathbf{H}\, (x < y)$
hd	$\exists x \in \mathbf{G}\, \forall y \in \mathbf{G}\, (x = y) \land \forall x \in \mathbf{G}\, \forall y \in \mathbf{H}\, (x < y)$
hL	$\exists x \in \mathbf{H}\, \forall y \in \mathbf{H}\, (x = y) \land \forall x \in \mathbf{G}\, \forall y \in \mathbf{H}\, (x < y)$

Algebraic properties of a single function.

Now we go into more detail on the properties of a single function which we shall investigate. From the point of view of general algebra, the main questions are: what happens to the cardinal function k under the passage to subalgebras, homomorphic images, products, and free products? There are natural problems too about more special operations on algebras in general, or on Boolean algebras in particular: what happens to k under weak products, amalgamated free products, unions of well-ordered chains of subalgebras, ultraproducts, dense subalgebras, subdirect products, global sections of sheaves, Boolean products, Boolean powers, set products, one-point gluing, Aleksandroff duplication, and the exponential? The mentioned operations which are not discussed in the Handbook will be explained in Chapter 1. Several of these operations are discussed for many of our functions, but only in the case of cellularity do we discuss all of them somewhat thoroughly.

One may notice that several of the above functions, such as depth and spread, are defined as supremums of the cardinalities of sets satisfying some property P. So, a natural question is whether such sups are *attained*, that is, with depth as an example, whether for every BA A there always is a subset X well-ordered by the Boolean ordering, with $|X| = \text{Depth}A$. Of course, this is only a question in case $\text{Depth}A$ is a limit cardinal. For such functions k defined by sups, we can define a closely related function k'; $k'A$ is the least cardinal such that there is no subset of A with the property P. So $k'A = (kA)^+$ if k is attained, and $k'A = kA$ otherwise.

Derived operations.

From a given cardinal function one can define several others; part of our work is to see what these new cardinal functions look like; frequently it turns out that they coincide with another of our basic 21 functions, but sometimes we arrive at a new function in this way:

$$k_{H+}A = \sup\{kB : B \text{ is a homomorphic image of } A\}.$$
$$k_{H-}A = \inf\{kB : B \text{ is an infinite homomorphic image of } A\}.$$
$$k_{S+}A = \sup\{kB : B \text{ is a subalgebra of } A\}.$$
$$k_{S-}A = \inf\{kB : B \text{ is an infinite subalgebra of } A\}.$$
$$k_{h+}A = \sup\{kY : Y \text{ is a subspace of Ult} A\}.$$
$$k_{h-}A = \inf\{kY : Y \text{ is an infinite subspace of Ult} A\}.$$
$$_d k_{S+}A = \sup\{kB : B \text{ is a dense subalgebra of } A\}.$$
$$_d k_{S-}A = \inf\{kB : B \text{ is a dense subalgebra of } A\}.$$

Note that $k_{h+}A$ and $k_{h-}A$ make sense only if k is a function which naturally applies to topological spaces in general as well as BAs. Any infinite Boolean space has a denumerable discrete subspace, and frequently k_{h-} will take its value on such a subspace.

Given a function defined in terms of ultrafilters, like character above, there is usually an associated function l assigning a cardinal number to each ultrafilter on A. Then one can introduce two cardinal functions on A itself:

$$l_{\sup}A = \sup\{lF : F \text{ is an ultrafilter on } A\}.$$
$$l_{\inf}A = \inf\{lF : F \text{ is a non-principal ultrafilter on } A\}.$$

Another kind of derived function applies to cases where the function is defined as the sup of cardinalities of sets X with a property P, where P is such that maximal families with the property P exist (usually seen by Zorn's lemma). For such a function k, we define

$$k_{mm}A = \min\{|X| : X \text{ is an infinite maximal family satisfying } P\}.$$

The derived functions so far mentioned are really cardinal functions. We also consider the following two spectrum functions, which assign to each BA a set of cardinal numbers:

$$k_{Hs}A = \{kB : B \text{ is an infinite homomorphic image of } A\}$$
$$\text{(the } homomorphic \ spectrum \text{ of } A)$$
$$k_{Ss}A = \{kB : B \text{ is an infinite subalgebra of } A\}$$
$$\text{(the } subalgebra \ spectrum \text{ of } A)$$

It is also possible to define a *caliber* notion for many of our functions, in analogy to the well-known caliber notion for cellularity. Given a property P associated with a cardinal function, a BA A is said to have κ, λ, P-*caliber* if among any set of λ elements of A there are κ elements with property P. It is hard to be very precise about this notion in general, but it has been extensively studied for cellularity, and studied somewhat for independence. The property P is not necessarily one used to define the function; thus for cellularity P is the finite intersection property, while for independence it is, indeed, independence.

Comparing two functions

Given two cardinal functions k and l, one can try to determine whether $kA \leq lA$ for every BA A or $lA \leq kA$ for every BA A. Given that one of these cases arises, it is natural to consider whether the difference can be arbitrarily large (as with cellularity and spread, for example), or if it is subject to restrictions (as with depth and length). If no general relationship is known, a counterexample is needed, and again one can try to find a counterexample with an arbitrarily large difference between the two functions. Of course, the known inequalities between our functions help in order to limit the number of cases that need to be considered for constructing such counterexamples; here the diagrams in Chapter 25 are sometimes useful. For example, knowing that $\pi\chi$ can be greater than c, we also know that χ can be greater than c.

Other considerations

In addition to the above systematic goals in discussing cardinal functions, there are some more ideas which we shall not explore in such detail. One can compare several cardinal functions, instead of just two at a time. Several deep theorems of this sort are known, and we shall mention a few of them. There are also a large number of relationships between cardinal functions which involve cardinal arithmetic; for example, Length$A \leq 2^{\mathrm{Depth}A}$ for any BA A. We mention a few of these as we go along.

One can compare two cardinal functions while considering algebraic operations; for example, comparing functions k, l with respect to the formation of subalgebras. We shall investigate just two of the many possibilities here:

$k_{\mathrm{Sr}}A = \{(\kappa, \lambda)$: there is an infinite subalgebra B of A such that $|B| = \lambda$ and $kB = \kappa\}$.

$k_{\mathrm{Hr}}A = \{(\kappa, \lambda)$: there is an infinite homomorphic image B of A such that $|B| = \lambda$ and $kB = \kappa\}$.

These are called, respectively, the *subalgebra k relation* and the *homomorphic k relation*. For each function k, it would be nice to be able to characterize the possible relations k_{Sr} and k_{Hr} in purely cardinal number terms.

Another general idea applies to several functions that are defined somehow in terms of finite sets; the idea is to take bounded versions of them. For example, independence has bounded versions: for any positive integer n, a subset X of a BA

A is called *n-independent* if for every subset Y of X with at most n elements and every $\varepsilon \in {}^Y 2$ we have $\prod_{y \in Y} y^{\varepsilon y} \neq 0$. (Here $x^1 = x$, $x^0 = -x$ for any x.) And then we define $\operatorname{Ind}_n A = \sup\{|X| : X \text{ is } n\text{-independent}\}$. It is interesting to investigate this notion and its relationship to actual independence; and similar things can be done for various other functions.

Special classes of Boolean algebras

We are interested in all of the above ideas not only for the class of all BAs, but also for various important subclasses: complete BAs, interval algebras, tree algebras, and superatomic algebras, which are discussed in the Handbook. To a lesser extent we give facts about cardinal functions for other subclasses like all atomic BAs, atomless BAs, initial chain algebras, minimally generated algebras, pseudo-tree algebras, semigroup algebras, and tail algebras. In Chapter 2 we describe some properties of the special classes mentioned which are not discussed in the Handbook, partly to establish notation.

1. Special operations on Boolean algebras

We give the basic definitions and facts about several operations on Boolean algebras which were not discussed in the Handbook. We first make some additional comments on the Boolean algebra of global sections of sheaves.

Sheaves

If $\langle A_x : x \in X \rangle$ is a system of Boolean algebras, B is a subalgebra of $\prod_{x \in X} A_x$, and a topology is given on X, then we say that B, X has the *patchwork property* if the following holds:

For any two $f, g \in B$ and any clopen subset N of X, the function $(f \restriction N) \cup (g \restriction (X \backslash N))$ is in B.

Next, if $\mathscr{S} = \langle S, \pi, X, \langle B_p \rangle_{p \in X} \rangle$ is a sheaf of Boolean algebras, we denote by $\mathrm{Gs}\mathscr{S}$ the BA of global sections of $\mathscr{S}$.

Theorem 1.1. *If $\mathscr{S} = \langle S, \pi, X, \langle B_p \rangle_{p \in X} \rangle$ is a sheaf of Boolean algebras, then $\mathrm{Gs}\mathscr{S}, X$ has the patchwork property. Moreover $\mathrm{Clop}X$ is isomorphic to a subalgebra of $\mathrm{Gs}\mathscr{S}$.*

Proof. Assume the hypotheses of the definition of patchwork property for $\mathrm{Gs}\mathscr{S}, X$, and let $h = (f \restriction N) \cup (g \restriction (X \backslash N))$. Then for any open subset U of S we have $h^{-1}[U] = (N \cap f^{-1}[U]) \cup (g^{-1}[U] \backslash N)$, proving that $h^{-1}[U]$ is open. Thus $h \in \mathrm{Gs}\mathscr{S}$, proving the patchwork property. The second assertion of the theorem follows from the first part using characteristic functions. $\qquad\square$

Boolean products

For the notion of *Boolean products* see Burris, Sankappanavar [81], whose notation we follow. We recall the definition. A *Boolean product* of a system $\langle A_x : x \in X \rangle$ of BAs is a subdirect product B of $\langle A_x : x \in X \rangle$ such that X can be endowed with a Boolean topology so that the following conditions hold:

(1) For any two $f, g \in B$ the set $[\![f = g]\!] \overset{\mathrm{def}}{=} \{ x \in X : fx = gx \}$ is clopen in X.
(2) B, X has the patchwork property.

Theorem 1.2. *Up to isomorphism, Boolean products coincide with the global section algebras in which the index space is Boolean and the sheaf space is Hausdorff.*

Proof. First suppose that we are given a sheaf $\mathscr{S} = \langle S, \pi, X, \langle A_x \rangle_{x \in X} \rangle$ with X Boolean and S Hausdorff. Let B be the algebra of global sections of $\mathscr{S}$. By Theorems 8.13 and 8.15 of the Handbook, part I, B is a subdirect product of $\langle A_x : x \in X \rangle$ and (1) holds; by Theorem 1.1, B, X has the patchwork property.

The other direction takes more work. Let a Boolean product be given as in the definition above. Without loss of generality we may assume that the sets A_x are pairwise disjoint, and we let $S = \bigcup_{x \in X} A_x$. It is "merely" a matter of putting a Hausdorff topology on S so that we get a sheaf such that B coincides with the BA of global sections of the sheaf. For each $b \in B$ define $f_b : X \to S$ by $f_b x = b_x$. As a base for the desired topology we take

$$\{f_b[U] : U \text{ is clopen in } X, b \in B\}.$$

First we show:

(1) The above set is a base for a topology on S.

To show this, suppose that $a \in f_{b1}[U_1] \cap f_{b2}[U_2]$. Say $a \in A_x$ for a certain $x \in X$. Thus there exist $x_i \in U_i$ such that $a = f_{bi} x_i$ for $i = 1, 2$. Thus $a = (bi)_{x_i}$, and so $x_1 = x_2$. Let

$$V = U_1 \cap U_2 \cap [\![b1 = b2]\!].$$

Then clearly

$$a \in f_{b1}[V] \subseteq f_{b1}[U_1] \cap f_{b2}[U_2],$$

as desired. And for any $s \in S$, say $s \in A_x$, choose $b \in B$ with $b_x = s$. Then $s \in f_b[X]$. This proves (1).

Next, let π be as in Definition 8.14 of Part I of the BA handbook. To show that π is continuous, it suffices to note that for U open in X we have

$$\pi^{-1}[U] = \bigcup_{b \in B} f_b[U].$$

The following statements, easily verified, show that π is an open mapping:

(2) $\pi f_b p = p$.
(3) $pi[f_b[U]] = U$.

That π is a local homeomorphism also follows easily from (2): Given $s \in S$, say $s \in A_p$ and $b_p = s$, $b \in B$. Then $\pi \upharpoonright f_b[X]$ is one-one by (2), and $f_b[X]$ is a neighborhood of s.

We still need to check the dreaded condition 8.14(d'). We first note that the following simplified form of it implies 8.14(d') itself in an easy manner:

(4) Let $U \subseteq X$ be open, $f_1, \ldots, f_n$ sections over U, and $0 \le i \le n$. Then the set

$$\{p \in U : f_1 p \cdot \ldots \cdot f_i p \cdot -f_{i+1} p \cdot \ldots \cdot -f_n p = 0\}$$

is open.

To prove (4), note that

$$\{p \in U : f_1 p \cdot \ldots \cdot f_i p \cdot - f_{i+1} p \cdot \ldots \cdot - f_n p = 0\}$$

$$= U \cap \bigcup_{b1,\ldots,bn \in B} \{p \in X : f_i p = (bi)_p \ (i = 1, \ldots, n) \text{ and}$$

$$(b1)_p \cdot \ldots \cdot (bi)_p \cdot -(b(i+1))_p \cdot \ldots \cdot -(bn)_p = 0\}$$

$$= U \cap \bigcup_{b1,\ldots,bn \in B} \left(\bigcap_{1 \le i \le n} \{p \in X : f_i p = (bi)_p\} \cap \right.$$

$$\left. \{p \in X : [(b1) \cdot \ldots \cdot (bi) \cdot -(b(i+1)) \cdot \ldots \cdot -(bn)]_p = 0\} \right);$$

hence it suffices to show that each set $\{p \in X : f_i p = (bi)_p\}$ is open; but this is clear, since this set is $f_i^{-1}[f_{bi}[X]]$.

Thus we have a sheaf. Now we need to show that B is exactly the set of all global sections with respect to this sheaf. First take any $b \in B$. To show that b is continuous, also take a typical member $f_c[U]$ of the base for the topology on S. Then

$$b^{-1}[f_c[U]] = \{p \in X : b_p = f_c q \text{ for some } q \in U\}$$
$$= \{p \in X : b_p = c_q \text{ for some } q \in U\}$$
$$= \{p \in X : b_p = c_p \text{ and } p \in U\}$$
$$= [\![b = c]\!] \cap U.$$

On the other hand, suppose that $g \in \prod_{x \in X} A_x$ is continuous; we need to show that $g \in B$. By compactness (as we shall describe in detail below) it suffices to take any $x \in X$ and find a clopen neighborhood U of x and a $b \in B$ such that $g \upharpoonright U = b \upharpoonright U$. In fact, take $b \in B$ such that $b_x = gx$, and let $U = g^{-1}[f_b[X]]$. Then $gx = b_x = f_b x$, so $x \in U$. And if $y \in U$, then

$$gy = f_b z \text{ for some } z \in X$$
$$= b_z \text{ for some } z \in X$$
$$= b_y,$$

as desired. We may assume that U is clopen. By compactness, we get a finite sequence $\langle b_i : i < n \rangle$ of elements of B and $\langle U_i : i < n \rangle$ of clopen subsets of X such that $\bigcup_{i<n} U_i = X$ and $g \upharpoonright U_i = b_i \upharpoonright U_i$ for all $i < n$. We may assume that the U_i's are pairwise disjoint. Then the patchwork property yields that $g \in B$. Finally, S is Hausdorff by Theorem 8.13. □

Boolean powers

The notion of Boolean power is related to that of Boolean product. To some extent this notion reduces to more familiar notions. In fact, bounded Boolean powers coincide with free products, while the unbounded Boolean powers are in between the free product and its completion. We prove these facts here; the first one is due to Quackenbush [72].

First we recall the definitions of Boolean power and bounded Boolean power, from Burris [75]. Given two BAs A and B, with B complete, the *Boolean power* $A[B]$ consists of all $f \in {}^A B$ such that the following two conditions hold:

(1) If $a_0, a_1 \in A$ with $a_0 \neq a_1$ then $fa_0 \cdot fa_1 = 0$;
(2) $\sum_{a \in A} fa = 1$.

The Boolean operations on $A[B]$ are defined like this:

$$(f + g)a = \sum_{b+c=a} (fb \cdot gc);$$

$$(f \cdot g)a = \sum_{b \cdot c = a} (fb \cdot gc);$$

$$(-f)a = f(-a);$$

$$0a = 0, \text{ if } a \neq 0;$$
$$0a = 1, \text{ if } a = 0;$$
$$1a = 0, \text{ if } a \neq 1;$$
$$1a = 1, \text{ if } a = 1.$$

It is easy to verify that $A[B]$ is a BA. The *bounded Boolean power* of A and B, denoted by $A[B]^*$, consists of those $f \in A[B]$ such that $\{a \in A : fa \neq 0\}$ is finite; in this case, we do not need to require that B is complete; and clearly $A[B]^*$ is a BA.

Theorem 1.3. *Let A and B be BAs. Then $C \stackrel{\text{def}}{=} \{f \in {}^{\text{Ult}B} A : f$ is continuous (with A having the discrete topology)$\}$ is a subalgebra of ${}^{\text{Ult}B} A$ which is isomorphic to $A[B]^*$. Moreover, C is a Boolean product of $\langle A : p \in \text{Ult}B \rangle$.*

Proof. Clearly the 0 function is in C. If $f \in C$, then $-f \in C$, since for any $X \subseteq A$,

$$(-f)^{-1}[X] = \{p \in \text{Ult}B : (-f)p \in X\}$$
$$= \{p \in \text{Ult}B : fp \in \{a \in A : -a \in X\}\}$$
$$= f^{-1}[\{a \in A : -a \in X\}],$$

and this last set is open. If $f, g \in C$, then $f + g \in C$, since for any $X \subseteq A$ we have

$$(f + g)^{-1}[X] = \{p \in \text{Ult}B : fp + gp \in X\}$$
$$= \bigcup_{a+b \in X} \{p \in \text{Ult}B : fp = a \text{ and } gp = b\}$$
$$= \bigcup_{a+b \in X} (f^{-1}[\{a\}] \cap g^{-1}[\{b\}]),$$

and the last set is open. This shows that C is a subalgebra of ${}^{\text{Ult}B} A$.

Note that for any $f \in C$ and $a \in A$ the set $\{p \in \mathrm{Ult}B : fp = a\}$ is clopen, since this set is equal to $f^{-1}[\{a\}]$, and its complement is equal to $\bigcup_{a' \in A \setminus \{a\}} f^{-1}[\{a'\}]$, which is open. Hence we may define $(Ff)a$ to be the unique $b \in B$ such that $\{p \in \mathrm{Ult}B : fp = a\} = \mathcal{S}b$. We now show that F is an isomorphism of C onto $A[B]^*$. First we note the following:

(1) if $f \in C$ and $a_0, a_1 \in A$ and $a_0 \neq a_1$, then $(Ff)a_0 \cdot (Ff)a_1 = 0$;
(2) if $f \in C$ then $\mathrm{Ult}B = \bigcup_{a \in A} \mathcal{S}((Ff)a)$, and $\sum_{a \in A}(Ff)a = 1$.

In fact, for any $p \in \mathrm{Ult}B$ we have $p \in \mathcal{S}((Ff)(fp))$, so $\mathrm{Ult}B = \bigcup_{a \in A} \mathcal{S}((Ff)a)$, and (2) follows.

(3) If $f \in C$, then $\{a \in A : (Ff)a \neq 0\}$ is finite.

This is clear by (1), (2), and the compactness of $\mathrm{Ult}A$.

Now we check that F is a homomorphism. For any $f \in C$ we have $(F(-f))a = (Ff)(-a)$ [hence $F(-f) = -(Ff)$], since for any $p \in \mathrm{Ult}B$ we have

$$
\begin{aligned}
p \in \mathcal{S}((F(-f))a) \text{ iff } (-f)p &= a \\
\text{iff } fp &= -a \\
\text{iff } p &\in \mathcal{S}((Ff)(-a)).
\end{aligned}
$$

Next, for any $f, g \in C$ and $a \in A$ we have $(F(f+g))a = \sum_{b+c=a}((Ff)b \cdot ((Ff)c))$ [hence $Ff + Fg = F(f+g)$], since for any $p \in \mathrm{Ult}\,B$ we have

$$
\begin{aligned}
p \in \mathcal{S}(F(f+g)a) \text{ iff } (f+g)p &= a \\
\text{iff } \exists b, c(fp = b \wedge gp = c \wedge b+c &= a) \\
\text{iff } \exists b, c(p \in \mathcal{S}((Ff)b) \wedge p \in \mathcal{S}((Fg)c) \wedge b+c &= a) \\
\text{iff } \exists b, c(p \in \mathcal{S}((Ff)b \cdot (Fg)c) \wedge b+c &= a) \\
\text{iff } p \in \mathcal{S}(\sum_{b+c=a} ((Ff)b \cdot (Fg)c)).&
\end{aligned}
$$

[The last equivalence holds since there are only finitely many nonzero summands.]
Next, F is one-one: if $f \in C \setminus \{0\}$, say $fp \neq 0$; then $p \in \mathcal{S}(((Ff)(fp))$, so $Ff \neq 0$. F is onto: given $g \in A[B]^*$, define $f : \mathrm{Ult}B \to A$ by setting $fp =$ the $a \in A$ such that $ga \in p$. [This is legal since $\sum_{a \in A} ga = 1$, $\langle ga : a \in A \rangle$ is a disjoint system, and $\{a \in A : ga \neq 0\}$ is finite.] If $X \subseteq A$, then

$$
\begin{aligned}
f^{-1}[X] &= \{p \in \mathrm{Ult}B : fp \in X\} \\
&= \{p \in \mathrm{Ult}B : g[X] \cap p \neq 0\} \\
&= \mathcal{S}(\sum_{a \in X, ga \neq 0} ga).
\end{aligned}
$$

Thus $f^{-1}[X]$ is clopen. So $f \in C$. For any $a \in A$ we have $(Ff)a = ga$, since for any $p \in \mathrm{Ult}B$ we have

$$p \in \mathcal{S}((Ff)a) \text{ iff } fp = a$$
$$\text{iff } ga \in p$$
$$\text{iff } p \in \mathcal{S}(ga).$$

Hence F is the desired isomorphism.

Finally, condition (2) in the definition of Boolean product clearly works for C. Now let $f, g \in C$. Then

$$[\![f = g]\!] = \bigcup_{a \in A} (f^{-1}[\{a\}] \cap g^{-1}[\{a\}]),$$

so $[\![f = g]\!]$ is open, and

$$\mathrm{Ult}B \backslash [\![f = g]\!] = \bigcup_{a,a' \in A, a \neq a'} (f^{-1}[\{a\}] \cap g^{-1}[\{a'\}]),$$

so $[\![f = g]\!]$ is clopen. $\square$

Theorem 1.4. *For any BAs A and B we have $A[B]^* = A \oplus B \leq A[B] \leq \overline{A \oplus B}$, where $\leq$ means "is a subalgebra of" and $\overline{A \oplus B}$ is the completion of $A \oplus B$.*

Proof. We define embeddings g of A into $A[B]^*$ and h of B into $A[B]^*$. For any $a \in A$, let

$$(ga)a' = \begin{cases} 1, & \text{if } a = a'; \\ 0, & \text{otherwise.} \end{cases}$$

It is straightforward to check that $ga \in A[B]^*$, and that, in fact, g is an isomorphic embedding of A into $A[B]^*$.

Next, define for any $b \in B$

$$(hb)a = \begin{cases} 0, & \text{if } a \neq 0, 1; \\ b, & \text{if } a = 1; \\ -b, & \text{if } a = 0. \end{cases}$$

Again, it is straightforward to check that $hb \in A[B]^*$, and that h is an isomorphic embedding of B into $A[B]^*$.

If $a \in A^+$ and $b \in B^+$, then $ga \cdot hb \neq 0$ since $(ga \cdot hb)a \geq b$. Hence to prove the theorem it suffices to prove the following: for any $\xi \in A[B]$, let $X = \{a \in A^+ : \xi a \neq 0\}$. Then

$$(1) \qquad\qquad \xi = \sum_{a \in X} ga \cdot h\xi a.$$

To prove this, it is convenient to prove the following fact first: for $a \in A^+$, $b \in B^+$, and $a' \in A$ we have

$$(ga \cdot hb)a' = \begin{cases} b, & \text{if } a = a'; \\ -b & \text{if } a' = 0; \\ 0, & \text{otherwise.} \end{cases}$$

We leave this proof to the reader. Then it is easy to show that ξ is an upper bound for $\{ga \cdot h\xi a : a \in X\}$. If η is any upper bound for this set, then one can prove the following two facts, valid for any $a \in A$:

$$\xi a = \sum_{a \leq c} \xi a \cdot \eta c;$$

$$\sum_{a \not\leq c} \xi a \cdot \eta c = 0.$$

From these two facts $\xi \leq \eta$ follows easily. Here are the details on the proofs of the two facts. They are clear for $a = 0$ and for $\xi a = 0$. Suppose $a \in X$. Then

$$\begin{aligned} \xi a &= (ga \cdot h\xi a)a \\ &= (ga \cdot h\xi a \cdot \eta)a \\ &= \sum_{b \cdot c = a} (ga \cdot h\xi a)b \cdot \eta c \\ &= \sum_{a \cdot c = a} \xi a \cdot \eta c, \end{aligned}$$

giving the first equality above. For the second, if $a \not\leq c$, then

$$\eta c \cdot \xi a = \sum_{a \leq d} \xi a \cdot \eta d \cdot \eta c = 0,$$

which yields the second equality above.　　　　　　　　　　　　　　　　　　□

For conditions under which $A[B]$ is equal to $\overline{A \oplus B}$ see Dwinger [82] and Takahashi [88].

Set products

This operation, due to Weese and Gurevich independently, is extensively studied in Heindorf [90]. Suppose that $\langle A_i : i \in I \rangle$ is a system of BAs; we assume that A_i is a field of subsets of some set J_i, and that the J_i's are pairwise disjoint. Furthermore, let B be an algebra of subsets of I containing all of the finite subsets of I. For each $b \in B$ let $\bar{b} = \bigcup_{i \in b} J_i$. Set $K = \bigcup_{i \in I} J_i$. For each $b \in B$, each finite $F \subset I$, and each $a \in \prod_{i \in F} A_i$, the set

$$\bar{b} \cup \bigcup_{i \in F} a_i$$

will be denoted by $h(b, F, a)$. It is easily checked that the set of all such elements $h(b, F, a)$ forms a field of subsets of K. This BA is the *set product of the A_i's over B*, and is denoted by $\prod_{i \in I}^{B} A_i$. Finco I is the BA of finite and cofinite subsets of I.

Theorem 1.5. *Suppose that $\langle A_i : i \in I \rangle$ is a system of BAs; each A_i is a field of subsets of some set J_i, the J_i's are pairwise disjoint.*

(i) *If* Finco $I \leq B \leq C \leq \mathscr{P}I$, *then* $\prod_{i \in I}^{B} A_i \leq \prod_{i \in I}^{C} A_i$.

(ii) *If* Finco $I \leq B \leq \mathscr{P}I$, *then* $\prod_{i \in I}^{B} A_i$ *can be embedded in* $\prod_{i \in I} A_i$.

(iii) $\prod_{i \in I}^{w} A_i \cong \prod_{i \in I}^{\text{Finco } I} A_i$.

Proof. (i) is clear. For (ii), define $(fx)_i = x \cap J_i$ for any $x \in \prod_{i \in I}^{B} A_i$ and each $i \in I$; it is easy to check that f is the desired embedding. In case $B = $ Finco I, this mapping is easily seen to be onto, proving (iii). $\qquad\square$

Theorem 1.6. *Assume the hypotheses of Theorem 1.5, and suppose that $L \overset{\text{def}}{=} \{i \in I : |A_i| > 2\}$ is infinite. Then $\prod_{i \in I}^{B} A_i$ is not complete.*

Proof. For each $i \in L$ choose $a_i \in A_i$ such that $0 \subset a_i \subset J_i$. Suppose that $\sum_{i \in L} a_i$ exists in $\prod_{i \in I}^{B} A_i$; say it is equal to $h(b, F, c)$, where we may assume that $b \cap F = 0$. Fix $i \in L \backslash F$. Then $a_i \subseteq h(b, F, c)$ implies that $i \in b$. But then $h(b \backslash \{i\}, F', c')$ is still an upper bound, where $F' = F \cup \{i\}$ and c' extends c with $c_i' = a_i$. Since $h(b \backslash \{i\}, F', c') \subset h(b, F, c)$, this is a contradiction. $\qquad\square$

It is clear that if each A_i is atomless, then so is $\prod_{i \in I}^{B} A_i$; similarly for each A_i atomic. Also note that B can be isomorphically embedded in $\prod_{i \in I}^{B} A_i$. It is somewhat less trivial to check that the set product preserves superatomicity:

Theorem 1.7. *If each A_i is superatomic and also B is superatomic, then $\prod_{i \in I}^{B} A_i$ is superatomic.*

Proof. For brevity write $C = \prod_{i \in I}^{B} A_i$. It suffices to show that if f is a homomorphism from C onto a nontrivial BA D, then D has an atom. We consider two cases.

Case 1. $f J_i \neq 0$ for some $i \in I$. Let $f u_i$ be an atom of $f[A_i]$; this is possible since A_i is superatomic. We claim that $f u_i$ is also an atom of D. For, suppose that $x \in C$ and $fx \cdot f u_i \neq 0$. Then $0 \neq x \cdot u_i \in A_i$, and so $0 \neq f(x \cdot u_i) \in f[A_i]$ and hence $f u_i = f(x \cdot u_i) \leq fx$, as desired.

Case 2. $f J_i = 0$ for all $i \in I$. Since D is nontrivial, $f[\{h(b, 0, 0) : b \in B\}] = D$. Then B being atomic yields the desired atom. $\qquad\square$

One-point gluing

Our next algebraic operation is one-point gluing. Suppose we are given a system $\langle A_i : i \in I \rangle$ of BAs, and a corresponding system $\langle F_i : i \in I \rangle$ of ultrafilters: F_i is an ultrafilter on A_i for each $i \in I$. The one-point gluing of the pair $(\langle A_i : i \in I \rangle, \langle F_i : i \in I \rangle)$ is the following subalgebra of the direct product $\prod_{i \in I} A_i$:

$$\{x \in \prod_{i \in I} A_i : \text{for all } i, j \in I(x_i \in F_i \text{ iff } x_j \in F_j)\}.$$

In the case of two factors A_i and A_j this amounts to identifying the two points F_i and F_j in the disjoint union of the Stone spaces; this is a special case of the following theorem.

Theorem 1.8. *Let $\langle A_i : i \in I \rangle$ be a system of BAs, and let $\langle F_i : i \in I \rangle$ be a system with F_i an ultrafilter on A_i for each $i \in I$. Let $C = \prod_{i \in I} A_i$, and let B be the one-point gluing of the pair $(\langle A_i : i \in I \rangle, \langle F_i : i \in I \rangle)$. Then for each $i \in I$ the set $F_i' \stackrel{\text{def}}{=} \{x \in C : x_i \in F_i\}$ is an ultrafilter on C. Further, let $K = \overline{\{F_i' : i \in I\}}$. Let X be the quotient of $\text{Ult}C$ obtained by collapsing K to a point. Then $\text{Ult}B$ is homeomorphic to X.*

Proof. The first assertion of the theorem is obvious. Now let π be the natural continuous mapping of $\text{Ult}C$ onto X. We now define g from $\text{Ult}B$ into X by setting $g(F \cap B) = \pi F$ for any ultrafilter F on C. To see that g is well-defined, suppose that $F \cap B = G \cap B$, where F and G are ultrafilters on C. If both F and G are in K, obviously $\pi F = \pi G$. Now suppose, say, that $F \notin K$. We claim then that $F = G$. Suppose to the contrary that $F \neq G$. Choose $u \in F \backslash G$. Now choose $x \in F$ such that $\mathcal{S}^C x \cap K = 0$. Here $\mathcal{S}^C$ is the Stone map associated with C. If $(x \cdot u)_i \in F_i$ for some $i \in I$, then $x \cdot u \in F_i'$, and hence $\mathcal{S}^C x \cap K \neq 0$, contradiction. Thus $(x \cdot u)_i \notin F_i$ for all $i \in I$, and consequently $x \cdot u \in B$. But $x \cdot u \in F$, so $x \cdot u \in G$ and so $u \in G$, contradiction.

g is one-one: suppose that F and G are ultrafilters on C and $\pi F = \pi G$; we want to show that $F \cap B = G \cap B$. We may assume that $F, G \in K$. Let $x \in B$; by symmetry we want to show that if $x \in F$ then $x \in G$. So, assume that $x \in F$. Thus $F \in \mathcal{S}^C x$ so, since $F \in K$, we can choose $i \in I$ so that $F_i' \in \mathcal{S}^C x$. Thus $x \in F_i'$ and so $x_i \in F_i$. If $x \notin G$, then $-x \in G$, and a similar argument gives $-x_j \in F_j$ for some $j \in I$. This contradicts the assumption that $x \in B$.

g maps onto X: let $F \in X$. If $F \notin K$, then $g(F \cap B) = F$. If F is the point which is the collapse of K, then for any $i \in I$ we have $g(F_i' \cap B) = F$.

g is continuous: suppose that U is open in X, and $F \cap B \in g^{-1}[U]$, where F is an ultrafilter on C. Thus $\pi F \in U$, and hence $F \in \pi^{-1}[U]$. So, choose $x \in C$ so that $F \in \mathcal{S}^C x \subseteq \pi^{-1}[U]$. *Case 1.* $F \notin K$. Choose $y \in C$ so that $F \in \mathcal{S}^C y$ and $\mathcal{S}^C y \cap K = 0$. Thus $F \in \mathcal{S}^C(x \cdot y)$, and clearly $x \cdot y \in B$. We claim that $F \cap B \in \mathcal{S}^B(x \cdot y) \subseteq g^{-1}[U]$. Obviously $F \cap B \in \mathcal{S}^B(x \cdot y)$. Suppose that $x \cdot y \in G \cap B$, where G is an ultrafilter on C. Then $G \in \mathcal{S}^C x$, and hence $g(G \cap B) = \pi G \in U$, as desired. *Case 2.* $F \in K$. Thus $K \subseteq \pi^{-1}[U]$. Choose $y \in C$ such that $K \subseteq \mathcal{S}^C y \subseteq \pi^{-1}[U]$. Thus $y \in B$. We claim that $F \in \mathcal{S}^B y \subseteq g^{-1}[U]$. Obviously $F \in \mathcal{S}^B y$. Now suppose that $G \cap B \in \mathcal{S}^B y$, where G is an ultrafilter on C. Then $G \in \mathcal{S}^C y$, and $g(G \cap B) = \pi G \in U$, as desired.

Finally, it is clear that X is Hausdorff. $\qquad\qquad\square$

The Aleksandroff duplicate

Given a BA A, its *Aleksandroff duplicate*, denoted by $\mathrm{Dup}A$, is the subalgebra of $A \times \mathscr{P}\mathrm{Ult}A$ whose set of elements is

$$\{(a, X) : a \in A, X \subseteq \mathrm{Ult}A, \text{ and } \mathcal{S}a \triangle X \text{ is finite}\}.$$

(It is easy to check that this is a subalgebra of $A \times \mathscr{P}\mathrm{Ult}A$; recall that $\mathcal{S}a = \{F \in \mathrm{Ult}A : a \in F\}$.)

We show now that this is equivalent to the usual definition. (See, for example, Gardner, Pfeffer [84].) That definition runs like this. Let X be a topological space. We put a topology on $X \times 2$ as follows: a base consists of all sets

- $F \times \{1\}$, F a finite subset of X;
- $(G \times 2) \backslash (F \times \{1\})$, G open in X, F a finite subset of X.

Theorem 1.9. *Let A be a BA. Under the above topology, $\mathrm{Ult}A \times 2$ is a Boolean space, and $\mathrm{Dup}A$ is isomorphic to $\mathrm{Clop}(\mathrm{Ult}A \times 2)$.*

Proof. The topology is clearly Hausdorff. Now we show that it is compact. Let $\mathcal{O}$ be a cover of $\mathrm{Ult}A \times 2$ by basic open sets. Then $\{G : (G \times 2) \backslash (F \times \{1\}) \in \mathcal{O}$ for some open $G \subseteq X$ and some finite $F \subseteq X\}$ is an open cover of $\mathrm{Ult}A$, so we can choose

$$(G_1 \times 2) \backslash (F_1 \times \{1\}), \ldots, (G_m \times 2) \backslash (F_m \times \{1\}) \in \mathcal{O}$$

such that $G_1, \ldots, G_m$ is a cover of $\mathrm{Ult}A$. There are only finitely many elements of $\mathrm{Ult}A \times 2$ remaining—some of the elements of $(F_1 \times \{1\}) \cup \ldots \cup (F_m \times \{1\})$— so $\mathcal{O}$ has a finite subcover.

Next we determine the clopen subsets of $\mathrm{Ult}A \times 2$. Clearly each set $F \times \{1\}$ is clopen, when F is a finite subset of $\mathrm{Ult}A$. And if G is a clopen subset of $\mathrm{Ult}A$ and F is a finite subset of $\mathrm{Ult}A$, then $(G \times 2) \backslash (F \times \{1\})$ is clopen, since its complement is $[(\mathrm{Ult}A \backslash G) \times 2] \cup (F \times \{1\})$. These two kinds of clopen subsets form a base for the topology, so every clopen set is a finite join of these two kinds.

So $\mathrm{Ult}A \times 2$ is a Boolean space. Now we define a function f which will extend to the desired isomorphism. For any $a \in A$ let $f(a, \mathcal{S}a) = \mathcal{S}a \times 2$, and for any ultrafilter F on A let $f(0, \{F\}) = \{F\} \times \{1\}$. So f maps a set of generators of $\mathrm{Dup}A$ onto a set of generators of $\mathrm{Clop}(\mathrm{Ult}A \times 2)$. An easy application of Sikorski's extension criterion shows that f extends to a one-one homomorphism, as desired. $\square$

We give some more simple facts about the Aleksandroff duplicate. Note that the duplicate is always atomic. The following result is easy.

Proposition 1.10. *For any BA A, define $g : A \to \mathrm{Dup}A$ by setting $ga = (a, \mathcal{S}a)$ for any $a \in A$. Then g is an isomorphism from A into $\mathrm{Dup}A$.* $\square$

The special nature of the duplicate is brought out by the following simple theorem. For any BA A, let I_{at}^A be the ideal of A generated by its atoms.

Theorem 1.11. *Let A be any BA. Then $A/I_{at}^A \cong \mathrm{Dup}A/I_{at}^{\mathrm{Dup}A}$.*

Proof. Let π be the natural homomorphism from $\mathrm{Dup}A$ onto $\mathrm{Dup}A/I_{at}^{\mathrm{Dup}A}$. We show that $\pi \circ g$ is a homomorphism from A onto $\mathrm{Dup}A/I_{at}^{\mathrm{Dup}A}$ with kernel I_{at}^A, where g is as in Proposition 1.10. If $(a, X) \in \mathrm{Dup}A$, then $(a, \mathcal{S}a) \triangle (a, X) = (0, \mathcal{S}a \triangle X) \in I_{at}^{\mathrm{Dup}A}$; thus $\pi \circ g$ maps onto $\mathrm{Dup}A/I_{at}^{\mathrm{Dup}A}$. Now for any $a \in A$, $(\pi \circ g)a = 0$ iff $(a, \mathcal{S}a) \in I_{at}^{\mathrm{Dup}A}$ iff $a \in I_{at}^A$, as desired. $\qquad\square$

A corollary of this theorem is that if A is superatomic, then so is $\mathrm{Dup}A$.

The exponential

Let X be any topological space. $\mathrm{Exp}\,X$ is the collection of all non-empty closed subspaces of X. We topologize it by taking the collection of sets of the following form as a base:

$$\mathscr{V}(U_1, \ldots, U_m) \overset{\mathrm{def}}{=} \{F \in \mathrm{Exp}X : F \subseteq U_1 \cup \ldots \cup U_m \text{ and } F \cap U_i \neq 0 \text{ for all } i\},$$

where $U_1, \ldots, U_m$ are open in X.

Theorem 1.12. *Let X be a compact Hausdorff space. Then $\mathrm{Exp}\,X$ is also Hausdorff and compact. Moreover, if X is a Boolean space, then so is $\mathrm{Exp}\,X$, and the set*

$$\{\mathscr{V}(U_1, \ldots, U_m) : \text{ each } U_i \text{ clopen}\}$$

is a collection of clopen sets forming a base for the topology on $\mathrm{Exp}\,X$.

Proof. Hausdorff: suppose that F and G are distinct non-empty closed sets. Say $x \in F\backslash G$. Let W and U be disjoint open sets such that $x \in W$ and $G \subseteq U$. Thus $G \in \mathscr{V}(U)$, $F \in \mathscr{V}(X, W)$, and $\mathscr{V}(U) \cap \mathscr{V}(X, W) = 0$.

Compact: First we note some facts. For each open U in X let $T_U = \{F \in \mathrm{Exp}X : F \cap U \neq 0\}$. This is an open set, since $T_U = \mathscr{V}(X, U)$. Next,

(1) $\{\mathscr{V}(U) : U \text{ open}\} \cup \{T_U : U \text{ open}\}$ is a subbase for the topology on $\mathrm{Exp}X$.

For, $\mathscr{V}(U_1, \ldots, U_m) = \mathscr{V}(U_1 \cup \ldots \cup U_m) \cap T_{U_1} \cap \ldots \cap T_{U_m}$.

Now to prove compactness of $\mathrm{Exp}X$, suppose that $\mathscr{O}$ is a cover of $\mathrm{Exp}X$ by subbase members in (1). *Case 1.* $\{W : T_W \in \mathscr{O}\}$ covers X. Choose $W_1, \ldots, W_m$ with each $T_{W_i} \in \mathscr{O}$ such that $X = W_1 \cup \ldots \cup W_m$. Then $\{T_{W_1}, \ldots, T_{W_m}\}$ covers $\mathrm{Exp}X$, as desired. *Case 2.* $\{W : T_W \in \mathscr{O}\}$ does not cover X. Let $Y = \bigcup_{T_W \in \mathscr{O}} W$. So Y is a proper open subset of X, and hence $X\backslash Y \in \mathrm{Exp}X$. Therefore $X\backslash Y \in \mathscr{V}(U)$ for some $\mathscr{V}(U) \in \mathscr{O}$. Thus $X\backslash Y \subseteq U$, so $X\backslash U \subseteq Y$. Since $X\backslash U$ is compact, there exist $W_1, \ldots, W_m$ with each $T_{W_i} \in \mathscr{O}$ such that $X\backslash U \subseteq W_1 \cup \ldots \cup W_m$. Hence $\{\mathscr{V}(U)\} \cup \{T_{W_1}, \ldots, T_{W_m}\}$ covers $\mathrm{Exp}X$, as is easily verified.

Now assume that X is a Boolean space. Then

(2) If $U_1, \ldots, U_m$ are clopen in X, then $\mathscr{V}(U_1, \ldots, U_m)$ is clopen in $\mathrm{Exp}X$.

For, suppose that $F \in \mathrm{Exp}X \backslash \mathscr{V}(U_1, \ldots, U_m)$. *Case 1.* $F \nsubseteq U_1 \cup \ldots \cup U_m$. Then for some $\Gamma \subseteq \{1, \ldots, m\}$ we have $F \in \mathscr{V}(X \backslash (U_1 \cup \ldots \cup U_m), \langle U_i \rangle_{i \in \Gamma})$, and this set is disjoint from $\mathscr{V}(U_1, \ldots, U_m)$. *Case 2.* $F \subseteq U_1 \cup \ldots \cup U_m$. Then for some $\Gamma \subset \{1, \ldots, m\}$ we have $F \in \mathscr{V}(\langle U_i \rangle_{i \in \Gamma})$, and this set is disjoint from $\mathscr{V}(U_1, \ldots, U_m)$.

(3) If $U_1, \ldots, U_m, W_1, \ldots, W_n$ are open and $F \in \mathscr{V}(U_1, \ldots, U_m) \cap \mathscr{V}(W_1, \ldots, W_n)$, then $F \in \mathscr{V}(\langle U_i \cap W_j : F \cap U_i \cap W_j \neq 0 \rangle)$.

(4) $\{\mathscr{V}(U_1, \ldots, U_m)$: each U_i clopen$\}$ is a base for the topology on $\mathrm{Exp}X$.

To prove (4), assume that $F \in \mathscr{V}(U_1, \ldots, U_m)$ with each U_i open; we want to find clopen $W_1, \ldots, W_n$ such that $F \subseteq \mathscr{V}(W_1, \ldots, W_n) \subseteq \mathscr{V}(U_1, \ldots, U_m)$. It suffices by (3) to show that the set

$$\{\mathrm{Exp}X \backslash \mathscr{V}(U_1, \ldots, U_m)\} \cup \{\mathscr{V}(W_1, \ldots, W_n) :$$

$$\text{each } W_i \text{ clopen and } F \in \mathscr{V}(W_1, \ldots, W_n)\}$$

has empty intersection. Suppose to the contrary that G is in each member of this set. *Case 1.* $G \nsubseteq U_1 \cup \ldots \cup U_m$. Thus $G \nsubseteq F$; say $x \in G \backslash F$. Let W be a clopen set such that $x \in W$ and $W \cap F = 0$. Thus $F \in \mathscr{V}(X \backslash W)$, so $G \in \mathscr{V}(X \backslash W)$, contradiction. *Case 2.* $G \subseteq U_1 \cup \ldots \cup U_m$. Since $G \in \mathrm{Exp}X \backslash \mathscr{V}(U_1, \ldots, U_m)$, it follows that $G \cap U_i = 0$ for some i. Say $x \in F \cap U_i$. Let W be clopen with $x \in W$ and $W \cap G = 0$. Then $F \in \mathscr{V}(W, X \backslash W)$ or $F \in \mathscr{V}(W)$, so $G \in \mathscr{V}(W, X \backslash W)$ or $G \in \mathscr{V}(W)$, contradiction. □

For any BA A, we denote by $\mathrm{Exp}A$ the Boolean algebra $\mathrm{Clop}(\mathrm{Exp}(\mathrm{Ult}(A)))$; this is called the *exponential* of A.

The following somewhat technical result will be useful.

Proposition 1.13. *For any BA A, $\mathrm{Exp}\,A$ is generated by $\{\mathscr{V}(\mathcal{S}a) : a \in A\}$.*

Proof. We already know from the proof of Theorem 1.12 that $\mathrm{Exp}A$ is generated by $\{\mathscr{V}(U_1, \ldots, U_m)$: each U_i clopen$\}$. So it suffices to see that each element of this set is generated by $\{\mathscr{V}(\mathcal{S}a) : a \in A\}$. This follows from:

$$\mathscr{V}(\mathcal{S}a_1, \ldots, \mathcal{S}a_m) = \mathscr{V}(\mathcal{S}(a_1 + \cdots + a_m) \backslash (\mathscr{V}(\mathcal{S}(-a_1)) \cup \ldots \cup \mathscr{V}(\mathcal{S}(-a_m))). \quad \square$$

Here is a useful example of a use of Proposition 1.13.

Proposition 1.14. *If A is an infinite BA and $0 < a < 1$, then there is a homomorphism f from $\mathrm{Exp}A$ onto $\mathrm{Exp}(A \restriction a) \oplus \mathrm{Exp}(A \restriction -a)$.*

Proof. For this proof we let $\mathscr{V}0 = 0$. Note that if $x, y \in A$ and $\mathscr{V}(\mathcal{S}^A x) = \mathscr{V}(\mathcal{S}^A y)$, then $x = y$. Hence there is a function f such that for all $x \in A$ we have $f\mathscr{V}(\mathcal{S}^A x) =$

$\mathscr{V}(\mathcal{S}^{A\restriction a}(x \cdot a)) \cdot \mathscr{V}(\mathcal{S}^{A\restriction -a}(x \cdot -a))$. To show that f extends to a homomorphism of $\mathrm{Exp}A$ into $\mathrm{Exp}\,(A \restriction a) \oplus \mathrm{Exp}\,(A \restriction -a)$, suppose that

$$(*) \quad \mathscr{V}(\mathcal{S}^A x_0) \cap \ldots \cap \mathscr{V}(\mathcal{S}^A x_{m-1}) \cap -\mathscr{V}(\mathcal{S}^A y_0) \cap \ldots \cap -\mathscr{V}(\mathcal{S}^A y_{n-1}) = 0.$$

Here we may assume that $m, n > 0$. It follows that $x_0 \cdot \ldots \cdot x_{m-1} \le y_i$ for some $i < n$, since otherwise $\mathcal{S}^A(x_0 \cdot \ldots \cdot x_{m-1})$ would be a member of the set in $(*)$. Then it easily follows that

$$f\mathscr{V}(\mathcal{S}^A x_0) \cap \ldots \cap f\mathscr{V}(\mathcal{S}^A x_{m-1}) \cap -f\mathscr{V}(\mathcal{S}^A y_0) \cap \ldots \cap -f\mathscr{V}(\mathcal{S}^A y_{n_1}) = 0,$$

as desired. Thus f extends to a homomorphism, still denoted by f.

To prove that f is onto, note that

$$\{\mathscr{V}(\mathcal{S}^{A\restriction a} x) \cdot 1 : x \in (A \restriction a)^+\} \cup \{1 \cdot \mathscr{V}(\mathcal{S}^{A\restriction -a} x) : x \in (A \restriction -a)^+\}$$

generates $\mathrm{Exp}\,(A \restriction a) \oplus \mathrm{Exp}\,(A \restriction -a)$. Now if $x \in (A \restriction a)^+$, then $f(x + -a) = \mathscr{V}(\mathcal{S}^{A\restriction a} x) \cdot 1$; similarly for the other desired elements. $\square$

Proposition 1.15. *If A is atomic, then so is $\mathrm{Exp}A$.*

Proof. For each atom a of A, let F_a be the principal ultrafilter determined by a. Suppose that $\langle a_0, \ldots, a_{m-1} \rangle$ is a system of distinct atoms of A. Then

$$\mathscr{V}(\mathcal{S}a_0, \ldots, \mathcal{S}a_{m-1}) = \{\{F_{a_0}, \ldots, F_{a_{m-1}}\}\},$$

and this is hence an atom of $\mathrm{Exp}A$. Now take any nonzero $x \in \mathrm{Exp}A$. To show that there is an atom below x it suffices to take the case in which x has the form $\mathscr{V}(\mathcal{S}b_0, \ldots, \mathcal{S}b_{m-1})$. Note that each b_i is nonzero, since $x \ne 0$, and each element of x has nonempty intersection with each $\mathcal{S}b_i$. Let a_i be an atom below b_i for each $i < m$. Then $\{\{F_{a_0}, \ldots, F_{a_{m-1}}\}\}$ is the desired atom. $\square$

Proposition 1.16. *If A is atomless, then so is $\mathrm{Exp}A$.*

Proof. Suppose that $0 \ne x \in \mathrm{Exp}A$; we want to find $0 < y < x$. We may assume that x has the form $\mathscr{V}(\mathcal{S}b_0, \ldots, \mathcal{S}b_{m-1})$. Note that each b_i is nonzero (see the proof of Proposition 1.15). And we clearly may assume that all the b_i are distinct from one another. Finally, we may assume that b_0 is minimal among them, i.e., if $b_i \le b_0$ then $i = 0$. Now choose $0 < a_0 < b_0$. We claim that the desired element is

$$y = \mathscr{V}(\mathcal{S}a_0, \mathcal{S}(b_1 \cdot -b_0), \ldots, \mathcal{S}(b_{m-1} \cdot -b_0)).$$

Clearly $y \subseteq x$. Let $u = a_0 + b_1 \cdot -b_0 + \cdots + b_{m-1} \cdot -b_0$. Then $\mathcal{S}u$ is a member of y, since if $b_i \cdot -b_0 = 0$ with $i > 0$ we get $b_i \le b_0$, contradiction. So $y \ne 0$. Let $v = b_0 \cdot -a_0 + b_1 \cdot -b_0 + \cdots + b_{m-1} \cdot -b_0$. Clearly $\mathcal{S}v$ is a member of x. Since clearly $v \cdot a_0 = 0$, it is not a member of y. So $y < x$. $\square$

From this proposition it follows that the free BA on ω free generators is isomorphic to its own exponential. Sirota [68] proved that also the free BA on ω_1 free generators is isomorphic to its own exponential. But Shapiro [76a], [76b] showed that this does not extend to higher cardinals.

To make this discussion of the exponential more concrete, we describe the exponential for A the finite-cofinite algebra on an infinite cardinal κ. For each $\alpha < \kappa$ let F_α be the ultrafilter of all $\Gamma \subseteq \kappa$ such that $\alpha \in \Gamma$. Let G be the ultrafilter of all finite subsets of κ. Thus $\mathrm{Ult}A = \{G\} \cup \{F_\alpha : \alpha < \kappa\}$. Now $\{Sa : a \in A\}$ is a basis for $\mathrm{Ult}A$. Note:

$$a \text{ finite} \quad \Rightarrow \quad Sa = \{F_\alpha : \alpha \in a\};$$
$$a \text{ cofinite} \quad \Rightarrow \quad Sa = \{G\} \cup \{F_\alpha : \alpha \in a\}.$$

Thus each F_α is isolated. The open subsets of $\mathrm{Ult}A$ are

$$\{F_\alpha : \alpha \in \Gamma\} \quad \text{for any} \quad \Gamma \subseteq \kappa;$$
$$\{G\} \cup \{F_\alpha : \alpha \in \Gamma\} \quad \text{for any cofinite } \Gamma \subseteq \kappa.$$

Hence the closed sets are

$$y_\Gamma \overset{\text{def}}{=} \{G\} \cup \{F_\alpha : \alpha \in \Gamma\} \quad \text{for any} \quad \Gamma \subseteq \kappa,$$
$$z_\Gamma \overset{\text{def}}{=} \{F_\alpha : \alpha \in \Gamma\} \quad \text{for any finite} \quad \Gamma \subseteq \kappa.$$

Hence
$$\mathrm{Exp}A = \{y_\Gamma : \Gamma \subseteq \kappa\} \cup \{z_\Gamma : 0 \neq \Gamma \text{ a finite subset of } \kappa\}.$$

Now we claim:

(1) Each z_Γ is isolated, Γ a finite nonempty subset of κ.

In fact, write $\Gamma = \{\alpha_0, \dots, \alpha_{m-1}\}$, $m > 0$. Then $\{z_\Gamma\} = \mathscr{V}(\{F_{\alpha_0}\}, \dots, \{F_{\alpha_{m-1}}\})$, as desired.

The following two statements are obvious:

(2) If $a_0, \dots, a_{m-1} \in A$ are all finite and $m > 0$, then

$$\mathscr{V}(Sa_0, \dots, Sa_{m-1}) = \{z_\Gamma : \Gamma \subseteq a_0 \cup \dots \cup a_{m-1} \text{ and } \Gamma \cap a_i \neq 0 \text{ for all } i < m\}.$$

(3) If $a_0, \dots, a_{m-1} \in A$ and some a_i is cofinite, then

$$\mathscr{V}(Sa_0, \dots, Sa_{m-1}) = \{y_\Gamma : \Gamma \subseteq a_0 \cup \dots a_{m-1} \text{ and } \Gamma \cap a_i \neq 0$$
$$\text{for all } i \text{ such that } a_i \text{ is finite}\}$$
$$\cup \{z_\Gamma : \Gamma \subseteq a_0 \cup \dots \cup a_{m-1}, \; \Gamma \text{ finite}, \; \Gamma \cap a_i \neq 0$$
$$\text{for all } i < m\}.$$

Next,

(4) No y_Γ is isolated.

For, suppose that $y_\Gamma \in \mathcal{V}(\mathcal{S}a_0, \ldots, \mathcal{S}a_{m-1})$, $m > 0$. Since $G \in y_\Gamma$, some a_i is cofinite. By the above, if

$$\bigcup \{a_i : a_i \text{ finite }\} \subseteq \Delta \subseteq a_0 \cup \ldots \cup a_{m-1},$$

then $y_\Delta \in \mathcal{V}(\mathcal{S}a_0, \ldots, \mathcal{S}a_{m-1})$. Thus $\mathcal{V}(\mathcal{S}a_0, \ldots, \mathcal{S}a_{m-1})$ has infinitely many members, as desired. We also proved:

(5) $\{y_\Gamma : \Gamma \subseteq \kappa\}$, as a subspace of $\mathrm{Ult}A$, is closed and has no isolated points.

The following is obvious:

(6) $\{\{z_\Gamma\} : \Gamma \text{ finite and non-empty}\}$ is the set of all atoms of $\mathrm{Exp}A$, which is atomic.

Let $x_\alpha = \mathcal{V}(\mathcal{S}(\kappa\backslash\{\alpha\}), \mathcal{S}\{\alpha\})$ for all $\alpha < \kappa$.

(7) $\langle x_\alpha/\text{fin} : \alpha < \kappa \rangle$ is a system of independent elements of $\mathrm{Exp}A/\text{fin}$.

To show this, suppose that Γ and Δ are finite disjoint subsets of κ; we want to show that $\bigcap_{\alpha \in \Gamma} x_\alpha \cap \bigcap_{\alpha \in \Delta} -x_\alpha$ is not a finite sum of atoms. Note by (3) that

$$x_\alpha = \{y_\Omega : \alpha \in \Omega\}$$
$$\cup \{z_\Omega : \alpha \in \Omega \text{ and } \Omega \neq \{\alpha\}, \ \Omega \text{ finite}\}.$$

It follows that if $\Gamma \subseteq \Omega$ and $\Delta \cap \Omega = 0$, then $y_\Omega \in \bigcap_{\alpha \in \Gamma} x_\alpha \cap \bigcap_{\alpha \in \Delta} -x_\alpha$. Hence (7) holds.

(8) $\langle x_\alpha/\text{fin} : \alpha < \kappa \rangle$ generates $\mathrm{Exp}A/\text{fin}$.

To prove this, by Proposition 1.13 it suffices to show that if a is cofinite then $\mathcal{V}(\mathcal{S}a)$ is generated by $\langle x_\alpha/\text{fin} : \alpha < \kappa \rangle$. So (8) follows from

(9) $\mathcal{V}(\mathcal{S}a)/\text{fin} = \bigcap_{\alpha \in \kappa\backslash a} -x_\alpha/\text{fin}$.

To prove this, first note that if $\alpha \in \kappa\backslash a$, then

$$\mathcal{V}(\mathcal{S}a) \cap x_\alpha = (\{y_\Gamma : \Gamma \subseteq a\} \cup \{z_\Gamma : 0 \neq \Gamma \subseteq a, \ \Gamma \text{ finite}\})$$
$$\cap (\{y_\Gamma : \alpha \in \Gamma\} \cup \{z_\Gamma : \alpha \in \Gamma, \ \Gamma \neq \{\alpha\}, \ \Gamma \text{ finite}\})$$
$$= 0.$$

This proves $\leq$ in (9). For the other direction, first note that

$$-x_\alpha = \{y_\Gamma : \alpha \notin \Gamma\} \cup \{z_\Gamma : (\alpha \notin \Gamma \text{ or } \Gamma = \{\alpha\}) \text{ and } \Gamma \text{ finite}\}.$$

Hence

$$\left(\bigcap_{\alpha\in\kappa\backslash a} -x_\alpha\right)\backslash\mathscr{V}(\mathcal{S}a) = (\{y_\Gamma : \Gamma\cap(\kappa\backslash a) = 0\}$$

$$\cup\{z_\Gamma : \forall\alpha\in\kappa\backslash a(\alpha\notin\Gamma \text{ or } \Gamma = \{\alpha\}) \text{ and } \Gamma \text{ finite}\})$$
$$\cap(\{y_\Gamma : \Gamma\nsubseteq a\}\cup\{z_\Gamma : \Gamma\nsubseteq a,\ \Gamma \text{ finite}\})$$
$$= \{z_{\{\alpha\}} : \alpha\in\kappa\backslash a\},$$

as desired.

Some further properties of the exponential will be developed in the discussion of semigroup algebras in the next chapter.

2. Special classes of Boolean algebras

We discuss several special classes of Boolean algebras not mentioned in the Handbook.

Semigroup algebras

The notion of a semigroup algebra is due to Heindorf [89b]. We give basic definitions and facts only. A subset H of a BA A is said to be *disjunctive* if $0 \notin H$, and $h, h_1, \ldots, h_n \in H$ and $h \leq h_1 + \cdots + h_n$ $(n > 0)$ imply that $h \leq h_i$ for some i.

If P is any partially ordered set, $M \subseteq P$, and $p \in P$, we define

$$M \uparrow p = \{a \in M : p \leq a\};$$
$$M \downarrow p = \{a \in M : a \leq p\}.$$

Proposition 2.1. *Let A be a BA and $H \subseteq A^+$. Then H is disjunctive iff for every $M \subseteq H$ there is a homomorphism f from $\langle H \rangle$ into $\mathscr{P}M$ such that $fh = M \downarrow h$ for all $h \in H$.*

Proof. $\Rightarrow$: In order to apply Sikorski's extension criterion, assume that $h_1, \ldots, h_m$, $k_1, \ldots, k_n \in H$ and $h_1 \cdot \ldots \cdot h_m \leq k_1 + \cdots + k_n$; we want to show that $(M \downarrow h_1) \cap \ldots \cap (M \downarrow h_m) \subseteq (M \downarrow k_1) \cup \ldots \cup (M \downarrow k_n)$. Let $x \in (M \downarrow h_1) \cap \ldots \cap (M \downarrow h_m)$. Then $x \leq h_1 \cdot \ldots \cdot h_m$, so $x \leq k_1 + \cdots + k_n$. Note that $n > 0$, since otherwise $x = 0$, contradicting $M \subseteq H \subseteq A^+$. Hence by disjunctiveness, $x \leq k_i$ for some i; so $x \in (M \downarrow k_i)$, as desired.

$\Leftarrow$: Suppose that $h, h_1, \ldots, h_m \in H$ and $h \leq h_1 + \cdots + h_m$ $(m > 0)$. Let $M = \{h\}$, and take the function f corresponding to M. Then

$$h \in (M \downarrow h) = fh \subseteq fh_1 \cup \ldots \cup fh_m = (M \downarrow h_1) \cup \ldots \cup (M \downarrow h_m),$$

so $h \leq h_i$ for some i. $\qquad\qquad\square$

A BA A is a *semigroup algebra* if it is generated by a subset H with the following properties: (1) $0, 1 \in H$; (2) H is closed under the operation $\cdot$ of A; (3) $H \backslash \{0\}$ is disjunctive. Here are three important examples of semigroup algebras:

A. *Tree algebras.* Let $A = \text{TreeAlg}\, T$. Without loss of generality T has only one root. Set $H = \{T \uparrow t : t \in T\} \cup \{0, 1\}$. The conditions for a semigroup algebra are easily verified.

B. *Interval algebras.* Let $A = \text{IntAlg}\, L$, where L is a linear ordering with first element 0_L. Let $H = \{[0_L, a) : a \in L\} \cup \{1\}$. Again the indicated conditions are easily checked.

C. *Free algebras.* Let A be freely generated by X, and set $H = \{x \in A : x \text{ is a finite product of members of } X\} \cup \{0, 1\}$. The indicated conditions clearly hold.

It is also useful to note that if A is a semigroup algebra, then so is $\mathrm{Dup}A$.

Proposition 2.2. *Suppose that A is a semigroup algebra with associated semigroup H, B is a BA, and f is a homomorphism from $(H, \cdot)$ into the semigroup $(B, \cdot)$ preserving 0 and 1. Then f has a unique extension to a homomorphism from A into B. Moreover, if f is onto, the extension is too. Finally, if B is a semigroup algebra on $(K, \cdot)$ and f is an isomorphism from $(H, \cdot)$ into $(K, \cdot)$ preserving 0 and 1, then the extension is an isomorphism into.*

Proof. In order to apply Sikorski's criterion, let $b_0, \ldots, b_{m-1}, c_0, \ldots, c_{n-1}$ be distinct elements of H and suppose that

$$b_0 \cdot \ldots \cdot b_{m-1} \cdot -c_0 \cdot \ldots \cdot -c_{n-1} = 0.$$

Without loss of generality, $m > 0$ and each c_i is different from 0. If $n = 0$, then

$$f b_0 \cdot \ldots \cdot f b_{m-1} = f(b_0 \cdot \ldots \cdot b_{m-1}) = f0 = 0,$$

as desired. Assume that $n > 0$. Then $b_0 \cdot \ldots \cdot b_{m-1} \leq c_0 + \cdots + c_{n-1}$, so $b_0 \cdot \ldots \cdot b_{m-1} \leq c_i$ for some i; hence $b_0 \cdot \ldots \cdot b_{m-1} \cdot c_i = b_0 \cdot \ldots \cdot b_{m-1}$ and

$$\begin{aligned}
f b_0 &\cdot \ldots \cdot f b_{m-1} \cdot -f c_0 \cdot \ldots \cdot -f c_{n-1} \\
&= f(b_0 \cdot \ldots \cdot b_{m-1}) \cdot -f c_0 \cdot \ldots \cdot -f c_{n-1} \\
&= f(b_0 \cdot \ldots \cdot b_{m-1} \cdot c_i) \cdot -f c_0 \cdot \ldots \cdot -f c_{n-1} \\
&= f b_0 \cdot \ldots \cdot f b_{m-1} \cdot f c_i \cdot -f c_0 \cdot \ldots \cdot -f c_{n-1} \\
&= 0,
\end{aligned}$$

as desired. Clearly if f is onto, then the extension is onto.

Assume the hypothesis of "Finally...". Let $b_0, \ldots, b_{m-1}, c_0, \ldots, c_{n-1}$ be distinct elements of H such that

$$f b_0 \cdot \ldots \cdot f b_{m-1} \cdot -f c_0 \cdot \ldots \cdot -f c_{n-1} = 0.$$

We want to show that $b_0 \cdot \ldots \cdot b_{m-1} \cdot -c_0 \cdot \ldots \cdot -c_{n-1} = 0$. Wlog $m, n > 0$. Thus either $b_0 \cdot \ldots \cdot b_{m-1} = 0$, as desired, or $f(b_0 \cdot \ldots \cdot b_{m-1}) \in K \backslash \{0\}$, and $f(b_0 \cdot \ldots \cdot b_{m-1}) \leq f c_0 + \cdots + f c_{n-1}$, so there is an $i < n$ such that $f(b_0 \cdot \ldots \cdot b_{m-1}) \leq f c_i$. Hence $f(b_0 \cdot \ldots \cdot b_{m-1}) = f(b_0 \cdot \ldots \cdot b_{m-1} \cdot c_i)$, so $b_0 \cdot \ldots \cdot b_{m-1} = b_0 \cdot \ldots \cdot b_{m-1} \cdot c_i$ since f is one-one, and the desired conclusion follows. $\square$

Corollary 2.3. *If A and B are semigroup algebras both with the same associated semigroup $(H, \cdot)$, then there is an isomorphism from A onto B which fixes H pointwise.* $\square$

Now we indicate the connection of the exponential of a BA with semigroup algebras.

Proposition 2.4. *For any BA A, $\mathrm{Exp}A$ is a semigroup algebra on a semigroup isomorphic to $(A, \cdot)$.*

Proof. For any $a \in A$ let $fa = \mathscr{V}(\mathcal{S}a)$, and let $H = f[A]$. We want to show that $\mathrm{Exp}A$ is a semigroup algebra on H and f is an isomorphism from $(A, \cdot)$ onto $(H, \cap)$. Clearly $f0 = 0$ and $f1 = 1$. If $a, b \in A$, then

$$f(a \cdot b) = \mathscr{V}(\mathcal{S}(a \cdot b)) = \mathscr{V}(\mathcal{S}a \cap \mathcal{S}b) = \mathscr{V}(\mathcal{S}a) \cap \mathscr{V}(\mathcal{S}b) = fa \cap fb.$$

If $a \neq b$, say $a \not\leq b$; then $\mathcal{S}a \in \mathscr{V}(\mathcal{S}a)$ but $\mathcal{S}a \notin \mathscr{V}(\mathcal{S}b)$; this shows that f is one-one. So we have checked that f is an isomorphism from $(A, \cdot)$ onto $(H, \cap)$ taking 0 to 0 and 1 to 1.

Note that H generates $\mathrm{Exp}A$ by Proposition 1.13. Finally, the disjunctive property follows like this: suppose that $\mathscr{V}(\mathcal{S}a) \subseteq \mathscr{V}(\mathcal{S}b_1) \cup \ldots \cup \mathscr{V}(\mathcal{S}b_m)$. Now $\mathcal{S}a \in \mathscr{V}(\mathcal{S}a)$, so $\mathcal{S}a \in \mathscr{V}(\mathcal{S}b_i)$ for some i, and hence $a \leq b_i$, as desired. □

The following result will also be useful.

Proposition 2.5. *For any BA A, $\mathrm{Exp}A$ embeds in $\prod_{n \geq 1} A^{*n}$, where A^{*n} denotes the free product of n copies of A.*

Proof. We use the notation of the proof of Proposition 2.4. For each $n \geq 1$ define $g_n : H \to A^{*n}$ as follows:

$$g_n fa = \prod_{i < n} h_i a,$$

where h_i is the natural embedding of A into the i-th factor of A^{*n}. Clearly g_n is a homomorphism from $(H, \cdot)$ into $(A^{*n}, \cdot)$ taking 0 to 0 and 1 to 1. Hence by Proposition 2.2 it extends to a homomorphism, still denoted by g_n, from $\mathrm{Exp}A$ into A^{*n}. For any $x \in \mathrm{Exp}A$ let $(kx)_n = g_n x$ for all $n \geq 1$. Clearly k is a homomorphism from $\mathrm{Exp}A$ into $\prod_{n \geq 1} A^{*n}$, so it suffices to show that k is one-one. We take an arbitrary non-zero member of $\mathrm{Exp}A$; we may assume that it has the form $\mathscr{V}(\mathcal{S}a_0, \ldots, \mathcal{S}a_{m-1})$, with each $a_i \neq 0$. Then

$$
\begin{aligned}
(k(\mathscr{V}(\mathcal{S}a_0, \ldots, \mathcal{S}a_{m-1}))_m &= g_m \mathscr{V}(\mathcal{S}a_0, \ldots, \mathcal{S}a_{m-1}) \\
&= g_m(\mathscr{V}(\mathcal{S}(a_0 + \cdots + a_{m-1})) \backslash (\mathscr{V}(\mathcal{S}(-a_0)) \cup \ldots \cup \mathscr{V}(\mathcal{S}(-a_{m-1}))) \\
&= \prod_{i<m} h_i(a_0 + \cdots + a_{m-1}) \cdot - \prod_{i<m} h_i(-a_0) \cdot \ldots \cdot - \prod_{i<m} h_i(-a_{m-1}) \\
&= \prod_{i<m} h_i(a_0 + \cdots + a_{m-1}) \cdot \sum_{i<m} h_i a_0 \cdot \ldots \cdot \sum_{i<m} h_i a_{m-1} \\
&\geq \prod_{i<m} h_i(a_0 + \cdots + a_{m-1}) \cdot \prod_{i<m} h_i a_i \\
&= \prod_{i<m} h_i a_i \neq 0. \quad \square
\end{aligned}
$$

Proposition 2.6. *For any BA, there is a homomorphism from $\mathrm{Exp}A$ onto A.*

Proof. By Proposition 2.4, $\mathrm{Exp}A$ is a semigroup algebra on a semigroup H, with an isomorphism f from $(H, \cdot)$ onto $(A, \cdot)$. Then by Proposition 2.2 there is an extension of f to a homomorphism from $\mathrm{Exp}A$ onto A.   □

Proposition 2.7. *If f is a homomorphism from A into B, then there is a homomorphism g from $\mathrm{Exp}A$ into $\mathrm{Exp}B$. Moreover, if f is onto, then g may be taken to be onto.*

Proof. By Proposition 2.4, the algebras $\mathrm{Exp}A$ and $\mathrm{Exp}B$ are semigroup algebras on semigroups H and K isomorphic to $(A, \cdot)$ and $(B, \cdot)$ respectively; hence f yields a homomorphism from H into K preserving 0 and 1, and the result follows by Proposition 2.2. The last statement of the proposition is obvious.   □

Proposition 2.8. *Any semigroup algebra can be isomorphically embedded in its exponential. Hence for any BA A, the algebra $\mathrm{Exp}\,A$ can be isomorphically embedded in $\mathrm{Exp}\,\mathrm{Exp}\,A$.*

Proof. Let A be a semigroup algebra. By Proposition 2.4, there is an isomorphism f from $(A, \cdot)$ onto a semigroup $(K, \cdot)$ such that $\mathrm{Exp}A$ is a semigroup on $(K, \cdot)$. But A is a semigroup algebra on some semigroup $(H, \cdot)$, so there is an isomorphism of $(H, \cdot)$ into $(K, \cdot)$. By the final part of Proposition 2.2, our proposition follows.   □

Pseudo-tree algebras

A *pseudo-tree* is a partially ordered system $(T, \leq)$ such that for each $t \in T$ the set $T \downarrow t$ is simply ordered. Thus this notion generalizes that of a tree, where $T \downarrow t$ is required to be well-ordered. We define $\mathrm{Treealg}\,T$ to be the subalgebra of $\mathscr{P}T$ generated by $\{T \uparrow t : t \in T\}$; such algebras are called *pseudo-tree algebras*. Pseudo-tree algebras are treated thoroughly in Koppelberg, Monk [92].

Much of the theory of tree algebras described in §16 of the BA Handbook, Vol. 1, carries over to pseudo-tree algebras. In particular, the normal form theorem 16.3 holds for pseudo-tree algebras. (In 16.3(b), the assumption should be that T has a smallest element. Note that a pseudo-tree may have only one root while having elements with no roots below them.) The proof of 16.3 as given works for pseudo-tree algebras. Then 16.4 follows. 16.6 also holds, but its proof must be modified, since at one point the well-ordering is used. The change that should be made is as follows. Where w is chosen, at the bottom of page 259, choose w instead to be minimal among all of the (finitely many) elements of

$$(*) \qquad \{\tau\} \cup \Sigma \cup \bigcup_{i=1}^{n} (\{t(i) \cup S(i)\})$$

which are in $\varepsilon \backslash e_i$, or simply let w be any element of $\varepsilon \backslash e_i$ if there are no such elements. Condition (16) should then be changed to say that if $x \in [\tau, w]$, x is among the elements $(*)$, and $x \notin e_i$, then $x = w$. With these changes the proof works for pseudo-tree algebras.

Now we want to give an abstract characterization of pseudo-tree algebras before proceeding with our survey of tree-algebra results which carry over to pseudo-tree algebras. For this purpose we need some easy propositions. A subset R of a BA A is a *ramification set* provided that any two elements of R are either comparable or disjoint. Thus R is then a pseudo-tree under the inverse ordering of the BA.

Proposition 2.9. *Let $(P, \leq)$ be a partially ordered system. Then $\{P \uparrow p : p \in P\}$ is a disjunctive set in $\mathscr{P}P$.*

Proof. Obviously 0 is not in the indicated set. Now suppose that $p, p_1, \ldots, p_n \in P$, where $n > 0$, and assume that $P \uparrow p \subseteq (P \uparrow p_1) \cup \ldots \cup (P \uparrow p_n)$. Then $p \in (P \uparrow p)$, and hence $p \in (P \uparrow p_i)$ for some i, and hence $(P \uparrow p) \subseteq (P \uparrow p_i)$, as desired. $\square$

As a corollary of Proposition 2.9 we see that every pseudo-tree algebra is a semigroup algebra.

Proposition 2.10. *Let R be a disjunctive ramification set of non-zero elements and let X, Y be finite subsets of R. Then $\prod X \leq \sum Y$ iff one of the following three conditions holds:*
(1) $X = 0$ and $\sum Y = 1$;
(2) $x \cdot y = 0$ for some $x, y \in X$;
(3) $x \leq y$ for some $x \in X$, $y \in Y$.

Proof. Obviously any of (1)–(3) implies that $\prod X \leq \sum Y$. Now suppose that $\prod X \leq \sum Y$ and (1) and (2) do not hold. Note that if $X = 0$ then $\sum Y = 1$; hence $X \neq 0$. From the falsity of (2) it then follows that $\prod X \in X$ and $Y \neq 0$. Then disjunctiveness yields (3). $\square$

Proposition 2.11. *Let $R \subseteq A^+$ be a ramification set, and let S be a subset of R maximal among disjunctive subsets of R. Then $\langle S \rangle = \langle R \rangle$.*

Proof. We need only show that $R \subseteq \langle S \rangle$; so let $r \in R \backslash S$. Since $r \notin S$, the set $S \cup \{r\}$ is not disjunctive. There are then two cases:

    Case 1. $r \leq s_1 + \cdots + s_n$ for certain $s_1, \ldots, s_n \in S$ $(n > 0)$, but $r \not\leq s_i$ for all i. Let n be minimal such that this can happen. By the minimality, $r \cdot s_i \neq 0$ for all i, so $s_i \leq r$ and hence $r = s_1 + \cdots + s_n \in \langle S \rangle$, as desired.

    Case 2. $s_1 \leq r + s_2 + \cdots + s_n$ for certain $s_1, \ldots, s_n \in S$ $(n > 0)$, but $s_1 \not\leq r$ and $s_1 \not\leq s_i$ for all $i > 1$. Again, take n minimal for this situation. Note that $n > 1$, by the minimality of n the elements $r, s_2, \ldots, s_n$ are pairwise disjoint, and, as in Case 1, $s_1 = r + s_2 + \cdots + s_n$. Hence $r = s_1 \cdot -(s_2 + \cdots + s_n) \in \langle S \rangle$, as desired. $\square$

With these preliminaries over, we can now give our abstract characterization of pseudo-tree algebras. At the same time we can establish 16.7 of the BA Handbook for pseudo-tree algebras; it can also be proved directly.

Theorem 2.12. *For any BA A, the following conditions are equivalent:*
    (i) A is isomorphic to $\mathrm{Treealg}\, T$ for some pseudo-tree T with a minimum element;

(ii) A is isomorphic to Treealg T for some pseudo-tree T;

(iii) A is generated by a ramification set;

(iv) A is generated by a ramification set $S \subseteq A^+$ such that $1 \in S$ and S is disjunctive.

Proof. Obviously $(i) \Rightarrow (ii)$, and it is also clear that $(ii) \Rightarrow (iii)$. For $(iii) \Rightarrow (iv)$, suppose that R is a ramification set which generates A. We may assume that $0 \notin R$ and $1 \in R$. Then by Proposition 2.11 we get a ramification set S as desired in (iv).

Finally, we prove $(iv) \Rightarrow (i)$. Clearly S is a pseudo-tree with minimum element under the converse of the Boolean ordering; so it suffices to show that A is isomorphic to Treealg S. By Proposition 2.1, there is a homomorphism f from A into $\mathscr{P}S$ such that $fs = S \uparrow s$ for all $s \in S$; here $S \uparrow s$ is in the tree sense. Clearly f maps onto Treealg S. It is also one-one; we see this by using Sikorski's criterion: assume that

$$t_0 \cdot \ldots \cdot t_{m-1} \cdot -s_0 \cdot \ldots \cdot -s_{n-1} \neq 0,$$

where all t_i and s_i are in S and $m, n \in \omega$. Since $1 \in S$, we may assume that $m > 0$. Then $u \overset{\text{def}}{=} t_0 \cdot \ldots \cdot t_{m-1}$ is an element of S, and $u \not\leq s_i$ for all i. Hence

$$u \in ft_0 \cap \ldots ft_{m-1} \cap -fs_0 \cap \ldots \cap -fs_{n-1},$$

as desired. $\square$

We continue our survey of tree algebra results which extend to pseudo-tree algebras. Lemma 16.8 of the BA Handbook, Vol. 1, extends with no changes in its proof. Proposition 16.9 extends, with some changes in the proof: if T' has only finitely many roots and each element lies above a root, proceed as in Case 2 of the old proof; otherwise proceed as in Case 1. The description of atoms in 16.10 is the same for pseudo-tree algebas. The description of ultrafilters in 16.11 carries over, where an initial chain is required to be non-empty if T has only finitely many roots and each element is above a root. This brings us to 16.12, which requires an essentially new proof in the pseudo-tree case:

Theorem 2.13. *Every pseudo-tree algebra embeds into an interval algebra.*

Proof. Let T be a pseudo-tree algebra; we may assume by 16.7 that T has a minimum element 0_T. We consider a first-order language with a binary relation symbol $<$ and for each $t \in T \backslash \{0_T\}$ two individual constants $\mathbf{a}_t, \mathbf{b}_t$. Let Σ be a set of first-order sentences expressing that in any model $\mathfrak{A}$ of Σ the following hold:

A is a dense linear order with first element $0_L^{\mathfrak{A}}$.

$0_T^{\mathfrak{A}} < \mathbf{a}_t^{\mathfrak{A}} < \mathbf{b}_t^{\mathfrak{A}}$ for all $t \in T \backslash \{0_T\}$.

$\mathbf{a}_s < \mathbf{a}_t$ and $\mathbf{b}_t < \mathbf{b}_s$ for $0_T < t < s$ in T.

$\mathbf{b}_t < \mathbf{a}_s$ or $\mathbf{b}_s < \mathbf{a}_t$ if $s, t \in T \backslash \{0_T\}$ and s, t are incomparable.

Clearly if Σ has a model $\mathfrak{A}$ with universe A, then there is an embedding from Treealg T into Intalg A mapping $T \uparrow 0_T$ to A and $T \uparrow t$ to the interval $[a_t, b_t)$ for $t \in T \backslash \{0_T\}$.

To show that Σ has a model, take any finite subset Σ_0 of Σ. Then there is a finite subset T_0 of T such that only members of T_0 occur as indices in the members of Σ_0. T_0 under the ordering of T is an ordinary finite tree, which can be embedded in an interval algebra by 16.12—and this yields a model of Σ_0. $\qquad\square$

Recently it has been shown that the converse of Theorem 2.13 holds: any subalgebra of an interval algebra is isomorphic to a pseudo-tree algebra; see Purisch [94]. Proposition 16.17 of the BA Handbook, Vol. 1, extends with no changes in the proof to pseudo-tree algebras. The proof of Proposition 16.18 also extends; but we can give a shorter proof, which works also for tree algebras:

Proposition 2.14. *Every homomorphic image of a pseudo-tree algebra is isomorphic to a pseudo-tree algebra.*

Proof. Let A be a pseudo-tree algebra, and f a homomorphism from A onto some algebra B. Then by Theorem 2.12, A is generated by a ramification set R. Clearly $f[R]$ is also a ramification set, and it generates B. Hence by Theorem 2.12 again, B is isomorphic to a pseudo-tree algebra. $\qquad\square$

Simple extensions of Boolean algebras

Given BA's A and B, we call B a *simple extension* of A provided that A is a subalgebra of B and $B = \langle A \cup \{x\}\rangle$ for some $x \in B$; then we write $B = A(x)$. We recall that each element of $A(x)$ can be written in the form $a \cdot x + b \cdot -x$ with $a, b \in A$; or in the form $c + a \cdot x + b \cdot -x$ with a, b, c pairwise disjoint elements of A. We now introduce some important ideals for studying simple extensions. If A is a subalgebra of B and $x \in B$, we let $A \restriction x = \{a \in A : a \leq x\}$. This is a slight extension of the usual notion; x is not necessarily in A. Under the same assumptions we let

$$\mathrm{Smp}_x^A = \langle (A \restriction x) \cup (A \restriction -x)\rangle^{\mathrm{Id}},$$

the ideal in A generated by $(A \restriction x) \cup (A \restriction -x)$. The three ideals $A \restriction x$, $A \restriction -x$, and Smp_x^A are important for studying simple extensions.

Proposition 2.15. *Let $A(x)$ be a simple extension of A. Then $A = A(x)$ iff $x \in \mathrm{Smp}_x^A$.*

Proof. If $A = A(x)$, then $x \in A \restriction x \subseteq \mathrm{Smp}_x^A$, as desired. Conversely, suppose that $x \in \mathrm{Smp}_x^A$. Write $x = a + b$, with $a \in A \restriction x$ and $b \in A \restriction -x$. Clearly $b = 0$, so $x = a \in A$. $\qquad\square$

Proposition 2.16. *Let $A(x)$ be a simple extension of A, and let $a \in A$. Then the following conditions are equivalent:*

    (i) $a \in \mathrm{Smp}_x^A$;
    (ii) $a = b + c$ for some $b \in A \restriction x$ and $c \in A \restriction -x$;
    (iii) $a \cdot x \in A$;
    (iv) $a \cdot -x \in A$;
    (v) For all $y \in A(x)$, if $y \leq a$ then $y \in A$; that is, $A(x) \restriction a = A \restriction a$;
    (vi) For all $y \in A(x)$, if $y \leq a$ then $y \in \mathrm{Smp}_x^A$.

Proof. Clearly $(i) \Leftrightarrow (ii)$. Assume (ii). Then $a \cdot x = b \in A$, i.e., (iii) holds. Assume (iii). Then $a \cdot -x = a \cdot -(a \cdot x) \in A$, i.e., (iv) holds. Similarly $(iv) \Rightarrow (iii)$. If (iii) holds, then (iv) holds and $a = a \cdot x + a \cdot -x$, so (ii) holds. $(ii) \Rightarrow (v)$: Assume (ii), and suppose that $y \in A(x)$ and $y \leq a$. Write $y = u \cdot x + v \cdot -x$ with $u, v \in A$. Then

$$y \cdot x = a \cdot y \cdot x = a \cdot u \cdot x \in A$$

by (iii). Similarly $y \cdot -x \in A$, so $y \in A$. Clearly $(v) \Leftrightarrow (vi)$. $(v) \Rightarrow (iii)$: $x \cdot a \in A(x)$ and $x \cdot a \leq a$, so $x \cdot a \in A$. $\qquad\square$

Corollary 2.17. *If $A(x)$ is a simple extension of A, then Smp_x^A is an ideal of $A(x)$.* $\qquad\square$

Proposition 2.18. *Suppose that $A(x)$ and $A(y)$ are simple extensions of A. Then the following conditions are equivalent:*

(i) There is a homomorphism from $A(x)$ into $A(y)$ which is the identity on A and sends x to y.

(ii) $A \restriction x \subseteq A \restriction y$ and $A \restriction -x \subseteq A \restriction -y$.

And also the following two conditions are equivalent:

(iii) There is an isomorphism from $A(x)$ onto $A(y)$ which is the identity on A and sends x to y.

(iv) $A \restriction x = A \restriction y$ and $A \restriction -x = A \restriction -y$.

Proof. $(i) \Rightarrow (ii)$: obvious. $(ii) \Rightarrow (i)$: by Sikorski's extension criterion. Since the homomorphism of (i) is unique, the equivalence of (iii) and (iv) is clear. $\qquad\square$

Proposition 2.19. *Let A be a BA, and let I_0 and I_1 be two ideals of A such that $I_0 \cap I_1 = \{0\}$. Then there is a simple extension $A(x)$ of A such that $A \restriction x = I_0$ and $A \restriction -x = I_1$.*

Proof. Let $A(y)$ be a free extension of A by y, that is, let $A(y)$ be the the free product of A with a four-element BA B, where $0 < y < 1$ in B. Consider the following ideal K of $A(y)$:

$$K = \langle \{a \cdot -y : a \in I_0\} \cup \{a \cdot y : a \in I_1\} \rangle^{\mathrm{Id}}.$$

Let f be the natural mapping from A onto A/K. It suffices to prove the following things:

(1) $A \cap K = \{0\}$.

(2) $(A/K) \restriction (y/K) = f[I_0]$.

(3) $(A/K) \restriction -(y/K) = f[I_1]$.

For (1), if $a \in A \cap K$, then $a \leq b \cdot -y + c \cdot y$ for some $b \in I_0$ and $c \in I_1$. An easy argument using freeness then yields $a \leq b \cdot c = 0$, as desired.

For (2), first suppose that $a \in A$ and $a/K \leq y/K$. Thus $a \cdot -y \in K$, so $a \cdot -y \leq b \cdot -y + c \cdot y$ for some $b \in I_0$ and $c \in I_1$. An easy argument then yields

$a \leq b$, and hence $a \in I_0$, as desired. On the other hand, if $a \in I_0$, obviously $a \cdot -y \in K$, and hence $a/K \leq y/K$ and $a/K \in (A/K) \restriction (y/K)$, as desired.

A similar argument proves (3). $\square$

Minimal extensions of Boolean algebras

We say that B is a *minimal extension* of a BA A, in symbols $A \leq_m B$, if B *is* an extension of A and there is no subalgebra C of B such that $A \subset C \subset B$. Clearly then B is a simple extension of A. This notion is studied in Koppelberg [89b]. First we want to see what this means in terms of the ideal Smp_x^A:

Proposition 2.20. *Let $A(x)$ be a simple extension of A. Then the following conditions are equivalent:*

(a) $A \leq_m A(x)$.

(b) Smp_x^A *is either equal to A or is a maximal ideal of A.*

(c) $A = \langle \{a \in A : a \text{ is comparable with } x\} \rangle$.

(d) *There is a $G \subseteq A$ which generates A and consists exclusively of elements comparable with x.*

(e) *If $y \in A(x) \backslash A$, then $x \triangle y \in A$.*

Proof. Obviously $(c) \Leftrightarrow (d)$. $(a) \Rightarrow (b)$: assume that (b) fails. Then there is an element $a \in A$ such that neither a nor $-a$ is in Smp_x^A. Then, we claim, $A \subset A(a \cdot x) \subset A(x)$. In fact, $a \cdot x \notin A$ by Proposition 2.16. And if $x \in A(a \cdot x)$, then we can write $x = b \cdot a \cdot x + c \cdot -(a \cdot x)$ with $b, c \in A$. But then $x = b \cdot a \cdot x + c \cdot -a + c \cdot -x$, hence $c \cdot -x = 0$ and $x = b \cdot a \cdot x + c \cdot -a$. Therefore $-a \cdot x = c \cdot -a \in A$, and hence by Proposition 2.16, $-a \in \text{Smp}_x^A$, contradiction.

$(b) \Rightarrow (c)$: Assume (b). Then Smp_x^A generates A. Let $G = \{a \in A : a \text{ is comparable with } x\}$. Now $A \restriction x \subseteq G$, and if $a \in A \restriction -x$ then $-a \in G$. It follows that $A = \langle \text{Smp}_x^A \rangle \subseteq \langle G \rangle$, and hence G generates A.

$(d) \Rightarrow (b)$: Clearly $G \subseteq \text{Smp}_x^A \cup \{a : -a \in \text{Smp}_x^A\}$, and the latter set is a subalgebra of A; hence it is all of A, which means that (b) holds.

$(b) \Rightarrow (e)$: Assume (b), and suppose that $y \in A(x) \backslash A$. Write $y = a + b \cdot x + c \cdot -x$, where a, b, c are pairwise disjoint elements of A. If b and c are both elements of Smp_x^A, then $b \cdot x$ and $c \cdot -x$ are both elements of A by Proposition 2.16, and so $y \in A$, contradiction. Assume that $b \notin \text{Smp}_x^A$. Hence $-b \in \text{Smp}_x^A$ by (b). Now $y \cdot b = x \cdot b$, so $x \triangle y \leq -b$, and hence $x \triangle y \in A$ by Corollary 2.17. If $c \notin \text{Smp}_x^A$, we obtain $(-x) \triangle y \in A$ similarly; then note that $(-x) \triangle y = -(x \triangle y)$.

$(e) \Rightarrow (a)$: If $y \in A(x) \backslash A$, then $x = x \triangle y \triangle y \in A(y)$, so $A(y) = A(x)$. $\square$

Proposition 2.21. *If $A \leq B$, $x \in B$, and $A \restriction x$ is a maximal ideal in A, then $A \leq_m B$.*

Proof. By Proposition 2.20. $\square$

Proposition 2.22. *Let $A \leq B$, let $f : B \to Q$ be an epimorphism, and set $P = f[A]$. Then $A \leq_m B$ implies that $P \leq_m Q$.*

If, moreover, $\ker f \subseteq A$, then $A \leq_m B$ iff $P \leq_m Q$.

Proof. This is a result of universal-algebraic nonsense: the function assigning to each subalgebra C of Q the subalgebra $f^{-1}[C]$ of B is one-one, and it maps $\{C : P \leq C \leq Q\}$ into $\{D : A \leq D \leq B\}$. In case $\ker f \subseteq A$, it maps onto the latter set: the preimage of such a D is $f[D]$, and $P \leq f[D] \leq Q$. $\qquad\square$

Proposition 2.23. *Assume that $A \leq B \leq M \geq D$. Set $P = A \cap D$ and $Q = B \cap D$. Then $A \leq_m B$ implies that $P \leq_m Q$.*

Proof. Assume all the hypotheses. We may also assume that $P \neq Q$. Now take any $x \in Q \backslash P$; we want to show that $Q = P(x)$. To this end, take any $y \in Q$; we show that $y \in P(x)$. Now $x \in B$ since $x \in Q$. Now $x \notin A$ since $x \notin P$. It follows that $B = A(x)$. We may assume that $y \notin P$; hence $y \in B \backslash A$. Now by Proposition 2.20(e) we get $x \triangle y \in A$. Also, $x \in D$ and $y \in D$, so $x \triangle y \in D$; hence $x \triangle y \in P$. It follows that $y = (x \triangle y) \triangle x \in P(x)$, as desired. $\qquad\square$

Proposition 2.24. *Suppose that $A(x)$ is a proper minimal extension of A. Then the following conditions are equivalent:*

(i) $A \upharpoonright x$ and $A \upharpoonright -x$ are non-principal ideals.

(ii) A is dense in $A(x)$.

Proof. Assume (i). Take any non-zero element y of $A(x)$. Wlog we may assume that $y = a \cdot x$ for some $a \in A$. By Proposition 2.20 there are two cases.

Case 1. $a \in \mathrm{Smp}_x^A$. Say $a = b + c$ with $b \in A \upharpoonright x$ and $c \in A \upharpoonright -x$. Then $a \cdot x = b \in A$ and there is nothing to prove.

Case 2. $-a \in \mathrm{Smp}_x^A$. Say $-a = b + c$ with $b \in A \upharpoonright x$ and $c \in A \upharpoonright -x$. Choose $d \in A \upharpoonright x$ with $b < d$ (which we can do because $A \upharpoonright x$ is non-principal). Then

$$d \cdot -b \cdot -a = d \cdot -b \cdot (b + c) \leq d \cdot c = 0$$

since $c \in A \upharpoonright -x$. Hence $0 \neq d \cdot -b \leq a \cdot x$, as desired.

Now assume (ii). To show that $A \upharpoonright x$ is non-principal, let $a \in A \upharpoonright x$. Now $x \cdot -a \neq 0$, so we can choose a non-zero $b \in A$ such that $b \leq x \cdot -a$. Then $a < a + b \leq x$, as desired. Similarly, $A \upharpoonright -x$ is non-principal. $\qquad\square$

Proposition 2.25. *Suppose that $A(x)$ is a proper minimal extension of A, and $A \upharpoonright x$ is a principal ideal generated by an element a^*. Set $y = -a^* \cdot x$. Then:*

(i) $y \notin A$, and hence $A(x) = A(y)$;

(ii) for all $a \in A$, $y \leq a$ iff $-a \in \mathrm{Smp}_x^A$;

(iii) y is an atom of $A(x)$;

(iv) if D is dense in A, then $\langle D \cup \{y\} \rangle$ is dense in $A(x)$.

Proof. (i): Clearly $y \notin A$, since otherwise $y \leq a^*$, $y = 0$, $x \leq a^*$, and $x = a^* \in A$. So $A(x) = A(y)$ by minimality.

(ii). First assume that $y \leq a$. Thus $-a \leq a^* + -x$. Now $-a = -a \cdot a^* + -a \cdot -a^*$, and $-a \cdot -a^* \leq -x$, proving that $-a \in \mathrm{Smp}_x^A$.

Second, assume that $-a \in \mathrm{Smp}_x^A$. Say $-a = b + c$ with $b \in A \restriction x$ and $c \in A \restriction -x$. Then $-a^* \cdot x \cdot -a = -a^* \cdot b = 0$, showing that $y \leq a$.

(iii). Clearly $y \neq 0$, by (i). Suppose that $z \leq y$; we show that $z = 0$ or $z = y$. Say $z = a \cdot x + b \cdot -x$ with $a, b \in A$. Since $y \leq x$, we have $b \cdot -x = 0$. By Proposition 2.20, either $a \in \mathrm{Smp}_x^A$ or $-a \in \mathrm{Smp}_x^A$. If $-a \in \mathrm{Smp}_x^A$, then $y \leq a$ by (ii), hence $y = z$, as desired. Now suppose that $a \in \mathrm{Smp}_x^A$. Write $a = c + d$ with $c \in A \restriction x$ and $d \in A \restriction -x$. Then $c \leq a^*$, hence $a \cdot x = c \cdot x \leq a^* \cdot x$, but also $a \cdot x = z \leq y = -a^* \cdot x$, so $a \cdot x = 0$, as desired.

(iv): Assume that u is a non-zero element of $A(x)$. If $u \cdot y \neq 0$, then $y \leq u$ by (iii), as desired. So, assume that $u \cdot y = 0$. Then we can write $u = b \cdot -y$ with $b \in A$, by (i). Choose $d \in D^+$ so that $d \leq b$. If $d \cdot -y = 0$, then $d \leq y$, hence $d = y$, from which it follows that $y \in A$, contradicting (i). So $0 \neq d \cdot -y \leq b \cdot -y = u$, as desired. $\qquad\square$

Minimally generated Boolean algebras

Let A and B be BA's. A *representing chain* for B over A is a sequence $\langle C_\alpha : \alpha < \rho \rangle$ of BA's with the following properties:

(1) $\alpha < \beta < \rho$ implies that $C_\alpha \leq C_\beta$.
(2) If λ is a limit ordinal less than ρ, then $C_\lambda = \bigcup_{\alpha < \lambda} C_\alpha$.
(3) $C_0 = A$.
(4) $\bigcup_{\alpha < \rho} C_\alpha = B$.

B is *minimally generated over* A if there is a representing chain for B over A such that $C_\alpha \leq_{\mathrm{m}} C_{\alpha+1}$ whenever $\alpha + 1 < \rho$. And B is *minimally generated* if it is minimally generated over 2. We write $A \leq_{\mathrm{mg}} B$ to abbreviate that B is minimally generated over A. Finally, if $A \leq_{\mathrm{mg}} B$, then $\mathrm{len}(B : A)$ is the smallest ordinal ρ demonstrating the minimal generation of B over A, and if B is minimally generated, then $\mathrm{len}B = \mathrm{len}(B : 2)$. The notion of a minimally generated BA is due to S. Koppelberg [89b].

Proposition 2.26. (i) *Suppose that $A \leq B$ and f is a homomorphism from B onto Q. Let $P = f[A]$. Then:*

  (a) if $A \leq_{\mathrm{mg}} B$, then $P \leq_{\mathrm{mg}} Q$ and $\mathrm{len}(Q : P) \leq \mathrm{len}(B : A)$;

  (b) if A includes the kernel of f, then $A \leq_{\mathrm{mg}} B$ iff $P \leq_{\mathrm{mg}} Q$, and if one and hence both of these holds then $\mathrm{len}(Q : P) = \mathrm{len}(B : A)$.

  (ii) A homomorphic image Q of a minimally generated BA B is minimally generated. Moreover, $\mathrm{len}Q \leq \mathrm{len}B$.

Proof. by Proposition 2.22. $\qquad\square$

Proposition 2.27. (i) *Suppose that $A \leq B \leq M \geq D$ and $A \leq_{\mathrm{mg}} B$. Set $P = A \cap D$ and $Q = B \cap D$. Then $P \leq_{\mathrm{mg}} Q$, and $\mathrm{len}(Q : P) \leq \mathrm{len}(B : A)$.*

  (ii) Every subalgebra D of a minimally generated BA B is minimally generated. Moreover, $\mathrm{len}D \leq \mathrm{len}B$.

  (iii) If $A \leq_{\mathrm{mg}} B$ and $A \leq C \leq B$, then $A \leq_{\mathrm{mg}} C$.

Proof. (i) and (ii) are clear from Proposition 2.23. For (iii), let $D = C$ in (i). $\qquad\square$

Proposition 2.28. *Suppose that A is an atomless subalgebra of B and there is an element $u \in B$ which is independent over A, i.e., $a \cdot u \neq 0 \neq a \cdot -u$ for all $a \in A^+$. Then B is not minimally generated over A.*

Proof. Otherwise we would have $A \leq_{\mathrm{mg}} A(u)$ by Proposition 2.27(iii). Hence there is an $x \in A(u)$ such that $x \notin A$ and $A \leq_m A(x)$. Say $x = a + b \cdot u + c \cdot -u$ with a, b, c pairwise disjoint elements of A. From the independence of u over A it then follows that $A \restriction x = A \restriction a$, a principal ideal in A. In fact, if $v \leq x$, then $v \leq a + b + c$; if $v \cdot b \neq 0$, then $v \cdot b \cdot -u \neq 0$. But $v \cdot b \leq b \cdot u$, contradiction. So $v \cdot b = 0$, and similarly $v \cdot c = 0$. Similarly, $A \restriction -x$ is a principal ideal in A. So Smp_x^A is a principal ideal in A. By Proposition 2.20, this gives us an atom of A, contradiction. $\qquad\square$

Proposition 2.29. *If A and B are minimally generated, then so is $A \times B$.*

Proof. Let $\langle C_\alpha : \alpha < \rho \rangle$ be a representing sequence for A's minimal generation (thus with $C_0 = 2$ and $C_\alpha \leq_m C_{\alpha+1}$ for all $\alpha + 1 < \rho$), and let $\langle D_\alpha : \alpha < \sigma \rangle$ be similarly chosen for B. The desired sequence for $A \times B$ is $\langle 2 \rangle^\frown \langle E_\alpha : \alpha < \rho + \sigma \rangle$, where for $\alpha < \rho$ we set

$$E_\alpha = \{(a, b) : a \in C_\alpha \text{ and } b \in \{0, 1\}\},$$

and for $\alpha < \sigma$ we set

$$E_{\rho+\alpha} = \{(a, b) : a \in A \text{ and } b \in D_\alpha\}. \qquad\square$$

Proposition 2.30. *If $\langle A_i : i \in I \rangle$ is a system of minimally generated BA's, then so is $\prod_{i \in I}^{\mathrm{w}} A_i$.*

Proof. Wlog I is an infinite cardinal κ. For all $\beta < \kappa$ let $\langle C_{\beta\alpha} : 0 < \alpha < \rho_\beta \rangle$ be a representing sequence for A_β (for technical reasons starting at 1 rather than 0) such that $C_{\beta\alpha} \leq_m C_{\beta,\alpha+1}$ if $\alpha + 1 < \rho_\beta$. Let $\sigma = \sup_{\beta<\kappa} \rho_\beta$. For each $\xi < \sigma$ we define a subalgebra E_ξ of $\prod_{\beta<\kappa}^{\mathrm{w}} A_\beta$. Say $\delta < \kappa$ and $\mu \stackrel{\mathrm{def}}{=} \sum_{\beta<\delta} \rho_\beta \leq \xi < \sum_{\beta\leq\delta} \rho_\beta$; say $\xi = \mu + \varepsilon$ with $\varepsilon < \rho_\delta$. If $\varepsilon = 0$ we set

$$E_\xi = \{x \in \prod_{\beta<\kappa}^{\mathrm{w}} A_\beta : \forall \theta < \delta(x_\theta \in A_\theta) \text{ and}$$

$$[\forall \theta < \kappa(\delta \leq \theta \Rightarrow x_\theta = 1) \text{ or } \forall \theta < \kappa(\delta \leq \theta \Rightarrow x_\theta = 0)]\}.$$

(Note that this makes E_0 the two-element subalgebra of $\prod_{\beta<\kappa}^{\mathrm{w}} A_\beta$.) If $\varepsilon \neq 0$ we set

$$E_\xi = \{x \in \prod_{\beta<\kappa}^{\mathrm{w}} A_\beta : \forall \theta < \delta(x_\theta \in A_\theta) \text{ and } x_\delta \in C_{\delta\varepsilon} \text{ and}$$

$$[\forall \theta < \kappa(\delta < \theta \Rightarrow x_\theta = 1) \text{ or } \forall \theta < \kappa(\delta < \theta \Rightarrow x_\theta = 0)]\}.$$

Then $\langle E_\xi : \xi < \sigma \rangle$ is as desired. $\qquad\square$

Proposition 2.31. *Every interval algebra is minimally generated.*

Proof. Let A be an interval algebra. Then it is generated by a chain C. Enumerate C: $C = \{c_\alpha : \alpha < \rho\}$. For each $\alpha < 1 + \rho$ let $B_\alpha = \langle\{c_\beta : \beta < \alpha\}\rangle$. Then by Proposition 2.20 we have $B_\alpha \leq_m B_{\alpha+1}$ whenever $\alpha + 1 \leq 1 + \rho$, so this shows the minimal generation of A. $\qquad\square$

Proposition 2.32. *Every superatomic BA is minimally generated.*

Proof. Let B be superatomic. By Proposition 2.31 we may assume that B is infinite. Let $\langle I_\alpha : \alpha$ an ordinal$\rangle$ be the standard sequence of ideals associated with B (I_1 is the ideal generated by the atoms, etc.). For each α let R_α be a complete set of representatives of the atoms of B/I_α: thus for each $x \in R_\alpha$, x/I_α is an atom of B/I_α; $(x/I_\alpha) \cdot (y/I_\alpha) = 0$ for distinct $x, y \in R_\alpha$; and for each atom b of B/I_α there is an $x \in R_\alpha$ such that $b = x/I_\alpha$. Fix σ such that $I_\sigma = A$. It is easy to see that $X \overset{\text{def}}{=} \bigcup_{\alpha<\sigma} R_\alpha$ generates B. Now well-order X by levels: $X = \{x_\alpha : \alpha < \rho\}$, where if $x_\alpha \in R_\beta$ and $x_\gamma \in R_\delta$ and $\beta < \delta$, then $\alpha < \gamma$. For each $\alpha \leq \rho$ let $C_\alpha = \langle\{x_\beta : \beta < \alpha\}\rangle$ Clearly this gives a representing chain for B, so it just suffices to show that $C_\alpha \leq_m C_{\alpha+1}$ whenever $\alpha + 1 \leq \rho$. To prove this we need the following fact: for any $\alpha \leq \rho$,

$$\{x_\nu : \nu < \alpha\} \subseteq \mathrm{Smp}^{C_\alpha}_{x_\alpha}.$$

And to prove this, let $\nu < \alpha$. Say $x_\nu \in R_\beta$ and $x_\alpha \in R_\gamma$; thus $\beta \leq \gamma$. If $\beta < \gamma$, then $x_\nu \in I_\gamma$, hence $x_\nu \cdot x_\alpha \in I_\gamma$, and $I_\gamma \subseteq \langle\bigcup_{\delta<\gamma} R_\delta\rangle \subseteq C_\alpha$. So, by Proposition 2.16, $x_\nu \in \mathrm{Smp}^{C_\alpha}_{x_\alpha}$. If $\beta = \gamma$, then again $x_\nu \cdot x_\alpha \in C_\alpha$ and the desired conclusion follows.

From the fact and Proposition 2.20 it follows that $C_\alpha \leq_m C_{\alpha+1}$. $\qquad\square$

Proposition 2.33. *The free BA A on ω_1 free generators is not minimally generated.*

Proof. Suppose it is, and let $\langle B_\alpha : \alpha < \sigma\rangle$ be a representing chain which demonstrates this. Wlog $\sigma = \mathrm{len}A$, and $B_\alpha \subset B_{\alpha+1}$ whenever $\alpha + 1 < \sigma$. Clearly $\sigma \geq \omega_1$.

We claim that $\sigma = \omega_1$. Otherwise B_{ω_1} is a subalgebra of A, and it has a subalgebra C isomorphic to A. Hence by Proposition 2.27,

$$\omega_1 \geq \mathrm{len}B_{\omega_1} \geq \mathrm{len}C = \mathrm{len}A,$$

contradiction.

Let $\{x_\alpha : \alpha < \omega_1\}$ be the set of free generators of A, and for each $\alpha < \omega_1$ let $C_\alpha = \langle\{x_\beta : \beta < \alpha\}\rangle$. This gives another representing chain for A. Hence $K \overset{\text{def}}{=} \{\alpha < \omega_1 : B_\alpha = C_\alpha\}$ is a club in ω_1. Take any infinite member α of K. Now x_α is independent over B_α and A is minimally generated over B_α, which contradicts Proposition 2.28. $\qquad\square$

Proposition 2.34. *If an infinite BA A satisfies any of the following conditions then it is not minimally generated:*

    (1) A is complete;
    (2) A is σ-complete;
    (3) A is ω_1-saturated (in the sense of model theory);
    (4) A has the countable separation property;
    (5) A is the product of infinitely many non-trivial algebras.

Proof. Each of (1), (2), (3) implies (4), and (5) implies that A has an infinite CSP subalgebra. Hence it suffices to show that if A is CSP then it is not minimally generated (using Proposition 2.27(ii)). Now A has $\mathscr{P}\omega$ as a homomorphic image, and $\mathscr{P}\omega$ has an independent set of size ω_1, so the same is true of A, and the conclusion follows from Proposition 2.33 and Proposition 2.27(ii). $\square$

Theorem 2.35. *For every minimally generated BA B there is a dense subalgebra A of B such that A is isomorphic to a tree algebra and B is minimally generated over A.*

Proof. Fix a subset X of B generating B and a well-ordering $<_X$ of X such that if we let $B_x = \langle\{y : y <_X x\}\rangle$ then the chain $\langle B_x : x \in X\rangle$ demonstrates the minimal generation of B; we assume that $1 \in X$, and, moreover, that $x \notin B_x$ for all $x \in X\backslash\{1\}$. In particular, $0 \notin X$. Define $S = \{x \in X : x \neq 1,\ B_x$ is dense in $B_x(x)\}$, and set $T = X\backslash S$. Then:

(*) If $x \in T\backslash\{1\}$, then there is an ultrafilter F on B_x and an element $y \in B_x(x)$ such that $B_x(x) = B_x(y)$ and $\forall a \in B_x(y \leq a$ iff $a \in F)$.

In fact, by Proposition 2.24, one of $B_x \upharpoonright x$ and $B_x \upharpoonright -x$ is a principal ideal. Thus (*) follows from Proposition 2.25.

    By (*), wlog we may assume:

(**) If $x \in T\backslash\{1\}$ then there is an ultrafilter F_x on B_x such that $\forall a \in B_x(x \leq a$ iff $a \in F)$.

Next we claim that T is a tree under the inverse of the Boolean ordering. In fact, suppose $x, y, z \in T$ and $x < y, x < z$; we want to show that y and z are comparable. Say $y <_X z$. Then $y \in B_z$, so by the property we just got, $z \leq y$ or $z \leq -y$; and $z \leq -y$ is ruled out since $0 \neq x \leq y \cdot z$. Thus z and y are comparable. Moreover, if $x, y \in T$ and $x < y$, then $y <_X x$; otherwise $x <_X y$, hence $x \in B_y$, and so by this same property, $y \leq x$ or $y \leq -x$, both of which are false. So, T is a tree.

    Also note that for $u, v \in T$ we have that u and v are incomparable iff $u \cdot v = 0$. In fact, assume that u and v are incomparable; say $u <_X v$. Then by (**) it follows that $u \notin F_v$ and hence $-u \in F_v$ and $v \leq -u$, as desired.

    Next, T is disjunctive. For, assume that $t, t_1, \ldots, t_n \in T$, where $n > 0$, and $t \leq t_1 + \cdots + t_n$, but $t \not\leq t_i$ for all i. Wlog the t_i's are pairwise disjoint and $t_i < t$ for each i. So, $t = t_1 + \cdots + t_n$. If t is $<_X$-maximum in $\{t, t_1, \ldots, t_n\}$, then $t \in B_t$, contradiction. Otherwise some t_i is $<_X$-maximum in $\{t, t_1, \ldots, t_n\}$, and $t_i = t \cdot \sum_{j \neq i} -t_j$, hence $t_i \in B_{t_i}$, contradiction. So, T is disjunctive.

Now by the proof of Theorem 2.12$(iv) \Rightarrow (i)$ it follows that $A \overset{\text{def}}{=} \langle T \rangle$ is isomorphic to Treealg T.

We claim that A is dense in B. To prove this, we show by $<_X$-induction on $x \in X$ that $\langle \{y \in T : y <_X x\} \rangle$ is dense in B_x for all $x \in X$. For x the smallest element of X the algebra B_x has only two elements, and the conclusion is obvious. Now suppose that the statement holds for $x \in X$, and $x' \in X$ is the immediate successor of x under $<_X$. Now $B_{x'} = B_x(x)$. If $x \in T$, then B_x is not dense in $B_{x'}$, and so one of $B_x \restriction x$ and $B_x \restriction -x$ is principal, by Proposition 2.24. Then by Proposition 2.25(iv) our desired result continues to hold. On the other hand, if $x \in S$, then B_x is dense in $B_x(x)$ and the desired conclusion is obvious.

Next, for x limit under $<_X$, the induction hypothesis clearly implies the desired conclusion. Hence our inductive statement holds, and then it is clear that A is dense in B.

If $T = X$, then $A = B$ and we are through. So assume that $S \neq 0$. We define a new well-ordering $\ll$ on X by putting T before S: for $x, y \in X$, we define $x \ll y$ iff $(x, y \in T$ and $x <_X y)$ or $(x, y \in S$ and $x <_X y)$ or $(x \in T$ and $y \in S)$. For each $x \in S$ let $C_x = \langle \{y \in X : y \ll x\} \rangle$. We claim that $\langle \{C_x : x \in S\} \rangle$ demonstrates that B is minimally generated over A. Since $C_s = A$ for s the least member of S under $\ll$, all we really need to prove is that $C_x \leq_m C_x(x)$ for all $x \in S$. To do this, it suffices to take any $y \ll x$ and show that $y \cdot x \in C_x$ or $-y \cdot x \in C_x$. In fact, if we can do this, then by Proposition 2.16, y or $-y$ will be in $\text{Smp}_x^{C_x}$, and since this will be true for each generator of C_x, it will follow that $\text{Smp}_x^{C_x}$ is either all of C_x or is a maximal ideal in C_x, so that $C_x \leq_m C_x(x)$ by Proposition 2.20.

If $y <_X x$, then $y \in B_x \leq_m B_x(x)$, and so by Propositions 2.16 and 2.20, one of $y \cdot x$ or $-y \cdot x$ is in B_x. Now $B_x \subseteq C_x$ since $x \in S$, so we obtain the desired conclusion. Assume, on the other hand, that $x <_X y$. This can only happen if $y \in T$. Hence by (**) either $y \leq x$ and hence $y \cdot x = y \in C_x$, or $y \leq -x$ and $y \cdot x = 0 \in C_x$, as desired. $\qquad \square$

We give a corollary of this theorem which depends on the notion of co-absolute, borrowed from topology. Two BA's A and B are *co-absolute* if their completions $\overline{A}$ and $\overline{B}$ are isomorphic.

Corollary 2.36. *Any minimally generated BA A is co-absolute with an interval algebra.*

Proof. By Theorem 2.35, A has a dense subalgebra B isomorphic to a tree algebra. From the Handbook we know that B can be embedded in an interval algebra, so let f be an isomorphism from B into an interval algebra C. Let I be an ideal of C maximal with respect to the property that $f[B] \cap I = \{0\}$. Then if $g : C \to C/I$ is the natural homomorphism, $g \circ f$ is still an embedding, and $g[f[B]]$ is dense in C/I; moreover, C/I is isomorphic to an interval algebra. Therefore we may assume in the original situation that $f[B]$ is dense in C. Now extend f to a homomorphism f^+ from $\overline{A}$ into $\overline{C}$. It is easy to see that f^+ is actually an isomorphism from $\overline{A}$ onto $\overline{C}$, as desired. $\qquad \square$

Tail algebras

For any partial order P, let the *tail algebra* of P be the subalgebra of $\mathscr{P}P$ generated by $\{P \uparrow p : p \in P\}$. Thus these algebras generalize tree algebras and pseudo-tree algebras. The notion is due to Gary Brenner. It is studied in Koppelberg, Monk [92], the main results being due to Koppelberg and Blass.

Theorem 2.37. *Every semigroup algebra is isomorphic to a tail algebra.*

Proof. Let A be a semigroup algebra, and choose a generating set H for A such that $0, 1 \in H$, H is closed under $\cdot$, and $P \stackrel{\text{def}}{=} H \backslash \{0\}$ is disjunctive. Let f be the homomorphism from A onto Tailalg (P^{-1}) given by Proposition 2.1: $fp = P \uparrow p$ for any $p \in P$. We show that f is one-one, which will finish the proof. By Sikorski's criterion, we have to show that $f(p_1) \cap \ldots \cap f(p_n) \subseteq f(q_1) \cup \ldots \cup f(q_m)$ (where $p_i, q_j \in P$) implies $p_1 \cdot \ldots \cdot p_n \leq q_1 + \ldots + q_m$. Without loss of generality, $p = p_1 \cdot \ldots \cdot p_n$ is nonzero and hence is in P. Now $p \in f(p_1) \cap \ldots \cap f(p_n)$. So $p \in f(q_j)$ for some j, $p \leq q_j$, and $p \leq q_1 + \ldots + q_m$, as desired. $\square$

We also need a set-theoretic lemma:

Lemma 2.38. *Let P be an infinite partially ordered set. Then: either P has a strictly ascending chain of type ω, or P has a strictly descending chain of type ω, or P is well-founded (with, say, P_α as its αth level) and there is some $n \in \omega$ such that P_n is infinite.*

Proof. Assume P has no descending chain of type ω (so P is well-founded) and no infinite level P_n ($n \in \omega$). For each $n \in \omega$, let

$$T_n = \{(p_0, \ldots, p_n) : p_i \in P_i \text{ for all } i \leq n, \text{ and } p_0 < \ldots < p_n\}.$$

So $T = \bigcup_{n \in \omega} T_n$ is a tree in which every level is finite and non-empty. But then T has an infinite branch, which yields an increasing chain of type ω in P. $\square$

An algebra which is generated by a disjunctive set is called *disjunctively generated*. Clearly tail algebras are disjunctively generated.

Theorem 2.39. *Every infinite disjunctively generated algebra has a countably infinite homomorphic image.*

Proof. Say $A = \langle P \rangle$, where P is an infinite disjunctive subset of A. We apply Lemma 2.38 to P^{-1}, and have three cases.

   Case 1. There is in P an ascending sequence $(p_n : n \in \omega)$. Let then $M = \{p_n : n \in \omega\}$, and consider the homomorphism f_M given by Proposition 2.1. Then f_M maps each $p \in P$ to an initial segment of M, and since $f_M(p_n) = \{p_0, \ldots, p_n\}$ and P generates A, it follows that the image of A under f_M is the finite-cofinite algebra on M, a countable algebra.

   Case 2. There is in P a descending sequence of type ω. This is similar to Case 1, again considering $M = \{p_n : n \in \omega\}$.

Case 3. P^{-1} is well-founded, and for some (minimal) $n \in \omega$, P_n is infinite. Consider $M = P_n$ and $f = f_M$ as in Proposition 2.1. Note that

1. $f(p) = \emptyset$ if $p \in P_\alpha, \alpha > n$
2. $f(p) = \{p\}$ for $p \in P_n$
3. $\{f(p) : p \in P_k, k < n\}$ is finite.

It follows that the image of A under f is superatomic, since its quotient under the ideal generated by the atoms is finite. It is well-known, and easy to check, that every superatomic algebra has a countable homomorphic image, giving the desired result. $\qquad\square$

Corollary 2.40. *No infinite Boolean algebra having the countable separation property is disjunctively generated.* $\qquad\square$

Theorem 2.41. *Every BA can be embedded into a tail algebra.*

Proof. This is trivial for a finite Boolean algebra B with, say, n atoms—just take a tree with n roots and no other points. So let B be an infinite Boolean algebra; we may assume that it is the algebra of clopen subsets of some Boolean space X.

For each $b \in B$, take two new points p_b, q_b such that the points p_b, q_b $(b \in B)$, are pairwise distinct and not in X. Then put

$$U = \{p_b, q_b : b \in B\}, \quad P = U \cup X$$

and define a partial order on P by setting $p_b < x$ and $q_b < x$ for all $x \in b$. Thus, for $b \in B$, $P \uparrow p_b = \{p_b\} \cup b$, $P \uparrow q_b = \{q_b\} \cup b$ and $b = (P \uparrow p_b) \cap (P \uparrow q_b) \in \text{Tailalg}\, P$.

We define a map e from B into the power set algebra of P by fixing a non-isolated point x^* of X and putting $e(b) = b$ if $x^* \notin b$ and $e(b) = U \cup b$ if $x^* \in b$. It is easily checked that e embeds B into the power set algebra of P and that $e(b) \in \text{Tailalg}\, P$ if $x^* \notin b$; hence e is an embedding from B into $\text{Tailalg}\, P$. $\qquad\square$

Initial chain algebras

Let T be a tree. The *initial chain algebra* of T, denoted by $\text{Init}\, T$, is the subalgebra of $\mathscr{P}T$ generated by $\{T \downarrow t : t \in T\}$. These algebras have been treated in the literature in a scattered fashion. More systematic studies have been made by Lynne Baur, Lutz Heindorf, and Monk (all unpublished). One can work with pseudo-trees too, but we restrict ourselves to trees. We prove just a few things about these algebras here. Call a tree T *limit-normal* if whenever u and v are distinct elements of T at the same limit level the sets $T \downarrow u$ and $T \downarrow v$ are distinct. The first theorem gives a simple normal form for elements of $\text{Init}\, T$ when T is limit-normal; the proof is obvious.

Theorem 2.42. *If T is limit-normal, then every monomial over $\text{Init}\, T$ has one of these three forms: (1) $T \downarrow t$ or (2) $(T \downarrow t) \backslash (T \downarrow s)$ or (3) $T \backslash \bigcup_{s \in F} (T \downarrow s)$ for some finite subset F of T.* $\qquad\square$

The next theorem, due to Lutz Heindorf, gives an abstract characterization of initial chain algebras. First we state two lemmas.

Lemma 2.43. *Let A be a BA generated by a set H with the following properties:*
    (i) $0 \notin H$;
    (ii) $H \cup \{0\}$ is closed under $\cdot$;
    (iii) $H \downarrow h$ is well-ordered, for all $h \in H$.
Then H is disjunctive.

Proof. Suppose that $h, h_1, \ldots, h_n \in H$, $n > 0$, and $h \leq h_1 + \cdots + h_n$. We may assume that $h \cdot h_i \neq 0$ for all i. Now $h = h \cdot h_1 + \cdots + h \cdot h_n$, and $h \cdot h_i \in (H \downarrow h)$ for each i, so by (iii) there is an i such that $h \cdot h_i \leq h \cdot h_j$ for all j. Hence $h = h \cdot h_i \leq h_i$, as desired.   □

Lemma 2.44. *Let A be a BA generated by a set H with the following properties:*
    (i) $0 \notin H$;
    (ii) $H \cup \{0\}$ is closed under $\cdot$;
    (iii) $H \downarrow h$ is well-ordered, for all $h \in H$;
    (iv) for every nonzero $a \in A$ there is an $h \in H$ such that $a \cdot h \neq 0$.
Then H is a tree under the Boolean ordering, and A is isomorphic to $\mathrm{Init}\, H$.

Proof. Obviously H is a tree under the Boolean ordering. By Lemma 2.43 and Proposition 2.1, there is a homomorphism f from A into $\mathscr{P}H$ such that $fh = H \downarrow h$ for all $h \in H$. Thus f maps onto $\mathrm{Init}\, H$. We need to show that f is one-one. Assume that

$$(H \downarrow h_1) \cap \ldots \cap (H \downarrow h_m) \cap [H \backslash (H \downarrow k_1)] \cap \ldots \cap [H \backslash (H \downarrow k_n)] = 0,$$

but $h_1 \cdot \ldots \cdot h_m \cdot -k_1 \cdot \ldots \cdot -k_n \neq 0$. By (iv) choose $h \in H$ such that $h \cdot h_1 \cdot \ldots \cdot h_m \cdot -k_1 \cdot \ldots \cdot -k_n \neq 0$. Now

$$h \cdot h_1 \cdot \ldots \cdot h_m \in (H \downarrow h_1) \cap \ldots \cap (H \downarrow h_n),$$

so $h \cdot h_1 \cdot \ldots \cdot h_m \in H(k_i)$ for some i. Thus $h \cdot h_1 \cdot \ldots \cdot h_m \cdot -k_i = 0$, contradiction.   □

Theorem 2.45. *For any BA A the following three conditions are equivalent:*
    (i) A is isomorphic to the initial chain algebra on some tree;
    (ii) A has a set of generators H such that the conditions of Lemma 2.43 hold.
    (iii) A has a set of generators H such that the conditions of Lemma 2.44 hold.

Proof. $(i) \Rightarrow (ii)$: let

$$H = \{\bigcap_{t \in F} (T \downarrow t) : F \text{ is a finite subset of } T\} \backslash \{0\}.$$

Clearly the conditions (i) and (ii) of Lemma 2.43 hold. As to condition 2.43(iii), note that the elements of H are initial chains of T; hence a strictly decreasing sequence in H would obviously yield a strictly decreasing sequence in T.

$(ii) \Rightarrow (iii)$: Suppose that (iv) of Lemma 2.44 fails to hold. Then there is some monomial x over H such that $x \cdot h = 0$ for all $h \in H$, with $x \neq 0$. Write $x = h_0 \cdot \ldots \cdot h_{m-1} \cdot -k_0 \cdot \ldots \cdot -k_{n-1}$ with all $h_i, k_j \in H$. Then $m = 0$ since $x \cdot h = 0$ for all $h \in H$. Let $H' = H \cup \{x\}$. Clearly H' satisfies the conditions of Lemma 2.44.

$(iii) \Rightarrow (i)$: by Lemma 2.44. □

Corollary 2.46. *For every tree T there is a limit-normal tree T' such that* $\operatorname{Init} T$ *is isomorphic to* $\operatorname{Init} T'$.

Proof. Let $H = \{\bigcap_{t \in F} (T \downarrow t) : F$ is a finite subset of $T\} \backslash \{0\}$. Clearly H is limit-normal and the conditions of Lemma 2.44 hold. □

Theorem 2.47. *Let T be a tree and $A = \operatorname{Init} T$. Then every homomorphic image of A is isomorphic to an initial chain algebra of a tree.*

Proof. By Corollary 2.46 we may assume that T is limit-normal. Let I be an ideal in A, and set $H = \{(T \downarrow t)/I : t \in T\} \backslash \{0/I\}$. It suffices to check the conditions of Lemma 2.43, and (i) and (ii) are clear. Now suppose that $t, s, r \in T$ and both $(T \downarrow s)/I$ and $(T \downarrow r)/I$ are $\leq (T \downarrow t)/I$; we want to show that they are comparable. Write $(T \downarrow s) \cap (T \downarrow t) = (T \downarrow s')$ and $(T \downarrow r) \cap (T \downarrow t) = (T \downarrow r')$. Then

$$(T \downarrow s)/I = [(T \downarrow s) \cap (T \downarrow t)]/I + [(T \downarrow s) \backslash (T \downarrow t)]/I$$
$$= (T \downarrow s')/I,$$

and similarly $(T \downarrow r)/I = (T \downarrow r')/I$. Now $r', s' \leq t$, so they are comparable; say $r' \leq s'$. Then $(T \downarrow r)/I \leq (T \downarrow s)/I$, as desired.

Finally, we need to show that $M \overset{\text{def}}{=} \{x \in H : x \leq (T \downarrow t)/I\}$ is well-ordered. Suppose on the contrary that $(T \downarrow t_0)/I > (T \downarrow t_1)/I > \cdots$ where $t_0 = t$. Write $(T \downarrow t_0) \cap \cdots \cap (T \downarrow t_m) = (T \downarrow s_m)$ for all $m \in \omega$. Then $s_0 > s_1 > \cdots$, contradiction. □

Theorem 2.48. *Every initial chain algebra on a tree is superatomic.*

Proof. By Theorem 2.47 it suffices to show that $\operatorname{Init} T$ is always atomic, for T a tree. Note that if $t \in T$ is not at a limit level, then $\{t\} \in A$. We may assume that T is limit-normal. Let x be a non-zero element of $\operatorname{Init} T$; we may assume that x is a monomial. By Theorem 2.42 we have three cases. *Case 1.* $x = (T \downarrow t)$ for some t. Let s be a root such that $s \leq t$. Then $\{s\}$ is the desired atom below x. *Case 2.* $x = (T \downarrow t) \backslash (T \downarrow s)$ for some s, t. Let $(T \downarrow t) \cap (T \downarrow s) = (T \downarrow r)$. Choose u at a successor level, $r < u \leq t$. Then $\{u\}$ is the desired atom below x. *Case 3.* $x = T \backslash \bigcup_{s \in F} (T \downarrow s)$ for some finite subset F of T. Since $x \neq 0$, choose $t \in x$. Then $(T \downarrow t) \backslash \bigcup_{s \in F} (T \downarrow s)$ reduces to Case 1 or Case 2, as desired. □

The following theorem will also be useful later.

Theorem 2.49. *Every initial chain algebra on a tree is a semigroup algebra.*

Proof. Let $A = \operatorname{Init} T$, T a tree. Wlog T is limit-normal. If there do not exist $t_1, \ldots, t_n \in T$ such that $T = (T \downarrow t_1) \cup \ldots \cup (T \downarrow t_n)$, then $\{T \downarrow t : t \in T\} \cup \{0, T\}$ works for the set H required in the definition of semigroup algebra (using limit normality to check closure under $\cdot$). Suppose now that there exist such elements, and choose elements so that n is minimum. The case $n = 1$ is clear, so assume that $n > 1$. Note that the elements t_i are exactly all of the maximal elements of T. Take $H = \{T \downarrow t : t \in T \backslash \{t_1\}\} \cup \{0, T\}$. All of the conditions for a semigroup algebra are clear except that H generates $\operatorname{Init} T$, and of course we just need to see that H generates $T \downarrow t_1$. If $(T \downarrow t_1) \cap (T \downarrow t_i) = 0$ for all $i = 2, \ldots, n$, then $(T \downarrow t_1) = T \backslash \bigcup_{2 \le i \le n} (T \downarrow t_i)$, as desired. Otherwise, let $M = \{i : 2 \le i \le n$ and $(T \downarrow t_i) \cap (T \downarrow t_1) \ne 0\}$. For each $i \in M$ write $(T \downarrow t_i) \cap (T \downarrow t_1) = (T \downarrow s_i)$, using the limit normality. Let $j \in M$ be such that $s_i \le s_j$ for all $i \in M$. Then

$$(T \downarrow t_1) = \left(T \backslash \bigcup_{2 \le k \le n} (T \downarrow t_k) \right) \cup (T \downarrow s_j),$$

as desired. $\square$

3. Cellularity

A BA A is said to satisfy the κ-*chain condition* if every disjoint subset of A has power $< \kappa$. Thus for κ non-limit, this is the same as saying that the cellularity of A is $< \kappa$. Of most interest is the ω_1-chain condition, called ccc for short (countable chain condition). We shall return to it below.

The attainment problem for cellularity is covered by two classical theorems of Erdös and Tarski: see Handbook, Part I, Theorem 3.10 and Example 11.14. Cellularity is attained for any singular cardinal, while for every weakly inaccessible cardinal there are examples of BAs with cellularity not attained.

If B is a subalgebra of A, then obviously $cB \leq cA$ and the difference can be arbitrarily large. If B is a dense subalgebra of A, then clearly $cA = cB$. If B is a homomorphic image of A, then cellularity can change either way from A to B. For example, if A is a free BA, then it has cellularity ω, while a homomorphic image of A can have very large cellularity. On the other hand, given any infinite BA A, it has a homomorphic image of cellularity ω: take a denumerable subalgebra B of A, and by Sikorski's theorem extend the identity mapping from B into the completion $\overline{B}$ of B to a homomorphism of A into $\overline{B}$.

By an easy argument, $c\left(\prod_{i \in I} A_i\right) = |I| + \sup_{i \in I} c(A_i)$, if all the A_i are non-trivial. The same computation holds for weak products.

Now we turn to chain conditions in free products, where there has been a lot of work done. Some partition theorems give results which clarify the situation:

(1) $c(A \oplus B) \leq 2^{cA \cdot cB}$ for infinite BAs A, B.

To see this, suppose that $\langle x_\alpha : \alpha < (2^{cA \cdot cB})^+ \rangle$ is a system of disjoint elements of $A \oplus B$. Without loss of generality we may assume that for each $\alpha < (2^{cA \cdot cB})^+$, the element x_α has the form $a_\alpha \times b_\alpha$, where $a_\alpha \in A$ and $b_\alpha \in B$ (we use $\times$ to make clear that the indicated product of elements is in the algebra $A \oplus B$). Thus for distinct α, β we have $a_\alpha \cdot a_\beta = 0$ or $b_\alpha \cdot b_\beta = 0$, and hence the Erdös-Rado partition theorem $(2^\kappa)^+ \rightarrow (\kappa^+)^2_\kappa$ implies that there is a subset Y of $(2^{cA \cdot cB})^+$ of power $(cA \cdot cB)^+$ such that either $a_\alpha \cdot a_\beta = 0$ for all distinct $\alpha, \beta \in Y$ or $b_\alpha \cdot b_\beta = 0$ for all distinct $\alpha, \beta \in Y$, which is impossible. (For this partition relation, see Erdös, Hajnal, Máté, Rado [84], pp. 98-100.)

Similarly, the partition theorem $(2^\kappa)^+ \rightarrow ((2^\kappa)^+, \kappa^+)^2$ (see the above book, Corollary 17.5) gives the following result:

(2) If $cA \leq 2^\kappa$ and $cB \leq \kappa$, then $c(A \oplus B) \leq 2^\kappa$.

Furthermore, if κ is strong limit, then $\kappa^+ \rightarrow (\kappa^+, \text{cf} \kappa)$ (see the above book, Theorem 17.1). Hence

(3) If κ is strong limit, $cA \leq \kappa$, and $cB < \text{cf} \kappa$, then $c(A \oplus B) \leq \kappa$.

The results (1) and (2) were first proved by Kurepa [62].

Under GCH, these results say the following: (1) For any BAs A and B, $cA \cdot cB \leq c(A \oplus B) \leq (cA \cdot cB)^+$; (2) For any BAs A and B, if $cB < \text{cf}(cA)$, then

$c(A \oplus B) = cA$. Thus even under GCH, there are two cases not covered by (1)-(3): when cA is limit with $\mathrm{cf}(cA) \leq cB < cA$, and when $cA = cB$.

About the first case, Shelah [94c] proved that if κ is a strong limit singular cardinal and $2^{\kappa} = \kappa^+$, then there are BAs A, B such that $cA = \kappa$, $cB < (2^{\mathrm{cf}\kappa})^+$, and $c(A \oplus B) = \kappa^+$. This gives a partial solution of the following problem:

Problem 1. *Is it true that for every singular cardinal κ there exist BAs A and B such that $cA = \kappa$, $\mathrm{cf}\kappa \leq cB < \kappa$, and $c(A \oplus B) > \kappa$?*

This is a modification of Problem 1 of Monk [90], which was solved in the form stated by the above result of Shelah. The problem is related to a result of Argyros and Tsarpalias. To formulate this result we need the notion of caliber. A BA A has *caliber* κ if for every system $\langle a_\alpha : \alpha < \kappa \rangle$ of elements of A^+ there is a $\Gamma \in [\kappa]^\kappa$ such that for all $\Delta \in [\Gamma]^\kappa$ the system $\langle a_\alpha : \alpha \in \Delta \rangle$ has the fip. Clearly if A has caliber κ and $A \oplus B$ has a system of κ disjoint elements, then B also has such a system. Thus in an example as in Problem 1, A must fail to have caliber κ. Now the result of Argyros and Tsarpalias is as follows (see Comfort, Negrepontis [82], Theorem 6.18): *If κ is a strong limit cardinal with $\mathrm{cf}\kappa = \omega$ and $2^\kappa = \kappa^+$, then there is a complete BA B of size at least κ^+ such that B has ccc and does not have caliber κ^+.* So an example as in Problem 1 yields a result in some respects stronger than the Argyros, Tsarpalias theorem.

The case $cA = cB$ has been intensively studied in the literature. That it is consistent to have a BA A such that $c(A \oplus A) > cA$ was essentially recognized quite early (probably at least implicitly by Kurepa); we give such an example shortly. Laver made a major advance by showing that CH suffices for such an example. The first example of such a phenomenon purely in ZFC was given by Todorčević, and we also give a simple case of his construction below. Working from the construction of Todorčević, Shelah has almost completely resolved the question concerning for which cardinals κ there is such an algebra of power κ in ZFC. To describe his results we introduce some terminology and prove some easy results. If λ is an infinite cardinal, we say that the λ-cc is *productive* iff for any BAs A and B, if they both satisfy the λ-cc, then so does $A \oplus B$. Obviously this is equivalent to saying that if $c'A, c'B \leq \lambda$, then also $c'(A \oplus B) \leq \lambda$. Note by the Erdös-Tarski theorem that if $c'A \leq \lambda$ and λ is singular, then $c'A < \lambda$. Hence λ-cc being productive is mainly interesting for λ regular.

Proposition 3.1. *Let λ be an infinite regular cardinal. Then the following conditions are equivalent:*
  (i) The λ-cc is productive.
  (ii) For all A, B, if $c'A = c'B = \lambda$, then $c'(A \oplus B) = \lambda$.
  (iii) For all A, if $c'A = \lambda$, then $c'(A \oplus A) = \lambda$.

Proof. Obviously $(i) \Rightarrow (ii) \Rightarrow (iii)$. Assume (iii). Suppose that A and B satisfy the λ-cc. Let C be any BA such that $c'C = \lambda$, and set $D = A \times B \times C$. Then $c'D = \lambda$. By (iii), $c'(D \oplus D) = \lambda$. Since $A \oplus B$ can be isomorphically embedded in $D \oplus D$, it follows that $A \oplus B$ satisfies the λ-cc. $\square$

Proposition 3.2. *Let κ be an infinite cardinal. Then the following conditions are equivalent:*

(i) The κ^+-cc is productive.

(ii) For all A, B, if $cA = cB = \kappa$, then $c(A \oplus B) = \kappa$.

(iii) For all A, if $cA = \kappa$, then $c(A \oplus A) = \kappa$.

Proof. Obviously $(i) \Rightarrow (ii) \Rightarrow (iii)$. Now assume (iii). We verify Proposition 3.1(iii). Let A be a BA with $c'A = \kappa^+$. Then $cA = \kappa$, and it is attained. It follows that $c(A \oplus A) = \kappa$, also attained. Thus $c'(A \oplus A) = \kappa^+$, as desired. $\square$

Now we can formulate most of the known results about productivity of λ-cc. Productivity of the ω_1-cc (also known as ccc) is independent of the axioms of set theory; we go into this in detail below. In Shelah [94b] 4.8, Appendix 1 of Shelah [94e], and Shelah [95b] it is shown that if κ is regular $> \omega_1$ then the κ^+-cc is not productive. In Shelah [94a] 4.2 it is shown that for κ singular, the κ^+-cc is not productive. In Shelah [94b] 4.8 it is also proved that if λ is uncountable inaccessible and not ω-Mahlo then the λ-cc is not productive. Finally, that the ω_2-cc is not productive is proved in Shelah [94f].

Perhaps the easiest proof concerning productivity of the λ-cc is as follows. Let T be a Suslin tree such that every element t has infinitely many successors, denote two of them by t_0 and t_1, and let A be the tree algebra on T. Now A satisfies ccc (see the description of cellularity for tree algebras at the end of this chapter), but $A \oplus A$ does not. To see this second fact, for each $t \in T$ consider the element $(T \uparrow t_0) \times (T \uparrow t_1)$ of $A \oplus A$. Suppose that $s, t \in T$, $s \neq t$, and $(T \uparrow t_0) \cap (T \uparrow s_0) \neq 0$. Then t_0 and s_0 are comparable; say $s_0 < t_0$. Clearly, then, t_1 and s_1 are not comparable, so $(T \uparrow t_1) \cap (T \uparrow s_1) = 0$, as desired.

We now give an example in ZFC of this kind of thing. We follow Todorčević [85], but we give only a simple form of his construction. To begin with we note that we can work with partial orders rather than Boolean algebras. Given a partial order $(P, \leq)$, we take $\{P \uparrow p : p \in P\}$ as a base for a topology on P, and we let ROP be the complete BA of regular open sets in this topology. For any $p \in P$ let $b_p = \text{int}(\text{cl}(P \uparrow p))$. Elements p, q of P are *compatible* if there is an $r \in P$ such that $p \leq r$ and $q \leq r$, i.e., if $(P \uparrow p) \cap (P \uparrow q) \neq 0$. We say that P satisfies the κ-cc if any collection of pairwise incompatible elements of P has fewer than κ elements.

Lemma 3.3. *Let P be a partial order.*

(i) Let $p, q \in P$. Then p and q are incompatible iff $b_p \cdot b_q = 0$.

(ii) For any infinite cardinal κ, the partial order P satisfies the κ-cc iff ROP satisfies the κ-cc.

Proof. (i) follows from the Handbook, Part I, p. 26 (10). (ii) follows from (i). $\square$

Given partial orders P and Q, the cartesian product $P \times Q$ is made into a partial order by defining $(p_1, q_1) \leq (p_2, q_2)$ iff $p_1 \leq p_2$ and $q_1 \leq q_2$.

Lemma 3.4. *Let P and Q be partial orders.*

(i) (p_1, q_1) is incompatible with (p_2, q_2) iff $b_{p_1} \cdot b_{q_1} \cdot b_{p_2} \cdot b_{q_2} = 0$ (in ROP $\oplus$ ROQ).

(ii) For any infinite cardinal κ, $P \times Q$ satisfies the κ-cc iff ROP $\oplus$ ROQ does.

Proof. (i): If (p_1, q_1) is compatible with (p_2, q_2), then p_1 and p_2 are compatible, and q_1 and q_2 are compatible, hence by Lemma 3.3, $b_{p_1} \cdot b_{q_1} \neq 0$ and $b_{p_2} \cdot b_{q_2} \neq 0$, and hence $b_{p_1} \cdot b_{q_1} \cdot b_{p_2} \cdot b_{q_2} \neq 0$. The converse is similar.

(ii): follows easily from (i). $\qquad\qquad\qquad\qquad\qquad\qquad\qquad\qquad\qquad\qquad\square$

Recall that $\beth_0 = \omega$, $\beth_{\alpha+1} = 2^{\beth_\alpha}$, and $\beth_\lambda = \sup_{\alpha < \lambda} \beth_\alpha$ for λ limit. We are going to construct BAs B and C which satisfy the $\beth_\omega^+$-cc while $B \oplus C$ does not; so $cB \leq \beth_\omega$, $cC \leq \beth_\omega$, and $c(B \oplus C) > \beth_\omega$. By the above lemmas, it suffices to work with partial orders.

A function λ is said to satisfy *condition* (α) provided that λ is an ω-termed sequence of infinite cardinals such that $\sup_{\xi < \omega} \lambda_\xi = \beth_\omega$ and

$$\left| \prod_{\eta < \xi} \lambda_\eta \right| < \lambda_\xi = \mathrm{cf} \lambda_\xi \text{ for all } \xi < \omega.$$

For the following definitions, assume that λ satisfies condition (α). For distinct $a, b \in \prod_{\xi < \omega} \lambda_\xi$ let $\rho(a, b) = \min\{\xi < \omega : a\xi \neq b\xi\}$. For $a, b \in \prod_{\xi < \omega} \lambda_\xi$ we define

$$a =^* b \text{ iff } \exists \xi < \omega \forall \eta \in (\xi, \omega)[a\eta = b\eta];$$
$$a \leq^* b \text{ iff } \exists \xi < \omega \forall \eta \in (\xi, \omega)[a\eta \leq b\eta];$$
$$a \lneq^* b \text{ iff } a \leq^* b \text{ and } a \neq^* b;$$
$$a <^* b \text{ iff } \exists \xi < \omega \forall \eta \in (\xi, \omega)[a\eta < b\eta].$$

A sequence $\langle a_\alpha : \alpha < \sigma \rangle \in {}^\sigma(\prod_{\xi < \omega} \lambda_\xi)$ is $\leq^*$-*increasing* if $\forall \alpha, \beta(\alpha < \beta < \sigma \Rightarrow a_\alpha \lneq^* a_\beta)$. For $I \subseteq \omega$, a subset $A \subseteq \prod_{\xi < \omega} \lambda_\xi$ is $\leq^*$-*unbounded on* I if there is no $b \in \prod_{\xi < \omega} \lambda_\xi$ such that $\forall a \in A \exists \xi < \omega \forall \eta \in I \cap (\xi, \omega)[a\eta \leq b\eta]$. For $A \subseteq \prod_{\xi < \omega} \lambda_\xi$ and $I \subseteq \omega$, let

$$\mathscr{P}_I A = \{p \in [A]^{<\omega} : (\forall \text{ distinct } a, b \in p)[\rho(a, b) \in I]\}.$$

We consider $\mathscr{P}_I A$ to be partially ordered by $\subseteq$. Suitable choices of I and A will give the partial orders we are after.

Lemma 3.5. *Let* λ *satisfy condition* (α). *Suppose that* $\langle a_\alpha : \alpha < \beth_\omega^+ \rangle$ *is* $\leq^*$-*increasing,* I *is an infinite subset of* ω, *and* $A \stackrel{\mathrm{def}}{=} \{a_\alpha : \alpha < \beth_\omega^+\}$ *is* $\leq^*$-*unbounded on* I. *Then* $\mathscr{P}_I A$ *satisfies the* $\beth_\omega^+$-cc.

Proof. Let $\langle p_\alpha : \alpha < \beth_\omega^+ \rangle$ be a sequence of distinct elements of $\mathscr{P}_I A$. We want to find distinct $\alpha, \beta < \beth_\omega^+$ such that p_α and p_β are compatible. Without loss of generality $\langle p_\alpha : \alpha < \beth_\omega^+ \rangle$ is a Δ-system, say with kernel q. It suffices to find distinct

α, β with $p_\alpha \backslash q$ and $p_\beta \backslash q$ compatible; so wlog the p_α's are pairwise disjoint. Also, we may assume that all the p_α's have the same size n. Since

$$\beth_\omega^+ = \bigcup_{\xi < \omega} \{\sigma < \beth_\omega^+ : (\forall \text{ distinct } a, b \in p_\sigma)[\rho(a,b) < \xi]\},$$

and $\omega < \beth_\omega^+$, we may assume that there is a $\xi_0 < \omega$ such that $\rho(a,b) < \xi_0$ for all $\sigma < \beth_\omega^+$ and all distinct $a, b \in p_\sigma$. For any $q \in A$ define $\Gamma q = \{\alpha < \beth_\omega^+ : a_\alpha \in q\}$. Then for $\sigma < \beth_\omega^+$ and $\xi < \omega$ let $b_\sigma \xi = \min\{a_\alpha \xi : \alpha \in \Gamma p_\sigma\}$. Then

(1) $\{b_\sigma : \sigma < \beth_\omega^+\}$ is $\leq^*$-unbounded on I.

For, suppose not: say $c \in \prod_{\xi < \omega} \lambda_\xi$ and $\forall \sigma < \beth_\omega^+ \exists \xi < \omega \forall \eta \in I \cap (\xi, \omega)[b_\sigma \eta \leq c\eta]$. Then $\{a_\alpha : \alpha < \beth_\omega^+\}$ is $\leq^*$-bounded on I by c (contradiction). For, let $\alpha < \beth_\omega^+$. Choose $\sigma < \beth_\omega^+$ such that $\alpha < \beta$ for each $\beta \in \Gamma p_\sigma$. (This is possible since the Γp_σ's are pairwise disjoint.) Then $a_\alpha \leq^* a_\beta$ for each $\beta \in \Gamma p_\sigma$, so there is a $\xi < \omega$ such that $\forall \beta \in \Gamma p_\sigma \forall \eta \in (\xi, \omega)[a_\alpha \eta \leq a_\beta \eta]$. Also, by the assumption on c choose $\xi' < \omega$ such that $\forall \eta \in I \cap (\xi', \omega)[b_\sigma \eta \leq c\eta]$. Without loss of generality $\xi' \leq \xi$. Hence $\forall \eta \in I \cap (\xi, \omega)[a_\alpha \eta \leq b_\sigma \eta \leq c\eta]$, as desired. Thus (1) holds.

By (1), there is an $\eta \in I \backslash \xi_0$ such that $\{b_\sigma \eta : \sigma < \beth_\omega^+\}$ is unbounded in λ_η. Now we define $\sigma \in {}^{\lambda_\eta}\beth_\omega^+$ by induction. Suppose it is defined for all $\alpha < \beta$, where $\beta < \lambda_\eta$. Then $\delta \overset{\text{def}}{=} \sup_{\alpha < \beta} \sup_{\gamma \in \Gamma p_{\sigma\alpha}} a_\gamma \eta < \lambda_\eta$, so there is a $\tau < \beth_\omega^+$ such that $b_\tau \eta$ exceeds δ, and we let $\sigma \beta$ be the least such τ. Let $C = \{\sigma \alpha : \alpha < \lambda_\eta\}$. Then

(2) If $\tau, \rho \in C$, $\tau < \rho$, $\alpha \in \Gamma p_\tau$, and $\beta \in \Gamma p_\rho$, then $a_\alpha \eta < a_\beta \eta$.

For each $\tau \in C$ write $\Gamma p_\tau = \{\alpha(\tau, 0), \ldots, \alpha(\tau, n-1)\}$, with $\alpha(\tau, 0) < \cdots < \alpha(\tau, n-1)$. Then

$$C = \bigcup \{\{\tau \in C : \forall i < n[a_{\alpha(\tau,i)} \restriction \eta = t_i]\} : t \in {}^n(\prod_{\xi < \eta} \lambda_\xi)\},$$

so, since $|\prod_{\xi < \eta} \lambda_\xi| < \lambda_\eta = \text{cf}\lambda_\eta$, wlog there is a $t \in {}^n(\prod_{\xi < \eta} \lambda_\xi)$ such that $a_{\alpha(\tau,i)} \restriction \eta = t_i$ for all $\tau \in C$ and all $i < n$. Thus if $i \neq j$ then $\rho(t_i, t_j) \in I \cap \eta$. Now if τ and ρ are distinct members of C and $a, b \in p_\tau \cup p_\rho$ it follows that $\rho(a,b) \in I$. For, say $a = a_{\alpha(\tau,i)}$ and $b = a_{\alpha(\rho,j)}$. If $i \neq j$, then $\rho(a,b) \in I$ from the above. If $i = j$, then $\rho(a,b) = \eta \in I$.

So p_τ and p_ρ are compatible for all $\tau, \rho \in C$, as desired. $\square$

Lemma 3.6. *Suppose that λ satisfies condition (α) and I and J are disjoint infinite subsets of ω. Let $A \subseteq \prod_{\xi < \omega} \lambda_\xi$ be of size $\beth_\omega^+$. Then $\mathscr{P}_I A \times \mathscr{P}_J A$ has a pairwise incompatible subset of size $\beth_\omega^+$.*

Proof. We claim that $\{(\{a\}, \{a\}) : a \in A\}$ is pairwise incompatible. Suppose that a and b are distinct elements of A, and $(\{a\}, \{a\})$ and $(\{b\}, \{b\})$ are compatible. So there is a $(p, q) \in \mathscr{P}_I A \times \mathscr{P}_J A$ such that $a, b \in p$ and $a, b \in q$. Then $\rho(a,b) \in I \cap J$, contradiction. $\square$

Lemma 3.7. *There is a λ satisfying condition (α) for which there is an $a \in$ $\beth_\omega^+ (\prod_{\xi<\omega} \lambda_\xi)$ such that a is $<^*$-increasing and $\{a_\alpha : a < \beth_\omega^+\}$ is $\leq^*$-unbounded on each infinite $I \subseteq \omega$.*

Proof. For each $\xi < \omega$ let $\mu_\xi = \beth_\xi^+$. We claim:

(1) If $b \in {}^{\beth_\omega}(\prod_{\xi<\omega} \mu_\xi)$ then there is a $c \in \prod_{\xi<\omega} \mu_\xi$ such that $b_\beta <^* c$ for all $\beta < \beth_\omega$.

In fact, let $c\eta$ be the least ordinal such that $b_\beta\eta < c\eta$ for all $\beta < \beth_\eta$. Now let $\beta < \beth_\omega$, in order to show that $b_\beta <^* c$. Choose $\xi < \omega$ so that $\beta < \beth_\xi$. Then if $\eta \in (\xi, \omega)$ we have $b_\beta\eta < c\eta$. Thus $b_\beta <^* c$, as desired.

    By (1), there is a $<^*$-increasing $b \in {}^{\beth_\omega^+}(\prod_{\xi<\omega} \mu_\xi)$. We now define a sequence $\langle c_\sigma : \sigma < \overline{\sigma} \rangle$ by induction on σ, each $c_\sigma \in \prod_{\xi<\omega}(\mu_\xi + 1)$. Let $c_0\xi = \mu_\xi$ for each $\xi < \omega$. Then we continue to define c_σ as long as possible, subject to the following two conditions:

(2) $b_\alpha <^* c_\sigma$ for all $\alpha < \beth_\omega^+$.
(3) If $\tau < \sigma < \overline{\sigma}$, then $c_\sigma \lneq^* c_\tau$.

Now note that $\overline{\sigma} < (2^\omega)^+$. In fact, otherwise we can write

$$[(2^\omega)^+]^2 = \bigcup_{\xi<\omega} \{\{\sigma,\tau\} : \sigma < \tau < (2^\omega)^+ \text{ and } c_\sigma\xi > c_\tau\xi\},$$

and the Erdös-Rado theorem $(2^\omega)^+ \to (\omega^+)_\omega^2$ would give an infinite decreasing sequence of ordinals. So, indeed, $\overline{\sigma} < (2^\omega)^+$. Next we claim

(4) $\overline{\sigma}$ is a successor ordinal.

Suppose not. For each $\xi < \omega$ let $B_\xi = \{c_\sigma\xi : \sigma < \overline{\sigma}\}$. Let $B = \prod_{\xi<\omega} B_\xi$. Now $|B_\xi| \leq 2^\omega$ since $\overline{\sigma} < (2^\omega)^+$, so $|B| \leq 2^\omega$. Hence:

(5) There is a $\gamma < \beth_\omega^+$ such that for all $d \in B$, if $b_\gamma <^* d$, then $b_\beta <^* d$ for all $\beta \in (\gamma, \beth_\omega^+)$.

In fact, if (5) fails, then there exist a $\Delta \in [\beth_\omega^+]^{\beth_\omega^+}$ and a $d \in B$ such that for all $\gamma \in \Delta$ we have $b_\gamma <^* d$ but $b_\beta \not<^* d$ for some $\beta \in (\gamma, \beth_\omega^+)$. But then $b_\gamma <^* d$ for all $\gamma < \beth_\omega^+$, since if $\gamma < \beth_\omega^+$, choose $\delta \in \Delta$ with $\gamma < \delta$; then $b_\gamma <^* b_\delta <^* d$. This is a contradiction. So (5) holds.

    Now define $d \in \prod_{\xi<\omega} B_\xi$ by setting

$$d_\xi = \begin{cases} c_0\xi & \text{if } \beta \leq b_\gamma\xi \text{ for all } \beta \in B_\xi, \\ \min\{\beta \in B_\xi : \beta > b_\gamma\xi\} & \text{otherwise,} \end{cases}$$

for each $\xi < \omega$. Then $b_\alpha <^* d \leq^* c_\sigma$ for all $\alpha < \beth_\omega^+$ and $\sigma < \overline{\sigma}$. In fact, obviously $b_\gamma <^* d$, so if $\alpha < \gamma$ then $b_\alpha <^* b_\gamma <^* d$, and if $\gamma \leq \alpha$ then $b_\alpha <^* d$ by (5). And $b_\gamma <^* c_\sigma$, so we can choose $\xi < \omega$ so that $b_\gamma\eta < c_\sigma\eta$ for all $\eta \in (\xi, \omega)$. Then for any $\eta \in (\xi, \omega)$ we have $c_\sigma \eta \in B_\eta$, and hence $d_\eta \leq c_\sigma \eta$; so $d \leq^* c_\sigma$. Now actually $d \lneq^* c_\sigma$ for all $\sigma < \overline{\sigma}$, since $\overline{\sigma}$ is a limit ordinal. But this contradicts the fact that $c_{\overline{\sigma}}$ is undefined. So (4) is proved.

Let $\overline{\sigma} = \sigma + 1$, and set $d = c_\sigma$. For each $\xi < \omega$ let $\lambda_\xi = \mathrm{cf}(d\xi)$, and pick a subset C_ξ of $d\xi$ of order type λ_ξ, so that $\sup C_\xi = \lambda_\xi$ if λ_ξ is a limit ordinal, and $C_\xi = \{\delta\}$ if $\lambda_\xi = \delta + 1$. For each $\alpha < \beth_\omega^+$ define $a_\alpha \in \prod_{\xi < \omega} \lambda_\xi$ by

$$a_\alpha \xi = \begin{cases} \text{order type of } C_\xi \cap b_\alpha \xi, & \text{if } b_\alpha \xi < d\xi, \\ 0 & \text{otherwise.} \end{cases}$$

(6) $\{a_\alpha : \alpha < \beth_\omega^+\}$ is $\leq^*$-unbounded in $\prod_{\xi < \omega} \lambda_\xi$ on any infinite $I \subseteq \omega$.

Proof of (6): Suppose that I is an infinite subset of ω and e is a $\leq^*$-bound in $\prod_{\xi < \omega} \lambda_\xi$ for $\{a_\alpha : \alpha < \beth_\omega^+\}$ on I. Thus

$$\forall \alpha < \beth_\omega^+ \exists \xi < \omega \forall \eta \in I \cap (\xi, \omega)[a_\alpha \eta \leq e\eta],$$

so there exist an $E \in [\beth_\omega^+]^{\beth_\omega^+}$ and a $\xi_0 < \omega$ such that for all $\alpha \in E$ and all $\eta \in I \cap (\xi_0, \omega)$ we have $a_\alpha \eta \leq e\eta$. And $b_\alpha <^* d$ for all $\alpha \in E$, so by a similar argument wlog $\forall \alpha \in E \forall \eta \in I \cap (\xi_0, \omega)[b_\alpha \eta < d\eta]$. Hence for all $\alpha \in E$ and all $\eta \in I \cap (\xi_0, \omega)$, $a_\alpha \eta$ is the order type of $C_\eta \cap b_\alpha \eta$. Define $f \in \prod_{\xi < \omega}(\mu_\xi + 1)$ by setting $f\xi = d\xi$ if $\xi \notin I \cap (\xi_0, \omega)$, and $f\xi = $ the $\delta \in C_\xi$ such that $e\xi$ is the order type of $C_\xi \cap \delta$ if $\xi \in I \cap (\xi_0, \omega)$. This is possible since $e\xi \in \lambda_\xi = \mathrm{cf}\lambda_\xi$ and $C_\xi \subseteq d\xi$ has order type λ_ξ. Now let $\alpha < \beth_\omega^+$. We claim that $b_\alpha <^* f$. For, choose $\beta \in E$ so that $\alpha < \beta$. Choose $\xi_1 < \omega$ such that $\forall \xi \in (\xi_1, \omega)[b_\beta \xi < d\xi]$, by (2). Suppose that $\xi \in (\max(\xi_0, \xi_1), \omega)$. If $\xi \notin I$, clearly $b_\beta \xi < f\xi$. Suppose that $\xi \in I$. Then $a_\beta \xi \leq e\xi$, $e\xi$ is the order type of $C_\xi \cap f\xi$, and $a_\beta \xi$ is the order type of $C_\xi \cap b_\beta \xi$, so $b_\beta \xi \leq f\xi$. This proves that $b_\beta \leq^* f$. Since $b_\alpha <^* b_\beta$, it follows that $b_\alpha <^* f$, as desired: we have proved that $b_\alpha <^* f$ for all $\alpha < \beth_\omega^+$.

Next, $f \lneq^* d$. In fact, if $\xi \notin I \cap (\xi_0, \omega)$, then $f\xi = d\xi$, while if $\xi \in I \cap (\xi_0, \omega)$ then $f\xi \in C_\xi$, hence $f\xi < \lambda_\xi = \mathrm{cf}(d\xi)$, and so $f\xi < d\xi$; since I is infinite, $f \neq^* d$. But this contradicts the fact that $c_{\overline{\sigma}}$ is not defined. Hence (6) holds.

(7) If $\alpha < \beta < \beth_\omega^+$, then $a_\alpha \leq^* a_\beta$.

In fact, choose $\xi < \omega$ such that $\forall \eta \in (\xi, \omega)[b_\alpha \eta < b_\beta \eta < d\eta]$. Then clearly $\forall \eta \in (\xi, \omega)[a_\alpha \eta \leq a_\beta \eta]$.

(8) If $\alpha < \beth_\omega^+$ then $\exists \gamma < \beth_\omega^+ \forall \beta \in (\gamma, \beth_\omega^+)[a_\alpha <^* a_\beta]$.

Suppose (8) fails for α. Then there is an $F \in [\beth_\omega^+]^{\beth_\omega^+}$ such that $a_\alpha \not<^* a_\beta$ for all $\beta \in F$, so by (7),

$$F = \bigcup_{I \in [\omega]^\omega} \{\beta \in F : I = \{\xi < \omega : a_\alpha \xi = a_\beta \xi\}\},$$

and hence wlog there is an infinite $I \subseteq \omega$ such that $\forall \beta \in F \forall \xi \in I[a_\alpha \xi = a_\beta \xi]$. Thus a_α is a $\leq^*$-bound for $\{a_\gamma : \gamma < \beth_\omega^+\}$ on I, which contradicts (6).

By (7) and (8), there is an increasing function $\delta \in {}^{\beth_\omega^+}\beth_\omega^+$ such that $a_{\delta\alpha} <^* a_{\delta\beta}$ for all α, β with $\alpha < \beta < \beth_\omega^+$. Let $a'_\alpha = a_{\delta\alpha}$ for all $\alpha < \beth_\omega^+$. Then a' is a one-one member of $\beth_\omega^+ \prod_{\xi<\omega} \lambda_\xi$, so $\sup_{\xi<\omega} \lambda_\xi = \beth_\omega$. Hence there is an increasing $\varepsilon \in {}^\omega\omega$ such that $\prod_{\eta<\xi} \lambda_{\varepsilon\eta} < \lambda_{\varepsilon\xi}$ for all $\xi < \omega$. Hence $\lambda' \stackrel{\text{def}}{=} \lambda \circ \varepsilon$ satisfies condition (α). For each $\alpha < \beth_\omega^+$ let $a''_\alpha \in \prod_{\xi<\omega} \lambda'_\xi$ be defined by: $a''_\alpha\xi = a'_\alpha\varepsilon\xi$ for all $\xi < \omega$. Clearly a'' is $<^*$-increasing. Suppose that $\{a''_\alpha : \alpha < \beth_\omega^+\}$ is $\leq^*$-bounded by g on an infinite subset I of ω. Now define $g' \in \prod_{\xi<\omega} \lambda'_\xi$ by

$$
g'\xi = \begin{cases} g\varepsilon^{-1}\xi, & \text{if } \xi \in \varepsilon[\omega], \\ 0, & \text{otherwise.} \end{cases}
$$

Note that if $\xi \in \varepsilon[\omega]$, then $g\varepsilon^{-1}\xi \in \lambda'_{\varepsilon^{-1}\xi} = \lambda_\xi$. So $g' \in \prod_{\xi<\omega} \lambda_\xi$. Now we claim that $\{a_\alpha : \alpha < \beth_\omega^+\}$ is $\leq^*$-bounded by g' on $\varepsilon[I]$, contradicting (6). In fact, given $\alpha < \beth_\omega^+$, choose $\beta < \beth_\omega^+$ so that $\alpha < \delta\beta$. Choose $\xi < \omega$ such that $\forall \eta \in (\xi, \omega)[a_\alpha\eta < a_{\delta\beta}\eta]$ and $\forall \eta \in I \cap (\xi, \omega)[a''_\beta\eta \leq g\eta]$. Then if $\nu \in \varepsilon[I] \cap (\varepsilon\xi, \omega)$, write $\nu = \varepsilon\eta$; then $\eta \in I \cap (\xi, \omega)$, so

$$
a_\alpha\nu < a_{\delta\beta}\nu = a_{\delta\beta}\varepsilon\eta = a''_\beta\eta \leq g\eta = g'\nu,
$$

as desired.   □

Theorem 3.8. *There exist BAs A and B such that $cA \leq \beth_\omega$, $cB \leq \beth_\omega$, and $c(A \oplus B) > \beth_\omega$.*   □

Corollary 3.9. *There is a BA C such that $cC = \beth_\omega$ while $c(C \oplus C) > \beth_\omega$.*   □

Another important and quite elementary fact about free products is that

$$
c(\oplus_{i\in I} A_i) = \sup\{c(\oplus_{i\in F} A_i) : F \in [I]^{<\omega}\}.
$$

In fact, $\geq$ is clear. Now let $\kappa = \sup\{c(\oplus_{i\in F} A_i) : F \in [I]^{<\omega}\}$, and suppose that X is a disjoint subset of $\oplus_{i\in I} A_i$ of size κ^+. For each $x \in X$ choose a finite $Fx \subseteq I$ such that $x \in \oplus_{i\in Fx} A_i$. We may assume that each $x \in X$ has the form $x = \prod_{i\in Fx} y_i^x$, where $y_i^x \in A_i$ for each $i \in Fx$. Without loss of generality, $\langle Fx : x \in X \rangle$ forms a Δ-system, say with kernel G. But then, by the free product property, $\langle \prod_{i\in G} y_i^x : x \in X \rangle$ is a disjoint system of elements of $\oplus_{i\in G} A_i$, contradiction.

As our final result on chain conditions in free products, we prove the folklore theorem that MA (Martin's axiom) + ¬CH implies that the free product of two ccc BAs is again ccc. This depends on the following lemma:

Lemma 3.10. (MA + ¬CH) *Suppose that $\langle x_\alpha : \alpha < \omega_1 \rangle$ is a system of elements in a ccc BA A. Then there is an uncountable $S \subseteq \omega_1$ such that $\langle x_\alpha : \alpha \in S \rangle$ has the finite intersection property.*

Proof. We may assume that A is complete. For each $\alpha < \omega_1$ let $y_\alpha = \sum_{\gamma>\alpha} x_\gamma$. Then, we claim,

(*) There is an $\alpha < \omega_1$ such that for all $\beta > \alpha$ we have $y_\beta = y_\alpha$.

Otherwise, since clearly $\alpha < \beta \to y_\alpha \geq y_\beta$, we easily get an increasing sequence $\langle \beta(\xi) : \xi < \omega_1 \rangle$ of ordinals less than ω_1 such that $y_{\beta(\xi)} > y_{\beta(\eta)}$ whenever $\xi < \eta < \omega_1$. But then $\langle y_{\beta(\xi)} \cdot -y_{\beta(\xi+1)} \rangle$ is a disjoint family of power ω_1, contradiction.

Thus (*) holds, and we fix an α as indicated there. The partial ordering P that we want to apply Martin's axiom to is $\{x \in A : 0 \neq x \leq y_\alpha\}$ under $\geq$. It is a ccc partial ordering since A is a ccc BA. Now for the dense sets. For each $\beta < \omega_1$ let

$$D_\beta = \{p \in P : \text{there is a } \gamma > \beta \text{ such that } p \leq x_\gamma\}.$$

To see that D_β is dense in P, let $p \in P$ be arbitrary. Choose $\delta \in \omega_1$ with $\delta > \alpha, \beta$. Then $y_\alpha = y_\delta$, so from $0 \neq p \leq y_\alpha$ we infer that there is a $\gamma > \delta$ such that $p \cdot x_\gamma \neq 0$. Thus $p \cdot x_\gamma$ is the desired element of D_β which is $\leq p$.

Now let G be a filter on P intersecting each dense set D_β for $\beta < \omega_1$, by MA + ¬CH. Then it is easy to see that $S \overset{\text{def}}{=} \{x_\gamma : \gamma < \omega_1, \text{ and } p \leq x_\gamma \text{ for some } p \in G\}$ is the set desired in the lemma. □

Now we prove, using MA+¬CH, that the free product of ccc BAs A and B is again ccc. Let $\langle x_\alpha : \alpha < \omega_1 \rangle$ be a disjoint system of elements of $A \oplus B$. Without loss of generality we may assume that each x_α has the form $a_\alpha \times b_\alpha$ where $a_\alpha \in A$ and $b_\alpha \in B$. By the lemma, let S be an uncountable subset of ω_1 such that $\langle a_\alpha : \alpha \in S \rangle$ has the finite intersection property. But then, by the free product property, $\langle b_\alpha : \alpha \in S \rangle$ is a disjoint system in B, contradiction.

The argument just given generalizes easily to show that MA+¬CH implies that if X and Y are ccc topological spaces, then so is $X \times Y$.

We now turn to more special operations. The basic fact about cellularity for amalgamated free products is as follows:

$$c(A \oplus_C B) \leq 2^{cA \cdot cB \cdot |C|}.$$

To prove this, let $\kappa = cA \cdot cB \cdot |C|$, and suppose that $\langle c_\alpha : \alpha < (2^\kappa)^+ \rangle$ is a disjoint system in $A \oplus_C B$. We may assume that each c_α is non-zero, and has the form $a_\alpha \cdot b_\alpha$, with $a_\alpha \in A$ and $b_\alpha \in B$. Thus for all distinct $\alpha, \beta < (2^\kappa)^+$ there is a $c \in C$ such that $a_\alpha \cdot a_\beta \leq c$ and $b_\alpha \cdot b_\beta \leq -c$. Hence by the Erdös-Rado theorem there is a $\Gamma \in [(2^\kappa)^+]^{\kappa^+}$ and a $c \in C$ such that $a_\alpha \cdot a_\beta \leq c$ and $b_\alpha \cdot b_\beta \leq -c$ for all distinct $\alpha, \beta \in \Gamma$. Thus $(a_\alpha \cdot -c) \cdot (a_\beta \cdot -c) = 0$ and $(b_\alpha \cdot c) \cdot (b_\beta \cdot c) = 0$ for all distinct $\alpha, \beta \in (2^\kappa)^+$. Since $cA < \kappa^+$, it follows that there is a $\Delta \in [\Gamma]^\kappa$ such that $a_\alpha \cdot -c = 0$ for all $\alpha \in \Gamma \backslash \Delta$; and there is a $\Theta \in [\Gamma \backslash \Delta]^\kappa$ such that $b_\alpha \cdot c = 0$ for all $\alpha \in (\Gamma \backslash \Delta) \backslash \Theta$. But then for any $\alpha \in (\Gamma \backslash \Delta) \backslash \Theta$ we have $a_\alpha \cdot b_\alpha = 0$, contradiction.

The above inequality is best-possible, in a sense. To see this, consider $\mathscr{P}\omega \oplus_C \mathscr{P}\omega$, where C is the BA of finite and cofinite subsets of ω. Let $\langle \Gamma_\alpha : \alpha < 2^\omega \rangle$ be a system of infinite almost disjoint subsets of ω; and also assume that each Γ_α is not cofinite. For each $\alpha < 2^\omega$ let y_α be the element $\Gamma_\alpha \cdot (\omega \backslash \Gamma_\alpha)$ of $\mathscr{P}\omega \oplus_C \mathscr{P}\omega$.

These elements are clearly non-zero. For distinct $\alpha, \beta < 2^\omega$ let $F = \Gamma_\alpha \cap \Gamma_\beta$. Then $\Gamma_\alpha \cap \Gamma_\beta = F$ and $(\omega \backslash \Gamma_\alpha) \cap (\omega \backslash \Gamma_\beta) \subseteq (\omega \backslash F)$, which shows that the system is disjoint. This demonstrates equality above.

For free amalgamated products with infinitely many factors we have

$$c(\oplus_{i \in I}^C A_i) \leq 2^{|C|} \cdot 2^{\sup_{i \in I} cA_i}.$$

To prove this, let κ be the cardinal on the right, and suppose that $\langle y_\alpha : \alpha < \kappa^+ \rangle$ is a disjoint system of elements of $\oplus_{i \in I}^C A_i$. We may assume that each y_α has the form

$$y_\alpha = \prod_{i \in F_\alpha} a_i^\alpha,$$

where F_α is a finite subset of I and $a_i^\alpha \in A_i$ for all $i \in F_\alpha$. We may assume, in fact, that the F_α's form a Δ-system, say with kernel G; and that they all have the same size. Thus by a change of notation we may write

$$y_\alpha = \prod_{j < m} a_{i_j^\alpha}^\alpha \cdot \prod_{j < n} a_{k_j}^\alpha,$$

where $F_\alpha \backslash G = \{i_j^\alpha : j < m\}$ and $G = \{k_j : j < n\}$. For distinct $\alpha, \beta < \kappa^+$ there then exist $c_j \in C$ for $j < m$, $d_j \in C$ for $j < m$, and $e_j \in C$ for $j < n$ such that $a_{i_j^\alpha}^\alpha \leq c_j$ for all $j < m$, $a_{i_j^\beta}^\beta \leq d_j$ for all $j < m$, and $a_{k_j}^\alpha \cdot a_{k_j}^\beta \leq e_j$ for all $j < n$, such that

$$\prod_{j < m} c_j \cdot \prod_{j < m} d_j \cdot \prod_{j < n} e_j = 0.$$

Using the Erdös-Rado theorem again, we get $\Gamma \in [\kappa^+]^\lambda$ and $c, d \in {}^m C$, $e \in {}^n C$ such that the above holds for all distinct $\alpha, \beta \in \Gamma$, where $\lambda = (|C| \cdot \sup_{i \in I} cA_i)^+$. Arguing similarly to the case of a free product with amalgamation of two algebras, we then easily infer that there is an $\alpha \in \Gamma$ such that $a_{k_j}^\alpha \leq e_j$ for each $j < n$. But then $y_\alpha = 0$, contradiction.

Since $\mathscr{P}\omega \oplus_C \mathscr{P}\omega$ can be considered as a subalgebra of $\oplus_{i \in \omega}^C \mathscr{P}\omega$, with C as in the example for the free product of two factors, it follows that the above inequality is again best possible.

The behaviour of cellularity under unions of well-ordered chains is clear on the basis of cardinal arithmetic. We restrict ourselves, without loss of generality, to well-ordered chains of regular type. Actually, we can formulate a more general fact about increasing chains of BAs; this fact will apply to several of our cardinal functions, namely to the ordinary sup-functions (see the introduction).

Theorem 3.11. *Let κ and λ be infinite cardinals, with λ regular. Suppose that k is an ordinary sup-function with respect to P. Then the following conditions are equivalent:*

(i) $cf\kappa = \lambda$.

(ii) There is a strictly increasing sequence $\langle A_\alpha : \alpha < \lambda \rangle$ of BAs each satisfying the $\kappa - k-chain$ condition such that $\bigcup_{\alpha < \lambda} A_\alpha$ does not satisfy this condition.

Proof. (i)$\Rightarrow$(ii): Assume (i). Let $\langle \mu_\xi : \xi < \lambda \rangle$ be a strictly increasing sequence of *ordinals* with sup κ (maybe κ is a successor cardinal, so that we cannot take the μ_ξ to be *cardinals*). Let A be a BA of size κ with a set $X \in PA$ such that $|X| = \kappa$. Write $A = \{a_\alpha : \alpha < \kappa\}$. For each $\xi < \lambda$ let $B_\xi = \langle \{a_\alpha : \alpha < \mu_\xi\} \rangle$. Thus $B_\xi \subseteq B_\eta$ if $\xi < \eta$, and $|B_\xi| < \kappa$ for all $\xi < \lambda$. Hence a strictly increasing subsequence is as desired (since λ is regular).

(ii)$\Rightarrow$(i). Assume that (ii) holds but (i) fails. Let X be a subset of $\bigcup_{\alpha < \lambda} A_\alpha$ of power κ which is in PA. If $\lambda < \mathrm{cf}\kappa$, then the facts that $X = \bigcup_{\alpha < \lambda}(X \cap A_\alpha)$, $|X| = \kappa$, and $|X \cap A_\alpha| < \kappa$ for all $\alpha < \lambda$, give a contradiction.

So, assume that $\mathrm{cf}\kappa < \lambda$. Now for all $\alpha < \lambda$ there is a $\beta > \alpha$ such that $X \cap A_\alpha \subset X \cap A_\beta$, since otherwise some A_α would contain X. It follows that $\lambda \leq \kappa$, and so κ is singular in the case we are considering. Let $\langle \mu_\alpha : \alpha < \mathrm{cf}\kappa \rangle$ be a strictly increasing sequence of cardinals with sup κ. Since $\sup_{\alpha < \lambda}|X \cap A_\alpha| = \kappa$, for each $\alpha < \mathrm{cf}\kappa$ choose $\nu(\alpha) < \lambda$ such that $|X \cap A_{\nu(\alpha)}| \geq \mu_\alpha$. Let $\rho = \sup_{\alpha < \mathrm{cf}\kappa} \nu(\alpha)$. Then $\rho < \lambda$ since $\mathrm{cf}\kappa < \lambda$ and λ is regular. But $|X \cap A_\rho| = \kappa$, contradiction.   □

With regard to Theorem 3.11, see also the end of this chapter.

We do not have a complete description of what happens to cellularity under ultraproducts, but we can give quite a bit of information. Many of the things which we mention hold for other cardinal functions as well. First we consider countably complete ultrafilters; the main result we want to give here is that if F is a countably complete ultrafilter on an infinite set I and A_i is a ccc BA for each $i \in I$, then $\prod_{i \in I} A_i/F$ also satisfies ccc. An analogous statement holds for many of our other functions. The result follows from the following standard facts. If F is countably complete and non-principal, then there is an uncountable measurable cardinal, and $|I|$ is at least as big as the first such — call it κ. (See Comfort, Negrepontis [74], p. 196.) Also, F is κ-complete. To see this, suppose not, and let λ be the least cardinal such that F is not λ-complete. Thus $\omega_1 < \lambda \leq \kappa$. Then there exist a cardinal $\mu < \lambda$ and disjoint $a_\alpha \subseteq I$ for $\alpha < \mu$ such that $I \backslash a_\alpha \in F$ for all $\alpha < \mu$, while $\bigcup_{\alpha < \mu} a_\alpha \in F$. Let $G = \{S \subseteq \mu : \bigcup_{\alpha \in S} a_\alpha \notin F\}$. Then it is easy to check that G is a σ-complete non-principal maximal ideal on μ, which is a contradiction, since μ is less than κ.

Now we can give the simple BA argument from these set-theoretical facts. Suppose $\prod_{i \in I} A_i/F$ does not satisfy ccc. Let $\langle [a_\alpha] : \alpha < \omega_1 \rangle$ be a system of non-zero disjoint elements of the product; $[x]$ denotes the equivalence class of x under F. Since F is ω_2-complete, the sets $J_{\alpha\beta} \overset{\text{def}}{=} \{i \in I : (a_\alpha)_i \cdot (a_\beta)_i = 0\}$ for $\alpha \neq \beta$ and the sets $K_\alpha \overset{\text{def}}{=} \{i \in I : (a_\alpha)_i \neq 0\}$ have a non-zero intersection, since that intersection is in F. But this is obviously a contradiction.

Thus countably complete ultrafilters tend to preserve chain conditions; we skip trying to give a more general version of the above argument.

Next, if F is a countably incomplete ultrafilter on I and each algebra A_i is infinite, then $\prod_{i \in I} A_i / F$ never has ccc. This follows from the fact that the product is ω_1-saturated in the model-theoretic sense; see Chang, Keisler [73], p. 305.

Now we present some results of Douglas Peterson. They depend on some well-known notions and results. An ultrafilter F on an infinite set I is *regular* if there is a system $\langle a_i : i \in I \rangle$ of elements of F such that $\bigcap_{j \in J} a_j = 0$ for every infinite subset J of I. The following concept is useful. Let F be an ultrafilter on I, and let $\langle \alpha_i : i \in I \rangle$ be a system of ordinals. We define the *essential supremum* of $\langle \alpha_i : i \in I \rangle$ over F to be

$$\text{ess.sup}_{i \in I}^{F} \alpha_i = \min\{\sup_{i \in b} \alpha_i : b \in F\}.$$

Keisler, Prikry [74] show that if F is a regular ultrafilter on an infinite set I and $\langle \kappa_i : i \in I \rangle$ is a system of infinite cardinals, then $|\prod_{i \in I} \kappa_i / F| = (\text{ess.sup}_{i \in I}^{F} \kappa_i)^{|I|}$.

Given an infinite cardinal κ and an ultrafilter F on some set I, we call F κ-*descendingly incomplete* provided that there is a system $\langle a_\alpha : \alpha < \kappa \rangle$ of elements of F such that $a_\alpha \supseteq a_\beta$ whenever $\alpha < \beta < \kappa$, and $\bigcap_{\alpha < \kappa} a_\alpha = 0$. We need the following well-known fact:

(*) If F is a regular ultrafilter on an infinite set I and κ is an infinite cardinal such that $\kappa \leq |I|$, then F is κ-descendingly incomplete.

To prove this fact, let $\langle a_\alpha : \alpha < |I| \rangle$ be a system of elements showing the regularity of F. For each $\alpha < \kappa$, let $b_\alpha = \bigcup_{\beta > \alpha} a_\beta$. Clearly the sequence $\langle b_\alpha : \alpha < \kappa \rangle$ shows the κ-descending incompleteness of F.

Now we begin Peterson's results, with two useful lemmas.

Lemma 3.12. *Suppose that κ is an uncountable limit cardinal, I is a set such that $\text{cf}\kappa \leq |I|$, and F is a $\text{cf}\kappa$-descendingly incomplete ultrafilter on I. Then there is a sequence $\langle \lambda_i : i \in I \rangle$ of infinite cardinals such that $\lambda_i < \kappa$ for all $i \in I$ and $\text{ess.sup}_{i \in I}^{F} \lambda_i = \kappa$.*

Proof. By the $\text{cf}\kappa$-descending incompleteness of F let $\langle a_\alpha : \alpha < \text{cf}\kappa \rangle$ be a system of elements of F such that $a_\alpha \supseteq a_\beta$ whenever $\alpha < \beta < \text{cf}\kappa$, and $\bigcap_{\alpha < \text{cf}\kappa} a_\alpha = 0$. We may assume that $a_0 = I$ and $a_\lambda = \bigcap_{\alpha < \lambda} a_\alpha$ for λ limit. Let $\langle \mu_\delta : \delta < \text{cf}\kappa \rangle$ be a strictly increasing continuous sequence of infinite cardinals with supremum κ. Now we define the sequence $\langle \lambda_i : i \in I \rangle$. Let $i \in I$. Then there is a $\gamma < \text{cf}\kappa$ such that $i \in a_\gamma \backslash a_{\gamma+1}$, and we define $\lambda_i = \mu_\gamma$. Now to show that $\text{ess.sup}_{i \in I}^{F} \lambda_i = \kappa$, take $a \in F$ and $\delta < \text{cf}\kappa$; we shall show that $\sup\{\lambda_i : i \in a\} \geq \mu_\delta$. For any $i \in a_\delta$ we have $\lambda_i \geq \mu_\delta$. Now $a \cap a_\delta \neq 0$ since $a \cap a_\delta \in F$, and if we choose $i \in a \cap a_\delta$ we have $\mu_\delta \leq \lambda_i$, and so $\mu_\delta \leq \sup\{\lambda_i : i \in a\}$, as desired. $\square$

Lemma 3.13. *Suppose that F is a regular ultrafilter on an infinite set I, κ is a limit cardinal, $\langle \kappa_i : i \in I \rangle$ is a system of infinite cardinals, $\kappa = \text{ess.sup}_{i \in I}^{F} \kappa_i$, $\text{cf}\kappa \leq |I|$, and $\omega < \kappa_i < \kappa$ for all $i \in I$. Then there is a system $\langle \lambda_i : i \in I \rangle$ of infinite cardinals such that $\lambda_i < \kappa_i$ for all $i \in I$ and $\text{ess.sup}_{i \in I}^{F} \lambda_i = \kappa$.*

Proof. Let $\langle a_\alpha : \alpha < |I| \rangle$ be a system of elements of F showing the regularity of F, with $a_0 = I$. Let $\langle \delta_\alpha : \alpha < \mathrm{cf}\kappa \rangle$ be a strictly increasing continuous sequence of cardinals with supremum κ such that $\delta_0 = \omega$. Let $G = \{i \in I : \kappa_i = \delta_\alpha$ for some limit $\alpha\}$, and for each $i \in G$ let $\alpha(i)$ be such that $\kappa_i = \delta_{\alpha(i)}$. Set $a^i_\xi = \bigcup_{\xi \leq \alpha < \alpha(i)} a_\alpha$ for each $i \in G$ and $\xi < \alpha(i)$. Then the following conditions clearly hold:

(1) $\bigcap_{\xi < \alpha(i)} a^i_\xi = 0$ for each $i \in G$.

(2) If $i, j \in G$ are such that $\alpha(i) < \alpha(j)$, then for each $\xi < \alpha(i)$ we have $a^i_\xi \subseteq a^j_\xi$.

(3) If $i \in G$ and ξ is a limit ordinal $< \alpha(i)$, then $\bigcap_{\gamma < \xi} a^i_\gamma = a^i_\xi$.

Here is the proof of (3), for example. Suppose the hypotheses of (3) hold but $k \in \left(\bigcap_{\gamma < \xi} a^i_\gamma \right) \setminus a^i_\xi$. Then $k \notin \bigcup_{\xi \leq \alpha < \alpha(i)} a_\alpha$, so for each $\gamma < \xi$, k is in some a_α with $\gamma \leq \alpha < \xi$. This clearly implies that k is in infinitely many a_α's, contradiction.

Now let $i \in I$ be arbitrary. We will define λ_i by cases. *Case 1.* $i \notin G$. There is an $\alpha < \mathrm{cf}\kappa$ such that $\delta_\alpha \leq \kappa_i < \delta_{\alpha+1}$. If α is a successor ordinal $\beta + 1$. let $\lambda_i = \delta_\beta$. Otherwise α is a limit ordinal or 0, and by $i \notin G$ we have $\delta_\alpha < \kappa_i < \delta_{\alpha+1}$, so let $\lambda_i = \delta_\alpha$. Under either possibility we then have $\lambda_i < \kappa_i$. *Case 2.* $i \in G$. Then there is a $\xi < \alpha(i)$ such that $i \in a^i_\xi \setminus a^i_{\xi+1}$; let $\lambda_i = \delta_\xi$. Thus $\lambda_i = \delta_\xi < \delta_{\alpha(i)} = \kappa_i$.

In order to show that $\mathrm{ess.sup}^F_{i \in I} \lambda_i = \kappa$, suppose that $a \in F$ and $\omega \leq \rho < \kappa$; we show that $\sup\{\lambda_i : i \in a\} \geq \rho$. Choose $\alpha < \mathrm{cf}\kappa$ such that $\delta_\alpha \leq \rho < \delta_{\alpha+1}$. We consider two cases. *Case 1.* $G \notin F$. Let $a' = \{i \in I : \kappa_i > \delta_{\alpha+2}\}$. Then $a' \in F$ since $\mathrm{ess.sup}^F_{i \in I} \kappa_i = \kappa$. If $i \in a' \setminus G$, then $\lambda_i \geq \delta_{\alpha+1} > \rho$. Since $I \setminus G \in F$, there is a $j \in (a \cap a') \setminus G$. Then $\sup\{\lambda_i : i \in a\} \geq \lambda_j > \rho$. *Case 2.* $G \in F$. Since $\mathrm{ess.sup}_{i \in I} \kappa_i = \kappa$, for each $\gamma < \kappa$ we have $M_\gamma \overset{\mathrm{def}}{=} \{i \in G : \delta_{\alpha(i)} > \gamma\} \in F$. Hence choose $k \in G$ such that $\delta_{\alpha(k)} > \delta_{\alpha+1}$. Then choose $j \in M_{\delta_{\alpha(k)}} \cap a^k_{\alpha+1} \cap a$. Then $j \in a^j_{\alpha+1}$ since $a^j_{\alpha+1} \supseteq a^k_{\alpha+1}$ by (2). Choose ε such that $j \in a^j_\varepsilon \setminus a^j_{\varepsilon+1}$. Then $\varepsilon \geq \alpha + 1$, and so $\lambda_j = \delta_\varepsilon \geq \delta_{\alpha+1} > \rho$. Hence $\sup\{\lambda_i : i \in a\} \geq \lambda_j > \rho$. $\square$

We need three more simple results.

Theorem 3.14. *If F is a regular ultrafilter over a set I then there is a system $\langle n_i : i \in I \rangle$ of natural numbers such that $\left| \prod_{i \in I} n_i / F \right| = 2^{|I|}$.*

Proof. Let $\langle a_i : i \in I \rangle$ be a system showing that F is regular. For each $i \in I$ let $M_i = \{j \in I : i \in a_j\}$. Thus $|M_i| < \omega$. We will show that $2^{|I|} \leq \left| \prod_{i \in I} {}^{M_i}2 / F \right|$, proving the theorem. For each $g \in {}^I 2$ define $g' \in \prod_{i \in I} {}^{M_i}2$ by $g'i = g \upharpoonright M_i$. If $g, h \in {}^I 2$ and $g \neq h$, pick $j \in I$ such that $gj \neq hj$; then for any $i \in a_j$ we have $j \in M_i$, and hence $g'i \neq h'i$. This shows that $a_j \subseteq \{i \in I : g'i \neq h'i\}$, and hence $g'/F \neq h'/F$. $\square$

Recall that if k is a cardinal function defined by supremums with respect to a function P, then

$$k'A = \min\{\kappa : |X| < \kappa \text{ for all } X \in PA\}.$$

Proposition 3.15. *If k is an ultra-sup function, $\langle A_i : i \in I \rangle$ is a system of infinite BAs, with I infinite, F is an ultrafilter on I, and $\kappa_i < k'A_i$ for all $i \in I$. then $k\left(\prod_{i \in I} A_i / F\right) \geq \left|\prod_{i \in I} \kappa_i / F\right|$.* □

Corollary 3.16. *If k is an ultra-sup function, $\langle A_i : i \in I \rangle$ is a system of BAs with I infinite, and F is an ultrafilter on I, then $k\left(\prod_{i \in I} A_i / F\right) \geq \operatorname{ess.sup}_{i \in I}^{F} kA_i$.*

Proof. Let $\lambda = \operatorname{ess.sup}_{i \in I}^{F} kA_i$. If λ is a successor cardinal, then we may assume that $kA_i = \lambda$ for all $i \in I$, and we are done by Proposition 3.15. If λ is a limit cardinal, then for each regular cardinal $\gamma < \lambda$ we have $k\left(\prod_{i \in I} A_i / F\right) \geq \left|\gamma^I / F\right| \geq \gamma$, and the result follows. □

Now we are ready for the main result of Peterson:

Theorem 3.17. *Let k be an ultra-sup function with respect to P such that if A is an infinite BA then PA contains arbitrarily large finite sets. Suppose that $\langle A_i : i \in I \rangle$ is a system of BAs with I infinite and F is a regular ultrafilter on I. Then $k\left(\prod_{i \in I} A_i / F\right) \geq \left|\prod_{i \in I} kA_i / F\right|$.*

Proof. Let $\lambda = \operatorname{ess.sup}_{i \in I}^{F} kA_i$, and recall that $\lambda^{|I|} = \left|\prod_{i \in I} kA_i / F\right|$. We now consider several cases.

Case 1. $\lambda = \omega$. Then $\lambda^{|I|} = 2^{|I|}$, and by Theorem 3.14, $k\left(\prod_{i \in I} A_i / F\right) \geq 2^{|I|}$.

Case 2. $\omega < \lambda \leq |I|$. Then $k\left(\prod_{i \in I} A_i / F\right) \geq \left|{}^I \omega / F\right| = 2^{|I|} = \lambda^{|I|}$.

Case 3. $\operatorname{cf}\lambda \leq |I| < \lambda$, and $\{i \in I : kA_i = \lambda\} \in F$. Then we may assume that $kA_i = \lambda$ for all $i \in I$. By Lemma 3.12, let $\langle \kappa_i : i \in I \rangle$ be a system of infinite cardinals such that $\kappa_i < \lambda$ for all $i \in I$ and $\operatorname{ess.sup}_{i \in I}^{F} \kappa_i = \lambda$. Then by Proposition 3.15, $k\left(\prod_{i \in I} A_i / F\right) \geq \left|\prod_{i \in I} kA_i / F\right|$.

Case 4. $\operatorname{cf}\lambda \leq |I| < \lambda$, and $\{i \in I : kA_i < \lambda\} \in F$. This is like Case 3, except Lemma 3.13 is used.

Case 5. $|I| < \operatorname{cf}\lambda$. Then $\lambda^{|I|} = \sup\{\kappa^{|I|} : \kappa < \lambda\}$. If $\omega \leq \kappa < \lambda$, then $k\left(\prod_{i \in I} A_i / F\right) \geq \left|{}^I \kappa / F\right| = \kappa^{|I|}$. Hence $k\left(\prod_{i \in I} A_i / F\right) \geq \lambda^{|I|}$. □

Recall that c is an ultra-sup function. Thus Theorem 3.17 gives a lower bound for $c\left(\prod_{i \in I} A_i / F\right)$, at least for regular F. The following simple result gives an upper bound.

Theorem 3.18. *Let $\langle A_i : i \in I \rangle$ is a system of infinite BAs, with I infinite, and suppose that F is a uniform ultrafilter on I. Let $\kappa = \max(|I|, \operatorname{ess.sup}_{i \in I}^{F} cA_i)$. Then $c\left(\prod_{i \in I} A_i / F\right) \leq 2^{\kappa}$.*

Proof. Let $\lambda = \operatorname{ess.sup}_{i \in I}^{F} cA_i)$. We may assume that $cA_i \leq \lambda$ for all $i \in I$. In order to get a contradiction, suppose that $\langle f_\alpha / F : \alpha < (2^\kappa)^+ \rangle$ is a system of disjoint elements. We may assume that $f_\alpha i \neq 0$ for all $i \in I$ and $\alpha < (2^\kappa)^+$. Thus $[(2^\kappa)^+]^2 = \bigcup_{i \in I} \{\{\alpha, \beta\} : f_\alpha i \cdot f_\beta i = 0\}$, so by the Erdös-Rado theorem $(2^\kappa)^+ \rightarrow (\kappa^+)^2_\kappa$ we get a homogeneous set which gives a contradiction. □

There are two main results about ultraproducts and cellularity in Shelah [90]. It is shown there that $c\left(\prod_{i \in I} A_i / F\right)$ does not depend solely on cardinal arithmetic.

Moreover, an example is given in which $|\prod_{i\in\omega} cA_i/F| < c\left(\prod_{i\in\omega} A_i/F\right)$ for any uniform ultrafilter F on ω. We give this result, since the reader may have a difficult time filling in the details of the proof given in Shelah [90]. We do not give it in the general form of that paper, restricting ourselves to a simple case. The theorem depends on a combinatorial theorem stated in Shelah [90] whose proof uses "Todorčević walks". For the proof see Todorčević [87b], Shelah [88e], and Bekkali [91]. The basic properties of the walks are stated in Todorčević [87b], but the proofs there are extremely terse. The author found proofs for some of these properties in the other two references, and while he could not understand those proofs completely, they suggested the proofs which follow.

Let θ be a regular uncountable cardinal. A *T-sequence* for θ is a sequence $\langle C_\xi : \xi < \theta \rangle$ such that for each limit $\xi < \theta$ the set C_ξ is a closed unbounded subset of ξ, for each $\xi + 1 < \theta$ the set $C_{\xi+1}$ is $\{\xi\}$, and $C_0 = 0$. We will consider the following condition on a T-sequence $\langle C_\xi : \xi < \theta \rangle$:

(T) If C is a club in θ, then there is an $\alpha < \theta$ such that for all $\beta \in [\alpha, \theta)$ we have $C \cap \alpha \not\subseteq C_\beta$.

Condition (T) implies some somewhat stronger conditions (T_m):

(T_m) If $\langle \zeta_\xi^k : \xi < \theta, k < m \rangle$ is a system of elements of θ such that $\{\zeta_\xi^k : k < m\} \cap \{\zeta_\eta^k : k < m\} = 0$ for all $\xi \neq \eta$, and if C is a club in θ, then there is an $\alpha < \theta$ such that for all $\beta \geq \alpha$ there is a ξ such that $\beta \leq \zeta_\xi^k$ for each $k < m$ and

$$C \cap \alpha \not\subseteq \bigcup_{k<m} C_{\zeta_\xi^k}.$$

In fact, clearly (T) implies (T_1). To derive the other conditions we need the following lemma, which is also useful for other purposes.

Lemma 3.19. *Assume that (T_m) holds, and assume its hypotheses. Then there exists $\langle \gamma^k : k < m \rangle$ such that $\gamma^k < \theta$ for each $k < m$ and*

$$\forall y < \theta \exists y' \in [y, \theta) \forall x < \theta \exists x' \in [x, \theta) \forall k < m[y' < \zeta_{x'}^k,$$
$$\text{and } y' \text{ is limit and } \gamma^k = \sup(C_{\zeta_{x'}^k} \cap y') < y'].$$

Proof. We can form a model M_m whose universe is θ, with $2m + 1$ relations: $<$, and for each $i < m$, the relations

$$S_i = \{(\alpha, \xi) : \alpha \in C_{\zeta_\xi^i}\} \quad T_i = \{(\alpha, \xi) : \alpha < \zeta_\xi^i\}.$$

Let $\langle N_\delta^m : \delta < \theta \rangle$ be a continuous increasing sequence of elementary submodels of M_m with union M_m, each of size less than θ. Then the set $C' \overset{\text{def}}{=} \{\delta \in C : N_\delta^m = \delta\}$ is club in θ. Choose α such that for all $\beta \geq \alpha$ there is a ξ such that $\beta \leq \zeta_\xi^k$ for each $k < m$ and $C' \cap \alpha \not\subseteq \bigcup_{k<m} C_{\zeta_\xi^k}$. Thus there is a set $B \in [\theta]^\theta$ such that for all

$\xi \in B$ we have $\alpha \leq \zeta_\xi^k$ for each $k < m$ and there is a $\delta_\xi \in C' \cap \alpha \setminus \bigcup_{k<m} C_{\zeta_\xi^k}$. Then there is a $\delta \in C' \cap \alpha$ and a $B' \in [B]^\theta$ such that $\delta_\xi = \delta$ for all $\xi \in B$. Note that for each $\xi \in B$ and $k < m$ the ordinal $\gamma_\xi^k \overset{\text{def}}{=} \sup(C_{\zeta_\xi^k} \cap \delta)$ is less than δ. Hence there exist a $B'' \in [B']^\theta$ and for each $k < m$ a $\gamma^k < \delta$ such that $\gamma_\xi^k = \gamma^k$ for all $\xi \in B'$. Now let $\varphi(u, v)$ say that $u < \zeta_v^k$, u is limit, and $\gamma^k = \sup(C_{\zeta_v^k} \cap u) < u$ for all $k < m$; clearly this is a formula in the language of M_m with constants in $N_\delta^m = \delta$, and $M_m \models \varphi[\delta, \xi]$ for all $\xi \in B''$. We claim:

$(*_m)$ $M_m \models \forall y \exists y' > y \forall x \exists x' > x \varphi(y', x')$.

In fact, suppose not. Then $N_\delta^m \models \exists y \forall y' > y \exists x \forall x' > x[\neg \varphi(y', x')]$. Hence choose $\rho \in N_\delta^m = \delta$ such that $N_\delta^m \models \forall y' > \rho \exists x \forall x' > x[\neg \varphi(y', x')]$. Then M_m models the same thing, so $M \models \exists x \forall x' > x[\neg \varphi(\delta, x')]$. So, choose ε so that $M \models \forall x' > \varepsilon \neg \varphi(\delta, x')$. Choose $\xi \in B''$ with $\xi > \varepsilon$. Then $M \models \neg \varphi[\delta, \xi]$, contradiction. So the conclusion of the lemma holds. $\qquad \square$

Lemma 3.20. *(T) and (T_m) imply (T_{m+1}).*

Proof. Assume (T) and (T_m). Suppose that $\langle \zeta_\xi^k : \xi < \theta, k \leq m \rangle$ is a system of elements of θ such that $\{\zeta_\xi^k : k \leq m\} \cap \{\zeta_\eta^k : k \leq m\} = 0$ for all $\xi \neq \eta$ and C is club in θ. Apply Lemma 3.19 to $\langle \zeta_\xi^k : \xi < \theta, k < m \rangle$ and C to get $\langle \gamma^k : k < m \rangle$ as indicated. Let $\delta = \sup_{k<m}(\gamma^k + 1)$. Apply (T) to get $\alpha < \theta$ such that for all ξ with $\alpha \leq \zeta_\xi^m$ we have $(C \setminus \delta) \cap \alpha \not\subseteq C_{\zeta_\xi^m}$. Suppose, to verify the conclusion of (T_m) for α, that $\beta \in [\alpha, \theta)$. By the conclusion of Lemma 3.17, choose a limit $\varepsilon \in [\beta + 1, \theta)$ so that

$(*)$ $\forall \eta < \theta \exists \xi \in [\eta, \theta) \forall k < m[\varepsilon < \zeta_\xi^k$ and $\gamma^k = \sup(C_{\zeta_\xi^k} \cap \varepsilon) < \varepsilon]$.

Choose $\eta < \theta$ such that $\forall \xi \in [\eta, j) \forall k < m(\beta \leq \zeta_\xi^k)$. Then choose $\xi \in [\eta, \theta)$ by $(*)$. Then take $\delta' \in (C \setminus \delta) \cap \alpha \setminus C_{\zeta_\xi^m}$. Then there are no members of $C_{\zeta_\xi^k}$ between γ^k and ζ_ξ^k. Since $\gamma^k < \delta \leq \delta' < \alpha < \varepsilon < \zeta_\xi^k$, it follows that $\delta' \notin C_{\zeta_\xi^k}$. So $\delta' \in C \cap \alpha \setminus \bigcup_{k \leq m} C_{\zeta_\xi^k}$, as desired. $\qquad \square$

Next, suppose that θ is a regular uncountable cardinal and $\langle C_\xi : \xi < \theta \rangle$ is a T-sequence. We define $\rho_2 : \{(\alpha, \beta) : \alpha \leq \beta < \theta\} \to \omega$ by induction: $\rho_2(\alpha, \alpha) = 0$, and if $\alpha < \beta$, then

$$\rho_2(\alpha, \beta) = \rho_2(\alpha, \min(C_\beta \setminus \alpha)) + 1.$$

Lemma 3.21. *Let θ be a regular uncountable cardinal, $\langle C_\xi : \xi < \theta \rangle$ a T-sequence, and assume (T). Suppose $\langle {}_\varepsilon \zeta_\xi^k : \xi < \theta, k < m \rangle$ is given for $\varepsilon \in 2$ so that $\{{}_\varepsilon \zeta_\xi^k : k < m\} \cap \{{}_\varepsilon \zeta_\eta^k : k < m\} = 0$ for any $\varepsilon < 2$ and any $\xi \neq \eta$. Then for any $n \in \omega$ there exist $\xi, \eta < \theta$ such that ${}_0 \zeta_\xi^k < {}_1 \zeta_\eta^l$ and $\rho_2({}_0 \zeta_\xi^k, {}_1 \zeta_\eta^l) \geq n$ for all $k, l < m$.*

Proof. We proceed by induction on n. The case $n = 0$ is clear. Assume the lemma for n. Apply Lemma 3.19 to $\langle {}_1 \zeta_\xi^k : \xi < \theta, k < m \rangle$ to obtain $\langle \gamma^k : k < m \rangle$ as indicated. We can then obtain ε_ν and ξ_ν for each $\nu < \theta$ so that the following conditions hold for all $k < m$:

(1) ε_ν is a limit ordinal;
(2) $\varepsilon_\nu < {}_1\zeta^k_{\xi_\nu}$;
(3) $\gamma^k = \sup(C_{{}_1\zeta^k_{\xi_\nu}} \cap \varepsilon_\nu) < \varepsilon_\nu$;
(4) ${}_1\zeta^k_{\xi_\nu} < \varepsilon_{\nu'}$ if $\nu < \nu'$.

Define ${}_2\zeta^k_\nu = \min(C_{{}_1\zeta^k_{\xi_\nu}} \backslash \varepsilon_\nu)$ for each $\nu < \theta$. Then if $\nu < \nu'$ and $k, l < m$ we have

$$ {}_2\zeta^k_\nu < {}_1\zeta^k_{\xi_\nu} < \varepsilon_{\nu'} \leq {}_2\zeta^l_{\nu'}. $$

Hence for $\nu < \nu'$ we have $\{{}_2\zeta^k_\nu : k < m\} \cap \{{}_2\zeta^k_{\nu'} : k < m\} = 0$. Choose $\xi_0 < \theta$ such that $\gamma^k < {}_0\zeta^l_\xi$ for all $k, l < m$ and all $\xi \in [\xi_0, \theta)$. Let ${}_3\zeta^k_\xi = {}_0\zeta^k_{\xi_0 + \xi}$ for all $\xi < \theta$. Now we apply the induction hypothesis to $\langle {}_3\zeta^k_\xi : \xi < \theta, k < m \rangle$ and $\langle {}_2\zeta^k_\nu : \nu < \theta, k < m \rangle$ to obtain $\xi, \nu < \theta$ such that

(5) $\forall k, l < m [{}_3\zeta^k_\xi < {}_2\zeta^l_\nu$ and $\rho_2({}_3\zeta^k_\xi, {}_2\zeta^l_\nu) \geq n]$.

Take any $k, l < m$. By (3), there are no members of $C_{{}_1\zeta^l_{\xi_\nu}}$ between γ^l and ${}_2\zeta^l_\nu$. Now

$$ \gamma^l < {}_0\zeta^k_{\xi_0 + \xi} = {}_3\zeta^k_\xi < {}_2\zeta^l_\nu $$

and

$$ \gamma^l < \varepsilon_\nu \leq {}_2\zeta^l_\nu, $$

so $\min(C_{{}_1\zeta^l_{\xi_\nu}} \backslash {}_3\zeta^k_\xi) = \min(C_{{}_1\zeta^l_{\xi_\nu}} \backslash \varepsilon_\nu) = {}_2\zeta^l_\nu$. Hence

$$ \rho_2({}_0\zeta^k_{\xi_0 + \xi}, {}_1\zeta^l_{\xi_\nu}) = \rho_2({}_3\zeta^k_\xi, \min(C_{{}_1\zeta^l_{\xi_\nu}} \backslash {}_3\zeta^k_\xi)) + 1 $$
$$ = \rho_2({}_3\zeta^k_\xi, {}_2\zeta^l_\nu) + 1 \geq n + 1, $$

as desired. $\qquad\square$

The combinatorial theorem which we actually need now follows:

Theorem 3.22. (Shelah) *Let $\lambda = \theta^+$ with θ an infinite cardinal. Then there is a $d : [\lambda]^2 \to \omega$ such that for all $m, n \in \omega$, if $\langle \zeta_i : i < \lambda \rangle$ is a system of n-tuples of members of λ such that $\zeta^1_i < \cdots < \zeta^n_i$ for all $i < \lambda$ and $\zeta^n_i < \zeta^1_j$ if $i < j < \lambda$, then there exist $i, j \in \lambda$ with $i < j$ such that $d\{\zeta^k_i, \zeta^l_j\} \geq m$ for all $k, l = 1, \ldots, n$.*

Proof. By Lemma 3.21 we just need to see that λ has a T-sequence satisfying (T). For each limit ordinal $\alpha < \lambda$ let C_α be a closed unbounded subset of α of order type cfα. For $\alpha < \lambda$ let $C_{\alpha+1} = \{\alpha\}$, and let $C_0 = 0$. Now if C is a closed unbounded subset of λ, let α be a member of C such that the order type of $C \cap \alpha$ is $\theta + 1$. Then for all $\beta \in [\alpha, \lambda)$ we have $C \cap \alpha \not\subseteq C_\beta$, as desired. $\qquad\square$

Theorem 3.23. (Shelah). *Let $\lambda = \theta^+$ with θ an infinite cardinal. Then there is a system $\langle B_n : n \in \omega \rangle$ of BAs each satisfying the λ-cc such that for any non-principal ultrafilter D on ω, the ultraproduct $\prod_{n \in \omega} B_n/D$ does not satisfy the λ-cc.*

(If $\lambda = (2^\omega)^{++}$, then we get $|\prod_{n \in \omega} cB_n/D| \leq (2^\omega)^+ < (2^\omega)^{++} \leq c(\prod_{n \in \omega} B_n/D)$, as indicated in the remark following Theorem 3.18.)

Proof. Choose d by Theorem 3.22. Temporarily fix $n \in \omega$. Let C_n be freely generated by $\langle x_\alpha^n : \alpha < \lambda \rangle$. Let I_n be the ideal of C_n generated by the set

$$\{x_\alpha^n \cdot x_\beta^n : \alpha < \beta < \lambda \text{ and } d\{\alpha, \beta\} \leq n\}.$$

Let $y_\alpha^n = x_\alpha^n/I_n$ for each $\alpha < \lambda$. Set $B_n = C_n/I_n$. Then for $\alpha < \beta < \lambda$ we have $\{n \in \omega : y_\alpha^n \cdot y_\beta^n = 0\} \supseteq \{n \in \omega : d\{\alpha, \beta\} \leq n\}$, so $(y_\alpha^n/D) \cdot (y_\beta^n/D) = 0$. We claim that $y_\alpha^n/D \neq 0$; in fact, $y_\alpha^n \neq 0$ for all $n \in \omega$. Otherwise we would get

$$x_\alpha^n \leq x_{\gamma_1}^n \cdot x_{\delta_1}^n + \cdots + x_{\gamma_m}^n \cdot x_{\delta_m}^n$$

with $\gamma_i \neq \delta_i$ for all i. Mapping x_α^n to 1 and all other generators to 0 then extending to a homomorphism, we get a contradiction.

To show that each B_n satisfies λ-cc, assume that $\langle b_\alpha : \alpha < \lambda \rangle \in {}^\lambda C_n$ is such that $b_\alpha \cdot b_\beta \in I_n$ for all distinct $\alpha, \beta < \lambda$, while each $b_\alpha \notin I_n$; we want to get a contradiction. Without loss of generality, we may assume that each b_α has the following form:

$$b_\alpha = \prod_{\beta \in F_\alpha} (x_\beta^n)^{\varepsilon_\alpha \beta},$$

where F_α is a finite subset of λ and $\varepsilon_\alpha \in {}^{F_\alpha}2$. Without loss of generality we may assume that: $\langle F_\alpha : \alpha < \lambda \rangle$ forms a Δ-system, say with kernel G; $\varepsilon_\alpha \upharpoonright G$ is the same for all $\alpha < \lambda$; and $|F_\alpha \backslash G| = |F_\beta \backslash G|$ for all $\alpha, \beta < \lambda$. Cutting down further, we may assume that $F_\alpha \backslash G < F_\beta \backslash G$ if $\alpha < \beta$ (that is, $\gamma < \delta$ if $\gamma \in F_\alpha \backslash G$ and $\delta \in F_\beta \backslash G$). Now by Lemma 3.22 choose $\alpha < \beta < \lambda$ so that $d\{\varepsilon, \zeta\} \geq n + 1$ for all $\varepsilon \in F_\alpha \backslash G$ and $\zeta \in F_\beta \backslash G$. Now we can write

(1) $$b_\alpha \cdot b_\beta \leq x_{\gamma_1}^n \cdot x_{\delta_1}^n + \cdots + x_{\gamma_m}^n \cdot x_{\delta_m}^n$$

with $\gamma_i < \delta_i$ and $d\{\gamma_i, \delta_i\} \leq n$; moreover, we assume that m is minimal so that such an inequality holds. It follows that $\gamma_i, \delta_i \in F_\alpha \cup F_\beta$ for all i. If $\gamma_i \in F_\alpha$, then $\varepsilon_\alpha \gamma_i = 1$, since otherwise the summand $x_{\gamma_i}^n \cdot x_{\delta_i}^n$ could be dropped. Similarly $\gamma_i \in F_\beta$ implies that $\varepsilon_\beta \gamma_i = 1$, and similarly for the δ_i's. Now it follows, since $b_\alpha \notin I_n$, that we cannot have $\gamma_1, \delta_1 \in F_\alpha$. Similarly, $\gamma_1, \delta_1 \notin F_\beta$. It follows, then, that $\gamma_1 \in F_\alpha \backslash G$ and $\delta_1 \in F_\beta \backslash G$. But then $d\{\gamma_1, \delta_1\} \geq n + 1$, contradiction. □

According to a result of Donder [88], it is consistent that the lower bound in Theorem 3.17 always holds (since his result says that it is consistent that every uniform ultrafilter is regular). However, the following problem appears to be open.

Problem 2. *Is it consistent that there is a an infinite set I, a system $\langle A_i : i \in I \rangle$ of infinite BAs, and an ultrafilter F on I such that* $c\left(\prod_{i \in I} A_i/F\right) < \left|\prod_{i \in I} cA_i/F\right|$?

It may be that a solution can be found using methods of Magidor, Shelah [91]. Theorems in Rosłanowski, Shelah [94] say that in an example of this kind, cA_i is inaccessible for a set of i's in the ultrafilter.

Also note that there are obvious examples where the upper bound given in Theorem 3.18 is attained, and other examples where it is not attained.

We turn to the examination of cellularity under other operations on Boolean algebras. If A is a dense subalgebra of B, obviously $cA = cB$.

The situation with subdirect products is clear. Suppose that B is a subdirect product of BAs $\langle A_i : i \in I \rangle$; what is the cellularity of B in terms of the cellularity of the A_i's? Well, since a direct product is a special case of a subdirect product, we have the upper bound $cB \leq \sup_{i \in I} cA_i \cup |I|$. The lower bound ω is obvious. And that lower bound can be attained, even if the algebras A_i have high cellularity. In fact, consider the following example. Let κ be any infinite cardinal, let A be the free BA on κ free generators, and let B be the algebra of finite and cofinite subsets of κ. We show that A is isomorphic to a subdirect product of copies of B. To do this, it suffices to take any non-zero element $a \in A$ and find a homomorphism of A onto B which takes $-a$ to 0. In fact, $A \restriction a$ is still free on κ free generators, and so there is a homomorphism of it onto B. So our desired homomorphism is obtained as follows:

$$A \to (A \restriction -a) \times (A \restriction a) \to A \restriction a \to B.$$

If $\mathscr{S}$ is a sheaf of BAs with base space X, then $c(\mathrm{Clop}X) \leq c(\mathrm{Gs}\mathscr{S})$ by Theorem 1.1. Even for the more special case of Boolean products the difference can be large; and equality is also possible. This follows from Theorems 1.2–1.4.

Problems concerning cellularity properties of Boolean powers reduce to more familiar problems concerning the cellularity of free products, discussed above; see Chapter 1.

For set products we clearly have

$$c\left(\prod_{i \in I}^{B} A_i\right) = |I| + \sup_{i \in I} c(A_i).$$

Next, let B be obtained from algebras $\langle A_i : i \in I \rangle$ by one-point gluing, as described in Chapter 1. With respect to cellularity, clearly B behaves much like the full direct product: If B is infinite and all algebras A_i have at least four elements, then

$$cB = |I| + \sup_{i \in I} cA_i.$$

Our next algebraic operation is Aleksandroff duplication. Clearly $c(\mathrm{Dup}A) = |\mathrm{Ult}A|$.

We consider now the exponential of a given BA A. We give an example, assuming the existence of a Suslin tree, of a ccc BA A such that $\mathrm{Exp}A$ has cellularity

ω_1. Recall that $\mathcal{S}$ is the Stone isomorphism of a BA onto the clopen algebra of its Stone space. Now assume that T is a Suslin tree in which every element s has infinitely immediate successors, among which we pick out two, $s0$ and $s1$. Let A be the tree algebra on T. For each $s \in T$ let $T \uparrow s = \{t \in T : s \leq t\}$, and let $x_s = \mathscr{V}(\mathcal{S}(T \uparrow s0), \mathcal{S}(T \uparrow s1))$. We claim that $x_s \cap x_t = 0$ for distinct $s, t \in T$. For, say $s \not\leq t$, but suppose that $F \in x_s \cap x_t$. Choose $\varepsilon \in \{0,1\}$ so that $t\varepsilon$ is incomparable with s. Now $F \cap \mathcal{S}(T \uparrow t\varepsilon) \neq 0$; say $M \in F \cap \mathcal{S}(T \uparrow t\varepsilon)$. Since $F \subseteq \mathcal{S}(T \uparrow s0) \cup \mathcal{S}(T \uparrow s1)$, choose $\delta \in \{0,1\}$ such that $M \in \mathcal{S}(T \uparrow s\delta)$. Thus $(T \uparrow t\varepsilon) \cap (T \uparrow s\delta) \neq 0$, so $t\varepsilon$ and $s\delta$ are comparable, contradiction.

The example of Todorčević concerning free products which was described above adapts to the exponential. Namely, in ZFC there is a BA D such that $cD \leq \beth_\omega$ while $c(\mathrm{Exp}D) \geq \beth_\omega^+$. In fact, let λ and a be chosen by Lemma 3.7, and let I and J be infinite disjoint subsets of ω. Set $A = \{a_\alpha : \alpha < \beth_\omega^+\}$, $B = \mathrm{RO}(\mathscr{P}_I A)$, $C = \mathrm{RO}(\mathscr{P}_J A)$, and $D = B \times C$, using the notation of Lemmas 3.1–3.5. Thus $cD \leq \beth_\omega$. We claim that

$$\langle \mathscr{V}(\mathcal{S}(b_{\{a\}}, 0), \mathcal{S}(0, b_{\{a\}})) : a \in A \rangle$$

is pairwise disjoint in $\mathrm{Exp}D$. For, suppose that a and a' are distinct elements of A and

$$C \in \mathscr{V}(\mathcal{S}(b_{\{a\}}, 0), \mathcal{S}(0, b_{\{a\}})) \cap \mathscr{V}(\mathcal{S}(b_{\{a'\}}, 0), \mathcal{S}(0, b_{\{a'\}})).$$

Then $C \cap \mathcal{S}(b_{\{a\}}, 0) \neq 0$, $C \subseteq \mathcal{S}(b_{\{a'\}}, 0) \cup \mathcal{S}(0, b_{\{a'\}})$, and $\mathcal{S}(b_{\{a\}}, 0) \cap \mathcal{S}(0, b_{\{a'\}}) = 0$, so $C \cap \mathcal{S}(b_{\{a\}}) \cap \mathcal{S}(b_{\{a'\}}, 0) \neq 0$, hence $b_{\{a\}} \cap b_{\{a'\}} \neq a$, so $\{a\}$ and $\{a'\}$ are compatible and hence $\rho(a, a') \in I$. Similarly, $\rho(a, a') \in J$, which is impossible.

Note from the remark after Lemma 3.10, and Proposition 2.5, that under MA+¬CH, A ccc implies $\mathrm{Exp}A$ ccc.

Now we proceed to discuss the derived functions associated with cellularity. First we show that c_{H+} is the same as spread. For this, it is convenient to have an equivalent definition of spread. A subset X of a BA A is *ideal independent* if $x \notin \langle X \backslash \{x\} \rangle^{\mathrm{Id}}$ for every $x \in X$; recall that $\langle Y \rangle^{\mathrm{Id}}$ denotes the ideal generated by Y, for any $Y \subseteq A$.

Theorem 3.24. *For any infinite BA A, $sA = \sup\{|X| : X$ is an ideal independent subset of $A\}$.*

Proof. First suppose that D is a discrete subspace of $\mathrm{Ult}A$. For each $F \in D$, let $a_F \in A$ be such that $Sa_F \cap D = \{F\}$. Then $\langle a_F : F \in D \rangle$ is one-one and $\{a_F : F \in D\}$ is ideal independent. In fact, suppose that $F, G_0, \ldots, G_{n-1}$ are distinct members of D such that $a_F \leq a_{G_0} + \ldots + a_{G_{n-1}}$. Then $Sa_F \subseteq Sa_{G_0} \cup \ldots \cup Sa_{G_{n-1}}$, and so $F \in Sa_{G_0} \cup \ldots \cup Sa_{G_{n-1}}$, which is clearly impossible.

Conversely, suppose that X is an ideal independent subset of A. Then for each $x \in X$, $\{x\} \cup \{-y : y \in X \backslash \{x\}\}$ has the finite intersection property, and so is included in an ultrafilter F_x. Let $D = \{F_x : x \in X\}$. Then $Sx \cap D = \{F_x\}$ for each $x \in X$, so D is discrete and $|D| = |X|$, as desired. $\quad \square$

By the proof of Theorem 3.24, spread in the two senses given in the theorem has the same attainment properties.

Theorem 3.25. *For any infinite BA A, $c_{H+}A$ is equal to sA, the spread of A.*

Proof. First let f be a homomorphism from A onto a BA B, and let X be a disjoint subset of B^+. We show that $|X| \le sA$; this will show that $c_{H+}A \le sA$. For each $x \in X$ choose $a_x \in X$ such that $f a_x = x$. Then $\langle a_x : x \in X \rangle$ is one-one and $\{a_x : x \in X\}$ is ideal independent. In fact, suppose that $x, y(0), \ldots y(n-1)$ are distinct elements of X, and $a_x \le a_{y(0)} + \cdots + a_{y(n-1)}$. Applying the homomorphism f to this inequality we get $x \le y(0) + \cdots + y(n-1)$. Since the elements $x, y(0), \ldots, y(n-1)$ are pairwise disjoint, this is impossible.

For the converse, suppose that X is an ideal independent subset of A; we want to find a homomorphic image B of A having a disjoint subset of size $|X|$. Let $I = \langle \{x \cdot y : x, y \in X, x \ne y\} \rangle^{\mathrm{Id}}$. It suffices now to show that $[x] \ne 0$ for each $x \in X$. ($[u]$ is the equivalence class of u under the equivalence relation naturally associated with the ideal I). Suppose that $[x] = 0$. Then x is in the ideal I, and hence there exist elements $y_0, z_0, \ldots, y_{n-1}, z_{n-1}$ of X such that $y_i \ne z_i$ for all $i < n$, and $x \le y_0 \cdot z_0 + \ldots + y_{n-1} \cdot z_{n-1}$. Without loss of generality, $x \ne y_i$ for all $i < n$. But then $x \le y_0 + \cdots + y_{n-1}$, contradicting the ideal independence of X. $\square$

For later purposes it is convenient to note the following corollary to the proof of the previous two theorems.

Corollary 3.26. $c_{H+}A$ *and* sA *have the same attainment properties, in the sense that sA is attained (in either the discrete subspace or ideal independence sense) iff there exist a homomorphic image B of A and a disjoint subset X of B such that $|X| = c_{H+}A$.* $\square$

Note in this corollary that attainment of $c_{H+}A$ involves two sups, while attainment of sA involves only one. Thus if sA is not attained, there are still two possibilities according to Corollary 3.26: there can exist a homomorphic image B of A with $sA = cB$ but cB is not attained, or there is no homomorphic image B of A with $sA = cB$. Both possibilities are consistent with ZFC; we shall return to this shortly and indicate the examples.

It is easy to see that $c_{H-}A = \omega$ for any infinite BA A: let B be a denumerable subalgebra of A, and extend the identity homomorphism h of B into $\overline{B}$ to a homomorphism from A into $\overline{B}$; the image of A under h is a ccc BA. (We are using here Sikorski's extension theorem; recall that $\overline{B}$ is the completion of B.) It is obvious that $c_{S+}A = cA$ and $c_{S-}A = \omega$ for any infinite BA A. $c_{h+}A$ is equal to sA, since a disjoint family of open subsets of a subspace Y of $\mathrm{Ult}A$ gives a discrete subset of $\mathrm{Ult}A$ of the same size, so that $c_{h+}A \le sA = c_{H+}A \le c_{h+}A$. It is obvious that $c_{h-}A = \omega$, and an easy argument gives that $_d c_{S+}A = cA = {}_d c_{S-}A$.

Next, recall from the introduction the definition of $c_{mm}A$:

$$c_{mm}A = \min\{|X| : X \text{ is an infinite maximal disjoint subset of } A\}.$$

We note some easy facts about this function.

(1) $c_{mm} A \geq \min\{|Y| : Y \subseteq A, \sum Y = 1$, and $\sum Y' \neq 1$ for every finite subset Y' of $Y\}$.

If $A = \mathscr{P}\omega/\mathrm{fin}$, then in the notation of van Douwen [84], $c_{mm} A = \mathfrak{a}$ and the right-hand side of (1) is $\mathfrak{p}$. It is known to be consistent that $\mathfrak{p} < \mathfrak{a}$. For BAs in general an example with $>$ in (1) can be given in ZFC. Namely, for any infinite cardinal κ let B be the free BA with distinct free generators x_α for $\alpha < \kappa$, and let $A = \mathrm{Dup}\, B$. We claim that $c_{mm} A = 2^\kappa$, while the right-hand side of (1) is $\leq \kappa$. Note that $|A| = 2^\kappa$, so $c_{mm} A \leq 2^\kappa$. Suppose that $\langle (a_\alpha, X_\alpha) : \alpha < \lambda \rangle$ is a system of pairwise disjoint nonzero elements of A with sum 1, $\omega \leq \lambda < 2^\kappa$. Since $cB = \omega$, we may assume that $a_\alpha = 0$ for all $\alpha \in [\omega, \lambda)$; hence X_α is finite for all such α. If $\sum_{\alpha \in F} a_\alpha = 1$ for some finite $F \subseteq \omega$, then $\bigcup_{\alpha \in F} X_\alpha$ is a cofinite subset of $\mathrm{Ult} B$, and so $\sum_{\alpha \in G}(a_\alpha, X_\alpha) = 1$ for some finite $G \subseteq \lambda$, contradiction. It follows that $\{-a_\alpha : \alpha < \omega\}$ has fip. For each $\alpha < \omega$ there is a finite $\Gamma_\alpha \subseteq \kappa$ such that $-a_\alpha$ is generated by $\{x_\beta : \beta \in \Gamma_\alpha\}$. Let $\Delta = \bigcup_{\alpha < \omega} \Gamma_\alpha$. If $\varepsilon \in {}^{\kappa \setminus \Delta}2$, then $\{-a_\alpha : \alpha < \omega\} \cup \{x_\beta^{\varepsilon\beta} : \beta \in \kappa \setminus \Delta\}$ has fip. Thus there are 2^κ ultrafilters D such that $-a_\alpha \in D$ for all $\alpha < \omega$. but

$$\left| \bigcup_{\alpha < \omega} (X_\alpha \setminus Sa_\alpha) \cup \bigcup_{\omega \leq \alpha < \lambda} X_\alpha \right| < 2^\kappa,$$

contradiction. Thus $c_{mm} A = 2^\kappa$.

Next, we exhibit $Y \subseteq A$ of size κ showing that the right-hand side of (1) is $\leq \kappa$. Namely, let

$$Y = \{(x_\alpha, Sx_\alpha) : \alpha < \kappa\} \cup \{(0, \{D\})\},$$

where D is the ultrafilter on B such that $-x_\alpha \in D$ for all $\alpha < \kappa$. To show that $\sup Y = 1$ it suffices to take any nonzero $(a, X) \in A$ and find $y \in Y$ such that $(a, X) \cdot y \neq 0$. Case 1. $a \neq 0$. Then there is an $\alpha < \kappa$ such that $a \cdot x_\alpha \neq 0$, and so $(a, X) \cdot (x_\alpha, Sx_\alpha) \neq 0$. Case 2. $a = 0$. So X is finite and nonempty. Without loss of generality, $D \notin X$. Take any $F \in X$. Then there is an $\alpha < \kappa$ such that $x_\alpha \in F$, so $(0, X) \cdot (x_\alpha, Sx_\alpha) \neq 0$, as desired.

Clearly $\sum Y' \neq 1$ for all finite $Y' \subseteq Y$.

(2) If A is the finite-cofinite algebra on κ, then $c_{mm} A = \kappa$.

(3) If A is κ-saturated in the model-theoretic sense, then $c_{mm} A \geq \kappa$.

(4) $c_{mm}(A \times B) = \min(c_{mm} A, c_{mm} B)$.

To see this (valid only for *infinite* A and B), suppose without loss of generality that $c_{mm} A \leq c_{mm} B$. Let X be a maximal disjoint subset of A of size $c_{mm} A$. Then $\{(x, 0) : x \in X\} \cup \{(0, 1)\}$ is a maximal disjoint subset of $A \times B$. This proves $\leq$ in (4). On the other hand, let Z be a maximal disjoint subset of $A \times B$ of size $c_{mm}(A \times B)$. Let pr_i be the projection of $A \times B$ into the i-th coordinate, $i = 1, 2$.

Clearly $\mathrm{pr}_1[Z]$ is a maximal disjoint subset of A, and similarly for $\mathrm{pr}_2[Z]$ and B. One at least of these projections is infinite, and this yields $\geq$.

(5) For I infinite and each A_i non-trivial we have

$$c_{\mathrm{mm}}\left(\prod_{i\in I} A_i\right) = \min(|I|, \min\{c_{\mathrm{mm}}A_i : i \in I, \ A_i \text{ infinite}\}).$$

Proof of (5): An argument as in (4) shows that

$$\min\{c_{\mathrm{mm}}A_i : i \in I, \ A_i \text{ infinite}\} \geq c_{\mathrm{mm}}\left(\prod_{i\in I} A_i\right).$$

Now for each $i \in I$ let f_i be the member of $\prod_{i\in I} A_i$ such that $f_i j = 0$ if $j \neq i$ while $f_i i = 1$. This gives a partition of 1 in $\prod_{i\in I} A_i$ of size $|I|$, so $|I| \geq c_{\mathrm{mm}}(\prod_{i\in I} A_i)$. Thus the right side of (5) is $\geq$ the left side. Now suppose that Z is an infinite partition of 1 in $\prod_{i\in I} A_i$, but its size is less than the right side of (5); we assume that all members of Z are nonzero. Now $\{z_i : z \in Z\}\backslash\{0\}$ is a partition of 1 in A_i for each $i \in I$, so $\{z_i : z \in Z\}$ is finite for each $i \in I$. Since Z is disjoint, $\{z \in Z : z_i \neq 0\}$ is finite for all $i \in I$. Choose $z \in Z\backslash\bigcup_{i\in I}\{z \in Z : z_i \neq 0\}$. Then $z = 0$, contradiction. So, (5) holds.

(6) Given $\omega \leq \kappa < \lambda$, there is a BA A such that $cA = \lambda$ and $c_{\mathrm{mm}}A = \kappa$.

Proof: let $A = \mathrm{Finco}\,\kappa \times \mathrm{Finco}\,\lambda$.

(7) If A is σ-complete, then $c_{\mathrm{mm}}A = \omega$.

(8) If I is infinite and each A_i has at least four elements, then $c_{\mathrm{mm}}(\oplus_{i\in I}A_i) = \omega$. To prove this, without loss of generality say that $I = \kappa$, an infinite cardinal. For each $i \in \omega$, let $0 < a_i < 1$ in A_i. Then consider the elements $x_0, x_1, \ldots$ in the free product defined by

$$x_0 = a_0,$$
$$x_1 = -a_0 \cdot a_1,$$
$$x_2 = -a_0 \cdot -a_1 \cdot a_2,$$
$$\text{etc.}$$

It is easily verified that these elements form a partition of unity in the free product, yielding (8).

The homomorphic spectrum of cellularity is interesting. First, we can easily see that $[\omega, sA) \subseteq c_{\mathrm{Hs}}A \subseteq [\omega, sA]$ (for cardinals $\kappa < \lambda$, $[\kappa, \lambda)$ denotes the set of all cardinals μ such that $\kappa \leq \mu < \lambda$; similarly for $[\kappa, \lambda]$). This follows from the fact already proved that $sA = c_{\mathrm{H+}}A$: given a homomorphic image B of A and a disjoint subset X of B, one can use Sikorski's extension theorem to get a homomorphic image C of B such that $cC = |X|$.

It is more difficult to decide whether $sA \in c_{Hs}A$. This amounts to the following question: is there always a homomorphic image B of A such that $cB = sA$? In case sA is attained, this is true by Corollary 3.26. For sA not attained, there are three consistency results which clarify things here and with respect to the question raised above after Corollary 3.26. First, example 11.14 in Part I of the BA handbook shows that for each weakly inaccessible cardinal κ there is a BA A such that $|A| = cA = sA = \kappa$ and cA is not attained but sA is attained (as is easily checked). Second, the interval algebra of a κ-Suslin line, for κ strongly inaccessible but not weakly compact, gives an example of a BA A such that $|A| = cA = sA$, with neither cA nor sA attained; see Juhász [71], example 6.6 (V=L or something beyond ZFC is needed for the existence of a κ-Suslin line). Third, an example of Todorčević [86], Theorem 12, shows that it is consistent to have a BA A in which sA is not attained, while there is no homomorphic image B of A with $cB=sA$. This example involves some interesting ideas, and we shall now give it. It depends on the following lemma about the real numbers.

Lemma 3.27. *There exist disjoint subsets E_0 and E_1 of $[0,1]$ which are of cardinality 2^ω, are dense in $[0,1]$, and satisfy the following two conditions:*

(i) For any $\kappa < 2^\omega$ there is a strictly increasing function from some subset of E_0 of size κ into E_1.

(ii) There is no strictly monotone function from a subset of E_0 of size 2^ω into E_1.

Proof. The idea of the proof is to construct E_0 and E_1 in steps, "killing" all of the possible big strictly monotone functions as we go along. The very first thing to do is to see that we can list out in a sequence of length 2^ω all of the functions to be "killed".

For the empty set 0 we let sup0=0, inf0=1. For any subset W of $[0,1]$ we let cl$W$ be its topological closure in $[0,1]$, and we let $C_1W = \{f : f : W \to [0,1]$, and f is either strictly increasing or strictly decreasing$\}$. For $W \subseteq [0,1]$ and $f \in C_1W$ (say f strictly increasing) we define $f_{cl} : clW \to [0,1]$ by

$$
f_{cl}x = \begin{cases} fx \text{ if } x \in W, \\ \sup\{fy : x > y \in W\} \text{ if } x \notin W \text{ and } x = \sup\{y \in W : y < x\}, \\ \inf\{fy : x < y \in W\} \text{ if } x \notin W \text{ and } x \neq \sup\{y \in W : y < x\}. \end{cases}
$$

(A similar definition is given if f is strictly decreasing.) Note that if $x \in clW \backslash W$ then $x = \sup\{y \in W : y < x\}$ or $x = \inf\{y \in W : x < y\}$. Now f_{cl} is increasing. For, suppose that $x, x' \in clW$ and $x < x'$. If $x, x' \in W$, then $f_{cl}x = fx < fx' = f_{cl}x'$. Suppose that $x \notin W$ and $x' \in W$. If $x = \sup\{y \in W : y < x\}$, then $fy < fx'$ for all $y \in W$ with $y < x$, and so $f_{cl}x \leq fx' = f_{cl}x'$. If $x \neq \sup\{y \in W : y < x\}$, then $x = \inf\{y \in W : x < y\}$, hence $f_{cl}x \leq fx' = f_{cl}x'$. Other possibilities for x and x' are treated similarly. Now if $x, y \in clW \backslash W$, $x < y$, and $f_{cl}x = f_{cl}y$, then $x = \sup\{z \in W : z < x\}$, $y = \inf\{z \in W : y < z\}$, and $|(x, y) \cap W| \leq 1$. For, if $x \neq \sup\{z \in W : z < x\}$, then $x = \inf\{z \in W : x < z\}$, and so there are $u, v \in W$

with $x < u < v < y$, hence

$$f_{cl}x \leq f_{cl}u = fu < fv = f_{cl}v \leq f_{cl}y,$$

contradiction. A similar contradiction is reached if $y \neq \inf\{z \in W : y < z\}$. And if $|(x, y) \cap W| > 1$ a contradiction is easily reached.

Next, note that for each $z \in [0, 1]$ the set $f_{cl}^{-1}[\{z\}]$ has at most three elements. For, if it has four or more, at most one of them is in W; so this gives three elements $w < x < y$ of $clW\backslash W$ all with the same value under f_{cl}. Applying the previous remark, $x = \sup\{u \in W : u < x\}$, and this gives infinitely many elements of W between w and x, contradicting the above statement.

For any $W \subseteq [0, 1]$ let C_2W be the set of all functions $f : W \to I$ such that

(1) f is either increasing or decreasing,

(2) $f^{-1}[\{y\}]$ is finite for all $y \in [0, 1]$,

(3) $|\{x \in W : fx \neq x\}| = 2^\omega$.

Thus by the above, $f_{cl} \in C_2W$ whenever $f \in C_1W$ and $|W| = 2^\omega$.

Now let W be a closed subset of $[0, 1]$. Choose a countable dense subset F_1 of W (pick $w_{rs} \in (r, s) \cap W$ for each pair $r < s$ of rationals such that $(r, s) \cap W \neq 0$, and let F_1 be the set of all such elements w_{rs}). Furthermore, let

$$F_2 = \{x \in W : \sup\{fy : x \geq y \in F_1\} < \inf\{fy : x \leq y \in F_1\}\}.$$

If $x \in F_2$, then $x \notin F_1$. Hence if $x, y \in F_2$ and $x < y$, then there is a $z \in F_1$ with $x < z < y$. It follows that the sup and inf above determine an open interval U_x in $\mathbb{R}$ so that $U_x \cap U_y = 0$ for $x \neq y$. So F_2 is countable. Note that f is determined by its restriction to $F_1 \cup F_2$. From these considerations it follows that $|C_2W| \leq 2^\omega$. Also recall that there are just 2^ω closed sets, since every closed set is the closure of a countable dense subset. Hence the set

$$C = \bigcup\{C_2F : F \subseteq [0, 1], F \text{ closed}\}$$

has cardinality $\leq 2^\omega$. Let $\langle f_\alpha : \alpha < 2^\omega \rangle$ be an enumeration of C. Let h be a strictly decreasing function from $\mathbb{R}$ onto $(0, 1)$; thus h^{-1} is also strictly decreasing. (For example, let $hx = 1/(e^x + 1)$ for all $x \in \mathbb{R}$.) Moreover, fix a well-ordering of $\mathbb{R}$.

Now we construct by induction pairwise disjoint subsets A_α of $[0, 1]$ for $\alpha < 2^\omega$. At the end we will let E_0 be the union of the A_α with even α and E_1 be the union of the rest. We will carry along the inductive hypothesis that $|\alpha| \leq |A_\alpha| \leq |\alpha| + \omega$. Let A_0 and A_1 be denumerable disjoint subsets of $[0, 1]$ which are dense in $[0, 1]$.

Now suppose that A_α has been constructed for all $\alpha < \beta$, where $\beta \geq 2$. Let $B_\beta = \bigcup_{\alpha < \beta} A_\alpha$ and $B_\beta^* = B_\beta \cup \bigcup_{\alpha < \beta} f_\alpha[B_\beta] \cup \bigcup_{\alpha < \beta} f_\alpha^{-1}[B_\beta]$. Note by our assumptions that $|\beta| \leq |B_\beta^*| \leq |\beta| + \omega$. For every real number r, let

$$C_r = h[\{r + h^{-1}b : b \in B_\beta^*\}].$$

We claim that there is an $r \in \mathbb{R}$ such that $C_r \cap B_\beta^* = 0$. Suppose not. For every $r \in \mathbb{R}$ choose $b_r \in C_r \cap B_\beta^*$. Since $|B_\beta^*| < 2^\omega$, there exist a set $S \subseteq \mathbb{R}$ and an element $c \in B_\beta^*$ such that $|S| > B_\beta^*$ and $b_r = c$ for all $r \in S$. Say $c = hd$ with $d \in \{r + h^{-1}x : x \in B_\beta^*\}$ for all $r \in S$. Thus $h(d - r) \in B_\beta^*$ for all $r \in S$. So there exist distinct $r, s \in S$ such that $h(d - r) = h(d - s)$. This contradicts h being one-one.

Finally, let r be the least real number (in the well-ordering fixed above) such that $C_r \cap B_\beta^* = 0$, and let $A_\beta = C_r$. Clearly the inductive hypothesis remains true. This finishes the construction of the sets A_α, $\alpha < 2^\omega$.

Let $E_0 = \bigcup_{\alpha \text{ even}} A_\alpha$ and $E_1 = \bigcup_{\alpha \text{ odd}} A_\alpha$. So E_0 and E_1 are disjoint subsets of $[0,1]$, and both of them are dense in $[0,1]$. Since all of the sets A_α are non-empty, it is clear that both E_0 and E_1 are of power 2^ω.

Now suppose that f is a strictly monotone function from a subset of E_0 of power 2^ω into E_1. Say $f_{\mathrm{cl}} = f_a$. Now $|\bigcup_{\beta < \alpha} A_\alpha| < 2^\omega$, so choose $y \in \mathrm{ran} f$ such that $y \in A_\gamma$ for some $\gamma > \alpha$. Say $fx = y$ with $x \in A_\delta$. Now δ is even and γ is odd. If $\delta < \gamma$, then $y \in B_\gamma^*$, so $y \in A_\gamma$ is a contradiction. If $\gamma < \delta$, then $x \in B_\delta^*$, so $x \in A_\delta$ is a contradiction. Thus (ii) of the lemma has been verified.

If $\omega \leq \kappa < 2^\omega$, choose $\beta < 2^\omega$ odd with $\beta > \kappa$. Say $A_\beta = C_r$, as in the definition. Now $b \mapsto h(r + h^{-1}b)$ is an increasing mapping from B_β^* into A_β, and $E_0 \cap B_\beta^*$ has at least κ elements. This verifies (i).   □

The example also depends upon the following lemma, which will also be useful later on.

Lemma 3.28. *Let A be the interval algebra on $\mathbb{R}$. Then there does not exist in A a strictly increasing sequence $\langle I_\alpha : \alpha < \omega_1 \rangle$ of ideals.*

Proof. Suppose that there is such a sequence. For each $\alpha < \omega_1$ define $r \equiv_\alpha s$ iff $r, s \in \mathbb{R}$ and either $r = s$ or else if, say, $r < s$, then $[r, s) \in I_\alpha$. Then $\equiv_\alpha$ is an equivalence relation on $\mathbb{R}$ and the equivalence classes are intervals. For each $r \in \mathbb{R}$ the left endpoints of the intervals $[r]_\alpha$ are decreasing for increasing α, and the right endpoints, increasing ($[r]_\alpha$ denotes the equivalence class of r under the equivalence relation $\equiv_\alpha$). Since there is no strictly monotone sequence of real numbers of type ω_1, there is an ordinal $\beta_r < \omega_1$ such that both the left and right endpoints of $[r]_\alpha$ are constant for $\alpha > \beta_r$. Let $\gamma = \sup\{\beta_r : r \text{ rational}\}$. Then all of the equivalence classes are constant for $\alpha > \gamma$, contradiction.   □

Corollary 3.29. *Let A be a subalgebra of the interval algebra on $\mathbb{R}$. Then A does not have an uncountable ideal independent subset.*

Proof. Suppose that X is an uncountable ideal independent subset of A. Let $\langle a_\alpha : \alpha < \omega_1 \rangle$ be a one-one enumeration of some elements of X. For each $\alpha < \omega_1$ let $I_\alpha = \langle \{a_\beta : \beta < \alpha\} \rangle^{\mathrm{Id}}$. Clearly then $\langle I_\alpha : \alpha < \omega_1 \rangle$ is a strictly increasing sequence of ideals in B, contradicting 3.28.   □

Finally, we are ready for the example. The main content of the example is from Todorčević [86], Theorem 12, as we mentioned.

Theorem 3.30. *There is a BA A of power 2^ω such that:*

(i) Ult$A$ *has, for each $\kappa < 2^\omega$, a discrete subspace of power κ, and A has an atomic homomorphic image B with κ atoms;*

(ii) Ult$A$ *has no discrete subspace of power 2^ω;*

(iii) If B is any homomorphic image of A, then there is a dense subset X of B such that there is a decomposition $X = W \cup \bigcup_{i \in \omega} Z_i$ with W the set of all atoms of B and for each $i \in \omega$, the set Z_i has the finite intersection property.

Proof. Let E_0 and E_1 be as in Lemma 3.27. Without loss of generality, $0, 1 \notin E_0 \cup E_1$. For $i < 2$ let K_i be the linearly ordered set obtained from $[0,1]$ by replacing each element $r \in E_i$ by two new points $r^- < r^+$. Taking the order topology on K_i, we obtain a Boolean space, as is easily verified. In fact, K_i is homeomorphic to the Stone space of the interval algebra on $E_i \cup \{0\}$. Namely, the following function f from Ult(Intalg($E_i \cup \{0\}$)) into K_i is the desired homeomorphism. Take any $F \in$ Ult(Intalg($E_i \cup \{0\}$)). Let $r = \inf\{a \in E_i : [0, a) \in F\}$; so $r \in [0, 1]$. If $r \in E_i$ and $[0, r) \in F$, let $fF = r^-$; if $r \in E_i$ and $[0, r) \notin F$, let $fF = r^+$; and if $r \notin E_i$ let $fF = r$. Clearly f is one-one and maps onto K_i. To show that it is continuous, first note that the following clopen subsets of K_i constitute a base for its topology:

$$\{[r^+, s^-) : r, s \in E_i, \ r < s\} \cup \{[0, s^-) : s \in E_i\} \cup \{[r^+, 1) : r \in E_i\}.$$

Then it is easy to check (with obvious assumptions) that

$$f^{-1}[[r^+, s^-)] = \{F : [r, s) \in F\};$$
$$f^{-1}[[0, s^-)] = \{F : [0, s) \in F\};$$
$$f^{-1}[[r^+, 1)] = \{F : [r, 1) \in F\}.$$

This completes the proof that f is a homeomorphism from Ult(Intalg($E_i \cup \{0\}$)) onto K_i.

By Corollary 3.29, neither K_0 nor K_1 has an uncountable discrete subspace. Also, $K_0 \times K_1$ is a Boolean space, and we let A be the BA of closed-open subsets of it.

First we check that for any $\kappa < 2^\omega$, $K_0 \times K_1$ has a discrete subset of power κ. Let f be a strictly increasing function from a subset of E_0 of power κ into E_1. Then we claim that $D \overset{\text{def}}{=} \{(r^-, (fr)^+) : r \in \text{dom}f\}$ is discrete. To show this, for each $r \in \text{dom}f$ let $a_r = [0, r^+) \times ((fr)^-, 1]$. Suppose $(s^-, (fs)^+) \in a_r$ and $s \neq r$. Thus $s^- < r^-$ and $(fr)^+ < (fs)^+$, contradiction.

From the proofs of 3.25 and 3.26 it now follows that A has a homomorphic image C which has a disjoint subset of power κ. By an easy application of the Sikorski extension theorem, A has an atomic homomorphic image B with κ atoms.

Next we prove (ii). Suppose that D is a discrete subspace of $K_0 \times K_1$ of size 2^ω. Now K_1 has no uncountable discrete subspace, so for each $x \in \text{dom}D$, the set $\{y : (x, y) \in D\}$ is countable. It follows that we may assume that D is a function. Similarly, we may assume that D is one-one.

For $(r,s) \in D$ let a_{rs} and b_{rs} be open intervals in K_0 and K_1 respectively such that $(a_{rs} \times b_{rs}) \cap D = \{(r,s)\}$. Let F_0 and F_1 be countable dense subsets of K_0 and K_1 respectively (in the sense that if $a < b$ in K_0 and $(a,b) \neq 0$ then there is a $c \in F_0$ such that $a < c < b$; similarly for K_1). Suppose that $\mathrm{dom}D \backslash (\{r^- : r \in E_0\} \cup \{r^+ : r \in E_0\})$ has power 2^ω. Then we may successively assume that $\mathrm{dom}D \cap (\{r^- : r \in E_0\} \cup \{r^+ : r \in E_0\}) = 0$, that each a_{rs} is an open interval with endpoints in F_0, and that all of the a_{rs} are equal, which implies that D has only one element, contradiction.

Thus we may assume that $\mathrm{dom}D \subseteq \{r^- : r \in E_0\} \cup \{r^+ : r \in E_0\}$, and similarly for $\mathrm{ran}D$. Hence we may assume that there are $\varepsilon, \delta \in \{-,+\}$ such that $\mathrm{dom}D \subseteq \{r^\varepsilon : r \in E_0\}$ and $\mathrm{ran}D \subseteq \{r^\delta : r \in E_1\}$.

Thus there are now four cases, which are very similar, and we treat only one of them: $\varepsilon = \delta = -$. We may assume that for each $(r^-, s^-) \in D$ the right endpoint of $a_{r^- s^-}$ is r^+ and that of $b_{r^- s^-}$ is s^+. Furthermore, we may assume that there exist $q_i \in F_i$, $i = 0,1$, such that $q_0 \in a_{r^- s^-}$ and $q_1 \in b_{r^- s^-}$ for each $(r^-, s^-) \in D$. Now we claim that the mapping $r \mapsto s$ for $(r^-, s^-) \in D$ is strictly decreasing (contradiction!). For, suppose that $(r^-, s^-) \in D$, $(u^-, v^-) \in D$, $r < u$, and $s < v$. Then it is clear that $(r^-, s^-) \in a_{u^- v^-} \times b_{u^- v^-}$, a contradiction (using the facts that $q_0 \in a_{r^- s^-} \cap a_{u^- v^-}$, and $q_1 \in b_{r^- s^-} \cap b_{u^- v^-}$).

Now we turn to the last part of the theorem. Suppose that B is a homomorphic image of A. Let F_i be a countable dense subset of E_i for $i = 0,1$. Let $E_i^+ = \{r^+ : r \in E_i\}$, $E_i^- = \{r^- : r \in E_i\}$ for $i = 0,1$. Now we are going to define some subsets $X_{\cdots}$ of B indexed by various objects in countable sets; each subset will satisfy the finite intersection property, and this will be obvious in each case. What is not so obvious is what these sets are good for. We show after defining them that their union with the set of atoms of B is dense in B, which is the desired conclusion of the theorem. It is convenient to work with the dual of B, which is some closed subspace Y of $K_0 \times K_1$.

Suppose that $p, q \in F_0$, $r, s \in F_1$, $p < q$, $r < s$, and $([p^+, q^-] \times [r^+, s^-]) \cap Y \neq 0$; then we set

$$X^1_{pqrs} = \{([p^+, q^-] \times [r^+, s^-]) \cap Y\}.$$

Next, suppose that $q \in F_0$, $r, s \in F_1$, and $r < s$. Then we set

$$X^2_{qrs} = \{([x, q^-] \times [r^+, s^-]) \cap Y : x \in E_0^+, \ x < q, \text{ and}$$
$$\exists y(r^+ < y < s^- \text{ and } (x,y) \in Y)\}.$$

The next three sets are similar to X^2_{qrs}. Suppose that $p \in F_0$, $r, s \in F_1$, and $r < s$. Set

$$X^3_{prs} = \{([p^+, x] \times [r^+, s^-]) \cap Y : x \in E_0^-, \ p < x,$$
$$\text{and } \exists y(r^+ < y < s^- \text{ and } (x,y) \in Y)\}.$$

Suppose that $p, q \in F_0$, $s \in F_1$, and $p < q$. Set

$$X^4_{pqs} = \{([p^+, q^-] \times [y, s^-]) \cap Y : y \in E_1^+, \ y < s,$$
$$\text{and } \exists x(p^+ < x < q^- \text{ and } (x,y) \in Y)\}.$$

Suppose that $p, q \in F_0$, $r \in F_1$, and $p < q$. Set

$$X^5_{pqr} = \{([p^+, q^-] \times [r^+, y]) \cap Y : y \in E_1^-,\ r < y,$$
$$\text{and } \exists x (p^+ < x < q^- \text{ and } (x, y) \in Y)\}.$$

Now suppose that $p, q \in F_0$, $r, s \in F_1$, $p < q$, and $r < s$. Set

$$X^6_{pqrs} = \{([x, q^-] \times [y, s^-]) \cap Y : x \in E_0^+,\ y \in E_1^+,\ x < p^-,$$
$$y < r^-,\ \text{and } ([p^+, q^-] \times [r^+, s^-]) \cap Y \neq 0\}.$$

The next three sets are similar to X^6_{pqrs}. For each of them we suppose that $p, q \in F_0$, $r, s \in F_1$, $p < q$, and $r < s$.

$$X^7_{pqrs} = \{([x, q^-] \times [r^+, y]) \cap Y : x \in E_0^+,\ y \in E_1^-,\ x < p^-,$$
$$s^+ < y,\ \text{and } ([p^+, q^-] \times [r^+, s^-]) \cap Y \neq 0\}.$$

$$X^8_{pqrs} = \{([p^+, x] \times [y, s^-]) \cap Y : x \in E_0^-,\ y \in E_1^+,\ q^+ < x,$$
$$y < r^-,\ \text{and } ([p^+, q^-] \times [r^+, s^-]) \cap Y \neq 0\}.$$

$$X^9_{pqrs} = \{([p^+, x] \times [r^+, y]) \cap Y : x \in E_0^-,\ y \in E_1^-,\ q^+ < x,$$
$$s^+ < y,\ \text{and } ([p^+, q^-] \times [r^+, s^-]) \cap Y \neq 0\}.$$

Next, if $p \in F_0$ and $r, s \in F_1$ with $r < s$, we set

$$X^{10}_{prs} = \{([p^+, x] \times [r^+, y]) \cap Y : x \in E_0^-,\ y \in E_1^-,\ s^+ < y,\ p < x,$$
$$\text{and there is a } v \text{ such that } (x, v) \in Y \text{ and } r^+ < v < s^-\}.$$

.

The other sets are similar to this one; with obvious assumptions,

$$X^{11}_{prs} = \{([p^+, x] \times [y, s^-]) \cap Y : x \in E_0^-,\ y \in E_1^+,\ y < r^-,\ p < x,$$
$$\text{and there is a } v \text{ such that } (x, v) \in Y \text{ and } r^+ < v < s^-\}.$$

$$X^{12}_{qrs} = \{([x, q^-] \times [r^+, y]) \cap Y : x \in E_0^+,\ y \in E_1^-,\ s^+ < y,\ x < q,$$
$$\text{and there is a } v \text{ such that } (x, v) \in Y \text{ and } r^+ < v < s^-\}.$$

$$X^{13}_{qrs} = \{([x, q^-] \times [y, s^-]) \cap Y : x \in E_0^+,\ y \in E_1^+,\ y < r^-,\ x < q,$$
$$\text{and there is a } v \text{ such that } (x, v) \in Y \text{ and } r^+ < v < s^-\}.$$

$$X^{14}_{pqs} = \{([x, q^-] \times [y, s^-]) \cap Y : x \in E_0^+,\ y \in E_1^+,\ x < p^-,\ y < s,$$
$$\text{and there is a } u \text{ such that } (u, y) \in Y \text{ and } p^+ < u < q^-\}.$$

$$X^{15}_{pqr} = \{([x, q^-] \times [r^+, y]) \cap Y : x \in E_0^+,\ y \in E_1^-,\ x < r^-,\ r < y,$$
$$\text{and there is a } u \text{ such that } (u, y) \in Y \text{ and } p^+ < u < q^-\}.$$

$$X_{pqs}^{16} = \{([p^+, x] \times [y, s^-]) \cap Y : x \in E_0^-, \ y \in E_1^+, \ q^+ < x, \ y < s,$$
$$\text{and there is a } u \text{ such that } (u, y) \in Y \text{ and } p^+ < u < q^-\}.$$

$$X_{pqr}^{17} = \{([p^+, x] \times [r^+, y]) \cap Y : x \in E_0^-, \ y \in E_1^-, \ q^+ < x, \ r < y,$$
$$\text{and there is a } u \text{ such that } (u, y) \in Y \text{ and } p^+ < u < q^-\}.$$

Now we show that the union of these sets with the set of atoms of B is dense in B. Suppose that U is a non-zero element of B; we may assume that U has the form $((a, b) \times (c, d)) \cap Y$, and that it is not $\geq$ any atom of B. Fix an element (x, y) of U. We consider various possibilities.

Case 1. $x \notin E_0^- \cup E_0^+$ and $y \notin E_1^- \cup E_1^+$. Then clearly there exist p, q, r, s such that $(x, y) \in X_{pqrs}^1 \subseteq U$.

Case 2. $x \in E_0^-$ and $y \notin E_1^- \cup E_1^+$. There are p, r, s such that $(x, y) \in X_{prs}^3 \subseteq U$.

Case 3. $x \in E_0^+$ and $y \notin E_1^- \cup E_1^+$. There are q, r, s such that $(x, y) \in X_{qrs}^2 \subseteq U$.

Case 4. $x \notin E_0^- \cup E_0^+$. Similar to above cases, using $X_{...}^1$, $X_{...}^4$, or $X_{...}^5$.

Case 5. $x \in E_0^-$ and $y \in E_1^-$. Then it is easy to find $p \in F_0$, $r \in F_1$ so that $(x, y) \in [p^+, x] \times [r^+, y] \cap Y \subseteq U$. Now there are two subcases. *Subcase 5.1.* There is a $(u, v) \in Y$ such that $p^+ < u < x$ and $r^+ < v < y$. Then it is easy to find q, s so that $(x, y) \in X_{pqrs}^9 \subseteq U$. *Subcase 5.2.* Otherwise, since we are assuming that U is not $\geq$ any atom of B, either there is a v such that $(x, v) \in Y$ and $r^+ < v < y$, or there is a u such that $(u, y) \in Y$ and $p^+ < u < x$. In the first instance there is an s such that $(x, y) \in X_{prs}^{10} \subseteq U$. In the second instance we use X^{17}.

Case 6. $x \in E_0^-$ and $y \in E_1^+$. This is like Case 5. We use X^8, X^{16}, and X^{11}.

Case 7. $x \in E_0^+$ and $y \in E_1^-$. This is like Case 5. We use X^7, X^{15}, and X^{12}.

Case 8. $x \in E_0^+$ and $y \in E_1^+$. This is like Case 5. We use X^6, X^{14}, and X^{13}. $\qquad\square$

Corollary 3.31. *Assume that 2^ω is a limit cardinal. Then there is a BA A of power 2^ω with spread 2^ω not attained, such that A has no homomorphic image B such that $cB = sA$.*

Proof. The first part of the conclusion follows immediately from the theorem. Now suppose that B is a homomorphic image of A such that $cB = sA$. Since sA is not attained, it follows from Corollary 3.26 that cB is not attained. Now let X, Y, etc., be as in (iii) of the theorem. Then $|Y| < 2^\omega$ since cB is not attained. Let W be a disjoint subset of B of power $|Y|^+$. For each $w \in W$ choose $x_w \in X$ such that $x_w \leq w$, and let $X' = \{x_w : w \in W\}$. Then there has to exist an $i \in \omega$ such that $|X' \cap Z_i| > 2$, which is a contradiction, since X' is disjoint and Z_i has the finite intersection property. $\qquad\square$

Returning to the program described in the introduction, we note that it is obvious that $c_{Ss}A = [\omega, cA]$. The caliber notion associated with cellularity has been worked on a lot. There are several variants of this notion. For a survey of results and problems, see Comfort, Negrepontis [82].

We shall compare c with other cardinal functions one-by-one in the discussion of those functions.

We turn to the relation c_{Sr}; see the end of the introduction. We do not have a purely cardinal number characterization of this relation (this problem was implicit in Monk [90]):

Problem 3. *Give a purely cardinal number characterization of* c_{Sr}.

Some restrictions to put on c_{Sr} are given in the following simple theorem:

Theorem 3.32. *For any infinite BA A the following conditions hold:*
(i) *If* $(\kappa, \lambda) \in c_{Sr}A$, *then* $\kappa \leq \lambda \leq |A|$ *and* $\kappa \leq cA$.
(ii) *For each* $\kappa \in [\omega, cA]$ *we have* $(\kappa, \kappa) \in c_{Sr}A$.
(iii) *If* $(\kappa, \lambda) \in c_{Sr}A$ *and* $\kappa \leq \mu \leq \lambda$, *then* $(\kappa, \mu) \in c_{Sr}A$.
(iv) *If* $(\lambda, (2^\kappa)^+) \in c_{Sr}A$ *for some* $\lambda \leq \kappa$, *then* $(\omega, (2^\kappa)^+) \in c_{Sr}A$.
(v) $(cA, |A|) \in c_{Sr}A$.
(vi) *If* $\omega \leq \lambda \leq |A|$ *then* $(\kappa, \lambda) \in c_{Sr}A$ *for some* κ. $\square$

The proof of this theorem is easy; for (iv), use Theorem 10.1 of Part I of the Handbook. To understand more about the possiblities for the relation $c_{Sr}A$, consider the following examples. If κ is an infinite cardinal and A is the finite-cofinite algebra on κ, then $c_{Sr}A = \{(\lambda, \lambda) : \lambda \in [\omega, \kappa]\}$. If A is the free algebra on κ free generators, then $c_{Sr}A = \{(\omega, \lambda) : \lambda \in [\omega, \kappa]\}$. If A is an infinite interval algebra and we assume GCH, then $c_{Sr}A$ does not have any gaps of size 2 or greater. That is, if $(\kappa, \lambda) \in c_{Sr}A$, then $\lambda = \kappa$ or $\lambda = \kappa^+$. This is seen by using Theorem 10.1 again: such a gap would imply the existence in A of an uncountable independent subset, which does not exist in an interval algebra. There are two deeper results:

(1) Todorčević in [87] shows that it is consistent (namely, it follows from V=L) to have each regular non-weakly compact cardinal κ a κ-cc interval algebra A of size κ such that any subalgebra or homomorphic image B of A of size $< \kappa$ has a disjoint family of size $|B|$. Applying this to subalgebras and to non-limit cardinals, this means in our terminology that is is consistent to have an algebra A with $c_{Sr}A = \{(\lambda, \lambda) : \lambda \in [\omega, \kappa]\} \cup \{(\kappa, \kappa^+)\}$.

(2) In models of Kunen [78] and Foreman, Laver [88], every ω_2-cc algebra of size ω_2 contains an ω_1-cc subalgebra of size ω_1. Thus in these models certain relations c_{Sr} are ruled out; cf. (1).

Now we survey what we know about c_{Sr} for small cardinals—those $\leq \omega_2$.

(3) $c_{Sr}A = \{(\omega, \omega)\}$ for any denumerable BA.

(4) $c_{Sr}A = \{(\omega, \omega), (\omega, \omega_1)\}$ for $A = Fr\omega_1$.

(5) $c_{Sr}A = \{(\omega, \omega), (\omega, \omega_1), (\omega_1, \omega_1)\}$ for $Fr\omega_1 \times Finco\,\omega_1$.

(6) $c_{Sr}A = \{(\omega, \omega), (\omega_1, \omega_1)\}$ for $A = Finco\,\omega_1$.

(7) $c_{Sr}A = \{(\omega, \omega), (\omega, \omega_1), (\omega, \omega_2)\}$ for $A = Fr\omega_2$.

(8) $c_{Sr}A = \{(\omega, \omega), (\omega_1, \omega_1), (\omega_1, \omega_2)\}$ for the algebra A of (1). Note that in the models mentioned in (2), such a value for c_{Sr} is not possible.

(9) $c_{\mathrm{Sr}}A = \{(\omega,\omega),(\omega,\omega_1),(\omega_1,\omega_1),(\omega_1,\omega_2)\}$ for the following BA A, assuming CH. Let

$$L = {}^{\omega_1}2\backslash\{f \in {}^{\omega_1}2 : \exists\alpha < \omega_1(f\alpha = 0 \text{ and } \forall\beta > \alpha(f\beta = 1))\},$$
$$M = \{f \in {}^{\omega_1}2 : \exists\alpha < \omega_1(f\alpha = 1 \text{ and } \forall\beta > \alpha(f\beta = 0))\}.$$

Clearly M is dense in L, and $|M| = \omega_1$ by CH. Let L' be a subset of L of size ω_2 which contains M. Then $A \stackrel{\text{def}}{=}$ Intalg (L') is as desired. For, by the denseness of M it has ω_2-cc, and it clearly has depth ω_1, and hence cellularity ω_1; so $(\omega_1,\omega_2) \in c_{\mathrm{Sr}}A$. We have $(\omega,\omega_2) \notin c_{\mathrm{Sr}}A$ by Theorem 10.1 of Part I of the Handbook. Obviously $(\omega_1,\omega_1) \in c_{\mathrm{Sr}}A$. The ordered set L'' constructed from ${}^\omega 2$ similarly to L' from ${}^{\omega_1}2$ has size ω_1 and a dense subset of size ω. Then Intalg (L'') is isomorphic to a subalgebra of Intalg (L') by Remark 15.2 of the BA Handbook, so $(\omega,\omega_1) \in c_{\mathrm{Sr}}A$

(10) $c_{\mathrm{Sr}}A = \{(\omega,\omega),(\omega,\omega_1),(\omega_1,\omega_1),(\omega,\omega_2),(\omega_1,\omega_2)\}$ for $A = \mathrm{Finco}\,\omega_1 \times \mathrm{Fr}\omega_2$.

(11) $c_{\mathrm{Sr}}A = \{(\omega,\omega),(\omega_1,\omega_1),(\omega_2,\omega_2)\}$ for $A = \mathrm{Finco}\,\omega_2$.

(12) $c_{\mathrm{Sr}}A = \{(\omega,\omega),(\omega,\omega_1),(\omega_1,\omega_1),(\omega_2,\omega_2)\}$ for $A = \mathrm{Finco}\,\omega_2 \times \mathrm{Fr}\omega_1$. To prove this, it suffices to show that any subalgebra B of A of size ω_2 has cellularity ω_2. Now $C \stackrel{\text{def}}{=} \{x \in \mathrm{Finco}\,\omega_2 : \exists y(x,y) \in B\}$ is a subalgebra of $\mathrm{Finco}\,\omega_2$ of size ω_2, and hence there is a system $\langle c_\alpha : \alpha < \omega_2\rangle$ of nonzero disjoint elements of C. Say $(c_\alpha, d_\alpha) \in B$ for all $\alpha < \omega_2$. Now there are only ω_1 possibilities for the d_α's, so wlog we may assume that they are all equal, and this easily gives rise to a disjoint subset of B of size ω_2.

(13) $c_{\mathrm{Sr}}A = \{(\omega,\omega),(\omega_1,\omega_1),(\omega_1,\omega_2),(\omega_2,\omega_2)\}$ for $A = B \times \mathrm{Finco}\,\omega_2$, where B is the algebra of (1), assuming V=L. For, $(\omega,\omega_2) \notin c_{\mathrm{Sr}}A$ by CH and Theorem 10.1 of the Handbook, Part I. So we just need to show that $(\omega,\omega_1) \notin c_{\mathrm{Sr}}A$. Suppose that D is a subalgebra of A of size ω_1. Let $E = \{b \in B : (b,c) \in A \text{ for some } c\}$ and $F = \{c \in \mathrm{Finco}\,\omega_2 : (b,c) \in A \text{ for some } b\}$. $Case\ 1.$ $|E| = \omega_1$. By the basic property of B, let $\langle e_\alpha : \alpha < \omega_1\rangle$ be a system of nonzero disjoint elements of E. Say $(e_\alpha, f_\alpha) \in D$ for all $\alpha < \omega_1$. If some f_β is cofinite, replace $\langle (e_\alpha, f_\alpha) : \alpha < \omega_1\rangle$ by $\langle (e_\alpha, f_\alpha \cdot -f_\beta) : \alpha < \omega_1, \alpha \neq \beta\rangle$; so wlog all f_α are finite. Wlog the f_α's form a Δ-system, say with kernel g. Pick distinct $\beta, \gamma < \omega_1$. Note that for $\alpha \neq \beta, \gamma$ we have $(e_\alpha, f_\alpha\backslash g) = (e_\alpha, f_\alpha) \cdot -[(e_\beta, f_\beta) \cap (e_\gamma, f_\gamma)]$; hence

$$\langle (e_\alpha, f_\alpha\backslash g) : \alpha < \omega_1, \alpha \neq \beta, \gamma\rangle$$

is a system of disjoint, nonzero elements of D, as desired. $Case\ 2.$ $|F| = \omega_1$ and $|E| < \omega_1$. This case is easy.

Note that in the models of (2), an algebra A of the sort just described is not possible.

(14) $c_{\mathrm{Sr}}A = \{(\omega,\omega),(\omega,\omega_1),(\omega_1,\omega_1),(\omega_1,\omega_2),(\omega_2,\omega_2)\}$ for $A = B \times \mathrm{Finco}\,\omega_2$, with B as in (9), assuming CH; the argument for this is easy.

(15) We do not know whether, under any set-theoretic assumptions, a BA A exists with $c_{\mathrm{Sr}}A = \{(\omega,\omega),(\omega,\omega_1),(\omega_1,\omega_1),(\omega,\omega_2),(\omega_2,\omega_2)\}$; this was Problem 2 in Monk [90]:

Problem 4. *Is there a BA A with $c_{Sr}A = \{(\omega,\omega), (\omega,\omega_1), (\omega_1,\omega_1), (\omega,\omega_2), (\omega_2,\omega_2)\}$? Equivalently, is there a BA A such that $|A| = \omega_2 = cA$, A has a ccc subalgebra of power ω_2, and every subalgebra of A of size ω_2 either has cellularity ω or ω_2?*

(16) $c_{Sr}A = \{(\omega,\omega), (\omega,\omega_1), (\omega_1,\omega_1), (\omega,\omega_2), (\omega_1,\omega_2), (\omega_2,\omega_2)\}$, where A is the algebra $B \times \text{Finco}\,\omega_2$, B the subalgebra of $\mathscr{P}\omega_1$ generated by the singletons and a set of ω_2 independent elements.

It is easy to check that (3)–(16) describe all the possibilities for c_{Sr} with the size of the algebra at most ω_2. We also mention the following problem, concerning (9) and (14); by (9) and (14), it is consistent that BAs of the sort indicated exist.

Problem 5. *Can one construct in ZFC BAs with c_{Sr} equal to the following relations?*
   (i) $\{(\omega,\omega), (\omega,\omega_1), (\omega_1,\omega_1), (\omega_1,\omega_2)\}$.
   (ii) $\{(\omega,\omega), (\omega,\omega_1), (\omega_1,\omega_1), (\omega_1,\omega_2), (\omega_2,\omega_2)\}$.

The relation $c_{Hr}A$ is similar to $c_{Sr}A$. We begin with a general theorem. Part (v) of this theorem is due to Piotr Koszmider.

Theorem 3.33. *For any infinite BA A the following conditions hold:*
   (i) If $(\kappa,\lambda) \in c_{Hr}A$, then $\kappa \le \lambda \le |A|$ and $\kappa \le sA$.
   (ii) For each $\kappa \in [\omega, sA)$ there is a $\lambda \le 2^\kappa$ such that $(\kappa,\lambda) \in c_{Hr}A$.
   (iii) If $(\lambda, (2^\kappa)^+) \in c_{Hr}A$, for some $\lambda \le \kappa$, then $(\omega, (2^\kappa)^+) \in c_{Hr}A$.
   (iv) $(cA, |A|) \in c_{Hr}A$.
   (v) If $(\kappa',\lambda') \in c_{Hr}A$, where κ' is a successor cardinal or a singular cardinal and $\kappa' < \text{cf}|A|$, then there is a $\kappa'' \ge \kappa'$ such that $(\kappa'', |A|) \in c_{Hr}A$.

Proof. Only (ii) and (v) need need proofs. For (ii), let $\kappa \in [\omega, sA)$. Take a homomorphic image B of A such that $cB > \kappa$; let C be a subalgebra of B generated by a disjoint set of power κ, and extend the identity on C to a homomorphism from B onto a subalgebra D of $\overline{C}$; then D is as desired.

Now we prove (v). For brevity let $\lambda = |A|$. There is nothing to prove if $\kappa' = \lambda$, so assume that $\kappa' < \lambda$. Let f be a homomorphism from A onto a BA B with $|B| = \lambda'$ and $cB = \kappa'$. By the Erdös-Tarski theorem, there is a system of $\langle b_\xi : \xi < \kappa' \rangle$ of nonzero disjoint elements in B. For each $\xi < \kappa'$ choose $a_\xi \in A$ such that $fa_\xi = b_\xi$. We now consider two cases.

Case 1. $|A \upharpoonright a_\xi| < \lambda$ for all $\xi < \kappa'$. Let J be the ideal in A generated by $\{a_\xi \cdot a_\eta : \xi < \eta < \kappa'\}$. Then

(*) $|J| < \lambda$.

In fact, $a \in J$ if and only if there is a finite set Γ of ordered pairs (ξ,η) with $\xi < \eta < \kappa'$ such that

$$a \le \sum_{(\xi,\eta)\in\Gamma} a_\xi \cdot a_\eta,$$

and the number of such sets Γ is κ'. Take any such set Γ. Write $\Gamma = \{(\xi_i, \eta_i) : i < n\}$. Define $c_i = a_{\xi_i} \cdot -\sum_{j<i} a_{\xi_j}$ for each $i < n$. Then $\sum_{i<n} c_i = \sum_{i<n} a_{\xi_i}$, and hence

$$\left| \left\{ a \in A : a \leq \sum_{i<n} a_{\xi_i} \cdot a_{\eta_i} \right\} \right| = \left| \prod_{i<n} A \restriction c_i \right| < \lambda,$$

which proves (*), since $\kappa' < \lambda$.

It is also clear that $a_\xi \notin J$ for all $\xi < \kappa'$. It follows that in A/J there is a system of κ' nonzero disjoint elements, and $|A/J| = \lambda$, as desired.

Case 2. There is a $\xi_0 < \kappa'$ such that $|A \restriction a_{\xi_0}| = \lambda$. Then if we take the homomorphism

$$A \cong (A \restriction a_{\xi_0}) \times (A \restriction -a_{\xi_0}) \to (A \restriction a_{\xi_0}) \times (B \restriction -b_{\xi_0})$$

determined by the identity and $f \restriction (A \restriction -a_{\xi_0})$, we get a homomorphism from A onto an algebra C of size λ and with κ' disjoint elements.   □

A related fact was noticed by P. Nyikos: if $(\omega_1, \omega_2) \in c_{\mathrm{Hr}}A$ and $(\omega, \omega_2) \notin c_{\mathrm{Hr}}A$, then $(\omega_1, \omega_1) \in c_{\mathrm{Hr}}A$.

Note in Theorem 3.33 (ii) that $\kappa = sA$ is not in general possible, by Corollary 3.26. The following examples shed some light on c_{Hr}. If A is complete and $(\kappa, \lambda) \in c_{\mathrm{Hr}}A$, then $\lambda^\omega = \lambda$. If A is the finite-cofinite algebra on an infinite cardinal κ, then $c_{\mathrm{Hr}}A = \{(\lambda, \lambda) : \omega \leq \lambda \leq \kappa\}$. If A is the free BA on κ free generators, κ infinite, then $c_{\mathrm{Hr}}A = \{(\lambda, \mu) : \omega \leq \lambda \leq \mu \leq \kappa\}$. If A an infinite interval algebra and GCH is assumed, then there is no gap of size 2 or greater in $c_{\mathrm{Hr}}A$, in the same sense as above. The algebra A of Todorčević [87] (assuming $V = L$) has $c_{\mathrm{Hr}}A = \{(\lambda, \lambda) : \lambda \in [\omega, \kappa]\} \cup \{(\kappa, \kappa^+)\}$. Another example is $\mathscr{P}\omega$. Under CH, its homomorphic cellularity relation is $\{(\omega, \omega_1), (\omega_1, \omega_1)\}$. If we assume that $2^\omega = \omega_2$ then we see that its homomorphic cellularity relation is $\{(\omega, \omega_2), (\omega_1, \omega_2), (\omega_2, \omega_2)\}$. Another relevant result is from Koppelberg [77]: assuming MA, if A is an infinite BA with $|A| < 2^\omega$, then A has a countable homomorphic image. And a special case of a result of Just, Koszmider [87] is that it is consistent to have $2^\omega = \omega_2$ with an algebra A having homomorphic cellularity relation $\{(\omega, \omega_1), (\omega_1, \omega_1)\}$. In Juhász [92] it is shown that if $\kappa > \omega$ and $|A| \geq \kappa$, then A has a homomorphic image of size λ for some λ with $\kappa \leq \lambda \leq 2^{<\kappa}$. Fedorchuk [75] constructed, assuming $\diamondsuit$, a BA A such that $c_{\mathrm{Hr}}A = \{(\omega, \omega_1)\}$. This example is described in Chapter 16. Koszmider (email message) has modified Fedorchuk's construction to give a model of ZFC $+ 2^\omega = \omega_2$ in which there are BAs A, B, C, and D with the following properties:

$$c_{\mathrm{Hr}}A = \{(\omega, \omega_2)\};$$
$$c_{\mathrm{Hr}}B = \{(\omega, \omega_1)\};$$
$$c_{\mathrm{Hr}}C = \{(\omega, \omega_2), (\omega_1, \omega_2)\};$$
$$c_{\mathrm{Hr}}D = \{(\omega, \omega_1), (\omega_1, \omega_1)\}.$$

R. Laver gave a forcing construction to show that it is consistent to have a system $\langle a_\alpha : \alpha < \omega_2 \rangle$ of almost disjoint subsets of ω such that if b is any subset of ω, then

$$\{\alpha < \omega_2 : b \cap a_\alpha \text{ is infinite}\} \text{ is infinite}$$

implies that

$$\{\alpha < \omega_2 : b \cap a_\alpha \text{ is infinite}\} \text{ is cocountable in } \omega_2.$$

P. Nyikos observed that one can then define a BA A such that

$$c_{Hr}A = \{(\omega, \omega), (\omega_1, \omega_1), (\omega, \omega_2), (\omega_2, \omega_2)\}.$$

The following elementary fact is also useful in constructing examples:

  C is a homomorphic image of $A \times B$ iff C is isomorphic to $A' \times B'$ for some homomorphic images A' and B' of A and B respectively.

Proof. $\Leftarrow$: obvious. $\Rightarrow$: suppose that f is a homomorphism from $A \times B$ onto C. It suffices to show that $C \restriction f(1,0)$ is a homomorphic image of A (similarly for B). Let I be a maximal ideal in A, and for any $a \in A$ let $ga = f(a, a/I) \cdot f(1,0)$. Clearly g is a homomorphism from A into $C \restriction f(1,0)$. To show that it is onto, let $x \in C \restriction f(1,0)$. Say $f(a,b) = x$. Then

$$ga = f(a, a/I) \cdot f(1,0) = f(a,b) \cdot f(1,0) = x. \qquad \square$$

Problem 6. *Describe in cardinal number terms the relation* c_{Hr}. (This problem was implicit in Monk [90].)

Now we consider small cardinals, like we did for c_{Sr}. There are many more problems here. The problems are of two sorts: cases in which we know that the existence of the appropriate BA is consistent but have no construction in ZFC, and cases in which we know that the existence of the appropriate BA is inconsistent, but have no proof of non-existence in ZFC. After stating the problems we shall systematically go through all of the possible cases of relations c_{Hr} for algebras of size at most ω_2.

  For each part of the following problem it is known to be consistent that there is a BA with the indicated relation.

Problem 7. *Can one prove in ZFC that BAs with the following relations* c_{Hr} *exist?*
  (i) $\{(\omega, \omega), (\omega, \omega_1), (\omega_1, \omega_1), (\omega_1, \omega_2)\}$. *See (H45).*
  (ii) $\{(\omega, \omega), (\omega, \omega_1), (\omega_1, \omega_1), (\omega_1, \omega_2), (\omega_2, \omega_2)\}$. *See (H60).*

Concerning the relations in the following problem, it is known to be consistent that no BA with that c_{Hr} relation exists.

Problem 8. *Is it consistent that BAs with the following relations* c_{Hr} *exist?*

(i) $\{(\omega, \omega_1), (\omega_1, \omega_1), (\omega_2, \omega_2)\}$. *See (H36).*
(ii) $\{(\omega, \omega_1), (\omega_1, \omega_1), (\omega_1, \omega_2)\}$. *See (H35).*
(iii) $\{(\omega, \omega_1), (\omega_1, \omega_1), (\omega_1, \omega_2), (\omega_2, \omega_2)\}$. *See (H55).*
(iv) $\{(\omega, \omega_1), (\omega_1, \omega_1), (\omega, \omega_2), (\omega_2, \omega_2)\}$. *See (H53).*

Possibilites for c_{Hr}. For the convenience of the reader we mention all of the 63 a priori possibilities. As a guide through these, we arrange the six possible pairs lexicographically and go through them in order of the number present.

(H1) $\{(\omega, \omega)\}$. Any countably infinite BA works.

(H2) $\{(\omega, \omega_1)\}$. The Fedorchuk example gives this, assuming $\diamondsuit$. If MA+$2^\omega > \omega_1$, it is ruled out by Koppelberg.

(H3) $\{(\omega, \omega_2)\}$. This is impossible under CH, by Juhász. If MA $+ 2^\omega > \omega_2$, it is ruled out by Koppelberg's Theorem. A consistent example is given by Koszmider's algebra A.

(H4) $\{(\omega_1, \omega_1)\}$. Any BA has a homomorphic image of countable cellularity, so this relation is impossible.

(H5) $\{(\omega_1, \omega_2)\}$. See (H4).

(H6) $\{(\omega_2, \omega_2)\}$. See (H4).

(H7) $\{(\omega, \omega), (\omega, \omega_1)\}$. A subalgebra of Intalg$\mathbb{R}$ of size ω_1 gives an example.

(H8) $\{(\omega, \omega), (\omega, \omega_2)\}$. This is impossible under CH, by Juhász. Assuming $2^\omega = \omega_2$, the algebra Intalg $\mathbb{R}$ works; see Theorem 9.4.

(H9) $\{\omega, \omega), (\omega_1, \omega_1)\}$. Finco$\omega_1$ works.

(H10) $\{(\omega, \omega), (\omega_1, \omega_2)\}$. This is impossible, by the result of Nyikos.

(H11) $\{(\omega, \omega), (\omega_2, \omega_2)\}$. Any BA of cellularity ω_2 has a homomorphic image of cellularity ω_1, so this relation is impossible.

(H12) $\{(\omega, \omega_1), (\omega, \omega_2)\}$. Not possible under CH, by Theorem 13.6. If MA $+ 2^\omega > \omega_1$, it is ruled out by Koppelberg's Theorem. A consistent example is given by $A \times B$, where A and B are Koszmider's algebras.

(H13) $\{(\omega, \omega_1), (\omega_1, \omega_1)\}$. Under CH, $\mathscr{P}\omega$ works; and also the Just, Koszmider example works. Koppelberg's Theorem indicates that it is not possible to have such an example in ZFC.

(H14) $\{(\omega, \omega_1), (\omega_1, \omega_2)\}$. This is ruled out by the result of Nyikos.

(H15) $\{(\omega, \omega_1), (\omega_2, \omega_2)\}$. Not possible: see (H11).

(H16) $\{(\omega, \omega_2), (\omega_1, \omega_1)\}$. This is not possible, since the homomorphic image of size ω_1 should have a homomorphic image which is ccc.

(H17) $\{(\omega, \omega_2), (\omega_1, \omega_2)\}$. This is impossible under CH, by Juhász. If MA $+ 2^\omega >$ ω_2, it is ruled out by Koppelberg's Theorem. A consistent example is given by Koszmider's algebra C.

(H18) $\{(\omega, \omega_2), (\omega_2, \omega_2)\}$. Not possible; see (H11).

(H19) $\{(\omega_1, \omega_1), (\omega_1, \omega_2)\}$. Not possible; see (H4).

(H20) $\{(\omega_1, \omega_1), (\omega_2, \omega_2)\}$. Not possible; see (H4).

(H21) $\{(\omega_1, \omega_2), (\omega_2, \omega_2)\}$. Not possible; see (H4).

(H22) $\{(\omega, \omega), (\omega, \omega_1), (\omega, \omega_2)\}$. Not possible under CH, since then there is an uncountable independent set. Under ¬CH, a product of certain interval algebras works.

(H23) $\{(\omega, \omega), (\omega, \omega_1), (\omega_1, \omega_1)\}$. A free algebra of size ω_1 works.

(H24) $\{(\omega, \omega), (\omega, \omega_1), (\omega_1, \omega_2)\}$. This is not possible, by the result of Nyikos.

(H25) $\{(\omega, \omega), (\omega, \omega_1), (\omega_2, \omega_2)\}$. Not possible; see (H11).

(H26) $\{(\omega, \omega), (\omega, \omega_2), (\omega_1, \omega_1)\}$. Ruled out by Theorem 3.33 (v).

(H27) $\{(\omega, \omega), (\omega, \omega_2), (\omega_1, \omega_2)\}$. This is impossible under CH, by Juhász. A consistent example is given by $C \times \text{Finco}\,\omega$, where C is Koszmider's algebra.

(H28) $\{(\omega, \omega), (\omega, \omega_2), (\omega_2, \omega_2)\}$. Not possible; see (H11).

(H29) $\{(\omega, \omega), (\omega_1, \omega_1), (\omega_1, \omega_2)\}$. Under V=L this is possible, by the result of Todorčević. This is not possible in the models of Kunen [78] and of Foreman and Laver; see (2) in the discussion of c_{Sr}. Namely, suppose that $c_{Hr}A$ is the indicated relation in one of the indicated models A. Let B be an ω_1-cc subalgebra of A of size ω_1. By the Sikorski extension theorem, there is a homomorphism from A onto some BA C such that $B \leq C \leq \overline{B}$. Thus C has ccc and size ω_1 or ω_2, contradiction.

(H30) $\{(\omega, \omega), (\omega_1, \omega_1), (\omega_2, \omega_2)\}$. Finco$\omega_2$ works.

(H31) $\{(\omega, \omega), (\omega_1, \omega_2), (\omega_2, \omega_2)\}$. This is ruled out by the result of Nyikos.

(H32) $\{(\omega, \omega_1), (\omega, \omega_2), (\omega_1, \omega_1)\}$. This is ruled out by Theorem 3.33(v).

(H33) $\{(\omega, \omega_1), (\omega, \omega_2), (\omega_1, \omega_2)\}$. This is impossible under CH, by Juhász. A consistent example is given by $B \times C$, both Koszmider's algebras.

(H34) $\{(\omega, \omega_1), (\omega, \omega_2), (\omega_2, \omega_2)\}$. This is not possible; see (H11).

(H35) $\{(\omega, \omega_1), (\omega_1, \omega_1), (\omega_1, \omega_2)\}$. If MA $+ 2^\omega > \omega_2$, this is ruled out by Koppelberg's Theorem. It is open whether an example is consistent (Problem 8(ii)).

(H36) $\{(\omega, \omega_1), (\omega_1, \omega_1), (\omega_2, \omega_2)\}$. If MA$+2^\omega > \omega_2$, this is ruled out by Koppelberg's theorem. It is open whether an example is consistent (Problem 8(i)).

(H37) $\{(\omega,\omega_1),(\omega_1,\omega_2),(\omega_2,\omega_2)\}$. This is ruled out by the result of Nyikos.

(H38) $\{(\omega,\omega_2),(\omega_1,\omega_1),(\omega_1,\omega_2)\}$. The homomorphic image of size ω_1 and cellularity ω_1 must have a homomorphic image of cellularity ω, contradiction.

(H39) $\{(\omega,\omega_2),(\omega_1,\omega_1),(\omega_2,\omega_2)\}$. Impossible; see (H38).

(H40) $\{(\omega,\omega_2),(\omega_1,\omega_2),(\omega_2,\omega_2)\}$. This is impossible under CH, by Juhász. Assuming $2^\omega = \omega_2$, $\mathscr{P}\omega$ has this relation. If MA$+2^\omega > \omega_2$, it is ruled out by Koppelberg's Theorem.

(H41) $\{(\omega_1,\omega_1),(\omega_1,\omega_2),(\omega_2,\omega_2)\}$. Not possible; see (H4).

(H42) $\{(\omega,\omega),(\omega,\omega_1),(\omega,\omega_2),(\omega_1,\omega_1)\}$. This is ruled out by Theorem 3.33(v).

(H43) $\{(\omega,\omega),(\omega,\omega_1),(\omega,\omega_2),(\omega_1,\omega_2)\}$. Not possible under CH, since then there must be an independent subset of size ω_2, and one of the pairs must be (ω_2,ω_2). A consistent example with this relation is $C \times E$, where C is Koszmider's example and E is a subalgebra of Intalg$\mathbb{R}$ of size ω_1.

(H44) $\{(\omega,\omega),(\omega,\omega_1),(\omega,\omega_2),(\omega_2,\omega_2)\}$. Not possible: see (H11).

(H45) $\{(\omega,\omega),(\omega,\omega_1),(\omega_1,\omega_1),(\omega_1,\omega_2)\}$. Under GCH the following algebra works: the standard interval algebra constructed from $^{\omega_1}2$; see (9) in the discussion of c_{Sr}. It is open whether it is consistent that there is no example of this sort (Problem 7(i)).

(H46) $\{(\omega,\omega),(\omega,\omega_1),(\omega_1,\omega_1),(\omega_2,\omega_2)\}$. Let B be a subalgebra of Intalg$\mathbb{R}$ of size ω_1, and set $A = B \times$ Finco ω_2. Then it suffices to show that A does not have a homomorphic image of size ω_2 with cellularity less than ω_2. But this is obvious by the above fact, since any homomorphic image of Finco ω_2 of size ω_2 is isomorphic to Finco ω_2.

(H47) $\{(\omega,\omega),(\omega,\omega_1),(\omega_1,\omega_2),(\omega_2,\omega_2)\}$. By the result of Nyikos this is impossible.

(H48) $\{(\omega,\omega),(\omega,\omega_2),(\omega_1,\omega_1),(\omega_1,\omega_2)\}$. Not possible under CH, since then there must be an independent subset of size ω_2, and one of the pairs must be (ω_2,ω_2). A consistent example is provided by $C \times$ Finco ω_1, where C is Koszmider's algebra.

(H49) $\{(\omega,\omega),(\omega,\omega_2),(\omega_1,\omega_1),(\omega_2,\omega_2)\}$. Not possible under GCH, since then there must be an independent subset of size ω_2, and one of the pairs must be (ω_1,ω_2). Laver's forcing example gives this relation.

(H50) $\{(\omega,\omega),(\omega,\omega_2),(\omega_1,\omega_2),(\omega_2,\omega_2)\}$. This is impossible under CH, by Juhász. Assuming that $2^\omega = \omega_2$, the algebra Finco$\omega \times \mathscr{P}\omega$ gives an example.

(H51) $\{(\omega,\omega),(\omega_1,\omega_1),(\omega_1,\omega_2),(\omega_2,\omega_2)\}$. This is possible under V=L: let B be the algebra of Todorčević, and set $A = B \times$ Finco ω_2. Then A has countable independence, and hence the pair (ω,ω_2) is ruled out. The elementary fact above rules out (ω,ω_1).

This relation is not possible in the models of Kunen and of Foreman and Laver.

(H52) $\{(\omega,\omega_1),(\omega,\omega_2),(\omega_1,\omega_1),(\omega_1,\omega_2)\}$. Not possible under CH, since then there is an independent subset of size ω_2, and one of the ordered pairs must be (ω_2,ω_2). Also ruled out by Koppelberg's theorem if $\mathrm{MA}+2^\omega>\omega_2$. A consistent example is provided by $C\times D$, where C and D are Koszmider's algebras.

(H53) $\{(\omega,\omega_1),(\omega,\omega_2),(\omega_1,\omega_1),(\omega_2,\omega_2)\}$. Not possible under GCH, since then there is an independent set of size ω_2, and one of the ordered pairs must be (ω_1,ω_2). Also ruled out by Koppelberg's theorem if $\mathrm{MA}+2^\omega>\omega_2$. It is open whether this relation is consistent (Problem 8(iv)).

(H54) $\{(\omega,\omega_1),(\omega,\omega_2),(\omega_1,\omega_2),(\omega_2,\omega_2)\}$. Not possible under CH since then there is an independent set of size ω_2, and hence (ω_1,ω_1) would have to be present. If $\mathrm{MA}+2^\omega>\omega_2$, this is ruled out by Koppelberg's Theorem. A consistent example is given by $B\times\mathscr{P}\omega$, where B is Koszmider's algebra.

(H55) $\{(\omega,\omega_1),(\omega_1,\omega_1),(\omega_1,\omega_2),(\omega_2,\omega_2)\}$. If $\mathrm{MA}+2^\omega>\omega_2$, this is ruled out by Koppelberg's Theorem. It is open to consistently give an example with this relation (Problem 8(iii)).

(H56) $\{(\omega,\omega_2),(\omega_1,\omega_1),(\omega_1,\omega_2),(\omega_2,\omega_2)\}$. This is impossible; a BA of size ω_1 has a ccc homomorphic image.

(H57) $\{(\omega,\omega),(\omega,\omega_1),(\omega,\omega_2),(\omega_1,\omega_1),(\omega_1,\omega_2)\}$. This is not possible under CH, since then there must be an independent subset of size ω_2, and one of the pairs must be (ω_2,ω_2). Assuming that $2^\omega>\omega_1$, we can take the standard linear order which is a subset of $^\omega 2$, take a subset L of size ω_2, containing a dense subset of size ω, and let $B=\mathrm{Intalg}\,L$ and $A=B\times\mathrm{Finco}\,\omega_1$.

(H58) $\{(\omega,\omega),(\omega,\omega_1),(\omega,\omega_2),(\omega_1,\omega_1),(\omega_2,\omega_2)\}$. This is not possible under CH, since then there must be an independent subset of size ω_2, and one of the pairs must be (ω_1,ω_2). A consistent example is given by $A\times B$, where A is Laver's algebra and B is a subalgebra of $\mathrm{Intalg}\mathbb{R}$ of size ω_1.

(H59) $\{(\omega,\omega),(\omega,\omega_1),(\omega,\omega_2),(\omega_1,\omega_2),(\omega_2,\omega_2)\}$. $\mathscr{P}\omega_2$ works, assuming GCH. This is ruled out by Koppelberg's theorem if $\mathrm{MA}+2^\omega>\omega_2$.

(H60) $\{(\omega,\omega),(\omega,\omega_1),(\omega_1,\omega_1),(\omega_1,\omega_2),(\omega_2,\omega_2)\}$. This is possible under CH: take the algebra B of (9) in the discussion of c_{Sr}, and let $A=B\times\mathrm{Finco}\,\omega_2$; (ω,ω_2) is ruled out by independence. It is open to give an example in ZFC (Problem 7(ii)).

(H61) $\{(\omega,\omega),(\omega,\omega_2),(\omega_1,\omega_1),(\omega_1,\omega_2),(\omega_2,\omega_2)\}$. Not possible under CH, since then there must be an independent subset of size ω_2, and one of the pairs must be (ω,ω_1). Assuming $2^\omega=\omega_2$, $\mathscr{P}\omega\times\mathrm{Finco}\omega_2$ works.

(H62) $\{(\omega,\omega_1),(\omega,\omega_2),(\omega_1,\omega_1),(\omega_1,\omega_2),(\omega_2,\omega_2)\}$. $\mathscr{P}\omega_2$ works, assuming GCH. Ruled out by Koppelberg's theorem if $\mathrm{MA}+2^\omega>\omega_2$.

(H63) $\{(\omega, \omega), (\omega, \omega_1), (\omega, \omega_2), (\omega_1, \omega_1), (\omega_1, \omega_2), (\omega_2, \omega_2)\}$. Many examples.

To conclude this chapter, we consider cellularity for special classes of BAs. For an atomic BA A, cA coincides with the number of atoms of A. Also note that some of the free product questions are trivial for atomic algebras; in particular, $c(A \oplus B) = \max\{cA, cB\}$ if A and B are atomic. There is one interesting result which comes up in considering cellularity and unions for complete BAs; this result is evidently due to Solovay, Tennenbaum [71]:

Theorem 3.34. *Let κ and λ be uncountable regular cardinals, and suppose that $\langle A_\alpha : \alpha < \lambda \rangle$ is an increasing sequence of complete BAs satisfying the κ−chain condition, such that A_α is a complete subalgebra of A_β for $\alpha < \beta < \lambda$, and for γ limit $< \lambda$, $\bigcup_{\alpha < \gamma} A_\alpha$ is dense in A_γ. Then $\bigcup_{\alpha < \lambda} A_\alpha$ also satisfies the κ−chain condition.*

Proof. By the proof of Theorem 3.11 we may assume that $\kappa = \lambda$. Let $B = \bigcup_{\alpha < \kappa} A_\alpha$. For each $\alpha < \kappa$ we define c_α mapping B into A_α by setting

$$c_\alpha x = \prod_{x \leq a \in A_\alpha} a.$$

(This function is a *cylindrification* on B, but we do not need to check that.)

Now, in order to get a contradiction, assume that X is a disjoint subset of B of size $\geq \kappa$. We may assume that X is maximal disjoint. Take any $\alpha < \kappa$. Now $\sum X = 1$, and hence $\sum \{c_\alpha x : x \in X\} = 1$. Since each A_α satisfies the κ−chain condition, choose $X_\alpha \subseteq X$ of size $< \kappa$ such that

(1) $\sum \{c_\alpha x : x \in X_\alpha\} = 1$.

Choose $\beta_\alpha < \kappa$ such that $X_\alpha \subseteq A_{\beta_\alpha}$; the ordinal β_α exists since $|X_\alpha| < \kappa$ and κ is regular. Finally, let γ be a limit ordinal $< \kappa$ such that $\beta_\alpha < \gamma$ for all $\alpha < \gamma$; the existence of γ is easy to see. We shall now prove that $X \subseteq A_\gamma$ (contradiction!).

Let $x \in X$ be arbitrary. Since $\bigcup_{\alpha < \gamma} A_\alpha$ is dense in A_γ, choose a non-zero $b \in \bigcup_{\alpha < \gamma} A_\alpha$ such that $b \leq c_\gamma x$. Say $b \in A_\alpha$ with $\alpha < \gamma$. By (1), choose $a \in X_\alpha$ such that $c_\alpha a \cdot b \neq 0$. If $b \cdot a = 0$, then $a \leq -b$ and hence $c_\alpha a \leq -b$ and so $c_\alpha a \cdot b = 0$, contradiction. Thus $b \cdot a \neq 0$, and so $c_\gamma x \cdot a \neq 0$. It follows that $x \cdot a \neq 0$, by the same argument as above. But both x and a are in X, so $x = a$. Thus $x \in X_\alpha \subseteq A_\gamma$, as desired. $\qquad\qquad\square$

There is a large literature on cellularity for BAs of the form $\mathscr{P}(\kappa)/I$; for a start, see Baumgartner, J., Taylor, A., Wagon, S. [82]. Usually BA terminology is not used in such investigations; *saturation* of ideals is the term used.

Note that $c(\mathscr{P}\kappa/\mathrm{fin}) = \kappa^\omega$, and this value is always attained; see the Handbook, v. 1, Lemma 17.15.

The cellularity of tree algebras has been described in Brenner [82]:

Theorem 3.35. *For $A = \mathrm{Treealg}\, T$, T a tree, cA is the maximum of $|\{t \in T : t \text{ has finitely many immediate successors}\}|$ and $\sup\{|X| : X$ is a collection of pairwise incomparable elements of $T\}$.*

Proof. If t has finitely many immediate successors, then $\{t\} \in A$. And if s and t are incomparable, then $(T \uparrow s) \cap (T \uparrow t) = 0$. Hence $\geq$ is clear. Now suppose that X is a collection of pairwise disjoint elements of A; we want to show that $|X|$ is $\leq$ the indicated maximum. Without loss of generality we may assume that each element $x \in X$ has the form $(T \uparrow t_x) \backslash \bigcup_{s \in F_x} (T \uparrow s)$, where F_x is a finite set of $s > t_x$. And we may assume that if t_x has only finitely many immediate successors, then $x = \{t_x\}$. Write $X = X_0 \cup X_1$, where X_0 is the set of singletons in X and $X_1 = X \backslash X_0$. Thus if $x \in X_1$, then t_x has infinitely many immediate successors. Therefore, if $x, y \in X_1$, $x \neq y$, then either t_x and t_y are incomparable, or $s \leq t_y$ for some $s \in F_x$, or $s \leq t_x$ for some $s \in F_y$. For each $x \in X_1$ let u_x be an immediate successor of t_x such that $u_x \not\leq s$ for all $s \in F_x$. Then it is easy to check that if x and y are distinct elements of X_1, then u_x and u_y are incomparable. This proves that $|X_1|$ is $\leq$ the sup mentioned in the theorem. $\square$

This characterization does not work for pseudo-tree algebras: for example, if L is a dense linear order of size ω_1 with an increasing subset of order type ω_1, then $c(\text{Treealg}\, L) = \omega_1$ (recall that for L a linear order, $\text{Treealg}\, L = \text{Intalg}\, L$). This gives rise to the following vague question.

Problem 9. *Describe cellularity for pseudo-tree algebras.*

4. Depth

Recall that Depth A is the supremum of cardinalities of subsets of A which are well-ordered by the Boolean ordering. There are two main references for results about this notion: McKenzie, Monk [82] and (implicitly) Grätzer, Lakser [69]. (Theorems 3.4.4 and 3.5.2 and their corollaries in McKenzie, Monk [82] were essentially already proved in Grätzer, Lakser [69].)

Some of the results which we shall present about depth depend on the following simple lemma.

Lemma 4.1. *Let A and B be BAs, and let X be a chain in $A \times B$ of infinite cardinality κ. Then the projections of X are chains, and at least one of them has cardinality κ. Furthermore, if X has order type κ, then X has a subset of order type κ on which one of the two projections is one-one.*

Proof. For any $z \in X$ write $z = (z_0, z_1)$. For $i = 0, 1$ write $z \equiv_i w$ iff $z, w \in X$ and $z_i = w_i$. Now note that

$$\{\{x\} : x \in X\} = \{a \cap b : a \in X/\equiv_0,\ b \in X/\equiv_1\}\backslash\{0\};$$

hence one of the two equivalence relations $\equiv_0, \equiv_1$ has κ equivalence classes, and the lemma follows. $\qquad\square$

Now we shall show that Depth A is attained if Depth A is a successor cardinal or a cardinal of cofinality ω; otherwise, there are counterexamples.

Theorem 4.2. *If $\mathrm{cf}(\mathrm{Depth}A) = \omega$, then $\mathrm{Depth}A$ is attained.*

Proof. Let $\kappa = \mathrm{Depth}A$. We may assume that κ is an uncountable limit cardinal. Let $\langle \lambda_i : i < \omega \rangle$ be a strictly increasing sequence of cardinals with supremum κ, and with $\lambda_0 = 0$ and λ_1 infinite. Now we call an element a of A an ∞-*element* if λ_i is embeddable in $A \upharpoonright a$ for all $i < \omega$. We claim

(*) If a is an ∞-element, and $a = b + c$ with $b \cdot c = 0$, then b is an ∞-element or c is an ∞-element.

In fact, by Lemma 4.1, for each $i < \omega$, λ_i is embeddable in $A \upharpoonright b$ or $A \upharpoonright c$, so (*) follows.

Using (*), we construct a sequence $\langle a_i : i < \omega \rangle$ of elements of A by induction. Suppose that a_j has been constructed for all $j < i$ so that $b \overset{\mathrm{def}}{=} \prod_{j<i} -a_j$ is an ∞-element. Let $\langle c(\alpha) : \alpha < \lambda_{i+1} \rangle$ be an isomorphic embedding of λ_{i+1} into b. By (*), one of the elements $c(\lambda_i)$ and $b \cdot -c(\lambda_i)$ is an ∞-element, while clearly λ_i is embeddable in both of these elements. So we can choose $a_i \leq b$ so that λ_i is embeddable in a_i, and $\prod_{j\leq i} -a_j$ is an ∞-element. This finishes the construction.

For each $i < \omega$ let $\langle b_{i\alpha} : \alpha < \lambda_i \rangle$ be an embedding of λ_i into a_i. Note that $a_i \cdot a_j = 0$ for $i < j < \omega$. Hence the following sequence $\langle d_\alpha : \alpha < \kappa \rangle$ is clearly the desired embedding of κ into A. Given $\alpha < \kappa$, there is a unique $i < \omega$ such that $\lambda_i \leq \alpha < \lambda_{i+1}$. We let $d_\alpha = a_0 + \cdots + a_i + b_{i+1,\alpha}$. $\qquad\square$

In order to see that Theorem 4.2 is "best possible", it is convenient to first discuss the depth of products.

Theorem 4.3. $\text{Depth}(\prod_{i \in I} A_i) = \max(|I|, \sup_{i \in I} \text{Depth} A_i)$.

Proof. Clearly $\geq$ holds. Suppose $=$ fails to hold, and let f be an order isomorphism of κ^+ into $\prod_{i \in I} A_i$, where $\kappa = \max(|I|, \sup_{i \in I} \text{Depth} A_i)$. For each $i \in I$ there is an ordinal $\alpha_i < \kappa^+$ such that $(f\alpha_i)_i = (f\beta)_i$ for all $\beta > \alpha_i$. Let $\gamma = \sup_{i \in I} \alpha_i$. Then for all $\delta > \gamma$ we have $f\delta = f\gamma$, contradiction. $\square$

Theorem 4.4. Let $\kappa = \sup_{i \in I} \text{Depth} A_i$, and suppose that κ is regular. Then the following conditions are equivalent:

(i) $\text{Depth}(\prod_{i \in I} A_i)$ is not attained.

(ii) $|I| < \kappa$, and for all $i \in I$, A_i has no chain of order type κ. $\square$

The proof of this theorem is very similar to that of Theorem 4.3. The case of singular cardinals is a little more involved:

Theorem 4.5. Let $\kappa = \sup_{i \in I} \text{Depth} A_i$, and suppose that κ is singular. Then the following conditions are equivalent:

(i) $\text{Depth}(\prod_{i \in I} A_i)$ is not attained.

(ii) These four conditions hold:

(a) $|I| < \kappa$.

(b) For all $i \in I$, A_i has no chain of type κ.

(c) $|\{i \in I : \text{Depth} A_i = \kappa\}| < \text{cf}\kappa$.

(d) $\sup\{\text{Depth} A_i : i \in I, \text{ Depth} A_i < \kappa\} < \kappa$.

Proof. Let $\langle \mu_\alpha : \alpha < \text{cf}\kappa \rangle$ be a strictly increasing continuous sequence of cardinals with supremum κ, with $\mu_0 = 0$. (i) $\Rightarrow$ (ii): (a) and (b) are clear. Suppose that (c) fails to hold; we show that (i) fails. Let i be a one-one function from $\text{cf}\kappa$ into $\{i \in I : \text{Depth} A_i = \kappa\}$. For each $\alpha < \text{cf}\kappa$ let $\langle a_{i\beta} : \mu_\alpha \leq \beta < \mu_{\alpha+1} \rangle$ be a strictly increasing sequence of elements of A_{i_α}. Now we define a sequence $\langle x_\beta : \beta < \kappa \rangle$ of elements of $\prod_{i \in I} A_i$. For each $\beta < \kappa$ choose $\alpha < \text{cf}\kappa$ so that $\mu_\alpha \leq \beta < \mu_{\alpha+1}$, and for any $j \in I$ set

$$x_{\beta j} = \begin{cases} 1 & \text{if } j = i_\gamma \text{ for some } \gamma < \alpha; \\ a_{i\beta} & \text{if } j = i_\alpha; \\ 0 & \text{otherwise.} \end{cases}$$

Clearly this sequence is as desired.

Next we show that if (d) fails then (i) fails. By induction we can define i_α for $\alpha < \text{cf}\kappa$ so that

$$\sup_{\beta < \alpha} (\text{Depth} A_{i_\beta} \cup \mu_\beta) < \text{Depth} A_{i_\alpha} < \kappa,$$

and then we can proceed as for (c).

$(ii) \Rightarrow (i)$: Assume (ii), and suppose that $\langle x_\alpha : \alpha < \kappa \rangle$ is strictly increasing in $\prod_{i \in I} A_i$. Define

$$J_i = \{\alpha < \kappa : x_\alpha i < x_{\alpha+1}i\} \text{ for } i \in I;$$
$$K = \{i \in I : \mathrm{Depth}A_i = \kappa\};$$
$$\lambda = \sup\{\mathrm{Depth}A_i : i \in I, \mathrm{Depth}A_i < \kappa\}.$$

Then by the above assumptions we have $\lambda < \kappa$, $|J_i| \leq \lambda$ for all $i \in I \backslash K$, $|K| < \mathrm{cf}\kappa$, and $|J_i| < \kappa$ for all $i \in K$. It follows that $|\bigcup_{i \in I} J_i| < \kappa$. But for any $\alpha \in \kappa \backslash \bigcup_{i \in I} J_i$ we have $x_\alpha = x_{\alpha+1}$, contradiction. $\qquad\square$

The above theorems completely describe the depth of products. The case of weak products is even simpler:

Theorem 4.6. *Let $\kappa = \sup_{i \in I} \mathrm{Depth}A_i$, and suppose that $\mathrm{cf}\kappa > \omega$. Then the following conditions are equivalent:*

   (i) $\prod_{i \in I}^{\mathrm{w}} A_i$ has no chain of order type κ.
   (ii) For all $i \in I$, A_i has no chain of order type κ.

Proof. $(i) \Rightarrow (ii)$ is clear. $(ii) \Rightarrow (i)$: Suppose that $\langle x_\alpha : \alpha < \kappa \rangle$ is strictly increasing in $\prod_{i \in I}^{\mathrm{w}} A_i$. For any $y \in \prod_{i \in I}^{\mathrm{w}} A_i$ let $Sy = \{i \in I : y_i \neq 0\}$.
   Case 1. Sx_α is finite for all $\alpha < \kappa$. Since $\mathrm{cf}\kappa > \omega$, it follows that there is an $\alpha < \kappa$ such that $Sx_\alpha = Sx_\beta$ whenever $\alpha < \beta < \kappa$. But then Lemma 4.1 easily gives a contradiction.
   Case 2. Otherwise we may assume that $\{i \in I : x_\alpha i \neq 1\}$ is finite for all $\alpha < \kappa$, and a contradiction is reached as in Case 1. $\qquad\square$

Corollary 4.7. $\mathrm{Depth}(\prod_{i \in I}^{\mathrm{w}} A_i) = \sup_{i \in I}\mathrm{Depth}A_i.$ $\qquad\square$

Theorem 4.6 enables us to easily show that Theorem 4.2 is best possible: if κ is a limit cardinal with $\mathrm{cf}\kappa > \omega$, then it is easy to construct a weak product B such that $\mathrm{Depth}B = \kappa$ but depth is not attained in B.

   If A is a subalgebra of B, then obviously $\mathrm{Depth}A \leq \mathrm{Depth}B$ and the difference can be arbitrarily large. If A is a homomorphic image of then depth can change either way from A to B; see the argument here for cellularity.

   For free products, we have $\mathrm{Depth}(\oplus_{i \in I} A_i) = \sup_{i \in I}\mathrm{Depth}A_i$. The proof is somewhat involved, and will be omitted; see McKenzie, Monk [82].

   We now briefly discuss depth and amalgamated free products. The following theorem is a special case of a theorem in McKenzie, Monk [82].

Theorem 4.8. *Let A be the BA of finite and cofinite subsets of ω. Then there exist $B, C \geq A$ both satisfying ccc such that $\mathrm{Depth}(B \oplus_A C) = \omega_1$.*

Proof. Let M be the collection of all even integers, N the set of all odd integers. Then we take two sequences $\langle a_\alpha : \alpha < \omega_1 \rangle$ and $\langle b_\alpha : \alpha < \omega_1 \rangle$ such that

(1) Each a_α is an infinite subset of M, and for $\alpha < \beta < \omega_1$ we have $a_\alpha \backslash a_\beta$ finite and $a_\beta \backslash a_\alpha$ infinite.

(2) Similarly for the b_α's, subsets of N.

Now let $B = C = A \times \mathscr{P}\omega$. For each $a \in A$ let $ga = (a, a)$. Then g is an isomorphism of A into $B = C$. So it is enough to prove that $\text{Depth}(B \oplus_{g[A]} C) = \omega_1$. For each $\alpha < \omega_1$ let $c_\alpha = (0, a_\alpha) \times (0, b_\alpha)$ [using $\times$ rather than $\cdot$ to indicate which one is in B and which one in C]. We claim that $\langle c_\alpha : \alpha < \omega_1 \rangle$ is as desired. Let $\alpha < \beta < \omega_1$. Then

$$c_\alpha \cdot -c_\beta = [((0, a_\alpha) \cdot (1, -a_\beta)) \times (0, b_\alpha)] + [(0, a_\alpha) \times ((0, b_\alpha) \cdot (1, -b_\beta))]$$
$$= [(0, a_\alpha \backslash a_\beta) \times (0, b_\alpha)] + [(0, a_\alpha) \times (0, b_\alpha \backslash b_\beta)].$$

Now $d \stackrel{\text{def}}{=} a_\alpha \backslash a_\beta$ is a finite subset of M, so $d \in A$. And $(0, a_\alpha \backslash a_\beta) \leq (d, d)$, while $(0, b_\alpha) \cdot (d, d) = (0, 0)$. Using a similar argument for the second summand, this shows that $c_\alpha \cdot -c_\beta = 0$.

Now suppose that $\alpha < \beta$ and $c_\beta \cdot -c_\alpha = 0$. So $(0, a_\beta \backslash a_\alpha) \times (0, b_\beta) = 0$, hence there is a $d \in A$ such that $(0, a_\beta \backslash a_\alpha) \leq (d, d)$ and $(0, b_\beta) \cdot (d, d) = (0, 0)$. Now $a_\beta \backslash a_\alpha$ is infinite, so d is cofinite. Hence $b_\beta \cap d \neq 0$, contradiction. $\qquad\square$

The following theorem solves Problem 2 of McKenzie, Monk [82]:

Theorem 4.9. *Let A be the BA of finite and cofinite subsets of ω, and let κ be an uncountable cardinal. Then there exist $B, C \geq A$ such that $|B| = |C| = \kappa$, $\text{Depth}B = \text{Depth}C = \omega$, and $\text{Depth}(B \oplus_A C) \geq \omega_1$.*

Proof. First we choose B and C as in the proof of Theorem 4.8. In particular, $B \oplus_A C$ has a chain of the form $\langle b_\alpha \times c_\alpha : \alpha < \omega_1 \rangle$, while B and C have size ω_1. Let $B' = B \times \text{Finco}\kappa$ and $C' = C \times \text{Finco}\kappa$. Thus $|B'| = |C'| = \kappa$. Set

$$A' = \{(a, 0) : a \in [\omega]^{<\omega}\} \cup \{(a, 1) : \omega \backslash a \in [\omega]^{<\omega}\}.$$

Clearly A' is isomorphic to A. To prove the theorem it suffices to show that $B' \oplus_{A'} C'$ has depth ω_1. Let $b'_\alpha = (b_\alpha, 0)$, $c'_\alpha = (c_\alpha, 0)$, and $d_\alpha = b'_\alpha \times c'_\alpha$ for all $\alpha < \omega_1$. We claim that $\langle d_\alpha : \alpha < \omega_1 \rangle$ is a chain in $B' \oplus_{A'} C'$, as desired. To prove this, suppose that $\alpha < \beta$. Choose $u, v \in A$ such that $b_\alpha \cdot -b_\beta \leq u$, $c_\alpha \cap u = 0$, $b_\alpha \cdot v = 0$, and $c_\alpha \cdot -c_\beta \leq v$. Now

$$d_\alpha \cdot -d_\beta = (b'_\alpha \cdot -b'_\beta) \times c'_\alpha + b'_\alpha \cdot (c'_\alpha \cdot -c'_\beta).$$

Note that $(b'_\alpha \cdot -b'_\beta) \times c'_\alpha = (b_\alpha \cdot -b_\beta, 0) \times (c_\alpha, 0)$. Then for some $\varepsilon \in \{0, 1\}$ we have $(u, \varepsilon) \in A'$, $(b_\alpha \cdot -b_\beta, 0) \leq (u, \varepsilon)$, and $(u, \varepsilon) \cdot (c_\alpha, 0) = (0, 0)$. Therefore $(b'_\alpha \cdot -b'_\beta) \times c'_\alpha = 0$. Similarly for the other summand, so $d_\alpha \cdot -d_\beta = 0$. Suppose that also $d_\beta \cdot -d_\alpha = 0$. This easily gives $(b_\beta \cdot -b_\alpha, 0) \times (c_\beta, 0) = (0, 0)$. Hence there is a $(w, \varepsilon) \in A'$ such that $(b_\beta \cdot -b_\alpha, 0) \leq (w, \varepsilon)$ and $(c_\beta, 0) \cdot (w, \varepsilon) = (0, 0)$. Hence $b_\beta \cdot -b_\alpha \leq w$ and $c_\beta \cdot w = 0$, contradiction. $\qquad\square$

The following variation on Problem 2 of McKenzie, Monk [82] remains open.

Problem 10. *Is it true that for every infinite BA A there is a cardinal κ such that if B and C are extensions of A with depth at least κ then $\text{Depth}(B \oplus_A C) = \max(\text{Depth}B, \text{Depth}C)$?*

A large cardinal κ might work here.

We also mention the following problem from McKenzie, Monk [82]:

Problem 11. *Is it true that for every infinite BA A there exist extensions B and C of A and an infinite cardinal κ such that B and C have no chains of order type κ but $B \oplus_A C$ does?*

Concerning unions, we note that Depth is an ordinary sup function with respect to the function P, where $PA = \{X \subseteq A : X$ is a well-ordered chain in $A\}$, and so Theorem 3.11 applies.

For ultraproducts the situation is similar to that for cellularity. The same argument as before shows that if F is a countably complete ultrafilter on an infinite set I and is a BA with depth ω for each $i \in I$, then $\prod_{i \in I} A_i/F$ has depth ω. And, as before, if F is a countably incomplete ultrafilter on I and each algebra is infinite, then $\prod_{i \in I} A_i/F$ has depth $> \omega$. This is easiest to see by recalling that $\prod_{i \in I} A_i/F$ is ω_1-saturated, and noting

(*) *If an infinite BA A is κ-saturated, then A has a chain of order type κ.*

To prove (*), we construct $a \in {}^\kappa A$ by recursion. Suppose that a_β has been defined for all $\beta < \alpha$, so that if β is a successor ordinal $\gamma + 1$, then $A \restriction -a_\gamma$ is infinite. If β is a successor ordinal, it is clear how to proceed in order to still have the indicated condition. If β is limit, consider the set

$$\{\mathbf{c}_{x_\alpha} < v_0 : \alpha < \beta\} \cup \{\text{ ``there are at least } n\text{''} v_1 (v_0 < v_1) : n \in \omega\}.$$

This set is finitely satisfiable in A, and so an element satisfying all of these formulas gives the desired element a_β.

Now we consider regular ultrafilters. The first result follows easily from a theorem of W. Hodges, that if F is a regular ultrafilter on I then in ${}^I\langle \omega, > \rangle/F$ there is a chain of order type $|I|^+$. We give a direct BA proof of the BA result:

Theorem 4.10. *Let F be a $|I|$-regular ultrafilter on I, and suppose that A_i is an infinite BA for every $i \in I$. Then in $\prod_{i \in I} A_i/F$ there is a chain of order type $|I|^+$.*

Proof. For brevity set $\kappa = |I|$. By the definition of regularity choose $E \subseteq F$ such that $|E| = \kappa$ and for all $i \in I$ the set $\{e \in E : i \in e\}$ is finite. Let G be a one-one function from E onto κ. For each $i \in I$ choose a strictly increasing sequence $\langle x_{ij} : j < \omega \rangle$ in A_i, and let $X_i = \{x_{ij} : j < \omega\}$. Then it suffices to show:

(*) If $g_\alpha \in \prod_{i \in I} X_i$ for all $\alpha < \kappa$, then there is an $f \in \prod_{i \in I} X_i$ such that $g_\alpha/F < f/F < 1$ for all $\alpha < \kappa$.

To define f, let $i \in I$. Let $e(1), \ldots, e(m)$ be all of the elements u of E such that $i \in u$. Then let fi be any element of X_i greater than all of the elements

$g_{Ge(1)}i, \ldots, g_{Ge(m)}i$. This defines f. Now if $\alpha < \kappa$ and $i \in G^{-1}\alpha$, we have $g_\alpha i < fi < 1$, as desired. $\square$

Theorem 4.11. *Let I be an infinite set, and suppose that A_i is an infinite BA for every $i \in I$. Then there is a proper filter G on I such that G contains all cofinite sets, and $\prod_{i \in I} A_i/F$ has a chain of order type $2^{|I|}$ for every ultrafilter F including G.*

Proof. Again let $\kappa = |I|$. Let $S \subseteq {}^\kappa\omega$ satisfy the following condition:

(1) $|S| = 2^\kappa$, and for every finite sequence $i_0, \ldots, i_{k-1}$ of natural numbers and every sequence $f_0, \ldots, f_{k-1}$ of distinct members of S of length k, there is an $\alpha < \kappa$ such that $f_t\alpha = i_t$ for all $t < k$.

For the existence of such a set, see Comfort, Negrepontis [74], pp. 75-77. Let $\langle f_\alpha : \alpha < 2^\kappa \rangle$ enumerate S without repetitions. For $\alpha < \beta < 2^\kappa$, let $J_{\alpha\beta} = \{\gamma < \kappa : f_\alpha\gamma < f_\beta\gamma\}$. From (1) it is clear that the intersection of any finite number of the sets $J_{\alpha\beta}$ is infinite. Hence

$$\{J_{\alpha\beta} : \alpha < \beta < 2^\kappa\} \cup \{\Gamma \subseteq \kappa : |\kappa \backslash \Gamma| < \omega\}$$

generates a proper filter G containing all cofinite sets. Clearly G is as desired. $\square$

Now we give some results of Douglas Peterson.

Theorem 4.12. *Suppose that $\langle A_i : i \in I \rangle$ is a system of infinite BAs, with I infinite, and F is an ultrafilter on I. Then $\mathrm{Depth}\left(\prod_{i \in I} A_i/F\right) \geq \mathrm{ess.sup}_{i \in I}^F \mathrm{Depth} A_i$.*

Proof. For any linearly ordered set L let $\mathrm{Depth}\, L$ be the supremum of the size of well-ordered subsets of L. Let $\lambda = \mathrm{ess.sup}_{i \in I}^F \mathrm{Depth} A_i$. If λ is a successor cardinal, then $\{i \in I : \mathrm{Depth} A_i = \lambda\} \in F$, and hence clearly $\mathrm{Depth}\left(\prod_{i \in I} A_i/F\right) \geq \mathrm{Depth}({}^I\lambda/F) \geq \lambda$, as desired. If λ is a limit ordinal, then by similar reasoning, $\mathrm{Depth}\left(\prod_{i \in I} A_i/F\right) \geq \kappa$ for every successor cardinal $\kappa < \lambda$, and so also $\mathrm{Depth}\left(\prod_{i \in I} A_i/F\right) \geq \lambda$. $\square$

Theorem 4.13. *Suppose that $\langle A_i : i \in I \rangle$ is a system of infinite BAs, with I infinite, and that F is a regular ultrafilter on I. Let $\lambda = \mathrm{ess.sup}_{i \in I}^F \mathrm{Depth} A_i$, and assume that $\mathrm{cf}\lambda \leq |I| < \lambda$. Then $\mathrm{Depth}\left(\prod_{i \in I} A_i/F\right) \geq \lambda^+$.*

Proof. *Case 1.* $\{i \in I : \mathrm{Depth} A_i = \lambda\} \in F$. We may assume that $\mathrm{Depth} A_i = \lambda$ for all $i \in I$. By Lemma 3.12 we get a system $\langle \kappa_i : i \in I \rangle$ of infinite cardinals such that $\kappa_i < \lambda$ for all $i \in I$, and $\mathrm{ess.sup}_{i \in I}^F \kappa_i = \lambda$. Let $\delta_i = \kappa_i^+$ for all $i \in I$. Then, using the notation in the proof of Theorem 4.12, $\mathrm{Depth}\left(\prod_{i \in I} A_i/F\right) \geq \mathrm{Depth}\left(\prod_{i \in I} \delta_i/F\right)$, so it suffices to show that $\mathrm{Depth}\left(\prod_{i \in I} \delta_i/F\right) \geq \lambda^+$. Suppose that $\{f_\alpha/F : \alpha < \lambda\}$ is a set of elements of $\prod_{i \in I} \delta_i/F$; we shall find an element $f \in \prod_{i \in I} \delta_i$ such that $f/F > f_\alpha/F$ for all $\alpha < \lambda$, and this will clearly finish the proof. Let $i \in I$. Then $\{f_\alpha i : \alpha < \kappa_i\}$ is not cofinal in δ_i, so we can let fi be an element of δ_i greater than each $f_\alpha i$, $\alpha < \kappa_i$. Then for any $\alpha < \lambda$ we have $\{i \in I : fi > f_\alpha i\} \supseteq \{i \in I : \kappa_i > \alpha\} \in F$, so $f/F > f_\alpha/F$, as desired.

Case 2. $\{i \in I : \mathrm{Depth}A_i < \lambda\} \in F$. Then we can assume that $\mathrm{Depth}A_i < \lambda$ for all $i \in I$. Then by Lemma 3.13 there is a system $\langle \kappa_i : i \in I \rangle$ of infinite cardinals such that $\kappa_i < \mathrm{Depth}A_i$ for all $i \in I$, and ess.sup$_{i \in I}^{F} \kappa_i = \lambda$. Let $\delta_i = \kappa_i^+$. Note that A_i has a well-ordered subset of size δ_i. Hence the rest of the proof of Theorem 4.12 goes through. $\square$

Theorem 4.14. (GCH) *Suppose that* $\langle A_i : i \in I \rangle$ *is a system of infinite BAs, with* I *infinite, and* F *is a regular ultrafilter on* I. *Then* $\mathrm{Depth}\left(\prod_{i \in I} A_i / F\right) \geq \left|\prod_{i \in I} \mathrm{Depth}A_i / F\right|$.

Proof. Let $\lambda = \mathrm{ess.sup}_{i \in I}^{F} \mathrm{Depth}A_i$. Then we consider three cases: *Case 1.* $\lambda \leq |I|$. Then $\lambda^{|I|} = 2^{|I|} = |I|^+$, and this case follows from Theorem 4.10. *Case 2.* $\mathrm{cf}\lambda \leq |I| < \lambda$. Then $\lambda^{|I|} = \lambda^+$, and the result follows from Theorem 4.13. *Case 3.* $|I| < \mathrm{cf}\lambda$. Then $\lambda^{|I|} = \lambda$ and we are through by Theorem 4.12. $\square$

We can get an upper bound as in the case of cellularity (see Theorem 3.18).

Theorem 4.15. *Let* $\langle A_i : i \in I \rangle$ *is a system of infinite BAs, with* I *infinite, let* F *be a uniform ultrafilter on* I, *and let* $\kappa = \max(|I|, \mathrm{ess.sup}_{i \in I}^{F} \mathrm{Depth}A_i)$. *Then* $\mathrm{Depth}\left(\prod_{i \in I} A_i / F\right) \leq 2^{\kappa}$. $\square$

So, again we have a lower and an upper bound. First consider the lower bound. It is consistent to have $\mathrm{Depth}\left(\prod_{i \in I} A_i / F\right) > \left|\prod_{i \in I} \mathrm{Depth}A_i / F\right|$ with F regular; see McKenzie,Monk [82], p. 158, for an example due to Laver. It is open to give such an example in ZFC.

Problem 12. *Is an example with* $\mathrm{Depth}\left(\prod_{i \in I} A_i / F\right) > \left|\prod_{i \in I} \mathrm{Depth}A_i / F\right|$ *possible in ZFC?*

But one can also consistently have inequality in the other direction; this is a result of Shelah [90] which also solves Problem 4 of Monk [90]. The proof is very similar to the proof of Theorem 1.5.8 in McKenzie, Monk [82] (also due to Shelah). Note that the theorem says that it is consistent to have a BA A such that $\mathrm{Depth}(^{\omega}A/F) < |^{\omega}\mathrm{Depth}A/F|$.

Theorem 4.16. *Suppose* $V \models CH$, *let* κ *be any uncountable cardinal in* V, *and let* P *be the partial order for adding* κ *Sacks reals side-by-side. Then in* V^P *there is a nonprincipal ultrafilter* F *on* ω *such that* $\mathrm{Depth}(^{\omega}A/F) = \omega_1$, *where* A *is the BA of finite and cofinite subsets of* ω.

(Since one can make the continuum large in this way, and cardinals are preserved, this does do the job.)

Proof. We shall use the notation in Jech [86]; in particular, we use the proof of Theorem 7.12 there. In fact, we need to give more details than were supplied for 7.12, so we give a proof of it here too.

A *perfect tree* is a nonempty subset T of $^{<\omega}2$ such that if $t \in T$ and m is smaller than the domain of t then $t \upharpoonright m \in T$, and such that for any $t \in T$ there is some $s \in T$ with $t \subseteq s$ such that $s0, s1 \in T$. We write $p \leq q$ in place of $p \subseteq q$

for perfect trees p, q. A *branching point* of p is a point $t \in p$ such that $t0 \in p$ and $t1 \in p$. An *nth branching point* is a branching point t such that there are exactly n branching points $< t$. Note that for any $s \in p$, if there are m branching points of p strictly less than s, then for any $n \geq m$ there is an nth branching point t of p such that $s \leq t$. For perfect trees p and q, $p \leq_n q$ means that $p \subseteq q$ and every nth branching point of q is a branching point of p. If $p \leq_n q$, then $p \leq_i q$ for every $i \leq n$. For, suppose that t is an ith branching point of q. By the above remark, choose an nth branching point u of q with $t \leq u$. Then $u \in p$, and hence $t \in p$. So $p \leq_i q$. Now it follows that every nth branching point of q is also an nth branching point of p.

We also note:

$(\star)$ If $p \leq q$ and $n \in \omega$, then there is an nth branching point t of q such that $t \in p$.

For, let s be an nth branching point of p. Then it is an mth branching point of q for some $m \geq n$. Let $t \leq s$ be an nth branching point of q. Thus $t \in p$, as desired.

Thus $p \leq_n q$ means that $p \subseteq q$, and any points of q thrown away to get p have more than n branching points strictly below them.

A *fusion sequence* is a sequence such that

$$p_0 \geq_0 p_1 \geq_1 p_2 \geq_2 \cdots \geq_{n-1} p_n \geq_n \cdots$$

Fusion Lemma 4.17. *If $\langle p_n : n \in \omega \rangle$ is a fusion sequence, then $p \stackrel{\text{def}}{=} \bigcap_{n \in \omega} p_n$ is a perfect tree, and $p \leq_n p_n$ for all $n \in \omega$.*

Proof. Let $n \in \omega$, and let s be an nth branching point of p_n. If $n \leq m$, then $p_n \geq_n p_m$, and so s is a branching point of p_m, so that $s, s0, s1 \in p_m$. Hence $s, s0, s1 \in p$, and s is a branching point of p.

Thus we just need to see that p is a perfect tree. If $t \in p$ and $m < \operatorname{dom}t$, then obviously $t \restriction m \in p$.

Now suppose that $s \in p$; we want to find $t \geq s$ such that $t0, t1 \in p$. Let $m = \operatorname{dom}s$. Now $s \in p_m$, and there are at most $m - 1$ branching points of $p_m < s$, since there are only that many elements of $^{<\omega}2$ which are $< s$. Choose an mth branching point t of p_m with $s \leq t$. By the first paragraph of this proof we have $t, t0, t1 \in p$, as desired. $\square$

If p is a perfect tree and $t \in p$, we define

$$p \restriction t = \{u \in p : u \text{ and } t \text{ are comparable}\}.$$

Now let p be a perfect tree, s an nth branching point of p, and t one of the immediate successors of s in p. Suppose that $q \leq p \restriction t$. Then

$$r \stackrel{\text{def}}{=} q \cup \{u \in p : u \text{ and } t \text{ are incomparable}\}$$

is a perfect tree called the *amalgamation* of q into p at t. Clearly $r \leq_n p$. Also note that $r \restriction t = q$.

Let Q be the collection of all perfect trees. Q is called *Sacks forcing*. Its greatest element is $^{<\omega}2$, the full binary tree of height ω. The partial order P that we are concerned with is the σ-product of κ copies of Q; it consists of all $p \in {}^{\kappa}Q$ such that $pi = 1$ for all but countably many $i \in \kappa$, where 1 is the full binary tree of height ω. The *support* of an element $p \in P$ is the set of all $i \in \kappa$ such that $pi \neq 1$; it is denoted by $\mathrm{Supp}(p)$. The essence of Theorem 7.12 of Jech [86] is the following lemma; we are interested not so much in the lemma itself as in its proof.

Lemma 4.18. *Let P be the σ-product of κ-many Sacks forcings, where κ is any infinite cardinal. Suppose that $B \in V$, $p \in P$, and $p \Vdash \dot{X} : \omega \to B$. Then there is a countable $A \in V$ and a $p_\infty \leq p$, $p_\infty \in P$, such that $p_\infty \Vdash \dot{X} : \omega \to A$.*

Proof. We assume given a well-ordering of all objects that play a role in this proof. This is so we can make the construction very definite, implicitly choosing the "first" object when we make an arbitrary choice. We construct a sequence $p = p_0 \geq p_1 \geq p_2 \geq \cdots$, and finite sets $A_0, A_1, \dots$. As soon as p_i is defined we let S_i be the support of p_i.

We need an auxiliary function $g : \omega \to \omega \times \omega$. Let $g0 = (0,0)$. If gn has been defined, say $gn = (i,j)$, let

$$g(n+1) = \begin{cases} (i+1, j-1), & \text{if } j \neq 0; \\ (0, i+1), & \text{otherwise.} \end{cases}$$

Then g maps onto $\omega \times \omega$, and if $gn = (i,j)$, then $i \leq n$.

If p_i has been defined, we let $\langle G_{ij} : j \in \omega \rangle$ be the first system of finite subsets of κ with union S_i. And let $F_i = \bigcup_{j \leq i} G_{gj}$. (Note that if $j \leq i$ and $gj = (k,l)$, then $k \leq j$, so G_{gj} has been defined already too.) So, $\langle F_i : i \in \omega \rangle$ will be an increasing sequence of finite sets with union $S \overset{\text{def}}{=} \bigcup_n S_n$.

Let $p_0 = p$ and $A_0 = 0$. Now suppose that p_{n-1} and A_{n-1} have been defined. For each $i \in F_{n-1}$ let E_{in} be the set of all successors of all nth branching points of the tree $p_{n-1}(i)$. Let $\sigma_{1n}, \dots, \sigma_{l_n n}$ be all of the functions σ on F_{n-1} such that $\sigma i \in E_{in}$ for all $i \in F_{n-1}$. We construct $q_{0n} \geq q_{1n} \geq \cdots \geq q_{l_n n}$ and $A_n = \{a_{1n}, \dots, a_{nl_n}\}$ as follows. Let $q_{0n} = p_{n-1}$. Assume that q_{kn} has been defined so that $q_{kn}(i) \leq_n p_{n-1}(i)$ for all $i \in F_{n-1}$. Thus $\sigma_{kn}(i)$ is a successor of an nth branching point of $q_{kn}(i)$ if $i \in F_{n-1}$. Let

$$q'_{kn}(i) = \begin{cases} q_{kn}(i) \upharpoonright \sigma_{kn}(i), & \text{if } i \in F_{n-1}, \\ q_{kn}(i), & \text{otherwise.} \end{cases}$$

So $q'_{kn} \leq q_{kn}$. Hence there is an $r_{kn} \leq q'_{kn}$ and an $a_{nk} \in B$ such that $r_{kn} \Vdash \dot{X}n = a_{nk}$. Let

$$q_{(k+1)n}(i) = \begin{cases} \text{amalgamation of } r_{kn}(i) \text{ into } q_{kn}(i) & \text{if } i \in F_{n-1}, \\ r_{kn}(i), & \text{otherwise.} \end{cases}$$

Thus $q_{(k+1)n}(i) \leq_n q_{kn}(i)$ if $i \in F_{n-1}$. Let $p_n = q_{l_n n}$. Thus $p_n(i) \leq_n p_{n-1}(i)$ for all $i \in F_{n-1}$.

Hence for all $i \in S$, $p_\infty(i) \stackrel{\text{def}}{=} \bigcap_{n \in \omega} p_n(i)$ is a perfect tree, by the fusion lemma, since for $i \in F_n$ we have

$$p_0(i) \geq \cdots \geq p_{n-1}(i) \geq_n p_n(i) \geq_{n+1} p_{n+1}(i) \geq \cdots.$$

Let $p_\infty(i) = 1$ for $i \notin S$. Define $A = \bigcup_{n \in \omega} A_n$. Now we prove that $p_\infty \Vdash \dot{X} : \omega \to A$, which will finish the proof. And to do this it suffices to show that, for any $n \in \omega$,

$$p_\infty \Vdash \dot{X}n = a_{1n} \vee \ldots \vee \dot{X}n = a_{l_n n}.$$

In turn, to do this it suffices to take an arbitrary $q \leq p_\infty$ and find $\tilde{q} \leq q$ and k such that $\tilde{q} \Vdash \dot{X} = a_{kn}$. Choose $\bar{q} \leq q$ and $b \in B$ such that $\bar{q} \Vdash \dot{X}n = b$. Consider F_{n-1} and $\sigma_{1n} \ldots \sigma_{l_n n}$ as above. For each $i \in F_{n-1}$ let $\tau(i) \in E_{in} \cap \bar{q}(i)$; it exists since $\bar{q}(i) \leq q(i) \leq p_\infty(i) \leq_n p_{n-1}(i)$ (see $(\star)$). Say $\tau = \sigma_{kn}$. Then if r_{kn} is as above, we have $\bar{q}(i) \restriction \sigma_{kn}(i) \leq q_{(k+1)n}(i) \restriction \sigma_{kn}(i) = r_{kn}(i)$ for $i \in F_{n-1}$. Thus if $\tilde{q}(i) = \bar{q}(i) \restriction \sigma_{kn}(i)$ for $i \in F_{n-1}$ and $\tilde{q}(i) = \bar{q}(i)$ otherwise, then $\tilde{q} \leq r_{kn}, \bar{q}$. In fact, clearly $\tilde{q} \leq \bar{q}$, and $\tilde{q}(i) \leq r_{kn}(i)$ for $i \in F_{n-1}$. For $i \notin F_{n-1}$,

$$\tilde{q}(i) = \bar{q}(i) \leq q(i) \leq p_\infty(i) \leq p_n(i) \leq q_{(k+1)n}(i) = r_{kn}(i),$$

as desired. $\tilde{q} \Vdash \dot{X}n = a_{kn}$, as desired. $\qquad\square$

We now begin the proof of Theorem 4.16 itself:

For each $p \in P$ and each subset Γ of κ, let $p \restriction \Gamma$ be the function which agrees with p on Γ and is the 1 of P otherwise. By a result of Laver, let F' be a Ramsey ultrafilter in V which generates a Ramsey ultrafilter F in V^P. By Theorem 4.10, we only need to show that $^\omega A / F$ has no chain of type ω_2. So, arguing by contradiction, suppose that $p \in P$ and

$$p \Vdash \forall \alpha < \omega_2 (\dot{f}_\alpha \in {}^\omega A) \wedge (\langle \dot{f}_\alpha / F : \alpha < \omega_2 \rangle \text{ is strictly increasing}).$$

where $\dot{f}$ is a name. Thus for each $\alpha < \omega_2$ we have $p \Vdash \dot{f}_\alpha : \omega \to A$, so we can apply the *proof* of the Lemma to p. We thus obtain for each α certain *constructed objects* in V; with an obvious correspondence with that proof, they are, for all $n, j \in \omega$,

p_n^α, G_{nj}^α,
F_n^α, E_{in}^α for all $i \in F_{n-1}^\alpha$,
A_n^α, l_n^α,

and, for all $i = 1, \ldots, l_n^\alpha$,

σ_{in}^α, q_{in}^α,
$(q_{in}^\alpha)'$, r_{in}^α,
and a_{in}^α;

finally, we have p_∞^α. Now we claim

(1) $\forall \alpha < \omega_2 \forall u \le p_\infty^\alpha \forall n \in \omega \forall j \in \omega (u \Vdash j \in \dot{f}_\alpha n$ iff $u \upharpoonright \mathrm{Supp}(p_\infty^\alpha) \Vdash j \in \dot{f}_\alpha n)$ and $(u \Vdash j \notin \dot{f}_\alpha n$ iff $u \upharpoonright \mathrm{Supp}(p_\infty^\alpha) \Vdash j \notin \dot{f}_\alpha n)$.

Suppose that $\alpha < \omega_2$, $u \le p_\infty^\alpha$, $n \in \omega$, $j \in \omega$, $u \Vdash j \in \dot{f}_\alpha n$, and $u \upharpoonright \mathrm{Supp}(p_\infty^\alpha) \nVdash j \in \dot{f}_\alpha n$; we want to get a contradiction. Choose $v \le u \upharpoonright \mathrm{Supp}(p_\infty^\alpha)$ such that $v \Vdash j \notin \dot{f}_\alpha n$. For each $i \in F_{n-1}^\alpha$ let $\tau(i) \in E_{in}^\alpha \cap v(i)$; it exists since $v(i) \le u(i) \le p_\infty^\alpha(i) \le p_{n-1}^\alpha(i)$. Say $\tau = \sigma_{kn}^\alpha$. Thus $r_{kn}^\alpha \Vdash \dot{f}_\alpha n = a_{kn}^\alpha$, and $r_{kn}^\alpha \le (q_{kn}^\alpha)'$. Also, if $i \in F_{n-1}^\alpha$, then $v(i) \upharpoonright \sigma_{kn}^\alpha(i) \le q_{k+1,n}^\alpha \upharpoonright \sigma_{kn}^\alpha(i) = r_{kn}^\alpha(i)$. Let $\tilde{v}(i) = v(i) \upharpoonright \sigma_{kn}^\alpha(i)$ for $i \in F_{n-1}^\alpha$ and $\tilde{v}(i) = v(i)$ otherwise. Then $\tilde{v} \le r_{kn}^\alpha, v$, so $j \notin a_k^\alpha$. On the other hand, $u(i) \upharpoonright \sigma_{kn}^\alpha \le r_{kn}^\alpha(i)$ for $i \in F_{n-1}^\alpha$. Let $\tilde{u}(i) = u(i) \upharpoonright \sigma_{kn}^\alpha(i)$ for $i \in F_{n-1}^\alpha$ and $\tilde{u}(i) = u(i)$ otherwise. Then $\tilde{u} \le r_{kn}^\alpha, u$. So $j \in a_k^\alpha$, contradiction. The other part of (1) is similar.

Now we may assume that $\langle \mathrm{Supp}(p_\infty^\alpha) : \alpha < \omega_2 \rangle$ forms a Δ-system, say with kernel Δ. Note that for each $\alpha < \omega_2$, $p_\infty^\alpha \upharpoonright \Delta : \Delta \to \mathcal{P}(^{<\omega}2)$; the set of all such functions has, by CH in V, ω_1 elements. Hence we may assume that for all $\alpha, \beta < \omega_2$ we have $p_\infty^\alpha \upharpoonright \Delta = p_\infty^\beta \upharpoonright \Delta$. Next, for each $\alpha < \omega_2$, the set $\mathrm{Supp}(p_\infty^\alpha) \backslash \Delta$ has a certain countable order type. There are ω_1 countable order types, so we may assume that all such order types are the same. Thus for any $\alpha, \beta < \omega_2$ there is a unique order isomorphism $\pi_{\alpha\beta}$ from $\mathrm{Supp}(p_\infty^\alpha) \backslash \Delta$ onto $\mathrm{Supp}(p_\infty^\beta) \backslash \Delta$. We extend $\pi_{\alpha\beta}$ to a permutation of κ, still denoted by $\pi_{\alpha\beta}$, by letting it be $\pi_{\beta\alpha}$ $(= \pi_{\alpha\beta}^{-1})$ on $\mathrm{Supp}(p_\infty^\beta) \backslash \Delta$ and the identity elsewhere. Thus $\pi_{\alpha\beta} = \pi_{\beta\alpha}$. And this permutation $\pi_{\alpha\beta}$ extends to other objects; for example, if $p \in P$, then $\pi_{\alpha\beta}(p)$ is the member q of P such that $q(i) = p(\pi_{\alpha\beta}(i))$ for all $i \in \kappa$. Note here that if p has support Γ, then $\pi_{\alpha\beta}(p)$ has support $\pi_{\alpha\beta}[\Gamma]$. Now consider the objects

$\pi_{\alpha 0}(p_n^\alpha)$, $\pi_{\alpha 0}[G_{nj}^\alpha]$,
$\pi_{\alpha 0}[F_n^\alpha]$, $\langle E_{(\pi_{\alpha 0}i)n}^\alpha : i \in \pi_{\alpha 0}[F_{n-1}^\alpha] \rangle$,
A_n^α, l_n^α,

and, for all $i = 1, \ldots, l_n^\alpha$,

$\sigma_{in}^\alpha \circ \pi_{0\alpha}$, $\pi_{\alpha 0}(q_{in}^\alpha)$,
$\pi_{\alpha 0}((q_{in}^\alpha)')$, $\pi_{\alpha 0}(r_{in}^\alpha)$,
and a_{in}^α;

and, finally, $\pi_{\alpha 0}(p_\infty^\alpha)$. By CH, there are only ω_1 of these things, so we may assume that they are the same for all $\alpha \in \omega_2 \backslash \{0\}$. Now take any two distinct $\alpha, \beta \in \omega_2 \backslash \{0\}$. Thus, for example,

$$\begin{aligned}
(\pi_{\alpha\beta}(p_\infty^\alpha))(i) &= (\pi_{0\beta}(\pi_{\alpha 0}(p_\infty^\alpha)))(i) \\
&= (\pi_{\alpha 0}(p_\infty^\alpha))(\pi_{0\beta}(i)) \\
&= (\pi_{\beta 0}(p_\infty^\beta))(\pi_{0\beta}(i)) \\
&= p_\infty^\beta(i).
\end{aligned}$$

Hence $\pi_{\alpha\beta}(p_\infty^\alpha) = p_\infty^\beta$. Another useful fact now is that if $i \in F_{n-1}^\alpha$ then $E_{(\pi_{\alpha\beta}i)n} = E_{in}^\alpha$ for all $i \in F_{n-1}^\alpha$. In fact, $\pi_{\alpha 0}(i) \in \pi_{\alpha 0}[F_{n-1}^\alpha] = \pi_{\beta 0}[F_{n-1}^\beta]$ and $E_{(\pi_{\alpha\beta}i)n}^\beta = E_{(\pi_{\beta 0}\pi_{\alpha 0}i)n}^\beta = E_{(\pi_{\alpha 0}\pi_{\alpha 0}i)n}^\alpha = E_{in}^\alpha$.

Next,

(2) $\forall u \le p_\infty^\alpha \forall n \in \omega \forall j \in \omega(u \Vdash j \in \dot{f}_\alpha n$ iff $\pi_{\alpha\beta}u \Vdash j \in \dot{f}_\beta n)$.

For, suppose that $u \Vdash j \in \dot{f}_\alpha n$ but $\pi_{\alpha\beta}u \nVdash j \in \dot{f}_\beta n$. By (1) we may assume that $\mathrm{Supp}(u) \subseteq \mathrm{Supp}(p_\infty^\alpha)$. Choose $v \le \pi_{\alpha\beta}(u)$ such that $v \Vdash j \notin \dot{f}_\beta n$. Now if $i \in F_{n-1}^\beta$, then

$$v(i) \le (\pi_{\alpha\beta}(u))(i) \le (\pi_{\alpha\beta}(p_\infty^\alpha))(i) = p_\infty^\beta(i),$$

so there is a $\tau(i) \in E_{in}^\beta \cap v(i)$. Say $\tau = \sigma_{kn}^\beta$. Thus $r_{kn}^\beta \Vdash \dot{f}_\beta n = a_{kn}^\beta$. As above, let $\tilde{v}(i) = v(i) \restriction \sigma_{kn}^\beta(i)$ for $i \in F_{n-1}^\beta$ and $\tilde{v}(i) = v(i)$ otherwise. Then $\tilde{v} \le r_{kn}^\beta, v$, so $j \notin a_{kn}^\beta$. Now $\pi_{\beta\alpha}v \le u$. Furthermore, if $i \in F_{n-1}^\alpha$ then $\pi_{\alpha\beta}i \in F_{n-1}^\beta$, and so

$$\sigma_{kn}^\alpha(i) = (\pi_{\beta\alpha}(\sigma_{kn}^\beta))(i) = \sigma_{kn}^\beta(\pi_{\alpha\beta}i) \in E_{(\pi_{\alpha\beta}i)n}^\beta \cap v(\pi_{\alpha\beta}i) = E_{in}^\alpha \cap (\pi_{\beta\alpha}v)(i).$$

Now let $\tilde{w}(i) = (\pi_{\beta\alpha}v)(i) \restriction \sigma_{kn}^\alpha(i)$ for $i \in F_{n-1}^\alpha$ and $\tilde{w}(i) = (\pi_{\beta\alpha}v)(i)$ otherwise. Then $\tilde{w} \le r_{kn}^\alpha$ and $\tilde{w} \le \pi_{\beta\alpha}v \le u$, so $j \in a_k$, contradiction. This proves (2).

Now let s be the member of P which agrees with p_∞^α and p_∞^β on their supports and is 1 otherwise. Clearly $\pi_{\alpha\beta}(s) = s$. We may assume that $\alpha < \beta$. Then, using the fact that F' generates F,

(3) $s \Vdash \exists X \in F' \forall i \in X \forall j \in \omega(j \in \dot{f}_\alpha i \Rightarrow j \in \dot{f}_\beta i)$.

We claim that

(4) $s \Vdash \exists X \in F' \forall i \in X \forall j \in \omega(j \in \dot{f}_\beta i \to j \in \dot{f}_\alpha i)$.

This is a clear contradiction. So, it suffices to prove (4). By (3), there is a $u \le s$ and an $X \in F'$ such that

(5) $u \Vdash \forall i \in X \forall j \in \omega(j \in \dot{f}_\alpha i \to j \in \dot{f}_\beta i)$.

It suffices now to show

$$\pi_{\alpha\beta}u \Vdash \forall i \in X \forall j \in \omega(j \in \dot{f}_\beta i \to j \in \dot{f}_\alpha i).$$

So, let $v \le \pi_{\alpha\beta}u$, $i \in X$, $j \in \omega$, and assume that $v \Vdash j \in \dot{f}_\beta i$. Since $v \le s \le p_\infty^\beta$, from (2) we get $\pi_{\beta\alpha}v \Vdash j \in \dot{f}_\alpha i$. And $\pi_{\beta\alpha}v \le u$, so by (5) we get $\pi_{\alpha\beta}(v) \Vdash j \in \dot{f}_\beta i$. But $\pi_{\alpha\beta}(v) \le s \le p_\infty^\alpha$, so by (2) again, $v \Vdash j \in \dot{f}_\alpha i$, as desired. $\square$

Shelah has a more recent construction proving the above inequality $<$ for depth and ultraproducts, and this construction applies to some other functions too. To formulate this result, we need a definition.

Suppose that $\mathbf{O}$ is an operation on sequences of BAs, and inv is a cardinal invariant on BAs such that $|\mathrm{inv}B| \le |B|$ for every infinite BA B. Then we say that the property $\square_O$ holds provided that if μ is a cardinal and B_i is a BA for each $i < \mu^+$, then

$$\sup_{i<\mu^+} \mathrm{inv} B_i \leq \mathrm{inv} \left(\mathbf{O}_{i<\mu^+} B_i \right) \leq \mu + \sup_{i<\mu^+} \mathrm{inv} B_i.$$

Rosłanowski, Shelah [94] proved the following result:

Suppose that inv *is a cardinal invariant on BAs satisfying* $\square_\oplus$ *or* $\square_{\Pi^w}$; *suppose that* inv$B \leq |B|$ *for any BA B. Suppose that for each infinite cardinal χ there is a BA B such that $\chi < $ inv$B$ and there is no inaccessible cardinal in the interval $(\chi, |B|]$. Assume further that*

$\odot$ $\langle \lambda_i : i < \kappa \rangle$ *is a sequence of weakly inaccessible cardinals* $\lambda_i > \kappa^+$, D *is an* $\aleph_1$-*complete ultrafilter on* κ, *and* $\prod_{i<\kappa}(\lambda_i, <)/D$ *is* μ^+-*like.*

Then there exist BAs B_i for $i < \kappa$ such that

$$\mathrm{inv} B_i = \lambda_i \quad \text{and} \quad \mathrm{inv}\left(\prod_{i<\kappa} B_i/D\right) \leq \mu.$$

As a corollary of this and Magidor, Shelah [91] one has:

Suppose that inv *is a cardinal invariant on BAs such that the following three conditions hold:*
(i) inv$B \leq |B|$ *for all infinite BAs B.*
(ii) $\sup_{i<\mu^+} \mathrm{inv} B_i \leq \mathrm{inv}\left(\prod^w_{i<\mu^+} B_i\right) \leq \mu + \sup_{i<\mu^+} \mathrm{inv} B_i$ *for every system* $\langle B_i : i < \mu^+ \rangle$ *of BAs.*
Then it is consistent to have a system $\langle B_i : i < \kappa \rangle$ *of BAs such that* inv$\left(\prod_{i<\kappa} B_i/D\right) < \prod_{i<\kappa} \mathrm{inv} B_i/D$.

This corollary applies not only to depth, but also to length, independence, π-character, and tightness.

Observe that the upper bound of Theorem 4.15 is strict in some ultraproducts, and mere equality in others.

Some additional results on the connection between depth and ultraproducts of BAs can be found in Shelah [95a].

Note that if A is a dense subalgebra of B, then trivially Depth$A \leq$ Depth$B$. The difference can be arbitrarily large: take B to be an interval algebra on a large cardinal, and let A be the subalgebra of B generated by its atoms.

For subdirect products the situation is similar to that for cellularity, with essentially the same proof: there is a BA with depth ω which is a subdirect product of BAs having high depth. Depth of Boolean powers is described by our discussion of free products.

Depth of set products can easily be described using the arguments for products:

Theorem 4.19. $\mathrm{Depth}(\prod_{i\in I}^{B} A_i) = \max\{\mathrm{Depth}B, \sup_{i\in I}\mathrm{Depth}A_i\}.$

Proof. We use the notation introduced in Chapter 1 for set products. Clearly $\geq$ holds. Now let $\kappa = \max\{\mathrm{Depth}B, \sup_{i\in I}\mathrm{Depth}A_i\}$, and, to get a contradiction, suppose that $\langle h(b_\alpha, F_\alpha, a^\alpha) : \alpha < \kappa^+\rangle$ is a strictly increasing sequence. Without loss of generality $b_\alpha \cap F_\alpha = 0$ and $a_i^\alpha \neq J_i$ for all $\alpha < \kappa^+$ and $i \in I$. Then $b_\alpha \subseteq b_\beta$ for $\alpha < \beta < \kappa^+$. Hence there is a $\Gamma \in [\kappa^+]^{\kappa^+}$ such that $b_\alpha = b_\beta$ for all $\alpha, \beta \in \Gamma$. Then $a^\alpha < a^\beta$ for $\alpha < \beta$, both in Γ, and this easily gives a contradiction. $\square$

An easy argument shows that depth for one-point gluing behaves like arbitrary products (Theorem 4.3), if all algebras have more than two elements. Note that a one-point gluing of a product of two-element algebras still has just two elements. For the Aleksandroff duplicate it is clear that the following result holds:

$$\mathrm{Depth}(\mathrm{Dup}A) = \mathrm{Depth}A.$$

For the exponential we also have $\mathrm{DepthExp}A = \mathrm{Depth}A$, by Proposition 2.5, Theorem 4.3, and the above remarks on free products.

Next we discuss derived functions with respect to depth. The first result is that $\mathrm{Depth}_{\mathrm{H}+}$ is the same as tightness. To prove this, we need an equivalent form of tightness due to Arhangelskiĭ and Shapirovskiĭ. It involves the notion of a free sequence in a topological space. Let X be a topological space. A *free sequence* in X is a sequence $\langle x_\xi : \xi < \alpha\rangle$ (α an ordinal) of elements of X such that for all $\xi < \alpha$ we have $\overline{\{x_\eta : \eta < \xi\}} \cap \overline{\{x_\eta : \xi \leq \eta < \alpha\}} = 0$. For an arbitrary topological space X and a point $x \in X$, the *tightness* tx of x in X is, by definition, the least cardinal κ such that if $Y \subseteq X$ and $x \in \overline{Y}$, then there is a subset $Z \subseteq Y$ such that $|Z| \leq \kappa$ and $x \in \overline{Z}$. And the tightness tX of X itself is $\sup_{x\in X} tx$. Clearly this means that $tA = t(\mathrm{Ult}A)$ for any BA A. The equivalent form of tightness due to Arhangelskiĭ (based on proofs of Shapirovskiĭ) is given in the following theorem.

Theorem 4.20. *Let X be a compact Hausdorff space. Then $tX = \sup\{|\alpha| :$ there is a free sequence in X of order type $\alpha\}$.*

Proof. For $\geq$, suppose that $\langle x_\xi : \xi < \kappa\rangle$ is a free sequence, where κ is regular; we shall find a point $y \in X$ such that $ty \geq \kappa$. First note:

(1) There is a $y \in X$ such that $|U \cap \{x_\xi : \xi < \kappa\}| = \kappa$ for each neighborhood U of y.

In fact, otherwise for every $y \in X$ let $U(y)$ be an open neighborhood of y such that $|U(y) \cap \{x_\xi : \xi < \kappa\}| < \kappa$. Thus $\{U(y) : y \in X\}$ is an open cover of X. Let $U(y_0), \ldots, U(y_{n-1})$ be a finite subcover. Then

$$\{x_\xi : \xi < \kappa\} = \bigcup_{i<n}(U(y_i) \cap \{x_\xi : \xi < \kappa\}),$$

and the right side has cardinality $< \kappa$, contradiction. So (1) holds.

Take y as in (1). Assume that $ty < \kappa$. Now $y \in \overline{\{x_\xi : \xi < \kappa\}}$. Hence by the definition of tightness, choose a subset Γ of κ of power at $< \kappa$ such that $y \in \overline{\{x_\xi : \xi \in \Gamma\}}$. Let $\eta = \sup\Gamma + 1$. Hence $y \in \overline{\{x_\xi : \xi < \eta\}}$, so by freeness $y \notin \overline{\{x_\xi : \eta \leq \xi\}}$. So there is a neighborhood U of y such that $U \cap \{x_\xi : \eta \leq \xi\} = 0$. This contradicts (1). We have now proved $\geq$ in the theorem.

Now, for $\leq$, let $\kappa = tX$ and suppose that $1 \leq \lambda < \kappa$. We shall construct a free sequence of length λ^+. Choose $y \in X$ with $t(y) > \lambda$; say $Y \subseteq X$, $y \in \overline{Y}$, and for all $Z \subseteq Y$ with $|Z| \leq \lambda$, $y \notin \overline{Z}$. Set

$$Y' = \{x : \text{there is a } Z \subseteq Y \text{ such that } |Z| \leq \lambda \text{ and } x \in \overline{Z}\}.$$

Thus $Y \subseteq Y'$, so $y \in \overline{Y'}$. Note

(2) If $Z \subseteq Y'$ and $|Z| \leq \lambda$, then $y \notin \overline{Z}$;
(3) If $Z \subseteq Y'$, $|Z| \leq \lambda$, and $z \in \overline{Z}$, then $z \in Y'$.

We now construct x_ξ, F_ξ, U_ξ for $\xi < \lambda^+$ such that $x_\xi \in Y'$, $y \in F_\xi \subseteq U_\xi$ with U_ξ open and F_ξ a closed neighborhood of y, by recursion. Suppose these have been constructed for all $\eta < \xi$, where $\xi < \lambda^+$. Since $y \notin \overline{\{x_\eta : \eta < \xi\}}$, let U_ξ be an open neighborhood of y such that $U_\xi \cap \{x_\eta : \eta < \xi\} = 0$. Let F_ξ be a closed neighborhood of y such that $F_\xi \subseteq U_\xi$. Then we claim

(4) $Y' \not\subseteq \bigcup_{\eta \leq \xi} (X \backslash F_\eta) \cup \overline{\{x_\eta : \eta < \xi\}}$.

For, suppose not; then we show that $y \in \overline{\{x_\eta : \eta < \xi\}}$ (contradiction). For, let U be an open neighborhood of y and let F' be a closed neighborhood of y which is included in U. Let W be the closure of the set $\{F_\eta : \eta \leq \xi\} \cup \{F'\}$ under finite intersections. Since $y \in \overline{Y'}$, for all $H \in W$ choose $z_H \in Y' \cap H$. Then $H' \cap \{z_H : H \in W\} \neq 0$ for all $H' \in W$. Choose

$$t \in \bigcap_{H \in W} H \cap \overline{\{z_H : H \in W\}}.$$

By (3), $t \in Y'$. Now $t \in F_\eta$ for all $\eta \leq \xi$, so by the "suppose not" for (4), $t \in \overline{\{x_\eta : \eta < \xi\}}$. Since $t \in F' \subseteq U$, it follows that $U \cap \{x_\eta : \eta < \xi\} \neq 0$, as desired.

So (4) holds; choose x_ξ in the left side of (4) but not in the right side. This completes the construction.

Suppose $\xi < \lambda^+$ and $s \in \overline{\{x_\eta : \eta < \xi\}} \cap \overline{\{x_\eta : \xi \leq \eta < \lambda^+\}}$. Then $s \notin U_\xi$, so $s \notin F_\xi$. Thus $s \in X \backslash F_\xi$, which is open, so there is an η with $\xi \leq \eta < \lambda^+$ such that $x_\eta \in X \backslash F_\xi$, contradiction. $\square$

Note that the proof of Theorem 4.20 shows that if tX is regular and is attained in the free sequence sense then it is attained in the defined sense, i.e., there is a point y with tightness tX.

Theorem 4.21. *For any infinite BA A we have* $\text{Depth}_{H+} A = tA$.

Proof. For $\geq$, let $\langle F_\xi : \xi < \alpha \rangle$ be a free sequence; we produce a quotient A/I of A having a strictly increasing sequence of order type α. For brevity let $Y = \{F_\xi : \xi < \alpha\}$. For every $\xi < \alpha$ there is an element a_ξ of A such that $\{F_\eta : \eta < \xi\} \subseteq Sa_\xi$ and $Sa_\xi \cap \{F_\eta : \xi \leq \eta < \alpha\} = 0$. Consider the following ideal on A: $I = \{x \in A : Y \subseteq S(-x)\}$. Suppose $\xi < \eta < \alpha$. Then $S(a_\xi \cdot -a_\eta) \cap Y = 0$: if $F_\nu \in S(a_\xi \cdot -a_\eta)$, then $F_\nu \in Sa_\xi$, hence $\nu < \xi$, and $-a_\eta \in F_\nu$, hence $\eta \leq \nu$, so $\eta < \xi$, contradiction. This shows that $[a_\xi] \leq [a_\eta]$ for $\xi < \eta < \alpha$. Still suppose that $\xi < \eta < \alpha$. Then $F_\xi \in Sa_\eta \backslash Sa_\xi = S(a_\eta \cdot -a_\xi)$. Thus $Y \not\subseteq S(-a_\eta + a_\xi)$, so $a_\eta \cdot -a_\xi \notin I$, which means that $[a_\eta] < [a_\xi]$, as desired.

For $\leq$, let I be an ideal in A, and let $\langle [a_\xi] : \xi < \alpha \rangle$ be a strictly increasing sequence in A/I. For each $\xi < \alpha$, the set $\{x : -x \in I\} \cup \{a_{\xi+1}, -a_\xi\}$ has the finite intersection property, since $a_{\xi+1} \cdot -a_\xi \notin I$. Let F_ξ be an ultrafilter including this set. Then, we claim, $\langle F_\xi : \xi < \alpha \rangle$ is a free sequence. To prove this it suffices to show that for any $\xi < \alpha$ we have

(1) $\{F_\eta : \eta < \xi\} \subseteq Sa_\xi$ and $Sa_\xi \cap \{F_\eta : \xi \leq \eta < \alpha\} = 0$.

If $\eta < \xi < \alpha$, then $a_{\eta+1} \cdot -a_\xi \in I$, and hence $-a_{\eta+1} + a_\xi \in F_\eta$; but also $a_{\eta+1} \in F_\eta$, so $a_\xi \in F_\eta$ and so $F_\eta \in Sa_\xi$, proving the first part of (1). For the second part, suppose that $\xi \leq \eta < \alpha$ and $F_\eta \in Sa_\xi$. Now $a_\xi \cdot -a_\eta \in I$, so $-a_\xi + a_\eta \in F_\eta$; but also $-a_\eta \in F_\eta$, so $-a_\xi \in F_\eta$, contradiction. $\square$

Corollary 4.22. Depth_{H+} and t (for free sequences) have the same attainment properties, i.e., for any BA A and any infinite cardinal κ, A has a homomorphic image with a chain of order type κ iff $\mathrm{Ult}A$ has a free sequence of type κ. $\square$

Note that, as in the relation between spread and cellularity, Depth_{H+} involves two sups, while t for free sequences involves only one; we return to this below.

Since $\mathrm{Depth}A \leq cA$, it is clear that $\mathrm{Depth}_{H-}A = \omega$. It is also easy to see that $\mathrm{Depth}_{S+}A = \mathrm{Depth}A$ and $\mathrm{Depth}_{S-}A = \omega$. Depth_{h+} is a little more interesting:

Theorem 4.23. $\mathrm{Depth}_{h+}A = sA$ for any infinite BA A.

Proof. For $\geq$, suppose that Y is a discrete subspace of $\mathrm{Ult}A$; clearly Y, since it is discrete, has an increasing sequence of closed-open sets of order type $|Y|$. For $\leq$, suppose that Y is a subspace of $\mathrm{Ult}A$ and $\langle U_\alpha : \alpha < \kappa \rangle$ is a strictly increasing system of closed-open subsets of Y. For each $\alpha < \kappa$ choose $y_\alpha \in U_{\alpha+1} \backslash U_\alpha$. Clearly $\{y_\alpha : \alpha < \kappa\}$ is a discrete subspace of $\mathrm{Ult}A$. $\square$

The proof shows that $\mathrm{Depth}_{h+}A$ and sA have the same attainment properties.

Since $\mathrm{Depth}_{h-}A \leq \mathrm{Depth}_{H-}A$, we have $\mathrm{Depth}_{h-}A = \omega$ for any infinite BA A. Obviously $_d\mathrm{Depth}_{S+}A = \mathrm{Depth}A$ for any BA A.

The status of the derived function $_d\mathrm{Depth}_{S-}$ is not clear. Note that for A the interval algebra on a cardinal κ we have $_d\mathrm{Depth}_{S-}A = \omega$: this follows upon considering the subalgebra of A generated by $\{\{\alpha\} : \alpha$ a non-limit ordinal $< \kappa\}$. Also, Koppelberg and Shelah have independently observed that if A is atomless and λ-saturated (in the model-theoretic sense), then $_d\mathrm{Depth}_{S-}A \geq \lambda$. To show this, suppose that B is a dense subalgebra of A. By induction choose elements $a_\alpha \in A$ and $b_\alpha \in B$ for $\alpha < \lambda$ so that $\alpha < \beta$ implies that $a_\alpha > a_\beta > b_\beta > 0$; the a_α's can be chosen by λ-saturation, and the b_α's by denseness. So the sequence $\langle b_\alpha : \alpha < \lambda \rangle$ shows that the depth of B is at least λ.

Depth does not quite fit into the framework for discussing $\mathrm{Depth}_{\mathrm{mm}}$. But there is a closely related idea which has been extensively discussed for the Boolean algebra $\mathscr{P}\omega/\mathrm{fin}$, and for completeness we define it. A *tower* in a BA A is a sequence $\langle a_\alpha : \alpha < \kappa \rangle$ of elements of A such that $a_\alpha \leq a_\beta < 1$ if $\alpha < \beta < \kappa$, and $\sum_{\alpha < \kappa} a_\alpha = 1$. We define

$$\mathrm{tow}A = \min\{\kappa : \text{there is a tower in } A \text{ of length } \kappa\}.$$

Next, clearly $[\omega, \mathrm{t}A) \subseteq \mathrm{Depth}_{\mathrm{Hs}}A$, by an argument very similar to that used for the function c. And, of course, $\mathrm{Depth}_{\mathrm{Hs}} \subseteq [\omega, \mathrm{t}A]$. Like for cellularity, there is a problem whether $\mathrm{t}A \in \mathrm{Depth}_{\mathrm{Hs}}A$. This is trivially true if $\mathrm{t}A$ is a successor cardinal or a limit cardinal of cofinality ω by Corollary 4.22 and Theorem 12.2 below. For each singular cardinal κ with $\mathrm{cf}\kappa > \omega$ there is a BA A such that $|A| = \mathrm{Depth}A = \mathrm{t}A$ and $\mathrm{Ult}A$ has no free sequence of length κ, hence by Corollary 4.22 A has no homomorphic image B such that $\mathrm{Depth}B = \mathrm{t}A$ and $\mathrm{Depth}B$ is attained. Namely, let $\langle \mu_\alpha : \alpha < \mathrm{cf}\kappa \rangle$ be a strictly increasing sequence of infinite cardinals with sup κ, and let $A = \prod_{\alpha < \mathrm{cf}\kappa}^{\mathrm{w}} \mathrm{Intalg}\,\mu_\alpha$, and use Theorem 12.1. Nevertheless, we have:

Problem 13. *Is* $\mathrm{t}B \in \mathrm{Depth}_{\mathrm{Hs}}B$ *for every infinite BA* B?

This is problem 5 of Monk [90].

Clearly $\mathrm{Depth}_{\mathrm{Ss}} = [\omega, \mathrm{Depth}A]$ for any infinite BA A.

Next comes the relation $\mathrm{Depth}_{\mathrm{Sr}}$. It is easy to see that parts (i)-(iii) and (v)-(vi) of Theorem 3.32 hold with cellularity replaced by depth. We do not know if Theorem 3.32 (iv) holds for depth:

Problem 14. *Are there an infinite cardinal* κ *and a BA* A *such that* $(\kappa, (2^\kappa)^+) \in \mathrm{Depth}_{\mathrm{Sr}}A$, *while* $(\omega, (2^\kappa)^+) \notin \mathrm{Depth}_{\mathrm{Sr}}A$?

This is problem 6 of Monk [90]. Note that if $(\omega_1, \omega_2) \in \mathrm{Depth}_{\mathrm{Sr}}A$ then also $(\omega, \omega_1) \in \mathrm{Depth}_{\mathrm{Sr}}A$. In fact, let B be a subalgebra of A with depth ω_1. Then B has a disjoint subset of size ω_1, and hence has a subalgebra C isomorphic to the finite-cofinite algebra on ω_1. C has depth ω, as desired. This observation solves problem 7 of Monk [90].

Problem 15. *Characterize the relation* $\mathrm{Depth}_{\mathrm{Sr}}$.

The following theorem seems relevant to these problems:

Theorem 4.24. (GCH) *For every infinite cardinal* κ *there is an interval algebra* A *of power* κ^+ *such that every subalgebra of* A *of power* κ^+ *has depth* $\geq \kappa$.

Proof. Let μ be minimum such that $\omega^\mu > \kappa$. Let L be the linearly ordered set $^\mu\mathbb{Q}$ under lexicographic order, where $\mathbb{Q}$ is the set of all rationals in $[0,1)$. Set

$D = \{f \in {}^{\mu}Q : \text{there is an } \alpha < \mu \text{ such that } f\beta = 0 \text{ for all } \beta > \alpha\}$. It is clear that $|D| \leq \kappa$ and D is dense in L in the sense that if $f, g \in L$ and $f < g$ then there is an $h \in D$ such that $f < h < g$. Let M be a subset of L of size κ^{+} which includes D, and let A be the interval algebra on M. Suppose that B is a subalgebra of A of power κ^{+}. Let N be any subset of B with κ^{+} elements; we shall first show that B includes a simply ordered subset of size κ^{+}; here we follow closely the proof of Theorem 15.22 in Part I of the handbook. For each $x \in N$ write

$$x = [a(1,x), b(1,x)) \cup \ldots \cup [a(m_x, x), b(m_x, x)),$$

where $a(1,x), b(1,x), \ldots, a(m_x, x), b(m_x, x)$ are in $M \cup \{+\infty\}$ and $a(1,x) < b(1,x) < \cdots < a(m_x, x) < b(m_x, x)$. By going from x to $-x$ if necessary, we may assume that $a(1,x) = 0$ for all $x \in N$. We may assume that m_x does not depend on x, so we drop the subscript x. Now for each $x \in N$ we choose

$$c(1,x), \ldots, c(m,x), d(1,x), \ldots, d(m,x) \in D$$

so that $a(i,x) < c(i,x) < b(i,x) < d(i,x) < a(i+1,x)$ for all $i = 1, \ldots, m$ (omitting the term $a(i+1,x)$ for $i = m$, and also omitting $d(i,x)$ if $b(i,x) = \infty$). We may assume that the elements $c(i,x)$ and $d(i,x)$ do not actually depend on x; so we write simply c_i and d_i. Next, we may assume that for some k, $1 \leq k \leq m$, the elements $a(k,x)$, $x \in N$, are pairwise distinct (the argument below is similar if some elements $b(k,x)$, $x \in N$ are pairwise distinct). Note that $k > 1$. Now define a homomorphism f of B into the BA of all subsets of $L \cap [d_{k-1}, c_k)$ by setting $fu = u \cap [d_{k-1}, c_k)$ for all $u \in B$. Now by Theorem 15.18 of the BA handbook, Part I, there is an isomorphism from the range of f into B. But clearly f takes N onto a linearly ordered set of power κ^{+}, as desired.

Now by the Erdös-Rado theorem $(2^{\lambda})^{+} \to (\lambda^{+})^2_{\lambda}$ it follows in an obvious way that B has depth $\geq \kappa$. $\qquad\square$

We note the following two obvious facts about Depth_{Hr}:

(1) If $(\kappa, \lambda) \in \text{Depth}_{\text{Hr}}A$, then $\kappa \leq \lambda \leq |A|$ and $\kappa \leq tA$.

(2) If $\kappa \in [\omega, tA)$ then there is a $\lambda \leq 2^{\kappa}$ such that $(\kappa, \lambda) \in \text{Depth}_{\text{Hr}}A$.

Also, the following examples are relevant: if A is the finite-cofinite algebra on κ, then $\text{Depth}_{\text{Hr}}A = \{(\omega, \lambda) : \omega \leq \lambda \leq \kappa\}$; if A is free on κ, then $\text{Depth}_{\text{Hr}}A = \{(\lambda, \mu) : \omega \leq \lambda \leq \mu \leq \kappa\}$. A problem similar to Problem 14 for Depth_{Sr} is open (this is Problem 8 of Monk [90]):

Problem 16. *Are there an infinite cardinal κ and a BA A such that $(\kappa, (2^{\kappa})^{+}) \in \text{Depth}_{\text{Hr}}A$, while $(\omega, (2^{\kappa})^{+}) \notin \text{Depth}_{\text{Hr}}A$?*

Problem 17. *Characterize the relation Depth_{Hr}.*

Concerning special classes of BAs, first notice that Depth is the same as cellularity for complete BAs. It is possible to have $\text{Depth}A < cA$ for an interval algebra. For example, let τ be the order type of the real numbers, let L be an ordered set of type $0 + (\omega + \omega^*) \cdot \tau$, and let A be the interval algebra on L. It is easily seen that $\mathrm{Depth} A = \omega$ while $cA = 2^\omega$. By Proposition 16.20 in the Handbook, if T is an infinite tree then $\mathrm{Depth}(\mathrm{Treealg}\, T)$ is equal to $\max(\sup\{|C| : C$ is a chain in $T\}, \omega)$.

We finish this chapter by giving two theorems concerning depth in the algebra $\mathscr{P}\omega/\mathrm{fin}$. The first theorem is due to Hechler [72].

Theorem 4.25. *Under MA,* $\mathrm{Depth}(\mathscr{P}\omega/\mathrm{fin}) = 2^\omega$.

Proof. It clearly suffices to prove the following statement:

(1) If $\langle x_\alpha : \alpha < \gamma \rangle$ is a system of infinite subsets of ω such that (a) $\gamma < 2^\omega$, (b) $\alpha < \beta < \gamma$ implies that $x_\alpha \backslash x_\beta$ is finite and $x_\beta \backslash x_\alpha$ is infinite, and (c) $\omega \backslash x_\alpha$ is infinite for all $\alpha < \gamma$, then there is an infinite subset x_γ of ω such that (d) $x_\alpha \backslash x_\gamma$ is finite for each $\alpha < \gamma$, (e) $x_\gamma \backslash x_\alpha$ is infinite for each $\alpha < \gamma$, and (f) $\omega \backslash x_\gamma$ is infinite.

To prove (1) we may assume that γ is nonzero, and we take two cases. *Case 1.* γ is a successor ordinal $\beta + 1$. Write $\omega \backslash x_\beta = y \cup z$, where y and z are infinite and disjoint. Let $x_\gamma = x_\beta \cup y$. Clearly this works. *Case 2.* γ is a limit ordinal. In this case we shall apply Theorem 2.15 of Chapter 2 in Kunen [80]. Let $\mathscr{A} = \{x_\alpha : \alpha < \gamma\}$ and $\mathscr{C} = \{\omega\}$. If F is a finite subset of γ with maximum element δ, then

$$\left(\omega \backslash \bigcup_{\alpha \in F} x_\alpha \right) / \mathrm{fin} = \prod_{\alpha \in F} -(x_\alpha / \mathrm{Fin})$$
$$= -(x_\delta / \mathrm{fin})$$
$$\neq 0,$$

by the assumption (c) of (1). This means that $\omega \backslash \bigcup_{\alpha \in F} x_\alpha$ is infinite, and verifies the hypothesis of Kunen 2.15. So we apply Kunen 2.15 and get a set $d \subseteq \omega$ such that $x_\alpha \cap d$ is finite for each $\alpha < \gamma$, and d itself is infinite. Let $x_\gamma = \omega \backslash d$. We proceed to check (d)–(f). (d) and (f) are clear. If (e) fails for a certain $\alpha < \gamma$, then

$$x_{\alpha+1} \backslash x_\alpha = (x_{\alpha+1} \cap d \backslash x_\alpha) \cup (x_{\alpha+1} \backslash d \backslash x_\alpha),$$

and the latter is finite, contradiction. $\square$

The second result is of a folklore nature.

Theorem 4.26. *There is a model of ZFC in which* $2^\omega > \omega_1$ *while* $\mathscr{P}\omega/\mathrm{fin}$ *has depth* ω_1. *In fact, we can take* $M[G]$, *where* M *satisfies CH and* G *adds Cohen reals.*

Proof. We use Boolean-valued forcing, as in Jech [78]. Assume that (in M) CH holds, and κ is a regular uncountable cardinal. Let $\mathscr{P} = \{p : p$ is a finite function with domain contained in κ and range contained in $2\}$. And let G be M-generic over $\mathscr{P}$. Let φ be the formula

$\langle T_\alpha : \alpha < \omega_2 \rangle$ is a sequence of subsets of ω,

and $\forall \alpha, \beta \in \omega_2 [\alpha < \beta \Rightarrow T_\alpha \backslash T_\beta$ is finite and $T_\beta \backslash T_\alpha$ is infinite$]$,

and suppose that $M[G] \models \varphi$; we want to get a contradiction. Choose $p \in G$ so that $p \Vdash \varphi$. From now on we work in M.

Temporarily fix $\alpha < \omega_2$ and $n \in \omega$. Now $p \Vdash n \in T_\alpha \vee n \notin T_\alpha$, so $\forall q \leq p \exists r \leq q(r \Vdash n \in T_\alpha \vee r \Vdash n \notin T_\alpha)$. Hence there is a maximal pairwise incompatible set $A_{\alpha n} \subseteq \{q : q \leq p\}$ such that $\forall q \in A_{\alpha n}(q \Vdash n \in T_\alpha \vee q \Vdash n \notin T_\alpha)$.

Now for any $\alpha \in \omega_2$ let

$$C_\alpha = \text{dmn} p \cup \bigcup_{n \in \omega} \bigcup_{q \in A_{\alpha n}} \text{dmn} q.$$

Thus C_α is countable. By CH there is an $X \in [\omega_2]^{\omega_2}$ such that $\langle C_\alpha : \alpha \in X \rangle$ is a Δ-system, say with kernel C. We may also assume that there is a $\gamma < \omega_1$ such that $C_\alpha \backslash C$ has order type γ for all $\alpha \in X$. For all $\alpha, \beta \in X$, let $j_{\alpha\beta}$ be the permutation of $C_\alpha \cup C_\beta$ such that $j_{\alpha\beta}$ is the identity on C, and is the unique order preserving map from $C_\alpha \backslash C$ onto $C_\beta \backslash C$ and from $C_\beta \backslash C$ onto $C_\alpha \backslash C$. Thus $j_{\alpha\alpha}$ is the identity on C_α, and $j_{\alpha\beta} = j_{\beta\alpha}$. Extend each $j_{\alpha\beta}$ to a permutation $j'_{\alpha\beta}$ of κ by letting $j'_{\alpha\beta}$ be the identity outside of $C_\alpha \cup C_\beta$. Obviously then

(1) $(j'_{\beta\gamma} \circ j'_{\alpha\beta}) \upharpoonright C_\alpha = j'_{\alpha\gamma} \upharpoonright C_\alpha$ for any $\alpha, \beta, \gamma \in X$.

Now $j'_{\alpha\beta}$ naturally induces an automorphism $j''_{\alpha\beta}$ of $\mathscr{P}$, given by: $\text{dmn}(j''_{\alpha\beta}p) = j'_{\alpha\beta}[\text{dmn}p]$ and for any $\alpha \in \text{dmn}p$, $(j''_{\alpha\beta}p)(j'_{\alpha\beta}\alpha) = p\alpha$. And of course then $j''_{\alpha\beta}$ induces an automorphism $j'''_{\alpha\beta}$ of $\text{RO}\mathscr{P}$. Finally, $j'''_{\alpha\beta}$ induces a permutation $j^{iv}_{\alpha\beta}$ of $V^{\text{RO}\mathscr{P}}$, defined recursively by setting $\text{dmn}(j^{iv}_{\alpha\beta}x) = \{j^{iv}_{\alpha\beta}y : y \in \text{dmn}x\}$ and $(j^{iv}_{\alpha\beta}x)(j^{iv}_{\alpha\beta}y) = j'''_{\alpha\beta}(xy)$. A basic property of this process is:

(2) $j'''_{\alpha\beta}[\![\varphi(x_1, \ldots, x_n)]\!] = [\![\varphi(j^{iv}_{\alpha\beta}x_1, \ldots, j^{iv}_{\alpha\beta}x_n)]\!]$.

Now for each $\alpha \in X$ define $\mathscr{P}_\alpha = \{q \in \mathscr{P} : \text{dmn}q \subseteq C_\alpha$ and $q \leq p\}$. Then the following are easy to prove:

(3) $j''_{\alpha\beta}[\mathscr{P}_\alpha] = \mathscr{P}_\beta$ for any $\alpha, \beta \in X$.
(4) $(j''_{\beta\gamma} \upharpoonright \mathscr{P}_\beta) \circ (j''_{\alpha\beta} \upharpoonright \mathscr{P}_\alpha) = (j''_{\alpha\gamma} \upharpoonright \mathscr{P}_\alpha)$.

We define $h_\alpha : \mathscr{P}_\alpha \to {}^\omega 3$ by

$$(h_\alpha q)n = \begin{cases} 0, & \text{if } q \Vdash n \notin T_\alpha, \\ 1, & \text{if } q \Vdash n \in T_\alpha, \\ 2, & \text{otherwise.} \end{cases}$$

Define $\alpha \equiv \beta$ iff $\alpha, \beta \in X$ and $h_\beta \circ (j''_{\alpha\beta} \upharpoonright C_\alpha) = h_\alpha$. It is straightforward to check that $\equiv$ is an equivalence relation on X. Then:

(5) There are at most ω_1 equivalence classes under $\equiv$.

In fact, suppose that $\Gamma \in [X]^{\omega_2}$ consists of pairwise inequivalent ordinals. Fix $\alpha \in \Gamma$. Then for any distinct $\beta, \gamma \in \Gamma$ we have $h_\beta \circ (j''_{\alpha\beta} \restriction \mathscr{P}_\alpha) \neq h_\gamma \circ (j''_{\alpha\gamma} \restriction \mathscr{P}_\alpha)$, since otherwise

$$h_\beta \circ (j''_{\beta\gamma} \restriction \mathscr{P}_\gamma) = h_\beta \circ (j''_{\alpha\beta} \restriction \mathscr{P}_\alpha) \circ (j''_{\alpha\gamma} \restriction \mathscr{P}_\gamma) = h_\gamma \circ (j''_{\alpha\gamma} \restriction \mathscr{P}_\alpha) \circ (j''_{\alpha\gamma} \restriction \mathscr{P}_\gamma) = h_\gamma,$$

contradiction. But this gives $\aleph_2$ members of $\mathscr{P}_\alpha(^\omega 3)$, which by CH has cardinality $\aleph_1$, contradiction. Thus (5) holds.

By (5), let X' be an equivalence class with $\aleph_2$ elements. Fix $\alpha, \beta \in X'$ with $\alpha < \beta$. Now $p \Vdash (T_\beta \backslash T_\alpha$ is infinite), so by (2), since dmn$p \subseteq C$,

(6) $p \Vdash ((j^{iv}_{\alpha\beta}T)_\beta \backslash (j^{iv}_{\alpha\beta}T)_\alpha$ is infinite).

Next we claim:

(7) $p \Vdash \forall n \in \omega(n \notin T_\alpha \to n \notin (j^{iv}_{\alpha\beta}T)_\beta)$.

To prove this, suppose that $q \leq p$ and $q \Vdash n \notin T_\alpha$; we want to show that $q \Vdash n \notin (j^{iv}_{\alpha\beta}T)_\beta$. Suppose that this is not true. Then there is an $r \leq q$ such that $r \Vdash n \in (j^{iv}_{\alpha\beta}T)_\beta$. Since $r \leq q$, we also have $r \Vdash n \notin T_\alpha$, so there is a $t \in A_{\alpha n}$ such that r and t are compatible. Thus $t \in \mathscr{P}_\alpha$ and $(h_\alpha t)n = 0$. Let $s = j''_{\alpha\beta}t$. Since $\alpha \equiv \beta$, we have $h_\beta s = h_\beta j''_{\alpha\beta}t = h_\alpha t$. Hence $(h_\beta s)n = 0$, so $s \Vdash n \notin T_\beta$. Hence by (2), $t \Vdash n \notin (j^{iv}_{\alpha\beta}T)_\beta$. Since r and t are compatible and $r \Vdash n \in (j^{iv}_{\alpha\beta}T)_\beta$, this is a contradiction. So, (7) holds.

Similarly:

(8) $p \Vdash \forall n \in \omega(n \in T_\beta \to n \in (j^{iv}_{\alpha\beta}T)_\alpha$.

Next, since $p \Vdash (T_\alpha \backslash T_\beta$ is finite), choose $m \in \omega$ and $q \leq p$ so that

(9) $q \Vdash \forall n \geq m(n \in T_\alpha \to n \in T_\beta)$.

By (6), $q \Vdash \exists n \geq m(n \in (j^{iv}_{\alpha\beta}T)_\beta \wedge n \notin (j^{iv}_{\alpha\beta}T)_\alpha)$, so choose $n \geq m$ and $r \leq q$ so that $r \Vdash n \in (j^{iv}_{\alpha\beta}T)_\beta \wedge n \notin (j^{iv}_{\alpha\beta}T)_\alpha)$. By (7), $r \Vdash n \in T_\alpha$, so by (9), $r \Vdash n \in T_\beta$. Then by (8), $r \Vdash n \in (j^{iv}_{\alpha\beta}T)_\alpha$, contradiction. $\square$

5. Topological density

We begin with some equivalents of this notion. A set X of non-zero elements of a BA A is said to be *centered* provided that it satisfies the finite intersection property. And A is called κ-*centered* if $A\backslash\{0\}$ is the union of κ centered sets.

Theorem 5.1. *For any infinite BA A, dA is equal to each of the following cardinals:*

$\min\{\kappa : A$ *is isomorphic to a subalgebra of* $\mathscr{P}\kappa\}$;
$\min\{\kappa : A$ *is κ-centered*$\}$;
$\min\{\kappa : A\backslash\{0\}$ *is a union of κ proper filters*$\}$;
$\min\{\kappa : A\backslash\{0\}$ *is a union of κ ultrafilters*$\}$.

Proof. Call the five cardinals mentioned $\kappa_0, \ldots \kappa_4$ respectively, starting with dA itself. $\kappa_0 \leq \kappa_1$: Let g be an isomorphism of A into $\mathscr{P}\kappa$. For each $\alpha < \kappa$ let $F_\alpha = \{a \in A : \alpha \in ga\}$. Then, as is easily checked, F_α is an ultrafilter on A. Let $Y = \{F_\alpha : \alpha < \kappa\}$. We claim that Y is dense in Ult$A$. For, let U be a non-empty open set in Ult$A$. We may assume that $U = \mathcal{S}a$ for some $a \in A$. Thus $a \neq 0$, so choose $\alpha \in ga$. Then $a \in F_\alpha$, and so $F_\alpha \in Y \cap U$, as desired.

$\kappa_1 \leq \kappa_2$: Suppose that $A\backslash\{0\} = \bigcup_{\alpha<\lambda} X_\alpha$, where each X_α is centered. Extend each X_α to an ultrafilter F_α. For each $a \in A$ let $fa = \{\alpha < \lambda : a \in F_\alpha\}$. Clearly f is an isomorphism of A into $\mathscr{P}\lambda$, as desired.

Obviously $\kappa_2 \leq \kappa_3 \leq \kappa_4$.

$\kappa_4 \leq \kappa_0$: Let X be a dense subset of Ult$A$. Then obviously $A\backslash\{0\} = \bigcup_{F\in X} F$, as desired. $\qquad\square$

We begin the discussion of algebraic operations for d. If A is a subalgebra of B, then d$A \leq$ dB, and the difference can be arbitrarily large. If A is a homomorphic image of B, then d can change either direction in going from B to A. Thus if B is is a large free BA and A is a countable homomorphic image of B, then d goes down. On the other hand, if $B = \mathscr{P}\omega$, and $A = \mathscr{P}\omega/\text{fin}$, then d$B = \omega$ while d$A = 2^\omega$, since in A there is a disjoint set of size 2^ω. Next, d$(A \times B) = \max($d$A,$d$B)$ for infinite BAs A, B. To see this, note that $\geq$ is clear, since A and B are isomorphic to subalgebras of $A \times B$. For the other inequality, suppose that f (resp. g) is an isomorphism of A (resp. B) into $\mathscr{P}\kappa$ (resp. $\mathscr{P}\lambda$). Let

$$X = \{(0,\alpha) : \alpha < \kappa\} \cup \{(1,\alpha) : \alpha < \lambda\}.$$

We define h mapping $A \times B$ into $\mathscr{P}X$ by setting

$$h(a,b) = \{(0,\alpha) : \alpha \in fa\} \cup \{(1,\alpha) : \alpha \in gb\}$$

for all $(a,b) \in A \times B$. It is easily verified that h is an isomorphism of $A \times B$ into $\mathscr{P}X$, and this proves $\leq$. A similar idea works for products and weak products in general:

Theorem 5.2. *If $\langle A_i : i \in I \rangle$ is a system of non-trivial BAs, then*

$$\mathrm{d}\left(\prod_{i \in I} A_i\right) = \mathrm{d}\left(\prod_{i \in I}^{\mathrm{w}} A_i\right) = \sum_{i \in I} \mathrm{d}A_i.$$

Proof. First we work with the full product, showing that $\sum_{i \in I} \mathrm{d}A_i = \mathrm{d}(\prod_{i \in I} A_i)$. Clearly $\mathrm{d}A_i \leq \mathrm{d}(\prod_{i \in I} A_i)$ for each $i \in I$. Since $\prod_{i \in I} A_i$ has a system of $|I|$ disjoint elements, we also have $|I| \leq \mathrm{d}(\prod_{i \in I} A_i)$. This verifies $\leq$. The direction $\geq$ is proved as in the case of two factors, using the "disjoint union" of all of the algebras. And the argument for weak products is the same. $\qquad\square$

Concerning ultraproducts, we do not know the full story. The following is fairly clear, though. Let $\langle A_i : i \in I \rangle$ be a system of infinite BAs, and F an ultrafilter on I. Then $\mathrm{d}(\prod_{i \in I} A_i/F) \leq |\prod_{i \in I} \mathrm{d}A_i/F|$. To see this, let f_i be an isomorphism of A_i into $\mathscr{P}(\mathrm{d}A_i)$ for each $i \in I$. Then the desired isomorphism g of $\prod_{i \in I} A_i/F$ into $\mathscr{P}(\prod_{i \in I} \mathrm{d}A_i/F)$ is given as follows: for any $x \in \prod_{i \in I} A_i$,

$$g(x/F) = \{y/F : y \in \prod_{i \in I} \mathrm{d}A_i \text{ and } \{i \in I : y_i \in f_i x_i\} \in F\}.$$

(This is easily verified.)

Rosłanowski, Shelah [94] constructed a system of BAs in ZFC such that $\mathrm{d}\left(\prod_{i \in I} B_i/F\right) < \left|\prod_{i \in I} \mathrm{d}B_i/F\right|$; this is a positive solution of Problem 9 of Monk [90].

Now we give some results of Douglas Peterson. We need the following simple fact about essential suprema:

(*) If F is any ultrafilter on a set I and $\langle A_i : i \in I \rangle$ is a system of sets, then $\left|\prod_{i \in I} A_i/F\right| \leq (\mathrm{ess.sup}_{i \in I}^{F} |A_i|)^{|I|}$.

To prove this, say $\mathrm{ess.sup}_{i \in I}^{F} |A_i| = \sup\{|A_i| : i \in a\}$ with $a \in F$. Then

$$\left|\prod_{i \in I} A_i/F\right| = \left|\prod_{i \in a} |A_i|/F\right| \leq \left|\prod_{i \in a} |A_i|\right| \leq (\sup_{i \in I} |A_i|)^{|a|} \leq (\sup_{i \in a} |A_i|)^{|I|}.$$

Theorem 5.3. *Suppose that k is a cardinal function on BAs such that $kA \leq |A| \leq 2^{kA}$ for every infinite BA A. Suppose that $\langle A_i : i \in I \rangle$ is a system of infinite BAs and F is an ultrafilter on I. Set $\lambda = \mathrm{ess.sup}_{i \in I}^{F} kA_i$. Then $k\left(\prod_{i \in I} A_i/F\right) \leq 2^{\lambda \cdot |I|}$.*

Proof. We have

$$k\left(\prod_{i \in I} |A_i/F\right) \leq \left|\prod_{i \in I} A_i/F\right|$$
$$\leq (\mathrm{ess.sup}_{i \in I}^{F} |A_i|)^{|I|}$$
$$\leq (\mathrm{ess.sup}_{i \in I}^{F} (2^{kA_i}))^{|I|}$$
$$\leq (2^{\lambda})^{|I|}.$$

The last inequality holds since if $\lambda = \sup_{i \in a} kA_i$ with $a \in F$, then $2^{kA_i} \leq 2^{\sup_{i \in a} kA_i}$ for each $i \in a$, and hence

$$\text{ess.sup}_{i \in I}^{F}(2^{kA_i}) \leq \sup_{i \in a}(2^{kA_i}) \leq 2^{\sup_{i \in a} kA_i}. \qquad \square$$

Corollary 5.4. *If F is a regular ultrafilter on an infinite set I and $\text{ess.sup}_{i \in I}^{F} dA_i \leq |I|$, then $d\left(\prod_{i \in I} A_i/F\right) = 2^{|I|}$.*

Proof. Using Theorems 3.17 and 5.3 we have

$$2^{|I|} = \left(\text{ess.sup}_{i \in I}^{F} cA\right)^{|I|} = \left|\prod_{i \in I} cA_i/F\right| \leq c\left(\prod_{i \in I} A_i/F\right) \leq d\left(\prod_{i \in I} A_i/F\right) \leq 2^{|I|}.$$

$$\square$$

Clearly $d(A \oplus B) = \max(dA, dB)$: if f is an isomorphism of A into $\mathscr{P}\kappa$ and g is an isomorphism of B into $\mathscr{P}\lambda$, then the following function clearly extends to an isomorphism of $A \oplus B$ into $\mathscr{P}(\kappa \times \lambda)$: for $a \in A$ and $b \in B$, $ha = fa \times \lambda$ and $hb = \kappa \times gb$. For free products of several algebras there is a much more general topological result. To prove it, we need the following lemma.

Lemma 5.5. *Let κ be an infinite cardinal. Then the product space $^{\kappa}2\kappa$ has density $\leq \kappa$ (where κ has the discrete topology).*

Proof. Let $D = \{f \in {}^{\kappa}2\kappa :$ there is a finite subset M of κ such that for all $x, y \in {}^{\kappa}2$, if $x \restriction M = y \restriction M$, then $fx = fy\}$. We show that $|D| \leq \kappa$. First,

$$D = \bigcup_{M \in [\kappa]^{<\omega}} \{f \in {}^{\kappa}2\kappa : \text{ for all } x, y \in {}^{\kappa}2(\text{ if } x \restriction M = y \restriction M, \text{ then } fx = fy\}.$$

So, it suffices to take any finite $M \subseteq \kappa$ and show that $N \stackrel{\text{def}}{=} \{f \in {}^{\kappa}2\kappa :$ for all $x, y \in {}^{\kappa}2$, if $x \restriction M = y \restriction M$ then $fx = fy\}$ has power at most κ. For any $f \in N$, let $f' \in {}^{M}2\kappa$ be defined as follows: for any $x \in {}^{M}2$, choose any $y \in {}^{\kappa}2$ such that $x \subseteq y$ and let $f'x = fy$. Clearly the assignment $f \mapsto f'$ is one-one. So $|N| \leq \kappa$, as desired.

To show that D is dense in $^{\kappa}2\kappa$, let U be an open set in $^{\kappa}2\kappa$. We may assume that U has a very special form, namely that there is a finite subset F of $^{\kappa}2$ and a function g mapping F into κ such that

$$U = \{f \in {}^{\kappa}2\kappa : g \subseteq f\}.$$

Now let G be a finite subset of κ such that $f \restriction G \neq h \restriction G$ for distinct $f, h \in F$. Define $k \in {}^{\kappa}2\kappa$ in the following way: for any $x \in {}^{\kappa}2$, set $kx = gf$ if $x \restriction G = f \restriction G$ for some $f \in F$, otherwise let kx be 0. Clearly $k \in D \cap U$, as desired. $\qquad \square$

Theorem 5.6. *Let $\langle X_i : i \in I \rangle$ be a system of topological spaces each having at least two disjoint non-empty open sets. Then $\mathrm{d}(\prod_{i \in I} X_i) = \max(\lambda, \sup_{i \in I} \mathrm{d} X_i)$, where λ is the least cardinal such that $|I| \leq 2^\lambda$.*

Proof. Clearly $\mathrm{d} X_i \leq \mathrm{d}(\prod_{i \in I} X_i)$ for each $i \in I$. Suppose that D is dense in $\prod_{i \in I} X_i$ but $2^{|D|} < |I|$. Let U_i^0 and U_i^1 disjoint non-empty open sets in X_i for all $i \in I$. For each $i \in I$ let

$$V_i = \{x \in \prod_{i \in I} X_i : x_i \in U_i^0\}.$$

Then our supposition implies that there are distinct $i, j \in I$ such that $V_i \cap D = V_j \cap D$. Let $W = \{x : x_i \in U_i^0 \text{ and } x_j \in U_j^1\}$. Choose $x \in W \cap D$. Then $x \in V_i$ but $x \notin V_j$, contradiction.

Up to this point we have proved the inequality $\geq$. Now for each $i \in I$, let D_i be dense in X_i with $|D_i| = \mathrm{d} X_i$. Set $\kappa = \max(\lambda, \sup_{i \in I} |D_i|)$. Then for each $i \in I$ there is a function f_i mapping κ onto D_i. Since $|I| \leq 2^\lambda$, we then get a continuous function from ${}^\kappa 2 \kappa$ onto $\prod_{i \in I} D_i$. Namely, let g be a one-one function from I into ${}^\kappa 2$. For each $x \in {}^\kappa 2 \kappa$ and each $i \in I$ let $(hx)_i = f_i x_{gi}$. Then h is the desired continuous function. To see that h is continuous, let U be basic open in $\prod_{i \in I} D_i$. Then there is a finite $F \subseteq I$ such that $\mathrm{pr}_i[U]$ is open in D_i for all $i \in F$ and $\mathrm{pr}_i[U] = D_i$ for all $i \in I \backslash F$. Let $L = \{l \in {}^{g[F]}\kappa : \forall i \in F(f_i l_{gi} \in \mathrm{pr}_i U)\}$. For each $l \in L$ the set $W_l \stackrel{\text{def}}{=} \{k \in {}^\kappa 2 \kappa : l \subseteq k\}$ is open in ${}^\kappa 2 \kappa$, and $h^{-1}[U] = \bigcup_{l \in L} W_l$. So h is continuous. Clearly h is maps onto $\prod_{i \in I} D_i$.

Now Lemma 5.5 yields the desired result.   □

The second part of the following corollary was observed by Sabine Koppelberg.

Corollary 5.7. *Let A be a free BA on κ free generators. Then $\mathrm{d} A$ is the smallest cardinal λ such that $\kappa \leq 2^\lambda$. More generally, if B is an infinite subalgebra of A, then $\mathrm{d} B$ is the least cardinal μ such that $|B| \leq 2^\mu$.*

Proof. For each infinite cardinal ν let $\log_2 \nu$ be the least cardinal μ such that $\nu \leq 2^\mu$. For A itself the corollary is true directly by Theorem 5.6. Now let B be an infinite subalgebra of A. Note that B is a subalgebra of a subalgebra of A generated by $|B|$ free generators, and so $\mathrm{d} B \leq \log_2 |B|$. If $|B| = \omega$, the desired conclusion is obvious. If $\omega < |B|$ and $|B|$ is regular, the conclusion follows from Theorem 9.16 of the BA handbook. Finally, suppose that $|B|$ is a singular cardinal. Then for each regular $\nu < |B|$ we have $\log_2 \nu \leq \mathrm{d} B$ by Theorem 9.16. Since clearly $\log_2 |B| = \sup_{\nu < |B|} \log_2 \nu$, this case now follows too.   □

Next we treat the topological density of the union of a well-ordered chain.

Proposition 5.8. *Let $\langle B_\alpha : \alpha < \kappa \rangle$ be a strictly increasing sequence of BAs with union A. Then:*

(i) $\sup_{\alpha < \kappa} \mathrm{d} B_\alpha \leq \mathrm{d} A \leq \sum_{\alpha < \kappa} \mathrm{d} B_\alpha \leq \kappa \cdot \sup_{\alpha < \kappa} \mathrm{d} B_\alpha \leq (2^{\sup_{\alpha < \kappa} \mathrm{d} B_\alpha})^+$.
(ii) $\kappa \leq |A| \leq 2^{\mathrm{d} A}$.

Proof. Clearly $\sup_{\alpha<\kappa} \mathrm{d}B_\alpha \leq \mathrm{d}A$. Now for each $\alpha < \kappa$ let X_α be a set of ultrafilters on A such that $|X_\alpha| = \mathrm{d}B_\alpha$ and $\{F \cap B_\alpha : F \in X_\alpha\}$ is dense in $\mathrm{Ult}B_\alpha$. So

$$\mathrm{d}A \leq \left| \bigcup_{\alpha<\kappa} X_\alpha \right| \leq \sum_{\alpha<\kappa} \mathrm{d}B_\alpha \leq \kappa \cdot \sup_{\alpha<\kappa} \mathrm{d}B_\alpha.$$

Since $|B_\beta| \leq 2^{\mathrm{d}B_\beta} \leq 2^{\sup_{\alpha<\kappa} \mathrm{d}B_\alpha}$ for each $\beta < \kappa$, we must have $\kappa \leq (2^{\sup_{\alpha<\kappa} \mathrm{d}B_\alpha})^+$, since otherwise $(2^{\sup_{\alpha<\kappa} \mathrm{d}B_\alpha})^+ \leq |B_\beta| \leq 2^{\sup_{\alpha<\kappa} \mathrm{d}B_\alpha}$ with $\beta = (2^{\sup_{\alpha<\kappa} \mathrm{d}B_\alpha})^+$, contradiction. Since $\langle B_\alpha : \alpha < \kappa \rangle$ is strictly increasing, clearly $\kappa \leq |A|$. Obviously $|A| \leq 2^{\mathrm{d}A}$. $\qquad\square$

Corollary 5.9. *Let $\langle B_\alpha : \alpha < \kappa \rangle$ be a strictly increasing sequence of BAs with union A. Suppose that $\sup_{\alpha<\kappa} \mathrm{d}B_\alpha < \mathrm{d}A$. Then $\mathrm{d}A \leq \kappa$.* $\qquad\square$

A connection between cellularity and topological density is given in Shelah [80]; we can use it to get another result about unions. Shelah's result is as follows: *If $\lambda = \lambda^{<\kappa}$, B satisfies the κ-cc, $|B| = \lambda^+$, and κ is regular and uncountable, then $\mathrm{d}B \leq \lambda$.*

Corollary 5.10. *Let $\langle B_\alpha : \alpha < \kappa \rangle$ be a strictly increasing sequence of BAs whose union is A.*

Suppose that $\mathrm{d}B_\alpha \leq \mu$ for all $\alpha < \kappa$, $\nu^\mu = \nu$, $\kappa = \nu^+$, and $2^\mu = \mu^+$. Then A satisfies the μ^+-cc, $|A| = \nu^+$, and $\mathrm{d}A \leq \nu$.

In particular, if $\mathrm{d}B_\alpha = \omega$ for all $\alpha < \omega_2$ and CH holds, then A is ccc, $|A| = \omega_2$, and $\mathrm{d}A \leq \omega_1$.

Proof. Let $\alpha < \kappa$. Since $2^\mu = \mu^+$ and $\mathrm{d}B_\alpha \leq \mu$ we have $|B_\alpha| \leq \mu^+$. Also, since $\nu^\mu = \nu$ we have $\mu^+ \leq \nu$. Now $\nu^{<\mu^+} = \nu^\mu = \nu$, so by Shelah's theorem, $\mathrm{d}A \leq \nu$. $\qquad\square$

Also note the following example. Assume GCH, and let B be a free BA on free generators $\{x_\alpha : \alpha < \omega_2\}$, and for each $\alpha < \omega_2$ let A_α be the subalgebra of B generated by $\{x_\xi : \xi < \alpha\}$. Then $\mathrm{d}B = \omega_1$, while $\mathrm{d}A_\alpha = \omega$ for all $\alpha < \omega_2$.

We turn to derived operations for topological density. We shall show that $\mathrm{d}_{H+} = \mathrm{hd}$, but to do this we need two results about tightness and spread which are corollaries of Theorems 4.20 and 3.25.

Theorem 5.11. (Shapirovskiĭ) $\mathrm{t}A \leq \mathrm{s}A$ *for any infinite BA A.*

Proof. By Theorem 4.20 it suffices to note that if $\langle F_\xi : \xi < \alpha \rangle$ is a free sequence, then $\langle F_\xi : \xi < \alpha \rangle$ is one-one and $\{F_\xi : \xi < \alpha\}$ is discrete. Let $\xi < \alpha$. There exist clopen sets Sa, Sb such that $\{F_\eta : \eta < \xi\} \cap Sa = 0$, $\{F_\eta : \xi \leq \eta < \alpha\} \subseteq Sa$, $\{F_\eta : \eta < \xi + 1 < \alpha\} \subseteq Sb$, and $\{F_\eta : \xi + 1 \leq \eta < \alpha\} \cap Sb = 0$. Clearly then $S(a \cdot b) \cap \{F_\eta : \eta < \alpha\} = \{F_\xi\}$, as desired. $\qquad\square$

Theorem 5.12. $\mathrm{s}A \leq \mathrm{d}_{H+}A$ *for any infinite BA A.*

Proof. Obviously $cB \leq dB$ for any infinite BA B. Hence by Theorem 3.25, $sA = c_{H+}A \leq d_{H+}A$. $\qquad\square$

Theorem 5.13. $d_{H+}A = d_{h+}A = hdA$ *for any infinite BA* A.

Proof. $d_{h+}A = hdA$ by definition, and $d_{H+}A \leq d_{h+}A$ since homomorphic images correspond to closed sets in $\text{Ult}A$. So it remains to show that $hdA \leq d_{H+}A$. Let $\kappa = d_{H+}A$, and suppose that $\kappa < hdA$. Choose $Y \subseteq \text{Ult}A$ such that $\kappa < dY$. Let Z be a dense subset of $\overline{Y}$ of size $\leq \kappa$. For each $z \in Z$ we have $z \in \overline{Y}$, and so $z \in \overline{W_z}$ for some $W_z \in [Y]^{\leq \kappa}$ by Theorems 5.11 and 5.12. We claim now that $\bigcup_{z \in Z} W_z$ is dense in Y; since $\left| \bigcup_{z \in Z} W_z \right| \leq \kappa$, this will be a contradiction. Let U be an open set in $\text{Ult}A$ such that $U \cap Y \neq 0$. Then $U \cap \overline{Y} \neq 0$, so choose $z \in U \cap Z$. Since $z \in \overline{W_z}$, we get $U \cap W_z \neq 0$, as desired. $\qquad\square$

Notice that attainment in the d_{H+} sense obviously implies attainment in the hd sense.

We have $d_{H-}A = d_{h-}A = \omega$ for infinite A, since by Sikorski's extension theorem there is a homomorphism of A onto an infinite subalgebra of $\mathscr{P}\omega$. Clearly $d_{S+}A = dA$, $d_{S-}A = \omega$, and $_d d_{S+}A = dA$ for any infinite BA A. Furthermore, $_d d_{S-}A = dA$: if B is a dense subalgebra of A and f is an isomorphism of B into $\mathscr{P}\kappa$, then f can be extended to an isomorphism of A into $\mathscr{P}\kappa$, as desired.

Concerning the spectrum function d_{Hs} we mention the following problem, Problem 10 in Monk [90].

Problem 18. *Is it true that* $[\omega, hdA) \subseteq d_{Hs}A$ *for every infinite BA* A?

Problem 19. *Completely describe* d_{Hs}.

Concerning d_{Ss} we have the following theorem and example, due to S. Koppelberg, solving Problem 11 in Monk [90].

Theorem 5.14. (GCH) $d_{Ss}A = [\omega, dA]$ *for every infinite BA* A.

Proof. Suppose that $\omega \leq \kappa < dA$; we want to find a subalgebra B of A such that $dB = \kappa$. If κ is a limit cardinal, then any subset of A of size κ will do, by GCH. So we may assume that κ is a successor cardinal, and hence is regular. If A has a disjoint subset of size κ, then the subalgebra generated by such a subset is isomorphic to $\text{Finco}\,\kappa$, which has topological density κ. So we may assume that A satisfies the κ-cc. Now $\mu^{<\kappa} < \kappa^+$ for every $\mu < \kappa^+$ by GCH. Hence by Theorem 10.1 of the BA Handbook, Part I, A has a free subalgebra B of size κ^+. By GCH, $dB = \kappa$, as desired. $\qquad\square$

The equality in Theorem 5.14 cannot be proved in ZFC. Namely, if for example $2^\omega = 2^{\omega_1} = \omega_2$ and $2^{\omega_2} = \omega_4$, then for A the free BA on ω_4 free generators we have $d_{Ss}A = \{\omega, \omega_2\}$ by Corollary 5.7.

From Theorem 5.1 the inequality $cA \leq dA$ for every infinite BA A is obvious. The difference between cA and dA can be arbitrarily large, for example in free BAs.

The equivalent definition of d using the notion of κ-centered set gives rise to bounded notions of d (see the introduction). A subset X of a BA A is said to have the *n-intersection property* (n a positive integer) if the product of at most n elements of X is always nonzero. And the *ω-intersection property* is the f.i.p. We set

$$d_n A = \sup\{|X| : X \subseteq A \text{ satisfies the } n\text{-intersection property}\}.$$

These notions were used in Roslanowski, Shelah [94] to give the example mentioned above concerning ultraproducts. The following proposition summarizes some easy facts.

Proposition 5.15. (i) If $1 \leq n < m$ and X has the m-intersection property, then X has the n-intersection property.

(ii) X has the finite intersection property iff it has the m-intersection property for every positive integer m.

(iii) If for each positive integer n the set X_n has the n-intersection property, then $\bigcap_{1 \leq n < \omega} X_n$ has the f.i.p.

(iv) If $n < m$, then $d_n B \leq d_m B$.

(v) $d_n B \leq dB$.

(vi) $dB \leq \prod_{1 \leq n < \omega} d_n B$.

Proof. Everything is trivial except (vi). To prove it, for each positive integer n write $B \setminus \{0\} = \bigcup_{i < d_n B} X_i^n$, where each X_i^n has the n-intersection property. For each $f \in \prod_{1 \leq n < \omega} d_n B$ define

$$Y_f = \bigcap_{1 \leq n < \omega} X_{fn}^n.$$

Then by (iii), each set Y_f has the f.i.p. Clearly $B = \bigcup_{f \in \prod_{1 \leq n < \omega} d_n B} Y_f$, so (v) follows. $\square$

Turning to topological density for special classes of BAs, note first that if A is atomic, then dA is the number of atoms of A. Hence if A is the finite-cofinite algebra on κ, then $d_{Sr} A = \{(\lambda, \lambda) : \omega \leq \lambda \leq \kappa\} = d_{Hr} A$. For interval algebras, we have one interesting inequality not true for BAs in general. It is actually true for linearly ordered spaces in general, and we give that general form, due to Kurepa [35]. This result has evidently been rediscovered by many people independently; see, e.g., Juhász [71].

Theorem 5.16. If L is an infinite linearly ordered space, then $dL \leq (cL)^+$.

Proof. Assume the contrary. Set $\kappa = (cL)^+$. Let $\prec$ be a well-ordering of L. Now we set

$$N = \{p \in L : p \text{ is the } \prec\text{-least element of some neighborhood of } p\}.$$

Clearly N is dense in L. Hence $|N| > \kappa$. Now for each $p \in N$ let I_p be the union of all open intervals having p as their $\prec$-first element. Then, we claim,

(1) If $p, p' \in N$ and $p \prec p'$, then $I_p \cap I_{p'} = 0$ or $I_{p'} \subset I_p$.

In fact, suppose $p \prec p'$ and $I_p \cap I_{p'} \neq 0$. This means that there exist an open interval U with $\prec$-first element p and an open interval U' with $\prec$-first element p' such that $U \cap U' \neq 0$; hence $U \cup U'$ is an open interval with both p and p' as members, and with p as $\prec$-first member. So, if V is any open interval with $\prec$-first element p', then $V \cup U \cup U'$ is an open interval with $\prec$-first element p, and hence $V \subseteq I_p$. This shows that $I_{p'} \subseteq I_p$. Since $p \in I_p \backslash I_{p'}$, (1) then follows.

Next, set

$$N_0 = \{p \in N : I_p \text{ is not contained in any other } I_{p'}\}.$$

Now $I_p \cap I_{p'} = 0$ for all distinct $p, p' \in N_0$, so $|N_0| < \kappa$. We continue inductively for all $\xi < \kappa$:

$$H_\xi = N \backslash \bigcup_{\eta < \xi} N_\eta;$$

$$N_\xi = \{p \in H_\xi : I_p \text{ is not contained in any other } I_{p'} \text{ for } p' \in H_\xi\}.$$

Note inductively that $|N_\xi| < \kappa$, and hence always $H_\xi \neq 0$. Hence $|\bigcup_{\xi < \kappa} N_\xi| \leq \kappa$, so there is a $p \in N \backslash \bigcup_{\xi < \kappa} N_\xi$. Thus $p \in H_\xi$ for all $\xi < \kappa$. But then for each $\xi < \kappa$ there is a $p(\xi) \in N_\xi$ such that $I_p \subset I_{p(\xi)}$. In fact, there is a $q \in H_\xi$ such that $I_p \subset I_q$. Taking the smallest such q under $\prec$, we get the desired $p(\xi)$. Hence for all $\xi, \eta < \kappa$ we have $I_{p(\xi)} \subset I_{p(\eta)}$ or $I_{p(\eta)} \subset I_{p(\xi)}$. By the partition relation $\kappa \to (\kappa, \omega)^2$ we may assume that $p(\xi) \prec p(\eta)$ whenever $\xi < \eta < \kappa$, and hence the sequence $\langle I_{p(\xi)} : \xi < \kappa \rangle$ is strictly decreasing. For each $\xi < \kappa$ choose $x_\xi \in I_{p(\xi)} \backslash I_{p(\xi+1)}$. Let

$$K^l = \{x_\xi : x_\xi < p(\xi+1)\}, \quad K^r = \{x_\xi : x_\xi > p(\xi+1)\}.$$

Now if $\xi < \eta$ and $x_\eta, x_\xi \in K^l$, then $x_\xi < x_\eta$: otherwise, note that x_ξ is less than all members of $I_{p(\xi+1)}$; so $x_\eta \leq x_\xi < p(\eta+1)$ and $x_\eta, p(\eta+1) \in I_{p(\eta)}$, so $x_\xi \in I_{p(\eta)}$, contradiction. Similarly, if $\xi < \eta$ and $x_\eta, x_\xi \in K^r$, then $x_\xi > x_\eta$. But this means that there are κ disjoint open intervals, contradiction. □

The interval algebra of a Suslin line gives an example of an interval algebra A in which $cA < dA$; on the other hand, Martin's axiom implies that for an interval algebra, $cA = \omega \Rightarrow dA = \omega$ (see any set theory book). In general, the existence of an interval algebra A such that $cA < dA$ is connected with the generalized Suslin problem.

Since interval algebras are retractive, it follows that if B is a homomorphic image of an interval algebra A, then $dB \leq dA$. Hence $dA = \text{hd}A$ for every interval algebra A.

For a minimally generated BA A we also have $dA \leq (cA)^+$. In fact, by Corollary 2.36 A is co-absolute with an interval algebra B. Hence

$$dA = d\overline{A} = d\overline{B} = dB \leq (cB)^+ = (cA)^+.$$

An example of a complete BA A for which $cA < dA$ can be obtained by taking A to be the completion of a large free BA.

Theorem 5.17. *For an infinite tree T we have $\pi(\text{Treealg } T) = d(\text{Treealg } T) = |T|$.*

Proof. Let $A = \text{Treealg } T$. Clearly $\pi B \leq dB$ for any BA B. Suppose that $\pi A < |T|$. Since clearly $cA \leq \pi A$, each level of T has at most πA elements. Hence T has $|T|$ levels. It also has height $|T|$, since an element of level $|T|$ would give a chain of order type $|T|$ and hence $|T|$ disjoint elements. If $|T|$ is singular, by considering chains in T we easily get $cA = |T|$, contradiction. Suppose that $|T|$ is regular. Let D be a dense subset of A of cardinality πA. We may assume that each element $d \in D$ has the form $(T \uparrow t_d) \setminus \bigcup_{s \in S_d} (T \uparrow s)$. Let u be an element of T at a level greater than the levels of all elements t_d for $d \in D$. Clearly no element of D is $\leq T \uparrow u$, contradiction. $\qquad\square$

6. π-weight

If A is a subalgebra of B, then πA can vary either way from πB; for clearly one can have $\pi A < \pi B$, and if we take $B = \mathscr{P}\omega$ and A the subalgebra of B generated by an independent subset of size 2^ω, then we have $\pi B = \omega$ and $\pi A = 2^\omega$. Similarly, if A is a homomorphic image of B: it is easy to get such A and B with $\pi A < \pi B$, and if we take $B = \mathscr{P}\omega$ and $A = B/\mathrm{Fin}$, then $\pi B = \omega$ while $\pi A = 2^\omega$ since A has a disjoint subset of size 2^ω. Turning to products, we have $\pi(\prod_{i\in I} A_i) = \max(|I|, \sup_{i\in I}\pi A_i)$ for any system $\langle A_i : i \in I\rangle$ of infinite BAs. For, $\geq$ is clear; now suppose D_i is a dense subset of A_i for each $i \in I$. Let

$$E = \left\{ f \in \prod_{i\in I}(D_i \cup \{0\}) : fi \neq 0 \text{ for only finitely many } i \in I \right\}.$$

Clearly E is dense in $\prod_{i\in I} A_i$, and $|E| = \max(|I|, \sup_{i\in I}\pi A_i)$, as desired. The equation $\pi(\prod^w_{i\in I} A_i) = \max(|I|, \sup_{i\in I}\pi A_i)$ is proved by the same argument.

Turning to ultraproducts, it is clear that $\pi(\prod_{i\in I} A_i/F) \leq |\prod_{i\in I}\pi A_i/F|$. In Koppelberg, Shelah [93] there is a forcing construction in which $<$ holds; this answers Problem 12 of Monk [90]. Several results about ultraproducts and π hold more generally for the sup-min functions defined in the introduction. These results are due to Douglas Peterson.

Theorem 6.1. *Let k be a sup-min function, $\langle A_i : i \in I\rangle$ a sequence of infinite BAs with I infinite, and F a regular ultrafilter on I. Suppose that $kA_i \geq \omega$ for all $i \in I$, $\lambda = \mathrm{ess.sup}^F_{i\in I} kA_i$, and $\mathrm{cf}\lambda \leq |I| < \lambda$. Then $k\left(\prod_{i\in I} A_i/F\right) \geq \lambda^+$.*

Proof. For brevity let $B = \prod_{i\in I} A_i/F$. *Case 1.* $\{i \in I : kA_i = \lambda\} \in F$. Then we may assume that $kA_i = \lambda$ for all $i \in I$. Now by Lemma 3.12 let $\langle \kappa_i : i \in I\rangle$ be a system of cardinals such that $\kappa_i < \lambda$ for all $i \in I$ and $\mathrm{ess.sup}^F_{i\in I}\kappa_i = \lambda$. Now fix $i \in I$. Since $kA_i = \lambda$, we can find $G_i \subseteq A_i$ such that $(A_i, G_i) \models \psi$ and $\min\{|P| : (A_i, G_i, P) \models \varphi\} > \kappa_i$. Let $H = \prod_{i\in I} G_i/F$. Thus $(B, H) \models \psi$. We claim that $\min\{|P| : (B, H, P) \models \varphi\} \geq \lambda^+$; this will prove the theorem. To prove the claim, suppose that $P = \{f_\alpha : \alpha < \lambda\} \subseteq B$ and $(B, H, P) \models \forall x \in \mathbf{P}(x \neq 0 \wedge \varphi')$; we shall show

$(*)$ $(B, H, P) \models \neg\forall x_0 \ldots x_{n-1} \in \mathbf{F} \exists y \in \mathbf{P}\varphi''$.

We may assume that $f_\alpha i \neq 0$ for all $\alpha < \lambda$ and $i \in I$. Now for any $\alpha < \lambda$ we have $(B, H) \models \varphi'[f_\alpha/F]$, and hence $\{i \in I : (A_i, G_i) \models \varphi'[f_\alpha i]\} \in F$. Hence we can assume that $(A_i, G_i) \models \varphi'[f_\alpha i]$ for all $\alpha < \lambda$ and $i \in I$. (If $(A_i, G_i) \not\models \varphi'[f_\alpha i]$, replace $f_\alpha i$ by a nonzero element a_i such that $(A_i, G_i) \models \varphi'[a_i]$; a_i exists by (3) of the definition of sup-min function.) Now fix $i \in I$ again. Then

$$(A_i, G_i, \{f_\alpha i : \alpha < \kappa_i\}) \models \forall x \in \mathbf{P}(x \neq 0 \wedge \varphi'),$$

so, since $(A_i, G_i, \{f_\alpha i : \alpha < \kappa_i\}) \not\models \varphi$, we can choose $a^i_0, \ldots, a^i_{n-1} \in G_i$ such that $(A_i, G_i) \models \neg\varphi''[a^i_0, \ldots, a^i_{n-1}, f_\alpha i]$ for all $\alpha < \kappa_i$. Now for any $\alpha < \lambda$ we have

$\{i \in I : \alpha < \kappa_i\} \in F$, and hence $(B, H) \models \neg\varphi''[a_0/F, \dots, a_{n-1}/F, f_\alpha/F]$, and this proves (*).

    Case 2. $\{i \in I : kA_i < \lambda\} \in F$. We may assume that $\omega < kA_i < \lambda$ for all $i \in I$. Then we can apply Lemma 3.13 to get a system $\langle \kappa_i : i \in I \rangle$ of infinite cardinals such that $\kappa_i < kA_i$ for all $i \in I$, and $\mathrm{ess.sup}\,^F_{i\in I}\kappa_i = \lambda$. Now we can proceed as in Case 1.   □

Theorem 6.2. *Suppose that k is a sup-min function, $\langle A_i : i \in I \rangle$ is a system of infinite BAs, with I infinite, and F is an ultrafilter on I. Then $k\left(\prod_{i\in I} A_i\right) \geq \mathrm{ess.sup}\,^F_{i\in I}kA_i$.*

Proof. Let $\lambda = \mathrm{ess.sup}\,^F_{i\in I}kA_i$ and $B = \prod_{i\in I} A_i/F$. Take any $\kappa < \lambda$. Then the set $K \overset{\mathrm{def}}{=} \{i \in I : kA_i > \kappa\} \in F$. For each $i \in K$ choose $G_i \subseteq A_i$ such that $(A_i, G_i) \models \psi$ and $\min\{|P| : (A_i, G_i, P) \models \varphi\} > \kappa$. For $i \in I\backslash K$ let $G_i = A_i$. Set $H = \prod_{i\in I} G_i/F$. Then $(B, H) \models \psi$. We claim that $\min\{|P| : (B, H, P) \models \varphi\} > \kappa$; this will prove the theorem. Suppose that $P = \{f_\alpha/F : \alpha < \kappa\} \subseteq B$ and $(B, H, P) \models \forall x \in \mathbf{P}(x \neq 0 \wedge \varphi')$. As in the proof of Theorem 6.1 we can assume that $(A_i, G_i) \models \varphi'[f_\alpha i]$ and $f_\alpha i \neq 0$ for all $\alpha < \lambda$ and $i \in I$. Fix $i \in K$. Then

$$(A_i, G_i, \{f_\alpha i : \alpha < \kappa\}) \models \forall x \in \mathbf{P}\,(x \neq 0 \wedge \varphi'),$$

so, since $(A_i, G_i, \{f_\alpha i : \alpha < \kappa\}) \not\models \varphi$, we can choose $a^i_0, \dots, a^i_{n-1} \in G_i$ such that $(A_i, G_i) \models \neg\varphi''[a^i_0, \dots, a^i_{n-1}, f_\alpha i]$ for all $\alpha < \kappa$. Then it follows that $(B, H) \models \neg\varphi''[a_0/F, \dots, a_{n-1}/F, f_\alpha/F]$, as desired.   □

Theorem 6.3. *Suppose that k is a sup-min function such that for every infinite BA A there is a $G \subseteq A$ such that $(A, G) \models \psi$ and $\min\{|P| : (A, G, P) \models \varphi\} \geq \omega$. Assume that $\langle A_i : i \in I \rangle$ is a system of infinite BAs, with I infinite, and F is a regular ultrafilter on I. Then $k\left(\prod_{i\in I} A_i/F\right) \geq |I|^+$.*

Proof. Let $B = \prod_{i\in I} A_i/F$. For each $i \in I$ choose $G_i \subseteq F_i$ such that $(A_i, G_i) \models \psi$ and $\min\{|P| : (A_i, G_i, P) \models \varphi\} \geq \omega$. Let $H = \prod_{i\in I} G_i/F$. Thus $(B, H) \models \psi$. We want to show that $\min\{|P| : (B, H, P) \models \varphi\} \geq |I|^+$; this will prove the theorem. To this end, suppose that $P = \{f_\alpha/F : \alpha < |I|\} \subseteq B$ and $(B, H, P) \models \forall x \in \mathbf{P}\,(x \neq 0 \wedge \varphi')$. As in the proof of Theorem 6.1 we may assume that $(A_i, G_i) \models \varphi'[f_\alpha i]$ and $f_\alpha i \neq 0$ for all $\alpha < |I|$ and $i \in I$. Let $\{a_\alpha : \alpha < |I|\}$ be a regular family for F. Now fix $i \in I$. Then the set $K_i \overset{\mathrm{def}}{=} \{\alpha < |I| : i \in a_\alpha\}$ is finite. Since $\min\{|Q| : (A_i, G_i, Q) \models \varphi\} \geq \omega$ and $P_i \overset{\mathrm{def}}{=} \{f_\alpha i : \alpha \in K_i\}$ is finite, it folows that $(A_i, G_i, P_i) \models \neg\varphi$. But $(A_i, G_i, P_i) \models \forall x \in \mathbf{P}\,(x \neq 0 \wedge \varphi')$, so we can choose $a^i_0, \dots, a^i_{n-1} \in G_i$ such that $(A_i, G_i) \models \neg\varphi''[a^i_0, \dots, a^i_{n-1}, f_\alpha i]$ for all $\alpha \in K_i$. Therefore $\{i \in I : (A_i, G_i) \models \neg\varphi''[a^i_0, \dots, a^i_{n-1}, f_\alpha i]\} \supseteq a_\alpha \in F$, so $(B, H) \models \neg\varphi''[a_0/F, \dots, a_{n-1}/F, f_\alpha/F]$, as desired.   □

Theorem 6.4. *Suppose that k is a sup-min function such that the formula ψ in the definition is $\forall x \mathbf{P} x$ and the formula φ' is $x = x$. Assume that $\langle A_i : i \in I \rangle$ is a system of infinite BAs, with I infinite, and F is an ultrafilter on I. Then $k\left(\prod_{i\in I} A_i/F\right) \leq |\prod_{i\in I} kA_i/F|$.*

Proof. For each $i \in I$ choose $P_i \subseteq A_i$ such that $|P_i| = kA_i$ and $(A_i, A_i, P_i) \models \varphi$. Then, we claim, $\left(\prod_{i \in I} A_i/F, \prod_{i \in I} A_i/F, \prod_{i \in I} P_i/F\right) \models \varphi$, which will prove the theorem. So, suppose that $x_0, \ldots, x_{n-1} \in \prod_{i \in I} A_i$. For each $i \in I$ choose $y_i \in P_i$ such that $(A_i, A_i) \models \varphi''[x_0 i, \ldots, x_{n-1} i, y_i]$. Then

$$\left(\prod_{i \in I} A_i/F, \prod_{i \in I} A_i/F\right) \models \varphi''[x_0/F, \ldots, x_{n-1}/F, y/F] \qquad \square$$

Theorem 6.5. (GCH) *Suppose that k is a sup-min function such that the formula ψ in the definition is $\forall x \mathbf{P} x$ and the formula φ' is $x = x$. We also suppose that for every infinite BA A there is a $G \subseteq A$ such that $(A, G) \models \psi$ and $\min\{|P| : (A, G, P) \models \varphi\} \geq \omega$. If $\langle A_i : i \in I \rangle$ is a system of infinite BAs, with I infinite, and F is a regular ultrafilter on I, then $k\left(\prod_{i \in I} A_i/F\right) = |\prod_{i \in I} kA_i/F|$.*

Proof. Let $\lambda = \operatorname{ess.sup}_{i \in I}^{F} kA_i$. We consider several cases. *Case 1.* $\lambda \leq |I|$. Then, using Theorem 6.4 we have

$$2^{|I|} = |{}^I \omega/F| \leq k\left(\prod_{i \in I} A_i/F\right) \leq \left|\prod_{i \in I} kA_i/F\right| = \lambda^{|I|} = 2^{|I|}.$$

Case 2. $\operatorname{cf}\lambda \leq |I| < \lambda$. Then by Theorem 6.1 we get $\lambda^+ \leq k\left(\prod_{i \in I} A_i/F\right) \leq |\prod_{i \in I} kA_i/F| = \lambda^{|I|} = \lambda^+$.

Case 3. $|I| < \operatorname{cf}\lambda$. Then by Theorem 6.2 we have $\lambda \leq k\left(\prod_{i \in I} A_i/F\right) \leq |\prod_{i \in I} kA_i/F| = \lambda^{|I|} = \lambda$. $\qquad \square$

By a result of Donder [88], $V = L$ implies that every uniform ultrafilter is regular, and hence that the equality in Theorem 6.5 always holds; thus this answers Problem 12 of Monk [90] in a different way from the solution of Koppelberg and Shelah mentioned at the outset.

An easy argument shows that $\pi(\oplus_{i \in I} A_i) = \max(|I|, \sup_{i \in I} \pi A_i)$ for any system $\langle A_i : i \in I \rangle$ of Boolean algebras. In fact, if D_i is dense in A_i for each $i \in I$, then

$$E \overset{\text{def}}{=} \{d_0 \cdot \ldots \cdot d_{n-1} : \exists \text{ distinct } i_0, \ldots, i_{n-1} \in I \text{ such that } \forall j < n(d_j \in D_{i_j})\}$$

is clearly dense in $\oplus_{i \in I} A_i$, and it has the indicated cardinality. On the other hand, suppose X is dense in $\oplus_{i \in I} A_i$. We may assume that each element of X is a product of members of $\bigcup_{i \in I} A_i$, with distinct factors coming from distinct A_i's. For each $i \in I$ let $Y_i = \{x \in X : x \leq a \text{ for some } a \in A_i\}$. For each $x \in Y_i$, let x_i^+ be obtained from x by replacing each factor of x which is not in A_i by 1. Clearly then $\{x_i^+ : x \in Y_i\} \subseteq A_i$ and this set is dense in A_i, so $\pi A_i \leq |\{x_i^+ : x \in Y_i\}| \leq |Y_i| \leq |X|$. It is also clear that $|I| \leq |X|$; so $|X| \geq \max(|I|, \sup_{i \in I} \pi A_i)$, as desired.

We turn to the discussion of unions. The following theorem takes care of some possibilities.

Theorem 6.6. *Suppose that* $\langle A_\alpha : \alpha < \kappa \rangle$ *is a strictly increasing sequence of BAs, with union* B, *where* κ *is regular. Let* $\lambda = \sup_{\alpha < \kappa} \pi A_\alpha$. *Then*
    (i) $\pi B \leq \sum_{\alpha < \kappa} \pi A_\alpha \leq \max(\kappa, \lambda)$.
    Now assume that, in addition, $A_\alpha = \bigcup_{\beta < \alpha} A_\beta$ *for all limit* $\alpha < \kappa$. *Then*
    (ii) $\kappa \leq 2^\lambda$,
    (iii) $\pi B \leq \lambda^+$.

Proof. For *(i)*, if X_α is dense in A_α for each $\alpha < \kappa$, then $\bigcup_{\alpha < \kappa} X_\alpha$ is dense in B. Now we make the additional assumption indicated, and prove *(ii)*. Assume that $\kappa \geq (2^\lambda)^+$. Let $\mu = (2^\lambda)^+$. Since clearly $|A_\alpha| \leq 2^\lambda$ for all $\alpha < \kappa$, we have $\kappa = \mu$. Let $S = \{\alpha < \mu : \text{cf}\alpha = \lambda^+\}$. Thus S is stationary in μ. For each $\alpha \in S$, A_α has a dense subset D_α of size $\leq \lambda$. Since $\text{cf}\alpha = \lambda^+$, it follows that there is an $f\alpha < \alpha$ such that $D_\alpha \subseteq A_{f\alpha}$. Now f is regressive on S, so f is constant on a stationary subset S' of S. Let β be the constant value of f on S'. Then D_β is dense in B. But $|B| \geq (2^\lambda)^+$, contradiction. So, *(ii)* holds. For *(iii)*, if $\pi B > \lambda^+$, then by *(i)* we have $\kappa > \lambda^+$, and we can use an argument similar to that for *(ii)*. $\qquad\square$

Note that the upper bound λ^+ mentioned in Theorem 6.6 can be attained: use a free algebra, as in the discussion of unions for topological density.

Turning to the functions derived from π, we first work toward proving that $\pi_{H+} = \pi_{h+} = \text{hd}$.

We call a sequence $\langle x_\xi : \xi < \kappa \rangle$ of elements of a space X *left-separated* provided that for every $\xi < \kappa$ there is an open set U in X such that $U \cap \{x_\eta : \eta < \kappa\} = \{x_\eta : \xi \leq \eta\}$. We now need the following important fact relating this notion to the function hd:

Theorem 6.7. *For any infinite Hausdorff space* X, $\text{hd}X$ *is the supremum of all cardinals* κ *such that there is a left-separated sequence in* X *of type* κ.

Proof. If $\langle x_\xi : \xi < \kappa \rangle$ is a left-separated sequence in X and κ is infinite and regular, then clearly the density of $\{x_\xi : \xi < \kappa\}$ is κ. Hence the inequality $\geq$ holds. Now suppose that Y is a subspace of X, and set $\text{d}Y = \kappa$. We construct a left-separated sequence $\langle x_\xi : \xi < \kappa \rangle$ as follows: having constructed $x_\eta \in Y$ for all $\eta < \xi$, where $\xi < \kappa$, it follows that $\{x_\eta : \eta < \xi\}$ is not dense in Y, and so we can choose $x_\xi \in Y \backslash \overline{\{x_\eta : \eta < \xi\}}$. This proves the other inequality. $\qquad\square$

Note that the proof of Theorem 6.7 shows that if $\text{hd}X$ is attained, then it is also attained in the left-separated sense, and conversely if $\text{hd}X$ is regular.

There is an algebraic version of left-separated sequences. We call a sequence $\langle a_\xi : \xi < \alpha \rangle$ of elements of A *left-separated* if for all $\xi < \alpha$ and all finite $F \subseteq \alpha$ such that $\xi < \beta$ for all $\beta \in F$ we have $a_\xi \cdot \prod_{\eta \in F} -a_\eta \neq 0$.

Lemma 6.8. A *has a left-separated sequence of length* α *iff* $\text{Ult}A$ *has a left-separated sequence of length* α.

Proof. $\Rightarrow$: Let $\langle a_\xi : \xi < \alpha \rangle$ be a left-separated sequence of elements of A. For each $\xi < \alpha$, let F_ξ be an ultrafilter on A containing the set $\{a_\xi\} \cup \{-a_\eta : \xi < \eta\}$. Clearly $\langle F_\xi : \xi < \alpha \rangle$ is a left-separated sequence in $\text{Ult}A$.

$\Leftarrow$: Let $\langle F_\xi : \xi < \alpha \rangle$ be a left-separated sequence in Ult$A$. For each $\xi < \alpha$ choose an element a_ξ such that $F_\xi \in \mathscr{S} a_\xi$ and $\mathscr{S} a_\xi \cap \{F_\eta : \eta < \alpha\} \subseteq \{F_\eta : \xi \le \eta < \alpha\}$. To check that $\langle a_\xi : \xi < \alpha \rangle$ is a left-separated sequence, suppose on the contrary that $a_\xi \le \sum_{\eta \in F} x_\eta$, where $\xi < \eta$ for all $\eta \in F$, F a finite subset of α. Since $a_\xi \in F_\xi$, it follows that $a_\eta \in F_\xi$ for some $\eta \in F$. This contradicts the choice of a_η. $\qquad\square$

The essential step in proving that $\pi_{H+} = $ hd is as follows; we follow the proof of van Douwen [89], 10.1.

Lemma 6.9. *Let A be an infinite BA. Then there exists a left-separated sequence $\langle x_\xi : \xi < \pi A \rangle$ such that $\{x_\xi : \xi < \pi A\}$ is dense in A.*

Proof. For brevity let $\pi = \pi A$. The major part of the proof consists in proving

(1) There is a sequence $\langle a_\xi : \xi < \pi \rangle$ of non-zero members of A such that $\{a_\xi : \xi < \pi\}$ is dense in A and for each $\eta < \pi$, $|\{\xi < \eta : a_\xi \cdot a_\eta \ne 0\}| < \pi(A \upharpoonright a_\eta)$.

To prove this, call an element $b \in A$ *π-homogeneous* provided that $\pi(A \upharpoonright c) = \pi(A \upharpoonright b)$ for every non-zero $c \le b$. Clearly the collection of all π-homogeneous elements of A is dense in A. Let $\mathcal{A}$ be a maximal disjoint collection of π-homogeneous elements of A. Let $\kappa = |\mathcal{A}|$; then $\kappa \le cA \le \pi A$. For each $b \in \mathcal{A}$ let M_b be a dense subset of $A \upharpoonright b$ of cardinality $\pi(A \upharpoonright b)$. Then $\bigcup\{M_b : b \in \mathcal{A}\}$ is dense in A. Now let $\langle N_b : b \in \mathcal{A}\rangle$ be a partition of π into disjoint subsets of power π. For each $b \in \mathcal{A}$ let f_b be a one-one function from M_b onto a subset of N_b of order type $\pi(A \upharpoonright b)$. Now for each $\xi < \pi$, let

$$a_\xi = \begin{cases} 0, & \text{if } \xi \notin \bigcup_{b \in \mathcal{A}} \mathrm{ran}(f_b); \\ f_b^{-1}\xi, & \text{if } \xi \in \mathrm{ran}(f_b),\ b \in \mathcal{A}. \end{cases}$$

Suppose that $\eta < \pi$ and $a_\eta \ne 0$. Say $\eta \in \mathrm{ran}(f_b)$. Then

$$|\{\xi < \eta : a_\xi \cdot a_\eta \ne 0\}| \le |\{\xi \in \mathrm{ran} f_b : \xi < \eta\}| < \pi(A \upharpoonright b).$$

Thus we have (1), except that some of the a_η's are zero. If we renumerate the non-zero a_η's in increasing order of their indices, we really get (1).

Now we construct a sequence $\langle b_\alpha : \alpha < \pi \rangle$ of non-zero elements of A so that the following conditions hold:

(2_α) $b_\alpha \le a_\alpha$ for all $\alpha < \pi$,

and

(3_α) for all $\xi < \alpha$ and every finite $F \subseteq (\xi, \alpha]$ we have $b_\xi \cdot \prod_{\eta \in F} -b_\eta \ne 0$ for all $\alpha < \pi$.

Assume that $\beta < \pi$, and b_α has been constructed for all $\alpha < \beta$ so that (2_α) and (3_α) hold. Then by (1) and the assumption that (2_α) holds for all $\alpha < \beta$ we see that the set $\Gamma \overset{\text{def}}{=} \{\alpha < \beta : b_\alpha \cdot a_\beta \neq 0\}$ has power $< \pi(A \restriction a_\beta)$. Hence there is a non-zero b_β in A such that $b_\beta \leq a_\beta$ and for all $\varphi \in \Gamma$ and all finite $G \subseteq \Gamma$, if $b_\varphi \cdot \prod_{\gamma \in G} -b_\gamma \neq 0$, then $b_\varphi \cdot \prod_{\gamma \in G} -b_\gamma \not\leq b_\beta$. Thus (2_β) and (3_β) hold, and the construction is complete.

It is clear from (2_α) and (3_α) that $\langle b_\alpha : \alpha < \pi \rangle$ is the desired dense sequence.
□

The following theorem is due to Shapirovskiĭ.

Theorem 6.10. $\pi_{\mathrm{H}+}A = \pi_{\mathrm{h}+}A = \mathrm{hd}A$ *for any infinite BA A.*

Proof. It is obvious that $\pi_{\mathrm{H}+}A \leq \pi_{\mathrm{h}+}A$. Now if $\mathcal{O}$ is a π-base for $Y \subseteq \mathrm{Ult}A$ with $|\mathcal{O}| = \pi Y$, without loss of generality $\mathcal{O}$ has the form $\{Sa \cap Y : a \in \mathcal{A}\}$ for some $\mathcal{A} \subseteq A$. Let $fx = Sx \cap Y$ for any $x \in A$. Then f is a homomorphism onto some algebra B of subsets of Y, and $\mathcal{O}$ is dense in B. This shows that $\pi_{\mathrm{h}+}A \leq \pi_{\mathrm{H}+}A$. It is also trivial that $\mathrm{hd}A \leq \pi_{\mathrm{h}+}A$, since if $S \subseteq \mathrm{Ult}A$ then $\mathrm{d}S \leq \pi S$. It remains just to show that $\pi_{\mathrm{H}+}A \leq \mathrm{hd}A$. Suppose that f is a homomorphism of A onto B, where B is infinite. Apply 6.9 to B to get a system $\langle b_\xi : \xi < \pi B \rangle$ of elements of B such that for any $\xi < \pi B$ and any finite subset G of $(\xi, \pi B)$ we have $b_\xi \cdot \prod_{\eta \in G} -b_\eta \neq 0$. For each $\xi < \pi B$ choose a_ξ so that $fa_\xi = b_\xi$. Clearly $\langle a_\xi : \xi < \pi B \rangle$ is a left-separated sequence of elements of A, which by Theorem 6.7 is as desired.   □

The proof of Theorem 6.10 shows that $\pi_{\mathrm{H}+}$ and $\pi_{\mathrm{h}+}$ have the same attainment properties; also, if $\pi_{\mathrm{H}+}$ is attained, then hd is attained in the left-separated sense. Also, if hd$A$ is attained in the defined sense then it is attained in the $\pi_{\mathrm{h}+}$ sense.

The cardinal function $\pi_{\mathrm{S}+}$ is of some interest, since it does not coincide with any of our standard ones. Obviously $\pi A \leq \pi_{\mathrm{S}+}A$ for any infinite BA A. Moreover, $\pi_{\mathrm{S}+}A \leq \pi_{\mathrm{H}+}A$; this follows from the following fact: for every subalgebra B of A there is a homomorphic image C of A such that $\pi B = \pi C$. To see this, by the Sikorski extension theorem extend the identity function from B into $\overline{B}$ to a homomorphism from A onto a subalgebra C of $\overline{B}$. Since $B \subseteq C \subseteq \overline{B}$, it is clear that $\pi B = \pi C$. Thus we have shown that $\pi A \leq \pi_{\mathrm{S}+}A \leq \pi_{\mathrm{H}+}A$ for any infinite BA A. It is possible to have $\pi A < \pi_{\mathrm{S}+}A$: let $A = \mathscr{P}\kappa$—then $\pi A = \kappa$, while $\pi_{\mathrm{S}+}A = 2^\kappa$, since A has a free subalgebra B of power 2^κ, and clearly $\pi B = 2^\kappa$. It is more difficult to come up with an example of an algebra where the other inequality is proper (this example is due to Monk):

Example 6.11. *There is an infinite BA A such that $\pi_{\mathrm{S}+}A < \pi_{\mathrm{H}+}A$.*

To see this, let B be the interval algebra on the real numbers, and let $A = B \oplus B$. Now, we claim, $\pi_{\mathrm{S}+}A = \omega$, while $\pi_{\mathrm{H}+}A = 2^\omega$. To prove that $\pi_{\mathrm{H}+}A = 2^\omega$, by 5.12, 5.13, and 6.10 it suffices to show that $sA = 2^\omega$. For each real number r let $c_r = b_r \cdot b'_r$, where $b_r = [-\infty, r)$ (as a member of the first factor of $B \oplus B$) and $b'_r = [r, \infty)$ (as a member of the second factor of $B \oplus B$). Note that we have adjoined

$-\infty$ as a member of $\mathbb{R}$ in order to fulfill the requirement for interval algebras that the ordered set in question always has a first element. To show that $\langle c_r : r \in \mathbb{R} \rangle$ is an ideal independent system of elements, suppost that $c_r \leq c_{s_1} + \cdots + c_{s_m}$ with $r \notin \{s_1, \ldots, s_m\}$. Let $\Gamma = \{i : s_i < r\}$ and $\Delta = \{i : r < s_i\}$. Then

$$c_r \cdot -c_{s_1} \cdot \ldots \cdot -c_{s_m} \geq b_r \cdot b_r' \cdot \prod_{i \in \Gamma} -b_{s_i} \cdot \prod_{i \in \Delta} -b_{s_i}' \neq 0,$$

contradiction.

To prove that $\pi_{S+} A = \omega$, we proceed as follows. Let C be any subalgebra of A. We want to show that $\pi C = \omega$. Now for each element c of C we choose a representation of c of the form

$$\sum_{i < m(c)} x_{0ic} \times x_{1ic},$$

where x_{0ic} and x_{1ic} are half-open intervals in B. Let $T = \{(m, r, s) : m \in \omega \backslash \{0\}$ and $r, s \in {}^m \mathbb{Q}\}$. An element (m, r, s) of T is a *frame* for $c \in C$ provided that $m(c) = m$, and $r_i \in x_{0ic}$, $s_i \in x_{1ic}$ for all $i \in m$. For each $(m, r, s) \in T$ let D_{mrs} be the set of all $c \in C$ with frame (m, r, s). Since C is the union of all sets D_{mrs}, it suffices to take an arbitrary $(m, r, s) \in T$ and find a countable subset of C dense in D_{mrs}.

For each $c \in D_{mrs}$ and each $i < m$ write

$$x_{0ic} = [a_{ic}, b_{ic}) \quad \text{and} \quad x_{1ic} = [d_{ic}, e_{ic}).$$

Thus $a_{ic} \leq r_i < b_{ic}$ and $d_{ic} \leq s_i < e_{ic}$ for all $c \in D_{mrs}$ and $i < m$. For each $i < m$ let N_{0i} be a countable subset of $\{a_{ic} : c \in D_{mrs}\}$ cofinal in that set. Similarly choose N_{1i} coinitial for the b_{ic}'s, N_{2i} cofinal for the d_{ic}'s, and N_{3i} coinitial for the e_{ic}'s. Let M be the set of all products

$$\prod_{i < m, j < 4} u_{ij}$$

with $u \in {}^{m \times 4} \bigcup_{i < m} (N_{0i} \cup N_{1i} \cup N_{2i} \cup N_{3i})$. Clearly all such products are nonzero. We claim that M is dense in D_{mrs}. For, let $c \in D_{mrs}$. For each $i < m$ choose $u_{i0} \in N_{0i}$ such that $a_{ic} \leq u_{i0}$; similarly for u_{ij}, $j = 1, 2, 3$. Then clearly $\prod_{i < m, j < 4} u_{ij} \leq c$, as desired. $\square$

Shelah [92b] showed that it is consistent to have a BA A with $\pi_{S+} A$ not attained; this answers Problem 13 in Monk [90]. But we do not know whether this can be done in ZFC:

Problem 20. *Can one find in ZFC a BA A such that $\pi_{S+} A$ is not attained?*

Clearly $\pi_{S-} A = \pi_{H-} A = \pi_{h-} A = \omega$ for any infinite BA A. Furthermore, $_d \pi_{S+} A = {}_d \pi_{S-} A = \pi A$ for any infinite BA A.

Concerning the function π_{Ss} we mention the following result from Shelah [92b]:

Theorem 6.12. *Let B be an infinite BA, and suppose that θ is an infinite regular cardinal less than $\pi_{\mathrm{S+}}B$. Then there is a subalgebra A of B such that $\pi A = \theta$.*

Proof. We may assume that $\omega < \theta < \pi B$. We define a sequence $\langle A_\alpha : \alpha < \theta \rangle$ of subalgebras of B, each of power less than θ, as follows. Let A_0 be any denumerable subalgebra of B. For α a limit ordinal $< \theta$, let A_α be the union of preceding algebras. If A_α has been defined, then, since it has fewer than πB elements, there is a nonzero element $b \in B$ such that for all $x \in A_\alpha^+$ we have $x \not\leq b$. Let A_α be the subalgebra of B generated by $A_\alpha \cup \{b\}$. Clearly $\bigcup_{\alpha < \theta} A_\alpha$ is as desired. $\square$

Thus $\pi_{\mathrm{Ss}}A$ contains all regular cardinals in the interval $[\omega, \pi_{\mathrm{S+}}A)$. But Shelah also showed in that paper that it is consistent to have a BA A with some of the singular cardinals in that interval not in $\pi_{\mathrm{Ss}}A$; and some special singular cardinals in that interval are always in $\pi_{\mathrm{Ss}}A$. These results answer problem 15 in Monk [90].

The following problem is open (this is Problem 14 in Monk [90]):

Problem 21. *Is it true that for every infinite BA A we have*

$$\pi_{\mathrm{Hs}}A = \begin{cases} [\omega, \mathrm{hd}A], & \text{if } \mathrm{hd}A \text{ is attained,} \\ [\omega, \mathrm{hd}A), & \text{otherwise?} \end{cases}$$

We have already observed that $\mathrm{d} \leq \pi$; the difference is small, though, since $\mathrm{d}A \leq \pi A \leq |A| \leq 2^{\mathrm{d}A}$ for any infinite BA A.

About π for special classes of BAs, note that $\pi A = \mathrm{d}A$ for any interval algebra A; in fact, πA is also equal to $\mathrm{hd}A$. To see this, note that $\mathrm{d}A = \mathrm{hd}A$ for A an interval algebra, since any interval algebra is retractive; then $\pi A = \mathrm{d}A$ by the above inequalities.

Another interesting fact about π and interval algebras was observed by Douglas Peterson: $\pi\,\mathrm{Intalg}\,L = \mathrm{d}L \cdot |M|$, where $\mathrm{d}L$ is the density of L as a topological space and M is the set of atoms of $\mathrm{Intalg}\,L$. To prove $\leq$, let X be dense in L with $|X| = \mathrm{d}L$; we show that $\{[x, y) : x, y \in X, x < y\} \cup M$ is dense in $\mathrm{Intalg}\,L$. Take any nonzero $a \in \mathrm{Intalg}\,L$. Wlog a has the form $[u, v)$. If $[u, v)$ is finite, then $b \leq [u, v)$ for some atom b. If $[u, v)$ is infinite, then there exist $x, y \in X$ with $u < x < y < v$, and so $[x, y) \subseteq [u, v)$. For $\geq$, suppose to the contrary that R is dense in $\mathrm{Intalg}\,L$ and $|R| < \mathrm{d}L \cdot |M|$. Clearly $M \subseteq R$. Wlog each member of R has the form $[a, b)$. Since $|M| \leq |R|$, we have $|R| < \mathrm{d}L$. Let $R' = \{a \in L : \exists b([a, b) \in R)\}$. Thus $L \backslash \overline{R'}$ is a non-empty open set. Say $w \in (u, v) \subseteq L \backslash \overline{R'}$. Then $[w, v) \in \mathrm{Intalg}\,L$, so $[a, b) \subseteq [w, v)$ for some $[a, b) \in R$; but then $a \in R'$, contradiction.

If A is a minimally generated algebra, then $\pi A = dA$. In fact, A is co-absolute with an interval algebra, and $\pi B = \pi \overline{B}$ and $dB = d\overline{B}$ for any BA B, so this follows from the interval algebra result.

For A atomic, clearly πA is the number of atoms of A. Also note that $\pi_{S+} A = |A|$ for A complete, and $\pi_{S+} A = hdA$ for A retractive.

If A is the completion of the free BA on ω_1 free generators, then $dA < \pi A$: clearly $\pi A = \omega_1$. The identity mapping from the free BA on ω_1 free generators into $\mathscr{P}\omega$ can be extended to a homomorphism f from A into $\mathscr{P}\omega$, and f must be one-one; so $dA = \omega$.

For tree algebras we recall from Theorem 5.17 that $\pi A = |A|$.

7. Length

Recall that Length$A$ is the sup of cardinalities of subsets of A which are simply ordered by the Boolean ordering. For references see the beginning of Chapter 4. The analysis of Length is similar to that for Depth; many of the proofs are similar, but there are some differences. To take care of the first problem, attainment of Length, we need two small lemmas about orderings:

Lemma 7.1. *Let L be a linear ordering of regular cardinality λ which has no strictly increasing or strictly decreasing sequences of length λ. Then there exist $a < b$ in L such that $|[a, b)| = \lambda$.*

Proof. Let $\langle a_\xi : \xi < \alpha \rangle$ and $\langle b_\xi : \xi < \beta \rangle$ be coinitial strictly decreasing and cofinal strictly increasing sequences in L, respectively. Then L, except for its greatest element, if it has such, is the union of all of the intervals $[a_\xi, b_\eta)$, and $\xi < \lambda$ and $\eta < \lambda$, so the conclusion is clear. □

Lemma 7.2. *Let L be a linear ordering with first element 0, and with cardinality κ^+, where κ is infinite. Then there exist $a < b$ in L such that $|[a, b)| \geq \kappa$ and $|L \backslash [a, b)| \geq \kappa$.*

Proof. Suppose not. Then clearly

(1) in L there is no strictly increasing or strictly decreasing sequence of length κ^+.

Define by induction a sequence $\langle [a_\xi, b_\xi) : \xi < \alpha \rangle$ of half-open intervals in L such that $[a_\eta, b_\eta) \subset [a_\xi, b_\xi)$ for $\xi < \eta$, $|[a_\xi, b_\xi)| = \kappa^+$, and $|L \backslash [a_\xi, b_\xi)| < \kappa$ for all $\xi < \alpha$, continuing as long as possible. We can start by Lemma 7.1. How long can we continue? Well, if $[a_\xi, b_\xi)$ has been defined, then $[a_{\xi+1}, b_{\xi+1})$ can be defined: choose c with $a_\xi < c < b_\xi$; then $|[a_\xi, c)| = \kappa^+$ or $|[c, b_\xi)| = \kappa^+$. Suppose that $[a_\xi, b_\xi)$ has been defined for all $\xi < \beta$, where β is a limit ordinal $< \kappa^+$. Then

$$|L \backslash \bigcap_{\xi < \beta} [a_\xi, b_\xi)| = |\bigcup_{\xi < \beta} L \backslash [a_\xi, b_\xi)| < \kappa^+,$$

so by (1) and Lemma 7.1 applied to $\bigcap_{\xi < \beta} [a_\xi, b_\xi)$, the interval $[a_\beta, b_\beta)$ can be defined. Thus $\alpha \geq \kappa^+$. Now $a_\xi \leq a_\eta$ and $b_\xi \geq b_\eta$ for $\xi < \eta$, so one of $\{a_\xi : \xi < \kappa^+\}$ and $\{b_\xi : \xi < \kappa^+\}$ contains a suborder of L of size κ^+. This contradicts (1). □

Theorem 7.3. *If $\mathrm{cf}(\mathrm{Length}A) = \omega$, then $\mathrm{Length}A$ is attained.*

Proof. The proof should be fairly clear, following the lines of the proof of 4.2. Some modifications: a is an ∞-*element* provided that for each $i \in \omega$, some ordering of size λ_i is embeddable in $A \restriction a$. When constructing a_i, Lemma 7.2 is used to obtain elements c, d such that $b = c + d$, $c \cdot d = 0$, and both $A \restriction c$ and $A \restriction d$ contain strictly increasing chains of length λ_i; then the new (*) is applied. □

The analog of 4.3 for Length does not hold. For example, if A is any denumerable BA, then $^\omega A$ has length 2^ω. This is because $\mathscr{P}\mathbb{Q}$ can be embedded in $^\omega A$, and $\mathbb{R}$

can be embedded in $\mathscr{P}\mathbb{Q}$: for each $r \in \mathbb{R}$, let $fr = \{q \in \mathbb{Q} : q < r\}$. To generalize this example, let us call a subset D of a linear order L *weakly dense in L* provided that if $a, b \in L$ and $a < b$, then there is a $d \in D$ such that $a \leq d \leq b$. Now for any infinite cardinal κ let $\mathrm{Ded}\kappa = \sup\{\lambda$: there is an ordering of size λ with a weakly dense subset of size $\kappa\}$. The following theorem from Kurepa [57] shows the connection of this notion to length in $\mathscr{P}\kappa$:

Theorem 7.4. *Let κ and λ be cardinals such that $\omega \leq \kappa \leq \lambda$. Then the follolwing two conditions are equivalent:*

(i) There is an ordering L of size λ with a weakly dense subset of size κ.

(ii) In $\mathscr{P}\kappa$ there is a chain of size λ.

Proof. $(i) \Rightarrow (ii)$. We may assume that $\kappa < \lambda$. Let D be weakly dense in L, with $|D| = \kappa$. Thus $|L\backslash D| = \lambda$. Let f be a one-one function from κ onto D. For each $a \in L\backslash D$ let $ga = \{\alpha < \kappa : f\alpha < a\}$. Clearly $a < b$ implies that $ga \subseteq gb$. Suppose $a < b$ with $a, b \in L\backslash D$; choose $x \in D$ so that $a \leq x \leq b$. Hence $a < x < b$, and so $f^{-1}x \in gb\backslash ga$ and $ga \neq gb$, as desired.

$(ii) \Rightarrow (i)$. Let L be a chain in $\mathscr{P}\kappa$ of size λ. For each $\alpha < \kappa$ let $x_\alpha = \bigcup\{a \in L : \alpha \notin a\}$. For any $\alpha, \beta < \kappa$ we have $x_\alpha \subseteq x_\beta$ or $x_\beta \subseteq x_\alpha$. For, suppose that $\gamma \in x_\alpha\backslash x_\beta$ and $\delta \in x_\beta\backslash x_\alpha$. Say $\gamma \in a \in L$ with $\alpha \notin a$ and $\forall b \in L(\beta \notin b \Rightarrow \gamma \notin b)$; and $\delta \in b \in L$ with $\beta \notin b$ and $\forall c \in L(\alpha \notin c \Rightarrow \delta \notin c)$. Say $a \subseteq b$. Then $\gamma \in b$, contradiction. For any $a \in L$ and $\alpha < \kappa$ we clearly have $a \subseteq x_\alpha$ or $x_\alpha \subseteq a$. Hence we may assume that $\{x_\alpha : \alpha < \kappa\} \subseteq L$. Let D be a subset of L of size κ such that $\{x_\alpha : \alpha < \kappa\} \subseteq D$. Now suppose that $a, b \in L$ and $a \subset b$. Choose $\alpha \in b\backslash a$. Then $a \subseteq x_\alpha$; and if $c \in L$ and $\alpha \notin c$, then $c \subseteq b$; so $x_\alpha \subseteq b$, as desired. $\qquad\square$

Because of this theorem, about all that we can say about the length of products is this:

$$\max(\mathrm{Ded}|I|, \sup_{i \in I}\mathrm{Length}A_i) \leq \mathrm{Length}\left(\prod_{i \in I} A_i\right) \leq \prod_{i \in I}\mathrm{Length}A_i.$$

Shelah [87] has shown that $\mathrm{Length}(\prod_{i \in I} A_i)$ cannot be calculated purely from $|I|$ and $\langle\mathrm{Length}A_i : i \in I\rangle$.

For weak products, we have the following analogs of 4.6 and 4.7:

Theorem 7.5. *Let $\kappa = \sup_{i \in I}\mathrm{Length}A_i$, and suppose that $\mathrm{cf}\kappa > \omega$. Then the following conditions are equivalent:*

(i) $\prod_{i \in I}^{\mathrm{w}} A_i$ has no chain of size κ.

(ii) For all $i \in I$, A_i has no chain of size κ.

Proof. For the non-trivial direction $(ii) \rightarrow (i)$, suppose that X is a chain in $\prod_{i \in I}^{\mathrm{w}} A_i$ of size κ. Wlog assume that for each $x \in X$, the set $M_x \overset{\mathrm{def}}{=} \{i \in I : x_i \neq 0\}$ is finite. Define $x \equiv y$ iff $M_x = M_y$. Then it is easy to see that $\equiv$ is a convex equivalence relation on X; there is an order induced on $X/\equiv$, and clearly that order is isomorphic to an interval of the ordered set ω. It follows from $\mathrm{cf}\kappa > \omega$ that some equivalence class has cardinality κ. Then Lemma 4.1 gives a contradiction. $\square$

Corollary 7.6. Length$\left(\prod_{i\in I}^{w} A_i\right) = \sup_{i\in I}\mathrm{Length}A_i$. $\square$

By 7.5 we see that 7.3 is best possible: if κ is a limit cardinal with $\mathrm{cf}\kappa > \omega$, then it is easy to construct a weak product B such that $\mathrm{Length}B = \kappa$ but the length of B is not attained.

If A is a subalgebra of B, then $\mathrm{Length}A \leq \mathrm{Length}B$, and the difference can be arbitrarily large. If A is a homomorphic image of B, then length can vary either way from B to A again, see the argument for cellularity.

Now we turn to ultraproducts, giving some results of Douglas Peterson. Since length is an ultra-sup function, Theorems 3.15–3.17 apply. Theorem 3.17 is especially to be noticed: Length $\left(\prod_{i\in I} A_i/F\right) \geq \left|\prod_{i\in I} \mathrm{Length}A_i/F\right|$ for F regular. Thus by Donder's theorem it is consistent that $\geq$ always holds. Magidor and Shelah have shown that it is consistent to have an example in which the length of an ultraproduct is strictly less than the size of the ultraproduct of their lengths; see Chapter 4 for details. It is consistent to have an example in which Length $\left(\prod_{i\in I} A_i/F\right) > \left|\prod_{i\in I} \mathrm{Length}A_i/F\right|$ with F regular; see the comments about depth. But it seems to be open whether this can be done in ZFC:

Problem 22. *Can one prove in ZFC that there exist a system $\langle A_i : i \in I\rangle$ of infinite BAs, I infinite, and an ultrafilter F on I such that* Length $\left(\prod_{i\in I} A_i/F\right) > \left|\prod_{i\in I} \mathrm{Length}A_i/F\right|$?

The following version of Theorem 3.18 holds:

Theorem 7.7. *Let $\langle A_i : i \in I\rangle$ be a system of infinite BAs, with I infinite, let F be a uniform ultrafilter on I, and let $\kappa = \max(|I|, \mathrm{ess.sup}_{i\in I}^{F}\mathrm{Length}A_i)$. Then* Length $\left(\prod_{i\in I} A_i/F\right) \leq 2^{\kappa}$.

Proof. Let $\lambda = \mathrm{ess.sup}_{i\in I}^{F}\mathrm{Length}A_i$. We may assume that $\mathrm{Length}A_i \leq \lambda$ for all $i \in I$. In order to get a contradiction, suppose that $\langle f_\alpha/F : \alpha < (2^{\kappa})^+\rangle$ is a system of distinct comparable elements. Thus $[(2^{\kappa})^+]^2 = \bigcup_{i\in I}\{\{\alpha,\beta\} : f_\alpha i$ and $f_\beta i$ are distinct comparable elements$\}$, so by the Erdös-Rado theorem $(2^{\kappa})^+ \rightarrow (\kappa^+)^2_\kappa$ we get a homogeneous set which gives a contradiction. $\square$

Concerning equality in Theorem 7.7, we note that it holds if F is regular and $\mathrm{ess.sup}|A_i| \leq |I|$, since then

$$2^{|I|} = (\mathrm{ess.sup}\,\mathrm{Length}A_i)^{|I|}$$

$$= \left|\prod_{i\in I}\mathrm{Length}A_i/F\right| \leq \mathrm{Length}\left(\prod_{i\in I} A_i/F\right)$$

$$\leq \left|\prod_{i\in I} A_i/F\right| \leq 2^{|I|}.$$

On the other hand, if $|A| = \text{Length} A = \kappa$ and $\kappa^\omega = \kappa$, then $\text{Length}\,({}^\omega A/F) < 2^\kappa$ for any nonprincipal ultrafilter F on ω.

For free products, we have $\text{Length}(\oplus_{i \in I} A_i) = \sup_{i \in I} \text{Length} A_i$; this result of Grätzer and Lakser was considerably generalized by McKenzie and Monk; but in any case the proof is too lengthy to include here. Bekkali [92] constructed a BA A of length $\aleph_{\omega_1}$ such that if L is a Suslin line, then $A \oplus \text{Intalg}\, L$ has no chain of size $\aleph_{\omega_1}$; this solves Problem 16 in Monk [90].

Length is an ordinary sup-function, so Theorem 3.11 applies.

We turn to derived functions for length. The function $\text{Length}_{H+} A$ seems to be new. Note just that $tA = \text{Depth}_{H+} A \leq \text{Length}_{H+} A$, using 4.21. It is possible to have $tA < \text{Length}_{H+} A$; this is true when A is the interval algebra on $\mathbb{R}$, since $tA = \omega$, while obviously $\text{Length}_{H+} A = 2^\omega$. To see that $tA = \omega$, one can use 3.24, 5.11, and 3.29.

Shelah has constructed an algebra A such that $\omega < \text{Length} A < |A|$ while A has no homomorphic image of power smaller that $|A|$, assuming $\neg$CH (email message of December 1990). This answers Problem 17 in Monk [90]. Since an infinite BA always has a homomorphic image of size $\leq 2^\omega$, the assumption $\neg$CH is needed here. We present this result here. It depends on the following notation. If $\langle a_n : n \in \omega \rangle$ is a system of elements of a BA A and $Y \subseteq \omega$, then $\{[x, a_n]^{\text{if } n \in Y} : n \in \omega\}$ denotes the following set of formulas:

$$\{a_n \leq x : n \in Y\} \cup \{a_n \cdot x = 0 : n \in \omega \backslash Y\}.$$

If L is a chain, then a *Dedekind cut* of L is a pair (M, N) such that $L = M \cup N$ and $u < v$ for all $u \in M$ and $v \in N$. If in addition L is a subset of a BA A, then an element $a \in A$ *realizes* the Dedekind cut (M, N) if $u \leq a \leq v$ for all $u \in M$ and $v \in N$.

For any BA A, $\text{Length}' A$ is the smallest infinite cardinal κ such that every chain in A has size less than κ.

Shelah's result will follow easily from the following lemma:

Lemma 7.8. *Let $\aleph_0 \leq \mu < \lambda \overset{\text{def}}{=} 2^{\aleph_0}$, and let A be a subalgebra of $\mathcal{P}\omega$ containing all singletons $\{i\}$, with $|A| = \mu$. Then there is a BA B of size $2^{\aleph_0}$ satisfying*

$\otimes_0$ *A is a dense subalgebra of B.*

$\otimes_1$ *If $\langle a_n : n < \omega \rangle$ is a system of pairwise disjoint elements of B^+, then for $2^{\aleph_0}$ subsets Y of ω there is an element $a_Y \in B$ realizing $\{[x, a_n]^{\text{if } n \in Y} : n \in \omega\}$.*

$\otimes_2$ *If $\langle a_n : n < \omega \rangle$ is a chain of members of B, then the number of Dedekind cuts of it realized in B is less than $2^{\aleph_0}$.*

Proof. First we obviously have:

(1) there is an enumeration $\langle \langle a_{\zeta n} : n < \omega \rangle : \zeta < \lambda \rangle$ of all of the ω-tuples of subsets of ω, each one repeated λ times.

Next we claim:

(2) There is a function $h : \lambda \to \lambda$ such that for all $\zeta < \lambda$ we have $h\zeta < \mu$ or $h\zeta < \zeta$, and the set

$$S_\zeta \overset{\text{def}}{=} \{\varepsilon < \lambda : h\varepsilon = \zeta\}$$

has power λ.

To see this, first choose a system $\langle D_\alpha : \alpha < \lambda \rangle$ of pairwise disjoint sets whose union is λ, each of power λ, with D_0 the set of all limit ordinals less than λ. Define

$$E_\alpha = (D_\alpha \cup \{\alpha + 1\}) \setminus \bigcup_{\beta < \alpha} E_\beta.$$

Then

(3) $D_\alpha \setminus E_\alpha \subseteq \alpha + 1$;
(4) $E_\alpha \setminus D_\alpha \subseteq \{\alpha + 1\}$.

For, (4) is obvious. For (3), suppose that $\zeta \in D_\alpha \setminus E_\alpha$. Then there is a $\beta < \alpha$ such that $\zeta \in E_\beta$. Now $\zeta \notin D_\beta$, so $\zeta = \beta + 1$ by (4). Thus (3) holds.

By (3), $|E_\alpha| = \lambda$ for all $\alpha < \lambda$. Next,

(5) $\bigcup_{\alpha < \lambda} E_\alpha = \lambda$.

For, suppose that $\zeta \notin \bigcup_{\alpha < \lambda} E_\alpha$. Since $E_0 = D_0 \cup \{1\}$, ζ is a successor ordinal $\alpha + 1$. Then $\zeta \in E_\alpha$, contradiction.

For each $\zeta < \lambda$ let $h\zeta$ be the α such that $\zeta \in E_\alpha$. Suppose $\mu \leq h\zeta$. Then $h\zeta \neq 0$, so ζ is a successor ordinal $\alpha + 1$. Clearly $\zeta \in \bigcup_{\beta \leq \alpha} E_\beta$, so $h\zeta < \zeta$. This proves (2).

Now we define a BA B_ε by induction on $\varepsilon < \lambda$ such that:

   a) B_ε is a subalgebra of $\mathscr{P}\omega$ containing all singletons, of cardinality $< \mu^+ + |\varepsilon|^+$.
   b) B_ε is increasing and continuous in ε.
   c) $B_0 = A$.
   d) If $\zeta < \varepsilon$ and $\langle a_{\zeta n} : n < \omega \rangle$ is a linearly ordered system of elements of B_ζ (no two equal), then every Dedekind cut of it which is not realized in B_ζ is also not realized in B_ε.
   e) Let Evens be the set of all even natural numbers. If $h\varepsilon = \zeta$ and $\langle a_{\zeta n} : n < \omega \rangle$ is a system of pairwise disjoint members of B_ε (some possibly zero) with union ω, then for some $Y_\varepsilon \subseteq$ Evens we have
     (i) $B_{\varepsilon+1}$ is generated by $B_\varepsilon \cup \{x_\varepsilon\}$, where $x_\varepsilon = \bigcup_{n \in Y_\varepsilon} a_{\zeta n}$.
     (ii) If $\psi < \varepsilon$ and $h\psi = \zeta$, then $Y_\psi \neq Y_\varepsilon$.
   If $\langle a_{\zeta n} : n < \omega \rangle$ is not such a system, then $B_{\varepsilon+1} = B_\varepsilon$.

The construction is determined for $\varepsilon = 0$ and for limit ε. At stage $\varepsilon \to \varepsilon + 1$, assume that $h\varepsilon = \eta$ and $\langle a_{\eta n} : n < \omega \rangle$ is a system of pairwise disjoint elements of B_ε with union ω, let $\kappa_\varepsilon = (\mu + |\varepsilon|)^+$, and let $\langle Y_i^\varepsilon : i < \kappa_\varepsilon \rangle$ be a sequence of almost disjoint infinite subsets of Evens; we try each of these as Y_ε and get $B_{\varepsilon i}$ by adjoining $x_{\varepsilon i}$ in order to satisfy e)(i); so the bad case is that one of the "demands" e)(ii) or d) fails. There are $< \kappa_\varepsilon$ demands, so we may assume that the same demand fails for all of them. It cannot be e)(ii), so it is d) for a certain $\zeta \leq \varepsilon$. There is then a term, wlog not depending on i, call it $t(x_{\varepsilon i}, b)$ with $b \in B_\varepsilon$, which realizes a Dedekind cut (M_i, N_i) of $\langle \alpha_{\zeta n} : n < \omega \rangle$ not realized in B_ζ; here $x_{\varepsilon i} = \bigcup_{n \in Y_i^\varepsilon} a_{\eta n}$. We can write $t(x_{\varepsilon i}, b) = b_0 + b_1 \cdot x_{\varepsilon i} + b_2 \cdot -x_{\varepsilon i}$ for some partition (b_0, b_1, b_2, b_3) of unity in B_ε.

If $(M_i, N_i) = (M_j, N_j)$ for two distinct $i, j < \kappa_\varepsilon$, we get a contradiction, as follows. For all $c \in M_i$ and $d \in N_i$ we have

$$c \cdot b_0 \leq b_0 \cdot t(x_{\varepsilon i}, b) = b_0 \leq b_0 \cdot d;$$
$$c \cdot b_1 \leq b_1 \cdot t(x_{\varepsilon i}, b) = b_1 \cdot x_{\varepsilon i} \leq b_1 \cdot d;$$
$$c \cdot b_2 \leq b_2 \cdot t(x_{\varepsilon i}, b) = b_2 \cdot -x_{\varepsilon i} \leq b_2 \cdot d;$$

it follows that $c \cdot b_1 \leq b_1 \cdot x_{\varepsilon i} \cdot x_{\varepsilon j} \leq b_1 \cdot d$; note also that $x_{\varepsilon i} \cdot x_{\varepsilon j} \in B_\zeta$. Similarly, $c \cdot b_2 \leq b_2 \cdot -x_{\varepsilon_i} + b_2 \cdot -x_{\varepsilon j} \leq b_2 \cdot d$, and $-x_{\varepsilon_i} + -x_{\varepsilon j} = -(x_{\varepsilon i} \cdot x_{\varepsilon j}) \in B_\zeta$. Hence the Dedekind cut (M_i, N_i) is realized by

$$b_0 + b_1 \cdot x_{\varepsilon i} \cdot x_{\varepsilon j} + b_2 \cdot -x_{\varepsilon_i} + b_2 \cdot -x_{\varepsilon j}$$

in B_ε, contradiction. Thus we have shown that distinct $i, j < \kappa_\varepsilon$ realize different Dedekind cuts.

Wlog the truth values of the following statements do not depend on i:

(6) $S_i^1 \overset{\text{def}}{=} \{n \in Y_i^\varepsilon : a_{\eta n} \cdot b_1 \neq 0\}$ is infinite;

(7) $S_i^2 \overset{\text{def}}{=} \{n \in Y_i^\varepsilon : a_{\eta n} \cdot b_2 \neq 0\}$ is infinite.

If both of these are false (for all i), take any i, and consider (M_i, N_i). Then, with $c \in M_i$ and $d \in N_i$,

$$c \cdot b_1 \leq b_1 \cdot x_{\varepsilon i} = b_1 \cdot \sum_{n \in S_i^1} a_{\eta n} \leq d \cdot b_1$$

and $b_2 \cdot x_{\varepsilon i} = b_2 \cdot \sum_{n \in S_i^2} a_{\eta n}$ and hence

$$c \cdot b_2 \leq b_2 \cdot -x_{\varepsilon i} = b_2 \cdot - \sum_{n \in S_i^2} a_{\eta n} \leq d \cdot b_2;$$

so (M_i, N_i) is realized by

$$b_0 + b_1 \cdot \sum_{n \in S_i^1} a_{\eta n} + b_2 \cdot - \sum_{n \in S_i^2} a_{\eta n}$$

in B_ε, contradiction.

Thus either (6) or (7) is true. Take distinct Dedekind cuts (M_i, N_i) and (M_j, N_j); say $M_i \subset M_j$. Choose $c \in M_j \backslash M_i$.

Case 1. (6) is true. Then in $B_{\varepsilon i}$ we have $t(x_{\varepsilon i}, b) \leq c$, so

(*) $x_{\varepsilon i} = t(x_{\varepsilon i}, b) \cdot b_1 \leq c \cdot b_1$.

And in $B_{\varepsilon j}$ we have $c \leq t(x_{\varepsilon j}, b)$, so

($\star$) $c \cdot b_1 \leq t(x_{\varepsilon j}, b) \cdot b_1 = x_{\varepsilon j} \cdot b_1$.

Now if $n \in S_i^1$, then $0 \neq a_{\eta n} \cdot b_1 \leq x_{\varepsilon i} \cdot b_1 \leq c \cdot b_1$ (by (*)) $\leq x_{\varepsilon j} \cdot b_1$ (by ($\star$)), so $n \in Y_i^{\varepsilon}$, hence $n \in S_j^1$. Thus $S_i^1 \subseteq S_j^1$, contradicting $S_i^1 \cap S_j^1 \subseteq Y_i^{\varepsilon} \cap Y_j^{\varepsilon}$ finite.

Case 2. (7) is true. Then in $B_{\varepsilon i}$ we have $t(x_{\varepsilon i}, b) \leq c$, so

(**) $-x_{\varepsilon i} \cdot b_2 = t(x_{\varepsilon i}, b) \cdot b_2 \leq c \cdot b_2$.

And

($\star\star$) $c \cdot b_2 \leq t(x_{\varepsilon j}, b) \cdot b_2 = -x_{\varepsilon j} \cdot b_2$.

If $n \in S_j^2$, then $a_{\eta n} \cdot b_2 \cdot -x_{\varepsilon j} = 0$, so by ($\star\star$), $a_{\eta n} \cdot b_2 \cdot c = 0$, hence by (**) $a_{\eta n} \cdot b_2 \cdot -x_{\varepsilon i} = 0$, so $n \in Y_i^{\varepsilon}$, hence $n \in S_i^2$. So $S_j^2 \subseteq S_i^2$, again giving a contradiction.

Thus the construction can be carried through. Let $B = \bigcup_{\varepsilon < \lambda} B_{\varepsilon}$. Clearly $\otimes_0$ holds, by c). Now suppose that $\langle a_n : n < \omega \rangle$ is a system of pairwise disjoint elements of B^+. Choose $\varepsilon < \lambda$ such that $a_n \in B_{\varepsilon}$ for all $n \in \omega$, extend $\{a_n : n \in \omega\}$ to a maximal disjoint set X in B_{ε}, and enumerate X as $\langle b_n : n < \omega \rangle$ so that $\{a_n : n < \omega\} = \{b_{2n} : n < \omega\}$. By (1), there is a $\zeta < \lambda$ such that $\langle b_n : n < \omega \rangle = \langle a_{\zeta n} : n < \omega \rangle$. Then by (2) and e) we get $2^{\aleph_0}$ subsets Y of ω such that $\{[x, a_n]^{\text{if } n \in Y} : n \in \omega\}$ is realized in B. Next, suppose that $\langle a_n : n < \omega \rangle$ is a chain of members of B^+. By (1), say $\langle a_n : n < \omega \rangle = \langle a_{\zeta n} : n < \omega \rangle$. Choose $\varepsilon < \lambda$ such that all a_n are in B_{ε}. Then by d), B realizes at most $|B_{\varepsilon}|$ Dedekind cuts of $\langle a_n : n < \omega \rangle$. □

Theorem 7.9. *If $\aleph_0 \leq \mu < 2^{\aleph_0}$ then there is a BA B such that:*
 (i) B is a subalgebra of $\mathscr{P}\omega$ containing all singletons, and hence $\pi B = \aleph_0$;
 (ii) $\mu^+ \leq \mathrm{Length}' B \leq 2^{\aleph_0}$;
 (iii) every infinite homomorphic image of B has size $2^{\aleph_0}$.

Proof. We apply the lemma to a subalgebra A of $\mathscr{P}\omega$ containing all singletons, of size μ and with length μ. We obtain a BA B as a result.

We check that every infinite homomorphic image of B has size $2^{\aleph_0}$. Let f be a homomorphism from B onto C with C infinite. Let $\langle c_n : n < \omega \rangle$ be a system of nonzero disjoint elements of C. Then there is a system $\langle b_n : n < \omega \rangle$ of non-zero disjoint elements of B such that $fb_n = c_n$ for all $n < \omega$. Let $\mathscr{D}$ be a collection of $2^{\aleph_0}$ subsets of ω such that for each $Y \in \mathscr{D}$ there is an element b_Y realizing $\{[x, b_n]^{\text{if } n \in Y} : n < \omega\}$. Clearly $\{fb_Y : Y \in \mathscr{D}\}$ is a subset of C of size $2^{\aleph_0}$, as desired.

Now suppose that J is a chain in B of size $2^{\aleph_0}$; we shall get a contradiction. For each $i \in \omega$ let $M_i = \{b \in J : i \notin b\}$. Clearly there is a countable subset K_i of

M_i cofinal in M_i. Let $I = \bigcup_{i \in \omega} K_i$. So, I is countable. Each element of J realizes a Dedekind cut of I, so we can contradict $\otimes_2$ of the Lemma by showing that any two distinct $u, v \in J$ realize distinct Dedekind cuts of I. Suppose that $u \subset v$ but $\{w \in I : u \subseteq w\} = \{w \in I : v \subseteq w\}$. Choose $i \in v \backslash u$. Choose $w \in K_i$ with $u \subseteq w$. Then $v \subseteq w$ and hence $i \in w$, contradiction. □

The function Length_{H-} is also new. Note that $\omega \leq \mathrm{Length}_{H-} A \leq 2^\omega$, by an easy argument using the Sikorski extension theorem. It is obviously possible to have $\omega = \mathrm{Length}_{H-} A$. Shelah has shown under $\Diamond$ that there is a BA A such that $\mathrm{Length}_{H-} A < \mathrm{Card}_{H-} A$. This answers Problem 18 in Monk [90]. Obviously $\mathrm{Length}_{S+} A = \mathrm{Length} A$ and $\mathrm{Length}_{S-} A = \omega$. The following is Problem 19 in Monk [90]:

Problem 23. *Is always* $\mathrm{Length}_{h-} A = \omega$?

Clearly $\mathrm{Length}_{h+} A \geq \mathrm{Depth}_{h+} A = sA$ by 4.23. And $\mathrm{Length}_{h+} A \geq \mathrm{Length}_{H+} A$; but it is possible to have $\mathrm{Length}_{h+} A > \mathrm{Length}_{H+} A$. This is true, for example, if A is the finite-cofinite algebra on an uncountable cardinal κ. For then $\mathrm{Length}_{h+} A = \mathrm{Ded}\kappa$, while $\mathrm{Length}_{H+} A = \omega$. That $\mathrm{Length}_{h+} A = \mathrm{Ded}\kappa$ is seen like this: $\mathrm{Ult}A$ has a discrete subspace S of size κ, and so Theorem 7.4 applies for the chains of subsets of S, since every subset is clopen.

Clearly $_d\mathrm{Length}_{S+} A = \mathrm{Length} A$. By the discussion of $_d\mathrm{Depth}_{S-}$ in Chapter 4 we see that if A is atomless and λ-saturated (in the model-theoretic sense), then $_d\mathrm{Length}_{S-} A \geq \lambda$. Thus Problem 20 of Monk [90] is answered.

Concerning the relationships of length to our previously treated functions, note that obviously $\mathrm{Depth} A \leq \mathrm{Length} A$ for any infinite BA A. Another clear relationship is $\mathrm{Length} A \leq 2^{\mathrm{Depth} A}$: if L is an ordered subset of A of power $(2^{\mathrm{Depth} A})^+$, let $\prec$ be a well-ordering of L; then by the Erdős-Rado partition relation $(2^\kappa)^+ \to (\kappa^+)^2_\kappa$ we get a well-ordered or inversely well-ordered subset of L of power $(\mathrm{Depth} A)^+$, contradiction.

Note that $\mathrm{Length} A > \pi A$ for $A = \mathscr{P}\omega$; and $cA > \mathrm{Length} A$ for A the finite-cofinite algebra on κ. If A is a tree algebra, then $\mathrm{Length} A = \mathrm{Depth} A$ by Proposition 16.20 of the Handbook. For A superatomic we have $\mathrm{Length} A = \mathrm{Depth} A$ by Rosenstein [82] Corollary 5.29.

8. Irredundance

Clearly $\mathrm{Irr}A \leq |A|$. If A is a subalgebra of B, then $\mathrm{Irr}A \leq \mathrm{Irr}B$, and Irr can change to any extent from B to A (along with cardinality). The same is true for A a homomorphic image of B. The following problem is open:

Problem 24. *Is* $\mathrm{Irr}(A \times B) = \max\{\mathrm{Irr}A, \mathrm{Irr}B\}$?

We prove a weak but useful result along the lines of this problem:

Theorem 8.1. *For any infinite BA A, $\mathrm{Irr}A = \mathrm{Irr}(A \times 2)$.*

Proof. Let $\kappa = \mathrm{Irr}A$, and suppose that $D \in [A \times 2]^{\kappa^+}$ is irredundant. Then $\{a \in A : (a, 0) \in D\}$ or $\{a \in A : (a, 1) \in D\}$ has size κ^+. By passing to $\{d : -d \in D\}$ if necessary, we may assume that the second set has size κ^+; and so we may assume that every element x of D has the form $(a_x, 1)$.

  Case 1. $\{a_x : x \in D\}$ does not have the fip. Say $z_0, \ldots, z_{m-1} \in D$ and $a_{z_0} \cdot \ldots \cdot a_{z_{m-1}} = 0$. Now $\{a_x : x \in D \setminus \{z_0, \ldots, z_{m-1}\}\}$ is redundant, so we can write

$$a_x = \sum_{i<p} \prod_{j<n} a_{y_j}^{\varepsilon_{ij}},$$

where $x, y_0, \ldots, y_{n-1}$ are distinct elements of $D \setminus \{z_0, \ldots, z_{m-1}\}$ and $\varepsilon_{ij} \in \{0, 1\}$. Then

$$a_x = \sum_{i<p} \prod_{j<n} a_{y_j}^{\varepsilon_{ij}} + a_{z_0} \cdot \ldots \cdot a_{z_{m-1}},$$

and hence

$$x = \sum_{i<p} \prod_{j<n} y_j + z_0 \cdot \ldots \cdot z_{m-1},$$

contradiction.

  Case 2. $\{a_x : x \in D\}$ has fip. Write

$$a_x = \sum_{i<m} \prod_{j<n} a_{y_j}^{\varepsilon_{ij}},$$

where $x, y_0, \ldots, y_{n-1}$ are distinct elements of D and $\varepsilon_{ij} \in \{0, 1\}$. If $\exists i < m \forall j < n(\varepsilon_{ij} = 1)$, then $x = \sum_{i<m} \prod_{j<n} y_j$, contradiction. Hence $\forall i < m \exists j < n(\varepsilon_{ij} = 0)$, and hence $a_{y_0} \cdot \ldots \cdot a_{y_{n-1}} \cdot a_x = 0$, contradiction. □

Corollary 8.2. *If A is infinite and B is finite, then $\mathrm{Irr}A = \mathrm{Irr}(A \times B)$.* □

Corollary 8.3. *If $|B| \leq \mathrm{Irr}A$, then $\mathrm{Irr}(A \times B) = \mathrm{Irr}A$.*

Proof. Let $\kappa = \mathrm{Irr}A$, and suppose that $X \in [A \times B]^{\kappa^+}$ is irredundant. Then there is a $b \in B$ such that $Y \overset{\text{def}}{=} \{(a, b) : (a, b) \in X\}$ has power κ^+. Thus $Y \subseteq A \times \{0, b, -b, 1\}$, so this contradicts Corollary 8.2. □

Irredundance is an ultra-sup function, so Theorems 3.15–3.17 of Peterson apply. By Theorem 3.17, Irr $\left(\prod_{i \in I} A_i/F\right) \geq \left|\prod_{i \in I} \mathrm{Irr} A_i/F\right|$ for regular F; hence by Donder's result it is consistent that this inequality always holds. But it seems to be open to actually find examples where equality fails to hold, giving two problems (under any set-theoretic assumptions):

Problem 25. *Is there an example of a system $\langle A_i : i \in I \rangle$ of infinite BAs, with I infinite, and a uniform ultrafilter F on I such that* Irr $\left(\prod_{i \in I} A_i/F\right) < \left|\prod_{i \in I} \mathrm{Irr} A_i/F\right|$?

Problem 25 may be solvable by the methods of Magidor, Shelah [91]. See also Rosłanowski, Shelah [94].

Problem 26. *Is there an example of a system $\langle A_i : i \in I \rangle$ of infinite BAs, with I infinite, and a uniform ultrafilter F on I such that* Irr $\left(\prod_{i \in I} A_i/F\right) > \left|\prod_{i \in I} \mathrm{Irr} A_i/F\right|$?

Concerning the derived operations, we note just the obvious facts that $\mathrm{Irr}_{\mathrm{S}+} A = \mathrm{Irr} A$, $\mathrm{Irr}_{\mathrm{S}-} A = \omega$, and $_{\mathrm{d}} \mathrm{Irr}_{\mathrm{S}+} A = \mathrm{Irr} A$.

Obviously any chain is irredundant; so $\mathrm{Length} A \leq \mathrm{Irr} A$. The difference can be large, e.g. in a free BA. By Theorem 4.25 of Part I of the BA handbook, $\pi A \leq \mathrm{Irr} A$. In particular, if $|A|$ is strong limit, then $|A| = \mathrm{Irr} A$, since then $\pi A = |A|$.

These trivial facts give the immediate results about irredundance. Deeper facts about it are that it is consistent that there is a BA with irredundance less than cardinality, and it is also consistent that every uncountable BA has uncountable irredundance (see Todorčević [90b]). We shall spend the rest of this chapter proving the first fact, in the form that under CH there is a BA of power ω_1 with countable irredundance. We give two examples for this. The first example is a compact Kunen line. We say "a" since there are various Kunen lines, and we say "compact" since the standard Kunen lines are only locally compact. For the Kunen lines, see Juhász, Kunen, Rudin [76]. The second construction uses considerably less than CH, and can be found in Todorčević [89]. For a forcing construction of an uncountable BA with countable irredundance, see Bell, Ginsburg, Todorčević [82]. A generalization of the main results about irredundance (to other varieties of universal algebras) can be found in Heindorf [89a] and Todorčević [90b].

The history of these results is complicated. I think that the first example of an uncountable BA with countable irredundance is due to Rubin [83] (the result was obtained several years before 1983). The papers with the constructions we give do not mention irredundance; their relevance for our purposes is due to a simple theorem of Heindorf [89a]. So, modulo the simple theorem of Heindorf, the first example with irredundance different from cardinality is a Kunen line.

Before beginning the examples we need the following topological lemma.

Lemma 8.4. *Suppose that X is a locally compact Hausdorff space, and Y is its one-point compactification. Then:*
(i) If the compact-open sets of X form a base, then Y is a Boolean space.

(ii) For every integer $k > 0$, if ^kX is hereditarily separable, then so is ^kY.

Proof. Recall that Y is obtained from X by adding one new point y, and declaring the topology on Y to consist of all open sets U of X together with all sets $\{y\} \cup U$ such that $U \subseteq X$ and $X \backslash U$ is compact in X. Note that if $U \subseteq X$ and $X \backslash U$ is compact in X then U is open in X. (i): We want to show that the clopen subsets of Y form a base. Any subset of X which is compact and open in X is clopen in Y. So it suffices to show that each "new" basic open set contains a new basic open set which is clopen. So, let W be a new basic open set—say $W = \{y\} \cup U$ where $U \subseteq X$ and $X \backslash U$ is compact in X. Then $X \backslash U \subseteq V$ for some compact open subset V of X, by the compactness of $X \backslash U$. Thus $\{y\} \cup (X \backslash V) \subseteq W$ and $\{y\} \cup (X \backslash V)$ is clopen in Y, as desired.

(ii) Fix $x \in X$. Assume that k is a positive integer and ^kX is hereditarily separable. Now suppose that S is a non-empty subspace of ^kY. For each $\Gamma \subseteq k$ let

$$S_\Gamma = \{z \in S : \forall i < k(z_i = y \text{ iff } i \in \Gamma)\};$$
$$S'_\Gamma = \{w \in {}^kX : \exists z \in S_\Gamma \forall i < k[(i \in \Gamma \Rightarrow w_i = x) \text{ and } (i \notin \Gamma \Rightarrow w_i = z_i)]\}.$$

Then for each $\Gamma \subseteq k$ let C'_Γ be a countable dense subset of S'_Γ. Next, let

$$C_\Gamma = \{z \in {}^kY : \exists w \in C'_\Gamma \forall i < k[(i \in \Gamma \Rightarrow z_i = y) \text{ and } (i \notin \Gamma \to z_i = w_i)\}.$$

We claim that $D \stackrel{\text{def}}{=} \bigcup_{\Gamma \subseteq k} C_\Gamma$ is dense in S (as desired). To this end, take an open set U such that $U \cap S \neq 0$. We may assume that U has the form $V_0 \times \cdots \times V_{k-1}$, where each V_i is open in Y. Say $U \cap S_\Gamma \neq 0$. Define, for $i < k$,

$$V'_i = \begin{cases} X, & \text{if } i \in \Gamma, \\ V_i \cap X, & \text{if } i \notin \Gamma. \end{cases}$$

Set $U' = V'_0 \times \cdots \times V'_{k-1}$. Then $U' \cap S'_\Gamma \neq 0$, so $U' \cap C'_\Gamma \neq 0$. Take $w \in U' \cap C'_\Gamma$. Define, for $i < k$,

$$z_i = \begin{cases} y, & \text{if } i \in \Gamma, \\ w_i, & \text{if } i \notin \Gamma. \end{cases}$$

Then $z \in U \cap C_\Gamma$, as desired. $\qquad\qquad\square$

Example 8.5. (CH) (*A compact Kunen line*). We construct a Boolean space making use of the topology on the real line; the resulting space is not linearly ordered, despite the name. We construct it by constructing a certain locally compact space, and then taking the one-point compactification to get the Boolean space we are interested in. Since we will be dealing with many topologies, we have to be precise about what we mean by a topology—for us, it is just the collection of all open sets. For any topology σ and any subset A of the space in question, $\bar{A}^\sigma$ denotes the closure of A with respect to the topology σ. Let $\langle x_\xi : \xi < \omega_1 \rangle$ be a one-one enumeration of $\mathbb{R}$. For each $\alpha \le \omega_1$ let $\mathbb{R}_\alpha = \{x_\xi : \xi < \alpha\}$. Let ρ be the usual topology on $\mathbb{R}$. Now we claim

(1) There is an enumeration $\langle S_\mu : \mu < \omega_1 \rangle$ of all of the countable subsets of $\mathbb{R} \times \mathbb{R}$ such that $S_\mu \subseteq \mathbb{R}_\mu \times \mathbb{R}_\mu$ for all $\mu < \omega_1$.

In fact, first let $\langle S'_\mu : \mu < \omega_1 \rangle$ be any old enumeration of the countable subsets of $\mathbb{R} \times \mathbb{R}$. We define $S_\mu = \mathbb{R}_\mu \times \mathbb{R}_\mu$ for all $\mu < \omega$. Now for $\omega \leq \mu < \omega_1$ let $S_\mu = S'_\nu$, where ν is minimum such that $S'_\nu \notin \{S_\eta : \eta < \mu\}$ and $S'_\nu \subseteq \mathbb{R}_\mu \times \mathbb{R}_\mu$. To see that this is the desired enumeration, suppose that S'_ν is not in the range of the function S, and choose ν minimum with this property. Then choose $\mu < \omega_1$ such that $\omega \leq \mu$, $S'_\rho \in \mathrm{Rng}(S \upharpoonright \mu)$ for each $\rho < \nu$, and $S'_\nu \subseteq \mathbb{R}_\mu \times \mathbb{R}_\mu$. Then the construction gives $S_\mu = S'_\nu$, contradiction.

Now we construct topologies τ_η for all $\eta \leq \omega_1$ so that the following conditions hold:

(2_η) τ_η is a topology on $\mathbb{R}_\eta$.

(3_η) $\tau_\xi = \{\mathbb{R}_\xi \cap U : U \in \tau_\eta\}$ for $\xi < \eta$.

(4_η) $\tau_\eta \supseteq \{\mathbb{R}_\eta \cap U : U \in \rho\}$.

(5_η) If $\xi, \xi' < \eta$, $\mu < \xi$ or $\mu < \xi'$, and $(x_\xi, x_{\xi'}) \in \bar{S}^\rho_\mu$, then $(x_\xi, x_{\xi'}) \in \bar{S}^{\tau_\eta}_\mu$.

(6_η) τ_η is first-countable.

(7_η) τ_η is Hausdorff.

(8_η) In τ_η, the compact open sets form a base.

For $\beta \leq \omega$ let τ_β be the discrete topology on $\mathbb{R}_\beta$. Then the conditions $(2_\beta) - (8_\beta)$ are clear; (5_β) holds since S_μ is finite under the indicated hypotheses.

Now assume that $\omega < \beta \leq \omega_1$ and τ_α has been constructed for all $\alpha < \beta$ so that $(2_\alpha) - (8_\alpha)$ hold. If β is a limit ordinal, let

$$\tau_\beta = \{U \subseteq \mathbb{R}_\beta : U \cap \mathbb{R}_\alpha \in \tau_\alpha \text{ for all } \alpha < \beta\}.$$

Then $(2_\beta) - (5_\beta)$ and (7_β) are clear. For (6_β), suppose that $\xi < \beta$; we want to find a countable neighborhood base for x_ξ. Let $\{U_n : n \in \omega\}$ be a countable neighborhood base for x_ξ in the topology $\tau_{\xi+1}$. If $V \in \tau_\beta$ and $x_\xi \in V$, then $V \cap \mathbb{R}_{\xi+1} \in \tau_{\xi+1}$, so there is an $n \in \omega$ such that $U_n \subseteq V \cap \mathbb{R}_{\xi+1} \subseteq V$, as desired. Finally, for (8_β), it suffices to notice that if $K \subseteq \mathbb{R}_\xi$ is compact in τ_ξ, where $\xi < \beta$, then it is compact in τ_β also.

Finally, suppose that β is an infinite successor ordinal $\alpha + 1$. If there is no $\mu < \alpha$ such that for some $\xi \leq \alpha$ we have $(x_\alpha, x_\xi) \in \bar{S}^\rho_\mu$ or $(x_\xi, x_\alpha) \in \bar{S}^\rho_\mu$, let τ_β be the topology with the base $\tau_\alpha \cup \{\{x_\alpha\}\}$. The conditions $(2_\beta) - (8_\beta)$ are easy to check.

Now suppose there is such a μ. Let T be the set of all ordered triples $(\gamma, \varepsilon, \mu)$ such that $\gamma, \varepsilon \leq \alpha$, $\gamma = \alpha$ or $\varepsilon = \alpha$, and $(x_\gamma, x_\varepsilon) \in \bar{S}^\rho_\mu$, where $\mu < \alpha$. Thus $0 < |T| \leq \omega$. Let $\langle (\xi_m, \eta_m, \mu_m) : m < \omega \rangle$ emumerate T, each element of T repeated infinitely many times. For each $\xi \leq \alpha$, let $\langle U^\xi_n : n < \omega \rangle$ be a decreasing sequence of open sets forming a neighborhood base for x_ξ in the usual topology ρ. Now for each $n < \omega$ choose $(p_n, q_n) \in S_{\mu_n} \cap (U^{\xi_n}_n \times U^{\eta_n}_n)$. Note that $p_n, q_n \in \mathbb{R}_{\mu_n}$ by (1). By (8_α)

we find compact open (in τ_α) $K_n \subseteq U_n^\alpha$ such that $p_n \in K_n$ if $\xi_n = \alpha$ and $q_n \in K_n$ if $\eta_n = \alpha$. Let τ_β be the topology on $\mathbb{R}_\beta$ having as a base the sets in τ_α together with all sets of the form $\{x_\alpha\} \cup \bigcup_{m>n} K_m$ for $n \in \omega$. We proceed to check $(2_\beta) - (8_\beta)$. (2_β) and (3_β) are clear. For (4_β), suppose that V is open in ρ. If $x_\alpha \notin V$, then $V \cap \mathbb{R}_\beta = V \cap \mathbb{R}_\alpha$ and so $V \cap \mathbb{R}_\alpha \in \tau_\alpha \subseteq \tau_\beta$. Suppose that $x_\alpha \in V$. For $x_\xi \in V$ with $\xi < \alpha$ we have $x_\xi \in V \cap \mathbb{R}_\alpha \in \tau_\alpha \subseteq \tau_\beta$. Choose $n \in \omega$ such that $U_n^\alpha \subseteq V$. Then $\bigcup_{m>n} K_m \subseteq U_n^\alpha \subseteq V$. Hence $V \cap \mathbb{R}_\beta = (V \cap \mathbb{R}_\alpha) \cup \{x_\alpha\} \cup \bigcup_{m>n} K_m \in \tau_\beta$, proving (4_β). For (5_β), assume that $\xi, \xi' < \beta$, $\mu < \xi$ or $\mu < \xi'$, and $(x_\xi, x_{\xi'}) \in \bar{S}_\mu^\rho$. We want to show that $(x_\xi, x_{\xi'}) \in \bar{S}_\mu^{\tau_\beta}$. To this end, take a neighborhood of $(x_\xi, x_{\xi'})$; we may assume that it has the form $W \times W'$ with W and W' open in τ_β. There are four possiblities. If $\xi = \alpha$ and $\xi' < \alpha$, we proceed as follows. We may assume that W has the form $\{x_\alpha\} \cup \bigcup_{m>n} K_m$ and W' has the form $U_n^{\xi'} \cap \mathbb{R}_\beta$ for some $n, a \in \omega$. Choose $r > n, a$ so that $(\xi_r, \eta_r, \mu_r) = (\xi, \xi', \mu)$. Then $(p_r, q_r) \in S_\mu \cap (W \times W')$, as desired. The possiblities $\xi < \alpha$ and $\xi' = \alpha$, and $\xi = \xi' = \alpha$ are treated similarly. The possibility $\xi, \xi' < \alpha$ follows easily from (5_α). So (5_β) is established. Condition (6_β) is obvious, as is (7_β). For (8_β), note that a set which is compact open in τ_α remains so in τ_β. Hence it suffices to show that $\{x_\alpha\} \cup \bigcup_{m>n} K_m$ is compact for each $n \in \omega$. Suppose that $\mathcal{O}$ is an open cover of this set. Choose $V \in \mathcal{O}$ such that $x_\alpha \in V$. Then there is a $p \in \omega$ such that $\{x_\alpha\} \cup \bigcup_{m>p} K_m \subseteq V$, and without loss of generality $n < p$. Since $\mathcal{O} \backslash \{V\}$ covers $\bigcup_{n<m\leq p} K_m$, which is compact, there is a finite subset of $\mathcal{O}$ which covers the desired set $\{x_\alpha\} \cup \bigcup_{m>n} K_m$. This finishes the construction of the topologies.

For brevity, let $\tau = \tau_{\omega_1}$. To proceed further, we need the following fact about the construction:

(9) If $A \subseteq \mathbb{R} \times \mathbb{R}$, then $|\bar{A}^\rho \backslash \bar{A}^\tau| \leq \omega$.

For, let B be countable and ρ-dense in A; thus $\bar{A}^\rho = \bar{B}^\rho$. Choose $\mu < \omega_1$ so that $B = S_\mu$. By condition (5_{ω_1}) we clearly have

$$\bar{A}^\rho \backslash \bar{A}^\tau \subseteq \bar{B}^\rho \backslash \bar{B}^\tau \subseteq \{x_\xi : \xi \leq \mu\} \times \{x_\xi : \xi \leq \mu\}.$$

and (9) follows.

(10) $(\mathbb{R}, \tau) \times (\mathbb{R}, \tau)$ is hereditarily separable.

To prove (10), let X be any subspace of $(\mathbb{R}, \tau) \times (\mathbb{R}, \tau)$. Let C be a countable subset of X which is ρ-dense in X. Then $C \cup (X \backslash \bar{C}^\tau) \subseteq C \cup (\bar{C}^\rho \backslash \bar{C}^\tau)$, so $C \cup (X \backslash \bar{C}^\tau)$ is countable by (9). It is τ-dense in X, since if $U, V \in \tau$ and $(U \times V) \cap X \neq 0$, then $(U \times V) \cap X \cap (X \backslash \bar{C}^\tau) = 0$ implies that $(U \times V) \cap X \subseteq \bar{C}^\tau$ and hence $(U \times V) \cap X \cap C \neq 0$, as desired.

Now we go to the final step in this example: let Y be the one-point compactification of $(\mathbb{R}, \tau)$. By Lemma 8.4, Y is a Boolean space. It is straightforward to check that the BA of closed-open sets is uncountable (new compact-open sets were introduced at each successor step). By (10) and Lemma 8.4, $Y \times Y$ is hereditarily separable. That the dual of Y has countable irredundance follows from the following result of Heindorf [89a] (upon noticing that $sA = c_{H+}A \leq d_{H+}A = hdA$ using 3.25 and 5.13). $\qquad\qquad\qquad\qquad\qquad\qquad\qquad\qquad\qquad\qquad\qquad$ $\square$

Theorem 8.6. *Let X be a Boolean space, and A its BA of closed-open sets. Then $\mathrm{Irr}A \leq s(X \times X)$.*

Proof. Suppose that I is an infinite irredundant subset of A; we will produce an ideal independent subset of $A \times A$ of power $|I|$ (as desired—see Theorem 3.24). Namely, take the set $\{a \times -a : a \in I\}$; it is as desired, for suppose that

$$a \times -a \subseteq (b_0 \times -b_0) \cup \ldots \cup (b_{m-1} \times -b_{m-1}),$$

where $a, b_0, \ldots, b_{m-1}$ are distinct elements of I. Now a is not in $\langle\{b_i : i < m\}\rangle$, so it follows that in that subalgebra, a splits some atom; this means that there is an $\varepsilon \in {}^m 2$ such that, if we set $d = \bigcap_{i<m} b_i^{\varepsilon i}$ then we have $d \cap a \neq 0 \neq d \cap -a$. Choosing $x \in d \cap a$ and $y \in d \cap -a$ it follows that $(x, y) \in a \times -a$ but $(x, y) \notin b_i \times -b_i$ for each $i < m$, giving the desired contradiction. $\qquad\qquad\qquad\qquad\qquad\qquad$ $\square$

This theorem gives rise to the following problem:

Problem 27. *Is it true that $\mathrm{Irr}A = s(A \oplus A)$ for every infinite BA A?*

Note that if $\mathrm{Irr}A = s(A \oplus A)$, then $\mathrm{Irr}(A \times A) = \mathrm{Irr}A$, since by 11.6(c) of the Boolean algebra handbook we have $(A \times A) \oplus (A \times A) \cong {}^4(A \oplus A)$, and hence

$$\mathrm{Irr}(A \times A) \leq s((A \times A) \oplus (A \times A))$$
$$= s({}^4(A \oplus A))$$
$$= s(A \oplus A) = \mathrm{Irr}A.$$

Example 8.7. This example, which as we mentioned is from Todorčević [89], constructs a topology on a certain subset of ${}^\omega\omega$. First, some notation: If A is a set with a linear order $<$ on it, and if $k \in \omega$, then $\langle A \rangle^k$ denotes the set of all $f \in {}^k A$ such that $f_i < f_j$ for all $i < j < k$. For $f, g \in {}^\omega\omega$ define $f <^* g$ if $\exists m \forall n \geq m(fn < gn)$. The BA we want will be constructed under the assumption that there is a subset A of ${}^\omega\omega$ of power ω_1 which is unbounded under $<^*$. This is an obvious consequence of CH, but is weaker.

Without loss of generality A has order type ω_1 under $<^*$ and all members of A are strictly increasing. In fact, take the A originally given, and write $A = \{f_\alpha : \alpha < \omega_1\}$. Then one can inductively define $\bar{f}_\alpha$ for $\alpha < \omega_1$ so that $\bar{f}_\beta <^* \bar{f}_\alpha$ for $\beta < \alpha$, $f_\alpha <^* \bar{f}_\alpha$, and $\bar{f}_\alpha$ is strictly increasing. Namely, let $\bar{f}_0$ be arbitrary. If $\bar{f}_\beta$ has been constructed for all $\beta < \alpha$, let $\langle g_n : n < \omega \rangle$ enumerate $\langle \bar{f}_\beta : \beta < \alpha \rangle$. Define $\bar{f}_\alpha(n)$ to be $> \bar{f}_\alpha(m)$ for all $m < n$, also $> f_\alpha(n)$, and also $> g_m(n)$ for all $m < n$. Clearly this works. The new set $\{\bar{f}_\alpha : \alpha < \omega_1\}$ (still denoted by A below) has the desired properties.

We will apply the above notation $\langle A \rangle^k$ to A under the ordering $<^*$. Let T be an Aronszajn subtree of $\{s \in {}^{<\omega_1}\omega : s$ is one-one$\}$. (See Kunen [80], p. 70.)

For each $\alpha < \omega_1$ let t_α be a member of T with domain α. Define $e : \langle A \rangle^2 \to \omega$ by $e(\bar{f}_\alpha, \bar{f}_\beta) = t_\beta \alpha$ for $\alpha < \beta$. Then the following conditions clearly hold:

(1) For all $b \in A$, the function $e_b \overset{\text{def}}{=} e(\cdot, b)$ is a one-one map from $A_b \overset{\text{def}}{=} \{a \in A : a <^* b\}$ into ω.

(2) For all $a \in A$, the set $\{e_b \upharpoonright A_a : b \in A, a <^* b\}$ is countable.

For distinct $a, b \in A$ let $\Delta(a, b)$ be the least $n < \omega$ such that $an \neq bn$. And let $\Delta(a, a) = \infty$. The following fact about this notation will be useful:

$(\star)$ If $\Delta(a, b) < \Delta(c, b)$ then $\Delta(a, c) = \Delta(a, b)$.

To see this, note that $a \upharpoonright \Delta(a, b) = b \upharpoonright \Delta(a, b) = c \upharpoonright \Delta(a, b)$, and $a\Delta(a, b) \neq b\Delta(a, b) = c\Delta(a, b)$, so $\Delta(a, c) = \Delta(a, b)$.
   Now we define $H : A \to \mathscr{P}A$ by

$$Hb = \{a \in A : a <^* b \text{ and } e(a, b) \leq b(\Delta(a, b))\}.$$

Note by (1) and the definition of H we have

(3)   for all $l < \omega$ and $b \in A$ the set $\{a \in Hb : \Delta(a, b) = l\}$ is finite.

Next we define Cb for $b \in A$ by recursion on b: $a \in Cb$ iff $a = b$ or

(4) $\exists c \in Hb(a \in Cc \text{ and } \forall d \in Hb(d \neq a \text{ and } d \neq c \Rightarrow \Delta(a, d) < \Delta(a, c)))$.

Note that

(5) $Hb \subseteq Cb$

for all $b \in A$ (if $a \in Hb$, take $c = a$ and note that $\Delta(a, a) = \infty$).
   For each $n \in \omega$ and $b \in A$ let $C_n b = \{a \in Cb : \Delta(a, b) \geq n\}$. Then

(6) $c \in Hb \Rightarrow \exists l(C_l c \subseteq Cb)$.

In fact, $\{x \in Hb : \Delta(x, b) = \Delta(c, b)\}$ is finite by (3). Choose $l > \Delta(x, c)$ for any $x \neq c$ which is in this set, and with $l > \Delta(c, b)$. Suppose that $d \in C_l c$. We claim that $d \in Cb$, and that the element c works to show this in (4). Indeed, suppose that $x \in Hb$, $x \neq d$, $x \neq c$, and $\Delta(d, x) \geq \Delta(d, c)$. Now $\Delta(c, b) < \Delta(d, c)$, so $\Delta(b, d) = \Delta(c, b)$ by $(\star)$, and $\Delta(b, d) < \Delta(d, x)$, so $\Delta(x, b) = \Delta(c, b)$ by $(\star)$. So x is in the indicated set, which gives $\Delta(x, c) < l \leq \Delta(c, d) \leq \Delta(d, x)$, so $\Delta(c, d) = \Delta(x, c)$ by $(\star)$, contradiction.

(7) $a \in Cb \Rightarrow \exists l(C_l a \subseteq Cb)$.

For, we may assume that $a \notin Hb$ by (6), and we proceed by induction on b. The conclusion is clear if $a = b$, so suppose that $a \neq b$. Choose c in accordance with

(4). Then $a \neq c$ since $a \notin Hb$. By the induction hypothesis, choose l such that $C_l a \subseteq Cc$. Without loss of generality, $l > \Delta(a, c)$. We claim that $C_l a \subseteq Cb$. To prove this, let $d \in C_l a$. So, $d \in Cc$. Suppose $x \in Hb$, $x \neq d$, and $x \neq c$ then $x \neq a$ since $a \notin Hb$. So $\Delta(a, x) < \Delta(a, c)$. Now $\Delta(a, c) < l \leq \Delta(a, d)$, so $\Delta(c, d) = \Delta(a, c)$ by $(\star)$. Also, $\Delta(a, x) < \Delta(a, d)$, so $\Delta(d, x) = \Delta(a, x) < \Delta(a, c) = \Delta(c, d)$, showing that $d \in Cb$.

From (7) we immediately get

$$(8) \qquad\qquad a \in C_m b \Rightarrow \exists l (C_l a \subseteq C_m b).$$

From (8) it follows that the collection of sets $\{C_m b : b \in A, m \in \omega\}$ forms a base for a topology on A. It is Hausdorff, since, given $a \neq b$, let $l = \Delta(a, b) + 1$; clearly $C_l a \cap C_l b = 0$. Also note that each set $Cb = C_0 b$ is open. Next,

$$(9) \qquad\qquad C_l b \text{ is closed in } Cb.$$

For, suppose that $x \in Cb \backslash C_l b$, and let $m = \Delta(x, b) + 1$. Then clearly $C_m x \cap Cb \subseteq Cb \backslash C_l b$, as desired.

$$(10) \qquad\qquad Cb \text{ is compact.}$$

We prove this by induction on b. So, assume that it is true for all $c <^* b$, and suppose that $Cb \subseteq \bigcup_{x \in X} C_{m(x)} x$. Then choose $y \in X$ such that $b \in C_{m(y)} y$. There is an l such that $C_l b \subseteq C_{m(y)} y$. Now we consider two cases:

Case 1. Hb is finite. In this case, we can easily show that Cb is closed: suppose that $a \in A \backslash Cb$. Hence $a \neq b$ and

$$(*) \quad \forall c \in Hb(a \notin Cc \text{ or } \exists d \in Hb(d \neq a \text{ and } d \neq c \text{ and } \Delta(a, d) \geq \Delta(a, c))).$$

If $c \in Hb$ and $a \notin Cc$, choose an open neighborhood U_c of a with the property that $U_c \cap Cc = 0$, using the inductive hypothesis. For $c \in Hb$ and $a \in Cc$, choose $d = d(a, c) \in Hb$ such that $d \neq a$, $d \neq c$, and $\Delta(a, d) \geq \Delta(a, c)$. Let

$$V = C_{\Delta(a,b)+1} a \cap \bigcap_{c \in Hb, a \notin Cc} U_c \cap \bigcap_{c \in Hb, a \in Cc} C_{\Delta(a,d(a,c))+1} a.$$

We claim that $V \cap Cb = 0$ (as desired, showing that Cb is closed). For, suppose that $x \in V \cap Cb$. Since $x \in C_{\Delta(a,b)+1} a$, we have $x \neq b$. Choose, then, $c \in Hb$ such that $x \in Cc$ and for all $d \in Hb$, if $d \neq x$ and $d \neq c$ then $\Delta(x, d) < \Delta(x, c)$. If $a \notin Cc$, then $x \in U_c \cap Cc$, contradiction. So $a \in Cc$. Set $d = d(a, c)$. Now $x \in C_{\Delta(a,d)+1} a$, so $x \neq d$ and $\Delta(a, x) > \Delta(a, d)$. Hence $\Delta(d, x) = \Delta(a, d)$ by $(\star)$. Now $\Delta(a, d) \geq \Delta(a, c)$, so $\Delta(a, c) < \Delta(a, x)$. Hence by $(\star)$, $\Delta(c, x) = \Delta(a, c) \leq \Delta(a, d) = \Delta(d, x)$, contradicting the choice of c.

Now for each $c \in Hb$ we have that $Cc \cap Cb$ is a closed subset of Cc, and hence the inductive hypothesis finishes this case. (Here one should note that $Cb = \{b\} \cup \bigcup_{c \in Hb}(Cc \cap Cb)$.)

Case 2. Hb is infinite. For all $c \in Hb$ let

$$D_c = \{a : a \notin Hb, \Delta(a,b) < l, \ a \in Cc, \text{ and}$$
$$\forall d \in Hb(d \neq a \text{ and } d \neq c \Rightarrow \Delta(a,d) < \Delta(a,c))\}.$$

Now

(**) If $c \in Hb$ and $a \in D_c$, then $\Delta(c,b) < l$.

For, assume otherwise. Now $\Delta(a,b) < \Delta(c,b)$, so $\Delta(a,c) = \Delta(a,b)$ by $(\star)$. Also, for all $d \in Hb \backslash \{a,c\}$ we have $\Delta(d,a) < \Delta(a,c) = \Delta(a,b)$, hence by $(\star)$ we get $\Delta(d,b) = \Delta(d,a) < \Delta(a,c)$. So Hb is finite by (3), contradiction. Thus (**) holds.

Let $Y = \{c \in Hb : \Delta(c,b) < l\}$. Note that Y is finite by (3). Now

(***) If $c \in Y$, $a \in D_c$, and $m = \Delta(a,c)$, then $C_m c \subseteq Cb$.

For, assume the hypotheses. If $\Delta(a,c) \leq \Delta(a,b)$, then $d \in Hb \backslash \{a,c\}$ implies that $\Delta(a,d) < \Delta(a,c) \leq \Delta(a,b)$, so $\Delta(d,b) = \Delta(a,d) < \Delta(a,c)$ by $(\star)$, hence Hb is finite by (3), contradiction. Thus $\Delta(a,c) > \Delta(a,b)$. Now suppose that $u \in C_m c$. Thus $\Delta(a,c) \leq \Delta(u,c)$. Suppose that $d \in Hb \backslash \{u,c\}$. Then $d \neq a$ since $a \notin Hb$. So $\Delta(a,d) < \Delta(a,c)$ since $a \in D_c$, so $\Delta(d,c) = \Delta(a,d) < \Delta(a,c) \leq \Delta(u,c)$ by $(\star)$, hence by $(\star)$ again, $\Delta(d,u) = \Delta(d,c) < \Delta(u,c)$. This shows that $u \in Cb$, and it proves (***).

For $c \in Y$ with $D_c \neq 0$ let $n(c) = \min\{\Delta(a,c) : a \in D_c\}$. Then

(****) $C_b = C_l b \cup Y \cup \bigcup_{c \in Y, D_c \neq 0} C_{n(c)} c$.

For, $\supseteq$ holds by (5) and (***). For $\subseteq$, suppose that $a \in Cb$, $a \notin C_l b$, $a \notin Y$. Since $a \notin C_l b$, we have $a \neq b$. Since $a \notin C_l b$ and $a \notin Y$, we have $a \notin Hb$. Since $a \in Cb$, choose $c \in Hb$ such that $a \in Cc$ and $\forall d \in Hb(d \neq a$ and $d \neq c \Rightarrow \Delta(a,d) < \Delta(a,c))$. So $a \in D_c$. By (**) we get $c \in Y$. Thus $a \in C_{n(c)} c$, as desired for $\subseteq$; (****) has been proved.

By the inductive hypothesis each $C_{n(c)} c$ is compact, so it follows that Cb is compact in Case 2.

So, we have proved (10).

From (9) and (10) we get

(11) $C_l b$ is compact; so A is locally compact.

(12) $a <^* b \Rightarrow Ca \neq Cb$.

This is true because $b \in Cb \backslash Ca$. So there are uncountably many compact open sets.

A subset $F \subseteq \langle A \rangle^k$ is *cofinal in* A provided that for all $a \in A$ there is an $f \in F$ such that $a <^* f_i$ for all $i < k$. Next we prove

(13) $\forall$ finite $k \geq 1$ and $\forall$ cofinal $F \subseteq \langle A \rangle^k \exists f, g \in F(f_i \in Hg_i$ for all $i < k)$.

This will take a while to prove, but it leads us close to the end of the matter. Since F is cofinal in A, we may assume that there is an enumeration $\langle h^\alpha : \alpha < \omega_1 \rangle$ of F such that $h_i^\alpha <^* h_j^\beta$ whenever $\alpha < \beta < \omega_1$ and $i, j < k$. Hence every uncountable subset of F is also cofinal in A; this is all we need the sequence $\langle h^\alpha : \alpha < \omega_1 \rangle$ for. Let $D \subseteq F$ be countable dense in F in the ordinary topology ($^\omega \omega$ has the product topology with ω having the discrete topology; $^k(^\omega \omega)$ gets the product topology too). Choose $c \in A$ such that $f_i <^* c$ for all $f \in D$ and $i < k$. Now

$$\forall f \in F \exists m \in \omega \forall n \geq m \forall i < j < k (f_i n < f_j n).$$

Hence there is an uncountable $F_0 \subseteq F$ and an m_0 such that

(14) $\forall f \in F_0 \forall n \geq m_0 \forall i < j < k (f_i n < f_j n).$

Next, let $F_1 = \{f \in F_0 : c <^* f_0\}$. Then

$$\forall f \in F_1 \exists m > m_0 \forall n \geq m (c_n < f_0 n).$$

Hence there is an uncountable $F_2 \subseteq F_1$ and an $m > m_0$ such that

(15) $\forall f \in F_2 \forall n \geq m (c_n < f_0 n).$

Now

$$F_2 = \bigcup \{\{f \in F_2 : \forall i < k (f_i \restriction m = s_i)\} : s \in {}^k(^m \omega)\}.$$

Hence there is an uncountable subset F_3 of F_2 and an $s \in {}^k(^m \omega)$ such that

(16) $\forall f \in F_3 \forall i < k (f_i \restriction m = s_i).$

Let $C = \{e_b \restriction A_c : b \in A\}$. Then

$$F_3 = \bigcup \{\{f \in F_3 : \forall i < k (e_{f_i} \restriction A_c = u^i)\} : u \in {}^k C\},$$

so, using (2), there is an uncountable $F_4 \subseteq F_3$ and a $u \in {}^k C$ such that

(17) $\forall f \in F_4 \forall i < k (e_{f_i} \restriction A_c = u^i).$

Now there is an $n \in \omega$ such that $\{f_0 n : f \in F_4\}$ is unbounded in ω, since otherwise, for each $n \in \omega$ let gn be greater than each $f_0 n$ for $f \in F_4$. Since F_4 is cofinal in A, it follows that g is an upper bound for A, contradiction. Take $m0$ to be the least such n. Then there is a $p \in \omega$ such that $\{f_0 \restriction m0 : f \in F_4\} \subseteq {}^{m0} p$; so there is a $t_0 \in {}^{m0} p$ and an infinite subset F_5 of F_4 such that $f_0 \restriction m0 = t_0$ for all $f \in F_5$ and $\{f_0 m0 : f \in F_5\}$ is unbounded in ω. Wlog $\langle f_0 m0 : f \in F_5 \rangle$ is one-one. Let $n = \max(m0, m_0)$. For any $i \in \omega$ choose $f \in F_5$ such that $i < f_0 m0$. then $i < f_0 n < f_1 n$ too. Thus $\{f_1 n : f \in F_5\}$ is unbounded. Let $m1$ be minimum such that $\{f_1 m1 : f \in F_5\}$ is unbounded. Continuing in this fashion, we get $m0, m1, \ldots, m(k-1), t_0, \ldots, t_{k-1}, F_5, \ldots, F_{4+k}$ such that $F_4 \supseteq F_5 \supseteq \cdots \supseteq F_{4+k}$, F_{4+k} is infinite, and the following conditions hold:

(18) $\qquad \forall f \in F_{4+k} \forall i < k (t_i \subseteq f_i),\ t_i \in {}^{mi}\omega.$

(19) $\qquad \langle f_i mi : f \in F_{4+k} \rangle$ is one-one for all $i < k$.

Now the open set in ${}^k({}^\omega \omega)$ determined by $\langle t_0, \ldots, t_{k-1} \rangle$ meets F, since F_{4+k} is contained in it; so by the denseness of D, choose $d \in D$ in this open set: $t_i \subseteq d_i$ for all $i < k$. By (19) there is an $f \in F_{4+k}$ such that

$$\forall i < k [f_i mi > \max\{u^j d_j : j < k\}].$$

Hence for all $i < k$ we have $f_i \restriction mi = t_i = d_i \restriction mi$, so $m_i \leq \Delta(d_i, f_i)$, and hence

$$e(d_i, f_i) = u^i d_i < f_i mi \leq f_i(\Delta(d_i, f_i)),$$

so d_i is in Hf_i for all $i < k$, as desired; we have proved (13)!

Next,

(21) For every positive integer k, the space ${}^k A$ does not have an uncountable discrete subspace.

We prove this by induction on k; suppose that it is true for all $k' < k$. Suppose that F is an uncountable discrete subspace of ${}^k A$. We may assume that there is an integer m such that $(C_m f_0 \times \cdots \times C_m f_{k-1}) \cap F = \{f\}$ for all $f \in F$. For all $f \in F$ define $\equiv_f$ on k by $i \equiv_f j$ iff $f_i = f_j$. Without loss of generality, $\equiv_f$ is the same for all $f \in F$. Hence by the induction hypothesis, $\equiv$ is the identity relation, so that each $f \in F$ is one-one. And then by a similar argument with permutations of k we may assume that $f_i <^* f_j$ whenever $f \in F$ and $i < j < k$. Next, we may assume that $\langle \text{rng} f : f \in F \rangle$ is a Δ-system, say with kernel G. For each $f \in F$, $\{i < k : f_i \in G\}$ is a finite subset of k; we may assume that this set is the same for all $f \in F$; call the set Γ. Thus $f_i = g_i$ for all $f, g \in F$ and all $i \in \Gamma$. So the set $F' \stackrel{\text{def}}{=} \{f \restriction (k \backslash \Gamma) : f \in F\}$ is still uncountable and discrete. Since $\langle \text{rng} f : f \in F' \rangle$ is a system of disjoint sets, F' is cofinal in A. And

$$F' = \bigcup \{\{f \in F : \forall i < k (f_i \restriction m = s_i)\} : s \in {}^k({}^m \omega)\},$$

so we may assume that $f_i \restriction m = g_i \restriction m$ for all $f, g \in F'$ and $i < k$. Now we apply (13) to get distinct $f, g \in F'$ such that $f_i \in Hg_i$ for all $i < k$. Since $Ha \subseteq Ca$ for all a, this clearly is a contradiction. So (21) holds.

The only remaining step is to take the one-point compactification A' of A. Lemma 8.1(i) says that A' is a Boolean space. An easy argument shows that ${}^k A'$ has no uncountable discrete subspace. $\qquad \square$

Problem 28. *Can one construct in ZFC a BA A such that $\mathrm{Irr}A < |A|$?*

This is Problem 21 in Monk [90].

There are reasonable finite versions of irredundance. For positive integers m, n, call a subset X of A *mn-irredundant* if for all $x \in X$ one cannot write

$$x = \sum_{i<m} \prod_{j<n} a_{ij},$$

with $a_{ij} \in X \backslash \{x\}$ or $-a_{ij} \in X \backslash \{x\}$ for all $i < m$, $j < n$. So, X is irredundant iff it is mn-irredundant for all m, n. And define

$$\mathrm{Irr}_{mn}A = \sup\{|X| : X \subseteq A, X \; mn - \text{irredundant}\}.$$

These function have not been studied. But note that in Rubin's algebra described in Chapter 18 we have $\mathrm{Irr}_{12}A = \omega$.

Since $\pi A = |A|$ for A a tree algebra, we also have $\mathrm{Irr}A = |A|$ for A a tree algebra.

9. Cardinality

We denote $|A|$ also by Card$A$. The behaviour of this function under algebraic operations is for the most part obvious. Note, though, that questions about its behaviour under ultraproducts are the same as the well-known and difficult problems concerning the cardinality of ultraproducts in general. Card$_{H-}$ is a non-obvious function. Clearly Card$_{H-}A \leq 2^\omega$ for every infinite BA, and Card$_{H-}A = \omega$ for many BAs, e.g. for free BAs and interval algebras. But Card$_{H-}A = 2^\omega$ for A satisfying CSP. W. Just and P. Koszmider [91] have shown that it is consistent to have a BA A such that $\omega_1 \leq$ Card$_{H-}A = |A| < 2^\omega$. Questions about Card$_{H-}$ are connected to some problems about cofinality and related cardinal functions which will not be considered here; see van Douwen [89]. The cardinal function Card$_{h+}$ is defined as follows:

$$\text{Card}_{h+}A = \sup\{|\text{Clop}X| : X \subseteq \text{Ult}A\}.$$

It is possible to have Card$_{h+}A > |\text{Ult}A|$: this is true, for example, with A the finite-cofinite algebra on an infinite cardinal κ, taking X to be the set of all principal ultrafilters on A, so that X is discrete and hence Clop$X = \mathscr{P}X$ and Card$_{h+}A = 2^\kappa$. On the other hand, Card$_{h-}$ coincides with Card$_{H-}$: obviously Card$_{h-}A \leq$ Card$_{H-}A$, and if X is any infinite subset of Ult$A$, then the function f such that $fa = Sa \cap X$ is a homomorphism from A onto an algebra B such that $|B| \leq$ Clop$X$; so Card$_{H-}A \leq$ Card$_{h-}A$. Clearly $_d$Card$_{S+}A = |A|$, and $_d$Card$_{S-}A = \pi A$.

We shall now go into some detail concerning the spectrum function Card$_{Hs}$, which seems to be another interesting derived function associated with cardinality. First we note some more-or-less obvious facts: (1) If A is an infinite free BA, then Card$_{Hs}A = [\omega, |A|]$; (2) If A is infinite and complete, then Card$_{Hs}A = [\omega, |A|] \cap \{\kappa : \kappa^\omega = \kappa\}$ (using in an essential way the Balcar-Franěk theorem); (3) if $\omega \leq \kappa \leq |A|$, then Card$_{Hs}A \cap [\kappa, 2^\kappa] \neq 0$; (4) if A has a free subalgebra of power $\kappa \geq \omega$, then Card$_{Hs}A \cap [\kappa, \kappa^\omega] \neq 0$. Now we prove a few more involved things.

Lemma 9.1. *If κ is an infinite cardinal, L is a linear ordering, the sequence $\langle a_\alpha : \alpha < \kappa \rangle$ is strictly increasing in L, and A is the interval algebra on L, then $[\omega, \kappa] \subseteq$ Card$_{Hs}A$.*

Proof. It suffices to show that $\kappa \in$ Card$_{Hs}A$. Define $x \equiv y$ iff $x, y \in L$ and $\forall \alpha < \kappa[(a_\alpha < x$ iff $a_\alpha < y)$ and $(x < a_\alpha$ iff $y < a_\alpha)]$. Then $\equiv$ is a convex equivalence relation on L with the equivalence classes of order type κ or $\kappa + 1$, and the desired homomorphism is easy to define. $\square$

Corollary 9.2. *If κ is an infinite cardinal and A is the interval algebra on κ, then Card$_{Hs}A = [\omega, \kappa]$.* $\square$

Corollary 9.3. *If κ is an infinite cardinal, L is a linear ordering of power $\geq (2^\kappa)^+$, and A is the interval algebra on L, then $[\omega, \kappa^+] \subseteq$ Card$_{Hs}A$.*

Proof. One can apply the Erdös-Rado theorem $(2^\kappa)^+ \to (\kappa^+)^2_\kappa$ to get a chain in L of order type κ^+ or $(\kappa^+)^*$. □

Theorem 9.4. *Let A be the interval algebra on $\mathbb{R}$. Then $\mathrm{Card}_{\mathrm{Hs}}A = \{\omega, 2^\omega\}$.*

Proof. The inclusion $\supseteq$ is obvious. Now suppose that f is a homomorphism of A onto an uncountable BA B; we want to show that $|B| = 2^\omega$. Notice that f is determined by a convex equivalence relation E on $\mathbb{R}$, where the number of E-equivalence classes is $|B|$. Now $L' \overset{\text{def}}{=} \bigcup\{k : k \text{ is an } E\text{-equivalence class with } |k| > 1\}$ is Borel, so $L'' \overset{\text{def}}{=} \mathbb{R}\backslash L'$ is also. There are only countably many E-equivalence classes k such that $|k| > 1$, so clearly $|L''| = |B|$. Hence $|B| = 2^\omega$ by the Aleksandroff-Hausdorff theorem (see Kuratowski [58] Theorem 3, p. 355). □

Theorem 9.5. $\mathrm{Card}_{\mathrm{Hs}}(A \times B) = \mathrm{Card}_{\mathrm{Hs}}A \cup \mathrm{Card}_{\mathrm{Hs}}B$.

Proof. The inclusion $\supseteq$ is obvious. For $\subseteq$, use the elementary fact given in the discussion of c_{Hr} in Chapter 3. □

Theorem 9.6. $\mathrm{Card}_{\mathrm{Hs}}(A \oplus B) = \mathrm{Card}_{\mathrm{Hs}}A \cup \mathrm{Card}_{\mathrm{Hs}}B$.

Proof. The inclusion $\supseteq$ is obvious. If f is a homomorphism from $A \oplus B$ onto C, then $f[A] \cup f[B]$ generates C, so the other inclusion follows. □

Corollary 9.7. *If $\omega \leq \kappa \leq 2^\omega$, then there is a BA A such that $\mathrm{Card}_{\mathrm{Hs}}A = [\omega, \kappa] \cup \{2^\omega\}$.*

Proof. Apply Theorem 9.5 to $A \times \mathscr{P}\omega$, where A is the free BA on κ free generators. □

The strongest result known about $\mathrm{Card}_{\mathrm{Hs}}$ is a special case of the following theorem of Juhász [92]:

Let κ be a regular uncountable cardinal and let X be a compact Hausdorff space of weight at least κ. Then there is a closed subspace $F \subseteq X$ such that the weight of F is in $[\kappa, 2^{<\kappa}]$ and

$$|F| \leq \sum_{\lambda < \kappa} 2^{2^\lambda}.$$

As a corollary, under GCH for every BA A the set $\mathrm{Card}_{\mathrm{Hs}}A$ contains all regular uncountable cardinals $\leq |A|$. As a special case, this solves Problem 22 of Monk [90].

Note that the relations $\mathrm{Card}_{\mathrm{Sr}}$ and $\mathrm{Card}_{\mathrm{Hr}}$ are trivial. In comparing cardinality with the cardinal functions so far introduced, we now note explicitly that $2^{\mathrm{d}A} \geq |A|$ for any infinite BA A. Finally, recall from Part I of the BA Handbook, Theorem 12.2, that $|A|^\omega = |A|$ for any infinite CSP algebra A, in particular for any (countably) complete infinite BA A.

10. Independence

There is a lot of information about independence in Part I of the Handbook. An even more extensive account is in Monk [83].

To treat the attainment problem, it is again convenient to first talk about independence in products.

Theorem 10.1. *If neither A nor B has an independent set of power $\kappa \geq \omega$ then $A \times B$ also does not.*

Proof. Let $\langle (a_\alpha, b_\alpha) : \alpha < \kappa \rangle$ be a system of elements of $A \times B$; we want to show that this system is dependent. Choose a finite subset Γ of κ and $\varepsilon \in {}^\Gamma 2$ so that $\prod_{\alpha \in \Gamma} a_\alpha^{\varepsilon\alpha} = 0$, and then choose a finite subset Δ of $\kappa \backslash \Gamma$ and $\delta \in {}^\Delta 2$ so that $\prod_{\alpha \in \Delta} b_\alpha^{\delta\alpha} = 0$. Let $\Theta = \Gamma \cup \Delta$ and $\theta = \delta \cup \varepsilon$. Then $\prod_{\alpha \in \Theta} (a_\alpha, b_\alpha)^{\theta\alpha} = 0$, as desired. $\qquad \square$

Corollary 10.2. $\mathrm{Ind}(A \times B) = \max(\mathrm{Ind}A, \mathrm{Ind}B)$ for infinite BAs A, B. $\qquad \square$

Corollary 10.3. *If $\langle A_i : i \in I \rangle$ is a system of BAs, κ is an infinite cardinal, and for every $i \in I$, the set A_i does not have an independent subset of power κ, then $\prod_{i \in I}^{\mathrm{w}} A_i$ also has no such subset.*

Proof. Suppose that X is an independent subset of $\prod_{i \in I}^{\mathrm{w}} A_i$ of power κ. Fix $x \in X$. We may assume that $F \stackrel{\mathrm{def}}{=} \{i \in I : x_i \neq 0\}$ is finite. Then

$$\langle y \restriction \prod_{i \in F} A_i : y \in X \backslash \{x\} \rangle$$

gives κ independent elements of $\prod_{i \in F} A_i$, contradicting Theorem 10.1. $\qquad \square$

Corollary 10.4. $\mathrm{Ind}(\prod_{i \in I}^{\mathrm{w}} A_i) = \sup_{i \in I} \mathrm{Ind}A_i$. $\qquad \square$

Corollary 10.3 enables us to take care of the attainment problem for independence. For each limit cardinal κ there is a BA A with independence κ not attained. For $\kappa = \omega$ we simply take for A any infinite superatomic BA. Now assume that κ is an uncountable limit cardinal. Let I be the set of all infinite cardinals $< \kappa$, and for each $\lambda \in I$ let B_λ be the free BA with λ free generators. Then $A \stackrel{\mathrm{def}}{=} \prod_{\lambda \in I}^{\mathrm{w}} B_\lambda$ is as desired, by Corollary 10.3.

It is perhaps surprising that the analog of Corollary 10.4 for arbitrary products is false. This follows from a theorem of L. Heindorf [92]; it was known earlier— see Cramer [74], but the construction there is rather ad hoc. Heindorf proves that $|A| \leq \mathrm{Ind}(\prod_{n \in \omega \backslash 1} A^{*n})$ for any infinite BA A, where A^{*n} is the free product of A with itself n times. For later purposes we need a modification of his theorem, as follows.

Theorem 10.5. *If A is an infinite BA and X is an infinite disjoint subset of A^+, then there is a function f mapping X into $\prod_{n \in \omega \backslash 1} A^{*n}$ such that if M and N are* *finite disjoint subsets of X then the set*

$$\left\{ n \in \omega\backslash 1 : \prod_{x \in M} f_x n \cdot \prod_{x \in N} -f_x n = 0 \right\}$$

is finite.

In particular, $cA \le \mathrm{Ind}(\prod_{n \in \omega\backslash 1} A^{*n})$.

Proof. For each $a \in X$ we define $f_a \in \prod_{n \in \omega\backslash 1} A^{*n}$ by setting $f_a n = g_0(-a) \cdot \ldots \cdot g_{n-1}(-a)$ for each $n \in \omega\backslash 1$, where g_i is the natural isomorphism of A onto the i-th free factor of A^{*n} for each $i < n$. Given finite disjoint subsets M and N of X, let $m = |N|+1$. We claim that $(\prod_{a \in M} f_a \cdot \prod_{a \in N} -f_a)n \ne 0$ for all $n \ge m$. For, extend N to a subset of X, still disjoint from M, of size n. Write $N' = \{a_0, \ldots, a_{n-1}\}$. Then

$$\left(\prod_{a \in M} f_a \cdot \prod_{a \in N'} -f_a \right) n = \prod_{a \in M} g_0(-a) \cdot \ldots \cdot \prod_{a \in M} g_{n-1}(-a) \cdot$$
$$\prod_{a \in N'} (g_0 a + \cdots + g_{n-1}a)$$
$$\ge \prod_{a \in M} g_0(-a) \cdot \ldots \cdot \prod_{a \in M} g_{n-1}(-a) \cdot$$
$$g_0 a_0 \cdot \ldots \cdot g_{n-1}a_{n-1} \ne 0,$$

as desired.   □

Now, given an infinite cardinal κ, let A be the finite-cofinite algebra on κ. Then each algebra A^{*n} is superatomic, hence has no infinite independent set, but the product $\prod_{n \in \omega\backslash 1} A^{*n}$ has independence at least κ. This shows a total failure of Corollary 10.4 for full direct products. Although this example takes care of the most obvious question about independence in products, there is another related question, namely whether an example of this sort can be done with an interval algebra (they always have independence ω too, just like superatomic algebras, although independence is attained for some interval algebras). The answer is no, and after several partial results by several mathematicians a complete solution was given by Shelah in December 1992; see Shelah [94d]:

If A_i is a non-trivial interval algebra for each $i \in I$, where I is infinite, then $\mathrm{Ind}(\prod_{i \in I} A_i) = 2^{|I|}$.

This answers Problem 23 in Monk [90].

We also give Heindorf's theorem itself:

Theorem 10.6. *If A is an infinite BA, then $|A| \le \mathrm{Ind}\left(\prod_{n \in \omega\backslash 1} A^{*n}\right)$.*

Proof. Let F be the set of all functions f such that f maps ${}^m 2$ into 2 for some $m \in \omega\backslash 1$; m is denoted by ρf. Let B be a free BA with free generators x_a for

$a \in A$. It suffices to isomorphically embed B into $\prod_{f \in F} A^{*\rho f}$. For each $f \in F$ and each $i < \rho f$ let g_i^f be the natural embedding of A into the i-th free factor of $A^{*\rho f}$. We define $G : B \to \prod_{f \in F} A^{*\rho f}$ by

$$(Gx_a)_f = \sum_{\varepsilon \in {}^{\rho f}2, f\varepsilon = 1} \left(\prod_{j < \rho f} (g_j^f a)^{\varepsilon j} \right),$$

extending G to a homomorphism. We want to show that G is one-one. To this end, let $a_0, \ldots, a_{n-1}$ be distinct elements of A and suppose that $\varepsilon \in {}^n 2$; we want to show that

$$y \overset{\text{def}}{=} (Gx_{a_0})^{\varepsilon 0} \cdot \ldots \cdot (Gx_{a_{n-1}})^{\varepsilon(n-1)} \neq 0.$$

Let $\Gamma = \{(i,j) : i < j < n\}$, and choose m and h so that h is a one-one function from m onto Γ. For each $k < m$ write $hk = (i,j)$, and let F_k be any ultrafilter on A such that $a_i \triangle a_j \in F_k$. For each $l < n$ we define $\delta_l \in {}^m 2$ by setting, for any $k < m$,

$$\delta_l k = \begin{cases} 1 & \text{if } a_l \in F_k, \\ 0 & \text{otherwise.} \end{cases}$$

Note that if $i < j < n$ then $\delta_i \neq \delta_j$, since if $k = h^{-1}(i,j)$ we have $\delta_i k \neq \delta_j k$. Hence there is an $f : {}^m 2 \to 2$ such that $f\delta_i = \varepsilon i$ for all $i < n$. Now we claim that $y_f \neq 0$, as desired. If $l < n$, then for $\varepsilon l = 1$ we have $f\delta_l = 1$, and so $\prod_{k<m}(g_k^f a_l)^{\delta_l k} \leq (Gx_{a_l})_f$; and for $\varepsilon l = 0$ we have $f\varepsilon_l = 0$ and so $\prod_{k<m}(g_k^f a_l)^{\delta_l k} \leq ((Gx_{a_l})_f)^0$; so in either case we have $\prod_{k<m}(g_k^f a_l)^{\delta_l k} \leq (Gx_{a_l})_f)^{\varepsilon l}$. It follows that $\prod_{l<n}\prod_{k<m}(g_k^f a_l)^{\delta_l k} \leq y_f$. Suppose that $l < n$ and $k < m$. Then $a_i^{\delta_l k} \in F_k$, so $\prod_{l<n} a_l^{\delta_l k} \in F_k$, so $\prod_{l<n} a_l^{\delta_l k} \neq 0$. Hence

$$\prod_{l<n} \prod_{k<m} (g_k^f a_l)^{\delta_l k} = \prod_{k<m} g_k^f \left(\prod_{l<n} a_l^{\delta_l k} \right) \neq 0,$$

as desired.   □

We turn to independence in ultraproducts. As in the case of cellularity, it is easy to see that if F is a countably complete ultrafilter on an index set I and each A_i has countable independence, then so does $\prod_{i \in I} A_i / F$. Namely, suppose that $\langle f_\alpha / F : \alpha < \omega_1 \rangle$ is a system of independent elements of $\prod_{i \in I} A_i / F$. Now for every $i \in I$ there exist finite disjoint subsets $M(i), N(i)$ of ω_1 such that

$$\prod_{\alpha \in M(i)} f_\alpha i \cdot \prod_{\alpha \in N(i)} -f_\alpha i = 0.$$

Hence

$$I = \bigcup_{M,N} \{i \in I : M = M(i) \text{ and } N = N(i)\},$$

with M and N ranging over finite subsets of ω_1, so, since F is ω_2-complete, there exist finite disjoint $M, N \subseteq \omega_1$ such that $\{i \in I : M = M(i) \text{ and } N = N(i)\} \in F$. But then $\prod_{\alpha \in M} f_\alpha / F \cdot \prod_{\alpha \in N} -f_\alpha / F = 0$, contradiction.

Further, if F is countably incomplete and each algebra A_i is infinite, then $\prod_{i \in I} A_i / F$ is ω_1-saturated, hence is CSP, from which it follows that $\prod_{i \in I} A_i / F$ has independence $\geq 2^\omega$. (See Part I of the BA handbook, Theorem 13.20.) Like with cellularity, if I is infinite, F is a $|I|$-regular ultrafilter on I, and A_i is an infinite BA for each $i \in I$, then $\mathrm{Ind}(\prod_{i \in I} A_i / I) \geq 2^{|I|}$. The proof is similar to that for cellularity: let E be a subset of F such that $|E| = |I|$ and each $i \in I$ belongs to only finitely many members of E; let Gi be the set of all $e \in E$ such that $i \in e$. With each $g \in {}^E 2$ we associate $g' \in \prod_{i \in I} A_i$ as follows. Let $\langle x_h : h \in {}^{Gi}2 \rangle$ be a system of independent elements of A_i. Then for any $i \in I$ we set $g'i = x_{g \restriction Gi}$. We claim that $\langle [g'] : g \in {}^E 2 \rangle$ is an independent system of elements of $\prod_{i \in I} A_i / I$. To see this, let $[(g0)'], \ldots, [(g(m-1))']$ be distinct elements of $\prod_{i \in I} A_i / I$ and let $\varepsilon \in {}^m 2$. Let H be a finite subset of E such that $(g0) \restriction H, \ldots, (g(m-1)) \restriction H$ are all distinct. Let $i \in \bigcap H$ be arbitrary. Now $H \subseteq Gi$, so $(g0) \restriction Gi, \ldots, (g(m-1)) \restriction Gi$ are all distinct. Hence

$$((g0)'i)^{\varepsilon 0} \cdot \ldots \cdot ((g(m-1))'i)^{\varepsilon(m-1)} = x_{(g0) \restriction Gi}^{\varepsilon 0} \cdot \ldots \cdot x_{(g(m-1)) \restriction Gi}^{\varepsilon(m-1)} \neq 0,$$

as desired.

An application of Theorem 10.5 shows that independence can jump greatly in an ultraproduct.

Independence is an ultra-sup function, so Theorems 3.15–3.17 of Peterson apply, Theorem 3.17 saying that $\mathrm{Ind}\left(\prod_{i \in I} A_i / F\right) \geq \left|\prod_{i \in I} \mathrm{Ind} A_i / F\right|$ for F regular. So by Donder's theorem it is consistent that $\geq$ always holds. The inequality can be strict, as is seen by Theorem 10.5. On the other hand, Magidor and Shelah have shown that is is consistent that there is an infinite set I, a system $\langle A_i : i \in I \rangle$ of infinite BAs, and an ultrafilter F on I such that $\mathrm{Ind}\left(\prod_{i \in I} A_i / F\right) < \left|\prod_{i \in I} \mathrm{Ind} A_i / F\right|$. See Rosłanowski, Shelah [94].

Independence in free products is treated in Part I of the BA handbook: $\mathrm{Ind}(A \oplus B) = \max(\mathrm{Ind} A, \mathrm{Ind} B)$, while if I is infinite and $|A_i| \geq 4$ for each $i \in I$, then $\mathrm{Ind}(\oplus_{i \in I} A_i) = \max(|I|, \sup_{i \in I} \mathrm{Ind} A_i)$; see Part I, Theorem 11.15. Under subalgebra and homomorphic image formation, the behaviour of independence is basically simple: if A is a subalgebra or homomorphic image of B, then $\mathrm{Ind} A \leq \mathrm{Ind} B$, and the difference can be arbitrarily large. Finally, independence is an ordinary sup-function, and so its behaviour with respect to unions of well-ordered chains is given by Theorem 3.11.

We turn to the functions derived from independence. Ind_{H+}, Ind_{S+}, and $_d\mathrm{Ind}_{S+}$ all coincide with Ind itself. Ind_{H-} appears to be a new function. Fedorchuk [75] has constructed, using $\diamond$, a BA A such that $\mathrm{Ind}_{H-}A = \mathrm{Ind} A = \omega$ and $\mathrm{Card}_{H-} A = \omega_1$; see also Nyikos [90]. Fedorchuk's construction is given in Chapter 16.

Problem 29. *Can one construct in ZFC a BA A with the property that* $\text{Ind}_{H-}A < \text{Card}_{H-}A$?

This is Problem 24 in Monk [90]. Clearly $\text{Ind}_{S-}A = \omega$ for any infinite BA A. We define

$$\text{Ind}_{h+}A = \sup\{|X| : X \subseteq \text{Clop}Y, \ X \text{ is independent}, \ Y \subseteq \text{Ult}A\}.$$

Then it is possible to have A superatomic, hence with $\text{Ind}A = \omega$, while $\text{Ind}_{h+}A > |\text{Ult}A|$; see the argument for Card. In fact, maybe it is always true that $\text{Ind}_{h+}A = \text{Card}_{h+}A$:

Problem 30. *Is* $\text{Ind}_{h+}A = \text{Card}_{h+}A$ *for every infinite BA A?*

This is Problem 25 in Monk [90]. Ind_{h-} is defined analogously. Again we do not know anything about this cardinal function; for example:

Problem 31. *Is* $\text{Ind}_{h-}A = \text{Ind}_{H-}A$ *for every infinite BA A?*

This is Problem 26 in Monk [90]. The function $_d\text{Ind}_{S-}$ appears to be interesting; if A is $\mathscr{P}X$ for some infinite X, then we have $_d\text{Ind}_{S-}A = \omega$ since the BA of finite and cofinite subsets of X is dense in A. On the other hand, if A is an infinite free BA of regular cardinality, then $_d\text{Ind}_{S-}A = |A|$; see Part I of the BA handbook, Theorem 9.16.

Concerning the spectrum function Ind_{Hs}, note that if $\text{Ind}_{H-}A \leq \mu \leq \text{Ind}A$, then A has a homomorphic image B such that $\mu \leq \text{Ind}B \leq \mu^\omega$. Moreover, if A has CSP, then this cannot be improved:

$$\text{Ind}_{Hs}A = \{\lambda : 2^\omega \leq \lambda \leq \text{Ind}A, \ \lambda^\omega = \lambda\}.$$

(These remarks are due to S. Koppelberg.)

The spectrum function Ind_{Ss} is trivial: $\text{Ind}_{Ss}A = [\omega, \text{Ind}A]$ for every infinite BA.

The comparison of independence with the cardinal functions already introduced is simple: $\text{Ind}A \leq \text{Irr}A$ for every infinite BA A, and the difference can be arbitrarily large, for example in an interval algebra; it is possible to have $\text{Ind}A$ bigger than πA, for example in $\mathscr{P}\kappa$. $\text{Depth}A$ can be much larger than $\text{Ind}A$, for example in the interval algebra on κ. Note that there are some close relationships between independence and cellularity, though. For example, if $(2^{cA})^+ \leq |A|$, then $(2^{cA})^+ \leq \text{Ind}A$ by Corollary 10.9 of Part I of the BA handbook. In particular, $|A| \leq 2^{\max(cA, \text{Ind}A)}$. And if $|A|$ is strong limit, then $|A| = \max(cA, \text{Ind}A)$. There are, however, some problems concerning the relationship of cellularity to independence. We give problems 7, 9, and 10 from Monk [83], where there is some background.

Problem 32. *Assume that* $\rho < \nu < \kappa \leq 2^\rho < \lambda \leq 2^\nu$ *with κ and λ regular. Is there a κ-cc BA A of power λ with no independent subset of power λ?*

Problem 33. *Can one prove the following in ZFC? Suppose that* $\mathrm{cf}\mu < \kappa < \mu <$ $\lambda \leq \mu^{\mathrm{cf}\mu} = \mu^{<\kappa}$ *and* $\forall \rho < \mu(\rho^{<\kappa} < \mu)$. *Then there is a BA of power* λ *satisfying the* κ-*cc with no independent subset of power* λ.

Problem 34. *Suppose that* κ *is uncountable and weakly inaccessible,* $2^\nu < \lambda$ *for all* $\nu < \kappa$, $2^{<\kappa} = \lambda$, *and* λ *is singular. Is there a* κ-*cc BA of power* λ *with no independent subset of power* λ?

A BA A has *free caliber* κ if $\forall X \in [A]^\kappa \exists Y \in [X]^\kappa (Y$ is independent$)$. Freecal$A$ is the set of all $\kappa \leq |A|$ such that A has free caliber κ. We mention some results and problems about this notion from Monk [83]. Problems 4 and 5 from Monk [83] are as follows.

Problem 35. *For all* $n \in \omega$ *let* A_n *be the free BA on* $\beth_n$ *free generators. Does* $\prod_{n \in \omega} A_n$ *have free caliber* $\beth_\omega^+$?

Problem 36. *Let* A *be free on a set of size* $\beth_{\omega+1}$. *Is* $\beth_{\omega+1} \in \overline{\text{Freecal}A}$?

Recall here that for any BA B, $\overline{B}$ is the completion of B.

    In Monk [83] it is observed that Freecal$(\overline{\text{Intalg}L})$ is empty for every linear ordering L with first element. This gives rise to the following problem, Problem 14 of Monk [83]:

Problem 37. *Is there for every* μ *a complete BA* A *of power* 2^μ *such that* Freecal$A = 0$?

The last problem of this sort that we mention is motivated by the following facts noted in Monk [83]:

(1) Assume GCH. Suppose that A is an infinite BA and

$$K \overset{\text{def}}{=} \{\kappa : \kappa \in \text{Freecal}A \text{ and } \kappa \text{ is regular}\}$$

is nonempty. Then the following conditions hold, where $\mu = \min K$ and $\nu = \sup K$:

(i) μ is uncountable.
(ii) For all $\lambda \in (\mu, \nu]$, if λ is regular and is not the successor of a singular cardinal, then $\lambda \in K$.
(iii) For all $\lambda \in (\mu, \nu]$, if $\lambda = \sigma^+$ for some singular σ with $\mu \leq \mathrm{cf}\sigma$, then $\lambda \in K$.

(2) Suppose that $\omega < \mu \leq \nu$ and μ is regular. Then there is a BA A such that

$$\text{Freecal}A = [\mu, \nu] \backslash \{\kappa : \mathrm{cf}\kappa = \omega\}.$$

(3) Assume GCH. Suppose that $\omega < \mu \leq \nu$ and μ is regular. Then there is a BA A such that

$$\{\kappa \in \text{Freecal}A : \kappa \text{ is regular}\} = \{\kappa \in (\mu, \nu] : \kappa \text{ is regular but}$$
$$\kappa \text{ does not have the form } \sigma^+ \text{ with } \sigma \text{ singular, } \mathrm{cf}\sigma < \mu\}.$$

Problem 38. *If K is a set of regular cardinals with $\mu = \min K$ and $\psi = \sup K$, and if K satisfies (1)(i)–(iii), is there a BA A such that K is the set of regular members of FreecalA?*

As mentioned in the introduction, there are bounded versions of independence. A set $X \subseteq A$ is *m-independent* (where m is a positive integer) if for every $Y \in [X]^m$ and every $\varepsilon \in {}^Y 2$ we have $\prod_{y \in Y} y^{\varepsilon y} \neq 0$. Then we set

$$\mathrm{Ind}_n A = \sup\{|X| : X \subseteq A \text{ and } X \text{ is } n\text{-independent}\}.$$

This notion is briefly studied in Monk [83], where the following problem is stated which is somewhat relevant to the notion:

Problem 39. *Can one prove the following in ZFC? For every $m \in \omega$ with $m \geq 2$ there is an interval algebra having a subset P of size ω_1 such that for all $Q \in [P]^{\omega_1}$, Q has m pairwise comparable elements and also m independent elements.*

The condition in this problem is shown to be consistent in Monk [83]. Rosłanowski, Shelah [94] consider the finite version of independence more extensively, proving the following results (and more):

(1) If $n \geq 2$ and λ is an infinite cardinal, then there is a BA A such that $\mathrm{Ind}_n A = \lambda = |B|$ and $\mathrm{Ind}_{n+1} A = \omega$.

(2) If λ is an infinite cardinal and n is an even integer > 2, then there is a BA A such that $\mathrm{Ind}_n A = \lambda$ and $\mathrm{Ind}(A \times A) = \omega$.

We close this chapter with some comments on independence for special kinds of BAs. By the Balcar-Franĕk theorem, $\mathrm{Ind} A = |A|$ for infinite and complete. For CSP algebras in general, all one can say is that $\mathrm{Ind} A = (\mathrm{Ind} A)^\omega$; see Part I of the BA handbook, Theorem 13.20. Finally, recall the important fact that interval algebras, tree algebras, and superatomic algebras have countable independence.

11. π-Character

First of all, note that if F is a non-principal ultrafilter on a BA A, then $\pi\chi F \geq \omega$. To see this, suppose that X is a finite set of non-zero elements of A which is dense in F. Choose $y \in F$ such that $y < \prod(X \cap F)$. Then choose $x \in X$ such that $x \leq y \cdot \prod\{z \in F : -z \in X\}$. This clearly gives a contradiction, whether $x \in F$ or not.

It can happen that A is a subalgebra of B and $\pi\chi A > \pi\chi B$: take $B = \mathscr{P}\omega$ and A an uncountable free subalgebra of B (see the description of $\pi\chi$ for free algebras below). A somewhat more complicated example works for A is a homomorphic image of B. Namely, let $B = \mathscr{P}\omega$, and using the fact that B has an independent set of size ω_1, obtain a homomorphism f from B onto an algebra A such that A is a subalgebra of the completion of the free algebra C on ω_1 free generators $\{x_\alpha : \alpha < \omega_1\}$, and C is a subalgebra of A. Then, we claim, $\pi\chi A = \omega_1$. For, suppose that F is an ultrafilter on A, and X is a countable subset of A. Then each element of X is a countable sum of monomials in the x_α's. If we take some α with x_α not in any of these monomials, then x_α (or $-x_\alpha$) is an element of F with no element of X below it.

We turn to products. Clearly $\pi\chi(A \times B) = \max(\pi\chi A, \pi\chi B)$ for any infinite BAs A and B. More generally, we have:

Theorem 11.1. $\pi\chi(\prod_{i \in I}^{w} A_i) = \sup_{i \in I} \pi\chi A_i$ for any system $\langle A_i : i \in I \rangle$ of infinite BAs.

Proof. We may assume that I is infinite. Since $\mathrm{Ult}(\prod_{i \in I}^{w} A_i)$ is the one-point compactification of the disjoint union of all of the spaces $\mathrm{Ult} A_i$, it suffices to prove the following:

(1) Let F be the ultrafilter on $\prod_{i \in I}^{w} A_i$ consisting of all $x \in \prod_{i \in I}^{w} A_i$ such that $\{i \in I : x_i \neq 1\}$ is finite. Then $\pi\chi F = \omega$.

To prove (1), let J be any denumerable subset of I. For each $j \in J$ we define an element x^j of $\prod_{i \in I}^{w} A_i$ by setting, for each $i \in I$,

$$x_i^j = \begin{cases} 0 & \text{if } j \neq i, \\ 1 & \text{if } j = i. \end{cases}$$

We claim that $\{x^j : j \in J\}$ is dense in F. To see this, take any $y \in F$. Then there is a $j \in J$ such that $y_j = 1$. So $x^j \leq y$, as desired. $\square$

Note that the proof of Theorem 11.1 shows that π-character is attained in $\prod_{i \in I}^{w} A_i$ iff there is an $i \in I$ such that $\pi\chi(\prod_{i \in I}^{w} A_i) = \pi\chi A_i$ and $\pi\chi A_i$ is attained. Using this remark, we can describe the attainment property of π-character: for each uncountable limit cardinal κ there is a BA A with π-character κ not attained: we take the weak product of free algebras of the obvious sizes. On the other hand, if $\pi\chi A = \omega$, then it is attained, since any non-principal ultrafilter has infinite π-character by our initial remark.

Turning to arbitrary products, we have:

Theorem 11.2. *If $\langle A_i : i \in I \rangle$ is a system of non-trivial BAs with $\prod_{i \in I} A_i$ infinite and $|I|$ regular, then $\pi\chi(\prod_{i \in I} A_i) \geq \max(|I|, \sup_{i \in I} \pi\chi A_i)$.*

Proof. If $i \in I$ and G is an ultrafilter on A_i, then the set $F \overset{\text{def}}{=} \{y \in \prod_{i \in I} A_i : y_i \in G\}$ is an ultrafilter on $\prod_{i \in I} A_i$, and a subset of $\prod_{i \in I} A_i$ dense in F clearly gives rise to a subset of A_i with no more elements which is dense in G. Hence $\pi\chi A_i \leq \pi\chi(\prod_{i \in I} A_i)$.

Next, assume that I is infinite; we show that $|I| \leq \pi\chi(\prod_{i \in I} A_i)$. For each subset J of I let x_J be the characteristic function of J, considered as a member of $\prod_{i \in I} A_i$. Let F be any ultrafilter on $\prod_{i \in I} A_i$ containing all elements $x_{I \setminus J}$ such that $|J| < |I|$. Then, we claim, $\pi\chi F \geq |I|$. In fact, suppose that $X \subseteq A^+$, X is dense in F, and $|X| < |I|$. For each $y \in X$ choose $i(y) \in I$ such that $y_{i(y)} \neq 0$. Let $J = \{i(y) : y \in X\}$. Then the element $x_{I \setminus J}$ of F is not $\geq$ any element of X, contradiction. $\qquad\square$

Actually, $\pi\chi$ can jump tremendously in a product. This follows in an obvious way from the following theorem, which is an observation of Douglas Peterson based on Theorem 10.5 and its proof.

Theorem 11.3. *If A is an infinite BA, then $cA \leq \pi\chi\left(\prod_{i \in \omega \setminus 1} A^{*i}\right)$ (with notation as in Theorem 10.5).*

Proof. Wlog $cA > \omega$. Let X and f be as in Theorem 10.5 and its proof, with X uncountable. Then by that proof, $\{-f_x : x \in X\}$ generates a proper filter in $\prod_{i \in \omega \setminus 1} A^{*i}$, and we extend it to an ultrafilter F. We claim that $\pi\chi F \geq |X|$; this will prove the Theorem. Suppose that $Y \subseteq \left(\prod_{i \in \omega \setminus 1} A^{*i}\right)^+$, $|Y| < |X|$, and Y is dense in F; we want to get a contradiction. There exist a $y \in Y$ and an uncountable $Z \subseteq X$ such that $y \leq -f_z$ for all $z \in Z$. Say $y_i \neq 0$. Wlog y_i has the form $a^0 \cdot a^1 \cdot \ldots \cdot a^{i-1}$, where a^j is in the j-th free factor of A^{*i}. Then $a^0 \cdot a^1 \cdot \ldots \cdot a^{i-1} \leq g_0 z + \cdots + g_{i-1} z$ for all $z \in Z$, so there is a $j < i$ such that $a^j \leq g_j z$. This being true for all $z \in Z$, and Z being infinite, it follows that there exist a $j < i$ and two distinct $z, w \in Z$ such that $a^j \leq g_j z$ and $a_j \leq g_j w$. Since $z \cdot w = 0$, it follows that $a^j = 0$, contradiction. $\qquad\square$

The possibility of doing the above with interval algebras, which naturally arose in Chapter 10, is not so interesting here, since interval algebras can have high π-character (see the end of this chapter).

We turn to ultraproducts, giving some results of Douglas Peterson. Since $\pi\chi$ is a sup-min function, Theorems 6.1–6.3 hold. An additional result of the sort described in these theorems, with a proof using independent matrices, is the following theorem of Peterson: *If $\langle A_i : i \in I \rangle$ is a system of infinite BAs, with I infinite, F is a regular ultrafilter on I, and $\text{ess.sup}_{i \in I}^F |A_i| \leq 2^{|I|}$, then $\pi\chi\left(\prod_{i \in I} A_i / F\right) \geq \text{cf}(2^{|I|})$.*

From 6.1–6.3 the following theorem follows, with a proof similar to that of Theorem 4.14:

Theorem 11.4. (GCH) *Suppose that* $\langle A_i : i \in I \rangle$ *is a system of infinite BAs, with* I *infinite, and* F *is a regular ultrafilter on* I. *Then* $\pi\chi \left(\prod_{i \in I} A_i / F \right) \geq \left| \prod_{i \in I} \pi\chi A_i / F \right|$.

As usual, the result of Donder shows that it is consistent to always have $\geq$. Peterson has shown that it is consistent to have $<$ in Theorem 11.4 in the absence of GCH. See also Chapter 4 for an independent solution by Shelah. For $>$ we have the following extension of Theorem 11.3, which shows that $\pi\chi$ can jump very much in an ultraproduct.

Theorem 11.5. *If* A *is an infinite BA and* F *is a nonprincipal ultrafilter on* ω, *then* $cA \leq \pi\chi \left(\prod_{i \in \omega \setminus 1} A^{*i} / F \right)$, *again with notation as in Theorem 10.5.*

Proof. Let X be as in Theorem 10.5, with X uncountable. By Theorem 10.5, if N is a finite subset of X then $\{n : \prod_{x \in N} -f_x n = 0\}$ is finite, and hence $\prod_{x \in N} -f_x / F \neq 0$. Thus $\{-f_x / F : x \in X\}$ has the finite intersection property, and we can let G be an ultrafilter on $\prod_{i \in \omega \setminus 1} A^{*i} / F$ containing this set. We claim that $\pi\chi G \geq |X|$, which will prove the theorem. To get a contradiction, suppose that $Y \subseteq \prod_{i \in \omega \setminus 1} A^{*i} / F$, $|Y| < |X|$, and Y is dense in G. Then there is a $y/F \in Y$ and an uncountable $X' \subseteq X$ such that $y/F \leq -f_x / F$ for all $x \in X'$. We may assume that $y_i \neq 0$ for all $i \in \omega$, and further that each y_i has the form $a_i^0 \cdot a_i^1 \cdot \ldots \cdot a_i^{i-1}$ with a_i^j from the j-th factor. Now for any $x \in X'$ we have $y/F \leq -f_x / F$, and so there is an $i \in \omega$ such that $y_i \leq -f_x i$. Hence there is an $i \in \omega$ and an uncountable $X'' \subseteq X'$ such that $y_i \leq -f_x i$ for all $x \in X''$. Now we proceed to a contradiction as in the proof of Theorem 11.3. $\qquad \square$

Next we describe π-character for free products:

Theorem 11.6. *If* $\langle A_i : i \in I \rangle$ *is a system of BAs each with at least 4 elements, then* $\pi\chi(\oplus_{i \in I} A_i) = \max(|I|, \sup_{i \in I} \pi\chi A_i)$.

Proof. For brevity let $B = \oplus_{i \in I} A_i$. First take any $i \in I$; we show that $\pi\chi A_i \leq \pi\chi B$. Let F be any ultrafilter on A_i, and extend F to an ultrafilter G on B. Suppose $X \subseteq B$ is dense in G. We may assume that each $x \in X$ has the form

$$(1) \quad x = \prod_{j \in Mx} y_j^x$$

for some finite subset Mx of I, where $y_j^x \in A_j$ for every $j \in Mx$. Now define $Y = \{y_i^x : x \in X, i \in Mx\}$. Then clearly Y is dense in F and $|Y| \leq |X|$. This proves that $\pi\chi A_i \leq \pi\chi B$.

Next, we show that $|I| \leq \pi\chi B$, where we assume that I is infinite. For each $i \in I$ choose $a_i \in A_i$ such that $0 < a_i < 1$. Let F be an ultrafilter on B such that $a_i \in F$ for each $i \in I$; clearly such an ultrafilter exists. Suppose $X \subseteq B$ is dense in F; we may assume that each $x \in X$ has the form (1) indicated above. Clearly then, by the free product property, we must have $|X| \geq |I|$.

Now let F be an ultrafilter on B. Then for each $i \in I$, $F \cap A_i$ is an ultrafilter on A_i, and so there is an $X_i \subseteq A_i$ of cardinality $\leq \pi\chi A_i$ which is dense in $F \cap A_i$. Let

$$Y = \{y : \text{there is a finite } J \subseteq I \text{ and a } b \text{ in } \textstyle\prod_{j \in J} X_j \text{ such that } y = \textstyle\prod_{j \in J} b_j\}.$$

Clearly $|Y| \leq \max(|I|, \sup_{i \in I} A_i)$ and Y is dense in F, as desired. $\qquad\square$

As a corollary, $\pi\chi A = \kappa$ if A is the free BA on κ generators.

Next we discuss the behaviour of $\pi\chi$ under unions.

Theorem 11.7. *Suppose that $\langle A_\alpha : \alpha < \kappa \rangle$ is a strictly increasing sequence of BAs with union B, where κ is regular. Let $\lambda = \sup_{\alpha < \kappa} \pi\chi A_\alpha$. Then $\pi\chi B \leq \sum_{\alpha < \kappa} \pi\chi A_\alpha \leq \max(\kappa, \lambda)$. Assume in addition that $A_\alpha = \bigcup_{\beta < \alpha} A_\beta$ for all limit $\alpha < \kappa$. Then $\pi\chi B \leq \lambda^+$.*

Proof. Let F be an ultrafilter on B. Choose $X_\alpha \subseteq A_\alpha$ which is dense in $F \cap A_\alpha$, with $|X_\alpha| = \pi\chi(F \cap A_\alpha)$, for each $\alpha < \kappa$. Then $\bigcup_{\alpha < \kappa} X_\alpha$ is dense in F, and $|\bigcup_{\alpha < \kappa} X_\alpha| \leq \sum_{\alpha < \kappa} \pi\chi A_\alpha$. So $\pi\chi B \leq \sum_{\alpha < \kappa} \pi\chi A_\alpha \leq \max(\kappa, \lambda)$.

Now we make the additional assumption indicated, and suppose that $\pi\chi B > \lambda^+$. Let F be an ultrafilter on B such that $\pi\chi F > \lambda^+$. Thus $\kappa > \lambda^+$ by the first part of this proof. Let $S = \{\alpha < \kappa : \mathrm{cf}\alpha = \lambda^+\}$. So, S is stationary in κ. For each $\alpha < \kappa$ let $X_\alpha \subseteq A_\alpha$ be dense in $F \cap A_\alpha$ with $|X_\alpha| \leq \lambda$. For $\alpha \in S$ we then have $X_\alpha \subseteq A_{f\alpha}$ for some $f\alpha < \alpha$. Therefore f is constant, say equal to β, on some stationary subset of S. So X_β is dense in F, contradicting $\pi\chi F > \lambda^+$. $\qquad\square$

In contrast to Theorem 6.6, we did not assert in 11.7 that $\kappa \leq 2^\lambda$. In fact, for any infinite cardinal κ there is a strictly increasing continuous sequence $\langle A_\alpha : \alpha < \kappa \rangle$ of BAs such that $\pi\chi A_\alpha = \omega$ for all $\alpha < \kappa$. Namely, take a strictly increasing continuous sequence of subalgebras of $\mathrm{Finco}\,\kappa$ with union $\mathrm{Finco}\,\kappa$; recall that if $A \leq \mathrm{Finco}\,\kappa$, then A is isomorphic to $\mathrm{Finco}\,\lambda$ for some $\lambda \leq \kappa$. (In Monk [90], $\kappa \leq 2^\lambda$ was mistakedly asserted.)

The upper bound λ^+ mentioned in Theorem 11.7 can be attained—take a sequence of free algebras.

Concerning the derived functions of π-character, the first result is that $tA = \pi\chi_{\mathrm{H}+}A = \pi\chi_{\mathrm{h}+}A$, where $\pi\chi_{\mathrm{h}+}A = \sup\{\pi\chi(F, Y) : F \in Y,\ Y \subseteq \mathrm{Ult}A\}$, and for any point x of any space X, $\pi\chi(x, X)$ is defined to be $\min\{|M| : M$ is a collection of non-empty open subsets of X and for every neighborhood U of x there is a $V \in M$ such that $V \subseteq U\}$. Such a set M is called a *local π-base* for x. It is also convenient for this proof to have an algebraic version of free sequences. Let A be a BA. A *free sequence* in A is a sequence $\langle x_\xi : \xi < \alpha \rangle$ of elements of A such that if $\xi < \alpha$ and F and G are finite subsets of ξ and $\alpha \setminus \xi$ respectively, then $\prod_{\eta \in F} x_\eta \cdot \prod_{\eta \in G} -x_\eta \neq 0$. Then A has a free sequence of length α iff $\mathrm{Ult}A$ has a free sequence (in the topological sense, defined in Chapter 4) of length α. In fact, first suppose that $\langle x_\xi : \xi < \alpha \rangle$ is a free sequence in A. For each $\xi < \alpha$ let F_ξ be an ultrafilter containing $\{x_\eta : \eta \leq \xi\} \cup \{-x_\eta : \xi < \eta < \alpha\}$. This is possible by the definition above. It is easy to check that $\langle F_\xi : \xi < \alpha \rangle$ is a free sequence in Ult$A$. Conversely, let $\langle F_\xi : \xi < \alpha \rangle$ be a free sequence in Ult$A$. Then by the definition of free sequences in spaces, for each $\xi < \alpha$ there is a $x_\xi \in A$ such that $\{F_\eta : \eta < \xi\} \subseteq \mathcal{S}(-x_\xi)$ and $\{F_\eta : \xi \leq \eta\} \subseteq \mathcal{S}x_\xi$. Then $\langle x_\xi : \xi < \alpha \rangle$ is a free sequence in A. This equivalence shows, in particular, that Ind$A \leq$ tA. Note that tightness in these two free sequence senses have the same attainment properties: one is attained iff the other is.

Theorem 11.8. (Shapirovskiĭ) *For any infinite BA* A *we have* t$A = \pi\chi_{\text{H+}}A = \pi\chi_{\text{h+}}A$.

Proof. First we show t$A \leq \pi\chi_{\text{h+}}A$. For brevity let $\kappa = \pi\chi_{\text{h+}}A$. Let F be an ultrafilter on A, and suppose that $Y \subseteq$ Ult$A$ and $F \subseteq \bigcup Y$; we want to find a subset Z of Y of size $\leq \kappa$ such that $F \subseteq \bigcup Z$. We may assume that $F \notin Y$. By the definition of $\pi\chi_{\text{h+}}A$, let M be a local π-base for F in $Y \cup \{F\}$ with $|M| \leq \pi\chi_{\text{h+}}A$. The assumption that $F \notin Y$ implies that F is not isolated in $Y \cup \{F\}$, and hence that $V \cap Y \neq 0$ for every $V \in M$. Taking a point from each such intersection, we get a subset Z of Y of power $\leq \kappa$ such that $V \cap Z \neq 0$ for every $V \in M$. Then clearly $F \subseteq \bigcup Z$, as desired. Clearly $\langle a_\alpha : \alpha < \kappa \rangle$ is a free sequence, as desired.

Next we show that $\pi\chi_{\text{h+}}A \leq \pi\chi_{\text{H+}}A$. Given $Y \subseteq$ Ult$A$, let $\overline{Y}$ be the closure of Y, and recall from the duality theory that $\overline{Y}$ corresponds to a homomorphic image of A. So, we just need to show that $\pi\chi Y \leq \pi\chi\overline{Y}$. Let $y \in Y$, and let M be a local π-base for y in $\overline{Y}$. Then $\{U \cap Y : U \in M\}$ is clearly a local π-base for y in Y. So, $\pi\chi Y \leq \pi\chi\overline{Y}$ follows.

Finally, we show that $\pi\chi_{\text{H+}}A \leq$ tA. Note that if Y is a closed subspace of X and $\langle x_\xi : \xi < \alpha \rangle$ is a free sequence in Y, then it is a free sequence in X also. Hence it suffices to show that if $F \in$ Ult$A$ and $\pi\chi F \geq \kappa$, then there is a free sequence of length κ in A, by Theorem 4.20. Thus we have:

(1) For every subset B of A^+ of power $< \kappa$ there is an $a \in F$ such that $b \cdot -a \neq 0$ for every $b \in B$.

We construct a sequence $\langle a_\alpha : \alpha < \kappa \rangle$ by induction. Choose a_0 arbitrary $\in F$. Now suppose that a_β has been defined for all $\beta < \alpha$, where $0 < \alpha < \kappa$. Let G_α be the set of all non-zero products $\prod_{\beta \in M} a_\beta \cdot \prod_{\beta \in N} -a_\beta$ such that M and N are finite disjoint subsets of α such that $M < N$ (meaning that $\forall\beta \in M \forall\lambda \in N(\beta < \lambda)$). By (1), choose $a_\alpha \in F$ such that $b \cdot -a_\alpha \neq 0$ for all $b \in G_\alpha$. Clearly $\langle a_\alpha : \alpha < \kappa \rangle$ is a free sequence, as desired. $\square$

Note from the proof of Theorem 11.8 that one of $\pi\chi_{\text{h+}}$ and $\pi\chi_{\text{H+}}$ is attained iff the other is; and if $\pi\chi_{\text{h+}}$ is attained, then so is t, in the free sequence sense.

It is possible to have $\pi\chi_{\text{S+}}A > \pi\chi A$; this is true, for example, for $A = \mathscr{P}\omega$, using the fact that $\mathscr{P}\omega$ has a free subalgebra of size 2^ω.

Clearly $\pi\chi_{\text{S-}}A = \pi\chi_{\text{H-}}A = \omega$. On the other hand, $\pi\chi_{\text{h-}}A = 1$ for any infinite BA A, since Ult$A$ has a denumerable discrete subspace. If B is dense in A, then $\pi\chi B \leq \pi\chi A$. In fact, if F is an ultrafilter on A, let $X \subseteq A$ be dense in F

with $|X| = \pi\chi F$. Wlog $X \subseteq B$. Hence X is dense in $F \cap B$, so $\pi\chi(F \cap B) \leq \pi\chi F$. This shows that, indeed, $\pi\chi B \leq \pi\chi A$. It is possible that $\pi\chi B < \pi\chi A$ when B is dense in A. For example, let A be the interval algebra on an uncountable cardinal κ and let B be Finco κ; see the description of $\pi\chi$ for interval algebras below. These comments show that $_d\pi\chi_{S+}A = \pi\chi A$, but there is an example with $_d\pi\chi_{S-}A < \pi\chi A$ (contradicting a statement in Monk [90]).

Recall from the introduction that for a cardinal function such as $\pi\chi$ we can define an associated function $\pi\chi_{\inf}$ as follows: $\pi\chi_{\inf}A = \inf\{\pi\chi F : F$ is an ultrafilter on $A\}$. And recall from Part I Theorem 10.16 the useful result of Shapirovskiĭ that $\text{Ind}A = (\pi\chi_{\inf})_{H+}A = \sup\{\pi\chi_{\inf}B : B$ is a homomorphic image of $A\}$, for A not superatomic. Moreover, $\pi\chi_{\inf}$ can be given a more elementary equivalent definition:

Theorem 11.9. *For any infinite BA, $\pi\chi_{\inf}A$ is the smallest cardinality of a subset D of A^+ such that for any finite partition of unity $\langle a_i : i < m \rangle$ in A there is a $d \in D$ and an $i < m$ such that $d \leq a_i$.*

Proof. Let $\pi\chi_{\inf}A = \pi\chi F$, where F is an ultrafilter on A. Let D be dense in F with $|D| = \pi\chi A$. Let $\langle a_i : i < m \rangle$ be a finite partition of unity in A. Then $a_i \in F$ for some $i < m$. Say $d \in D$ and $d \leq a_i$. This shows that D satisfies the indicated condition.

For the other direction, suppose that D satisfies the indicated condition, but $|D| < \pi\chi_{\inf}A$. For all $F \in \text{Ult}A$, D is not dense in F, so there is an $a_F \in F$ such that $d \not\leq a_F$ for all $d \in D$. Now $\{Sa_F : F \in \text{Ult}A\}$ covers $\text{Ult}A$. Let $\{Sa_{F_0}, \ldots, Sa_{F_{n-1}}\}$ be a finite subcover. So $a_{F_0} + \cdots + a_{F_{n-1}} = 1$, and $d \not\leq a_i$ for all $d \in D$ and $i < n$. Without loss of generality the a_i's are pairwise disjoint, and this gives a contradiction. $\square$

This theorem suggests another function related to $\pi\chi_{\inf}$: call a subset $D \subseteq A^+$ *weakly dense* if for all $a \in A$ there is a $d \in D$ such that $d \leq a$ or $d \leq -a$. Let $\text{wd}A = \min\{|D| : D$ is weakly dense in $A\}$. Then $\text{wd}A \leq \pi\chi_{\inf}A$ by Theorem 11.9. Balcar and Simon [91a], [91b] have shown that there are BAs where these two cardinals are different, although they are equal for all complete BAs and for all homogeneous BAs.

Clearly $\pi\chi A \leq \pi A$ for any infinite BA A. The difference between $\pi\chi$ and π can be large, for example in a finite-cofinite algebra: as in the proof of Theorem 11.1, $\pi\chi A = \omega$ for a finite-cofinite algebra A. $\pi\chi A > \text{d}A$ for some free algebras A; a free algebra also shows that $\pi\chi A$ can be greater than $\text{Length}A$. It is easy to construct an example where $\pi\chi$ is much smaller than Ind. In fact, let A be a free BA on κ free generators. Then we construct a sequence $\langle B_n : n \in \omega \rangle$ of algebras by recursion. Let $B_0 = A$. Having constructed B_n, let B_{n+1} be an extension of B_n obtained by adding for each ultrafilter F on B_n an element $0 \neq y_F^n$ such that $y_F^n \leq b$ for all $b \in F$; it is easy to see that this is possible. Let $C = \bigcup_{n \in \omega} B_n$. Then $\text{Ind}C \geq \kappa$, while $\pi\chi C = \omega$. For, let G be any ultrafilter on C. Then $\{y_{G \cap B_n}^n : n \in \omega\}$ is dense in G, showing that $\pi\chi G \leq \omega$.

$\pi\chi A > \text{Ind} A$ for A the interval algebra on an uncountable cardinal κ, and $\text{Depth} A > \pi\chi A$ for A the interval algebra on $1 + \omega^* \cdot (\kappa + 1)$; both of these results are clear on the basis of the description of $\pi\chi$ for interval algebras given at the end of this chapter.

There are two interesting positive results concerning the relationship of $\pi\chi$ with our earlier cardinal functions. The first of these is true for arbitrary non-discrete regular Hausdorff spaces, with no complications in the proof from the BA case:

Theorem 11.10. $dX \leq \pi\chi X^{cX}$ *for any non-discrete regular Hausdorff space* X.

Proof. By non-discreteness, $\pi\chi X \geq \omega$; this is easy to check, following the lines of the argument at the beginning of this chapter. For each $x \in X$ let $\mathcal{O}_x$ be a family of non-empty open subsets of X such that $|\mathcal{O}_x| \leq \pi\chi X$ and for every neighborhood U of x there is a $V \in \mathcal{O}$ such that $V \subseteq U$. Now we define subsets $Y_\alpha \subseteq X$ and collections $\mathscr{P}_\alpha$ of open sets for $\alpha < (cX)^+$ by induction so that the following conditions hold:

(1) $|Y_\alpha| \leq (\pi\chi X)^{cX}$;

(2) $|\mathscr{P}_\alpha| \leq (\pi\chi X)^{cX}$.

Fix $x_0 \in X$. Set $Y_0 = \{x_0\}$ and $\mathscr{P}_0 = \mathcal{O}_{x_0}$. Suppose that Y_β and $\mathscr{P}_\beta$ have been defined for all $\beta < \alpha$. If α is a limit ordinal, set $Y_\alpha = \bigcup_{\beta<\alpha} Y_\beta$ and $\mathcal{P}_\alpha = \bigcup_{\beta<\alpha} \mathscr{P}_\beta$. Now suppose that α is a successor ordinal $\beta + 1$. Set

$$Q_\alpha = \{\mathcal{R} : \mathcal{R} \subseteq \mathscr{P}_\beta, |\mathcal{R}| \leq cX, \overline{\bigcup \mathcal{R}} \neq X\}$$

Clearly $|Q_\alpha| \leq \pi\chi X^{cX}$. For every $\mathcal{R} \in Q_\alpha$ choose $\varphi_\mathcal{R} \in X \backslash \overline{\bigcup \mathcal{R}}$ and put

$$Y_\alpha = Y_\beta \cup \{\varphi_\mathcal{R} : \mathcal{R} \in Q_\alpha\},$$

$$\mathscr{P}_\alpha = \bigcup_{x \in Y_\alpha} \mathcal{O}_x.$$

This finishes the definition. Now we claim

(3) $L \stackrel{\text{def}}{=} \bigcup_{\alpha < (cX)^+} Y_\alpha$ is dense in X.

Since $|L| \leq (\pi\chi X)^{cX}$, (3) finishes the proof. To prove (3), suppose that it is not true. Then by regularity, there is an open U such that $\overline{L} \subseteq U \subseteq \overline{U} \neq X$. Set $\mathcal{P}^* = \bigcup_{x \in L} \mathcal{O}_x$, and $\mathcal{T} = \{V \in \mathcal{P}^* : V \subseteq U\}$. Let $\mathcal{R}$ be a maximal disjoint subset of $\mathcal{T}$. Then $L \subseteq \overline{\bigcup \mathcal{R}}$; for, if $x \in L \backslash \overline{\bigcup \mathcal{R}}$, then $x \in U \backslash \overline{\bigcup \mathcal{R}}$, which is open, so there is a $V \in \mathcal{O}_x$ such that $V \subseteq U \backslash \overline{\bigcup \mathcal{R}}$, and $\mathcal{R} \cup \{V\}$ contradicts the maximality of $\mathcal{R}$. Also, $\overline{\bigcup \mathcal{R}} \subseteq \overline{\bigcup \mathcal{T}} \subseteq \overline{U} \neq X$. Since $\mathcal{R} \subseteq \mathscr{P}_\beta$ for some $\beta < (cX)^+$, it follows that $\mathcal{R} \in Q_\beta$ for some $\beta < (cX)^+$, and hence we get $\varphi_\mathcal{R} \in X \backslash \overline{\bigcup \mathcal{R}} \subseteq X \backslash L$, contradiction. $\square$

Theorem 11.11. $dA \cdot \pi\chi A = \pi A$ *for any infinite BA A.*

Proof. We already know that $dA \leq \pi A$ and $\pi\chi A \leq \pi A$. Now let D be a dense subset of $\mathrm{Ult}A$ with $|D| = dA$, and for each $F \in D$ let X_F be a local base for F of size $\leq \pi\chi A$. Clearly $\bigcup_{F \in D} X_F$ is dense in A, as desired. $\qquad\square$

Concerning $\pi\chi$ for special classes of algebras, we first give a description of what happens for interval algebras. Let L be a linearly ordered set with first element 0, and let A be the interval algebra on L. The ultrafilters on A are in one-one correspondence with the final segments of L not containing 0; corresponding to the ultrafilter F is the segment $\{a \in L : [0, a) \in F\}$. Given a terminal segment T of L, let κ be the type of a shortest cofinal sequence in $L\backslash T$ and λ the type of a shortest coinitial sequence in T. If both κ and λ are infinite, then $\pi\chi F$ is the minimum of κ and λ. If one is infinite and the other is 1, then $\pi\chi F$ is the infinite one. If both are 1, then $\pi\chi F$ is 1. From this description it is easy to construct a linear order L such that if A is the interval algebra on L then $\pi\chi A < \chi A$, with the difference arbitrarily large: for example, let κ be any infinite cardinal, and let L be $0 + \omega^* \cdot \kappa + \omega^*$. The above description implies that $\pi\chi A = \omega$, while if F is the ultrafilter corresponding to the terminal segment ω^*, then $\chi F = \kappa$. In this example we also have $\pi\chi A < \mathrm{Depth}A$. The description of $\pi\chi$ also shows that $\pi\chi A \leq \mathrm{Depth}A$ for an interval algebra A.

If A is complete, then $cA \leq \pi\chi A$: in fact, suppose that $\pi\chi A < cA$. Let X be disjoint in A with $\sum X = 1$ and $|X| = (\pi\chi A)^+$. Let F be an ultrafilter on A such that $\sum(X\backslash Y) \in F$ for each $Y \subset X$ such that $|Y| < |X|$. Let Y be a π-base for F with $|Y| < |X|$. For each $y \in Y$ choose $x_y \in X$ such that $y \cdot x_y \neq 0$. Then $\{x_y : y \in Y\}$ is a π−base for $F \cap \langle X \rangle^{\mathrm{cm}}$ (where $\langle X \rangle^{\mathrm{cm}}$ is the complete subalgebra of A generated by X). But $-\sum_{y \in Y} x_y \in F \cap \langle X \rangle^{\mathrm{cm}}$, contradiction.

K. Bozeman [91] shows that under GCH we have $\pi A = \pi\chi A$ for A complete; this is a partial solution of Problem 27 of Monk [90]. We reformulate that problem:

Problem 40. *Can one show in ZFC that $\pi A = \pi\chi A$ for A complete?*

(Bozeman's results must be suitably analyzed to get the indicated result. First some notation. Let B be a BA, $X \subseteq B$, and $a \in B$. Then we set $X \restriction a = \{x \cdot a : x \in X\}$. We say that X is *hereditarily weakly dense* if $X \restriction a$ is weakly dense in $B \restriction a$ for all $a \in B^+$. Then we set

$$\mathrm{hwd}B = \min\{|X| : X \text{ is hereditarily weakly dense in } B\}.$$

Note that $\mathrm{wd}B \leq \mathrm{hwd}B$. If k is a cardinal function on Boolean algebras, we say that B is k-*homogeneous* if $k(B \restriction a) = kB$ for every $a \in B$. Two major results in Bozeman [91] are as follows:

(1) If B is complete and hwd-homogeneous, then $\mathrm{wd}B = \mathrm{hwd}B$. (This result is rather easy.)
(2) If B is complete and both π- and hwd-homogeneous, then $\pi B \leq 2^{<\mathrm{hwd}B}$. (The proof is rather involved.)

On the basis of these results, if B is complete and both π- and hwd-homogeneous, and if GCH holds, then

$$\text{wd}B \leq \pi\chi_{\inf}B \leq \pi\chi B \leq \pi B \leq \text{hwd}B \leq \text{wd}B.$$

Now assume GCH, and let B be any complete BA. It is easy to see that we can write $B \cong \prod_{i \in I} C_i$ with each C_i both π- and hwd-homogeneous. Then

$$\pi B = \max\{|I|, \sup_{i \in I} \pi C_i\}$$
$$= \max\{\text{c}B, \sup_{i \in I} \pi\chi C_i\}$$
$$\leq \pi\chi B \leq \pi B,$$

as desired.)

In Chapter 6 we gave an example of a complete algebra A with the property that $\text{d}A < \pi A$; hence by Theorem 11.11 we have $\text{d}A < \pi\chi A$ also.

$\pi\chi$ is characterized for tree algebras by the following theorem.

Theorem 11.12. *Let T be an infinite tree. Then $\pi\chi(\text{Treealg}\,T) = \sup\{\text{cf}\,C : C$ is an initial chain of T with finitely many immediate successors$\}$.*

Proof. We describe $\pi\chi F$ for each ultrafilter F on $\text{Treealg}\,T$. Recall that the ultrafilters on $\text{Treealg}\,T$ are in one-one correspondence with the initial chains of T, where if T has finitely many roots we exclude the empty chain (a correction of the description in the Handbook). Given an initial chain C, we let F_C be generated by

$$\{T \uparrow t : t \in C\} \cup \{T \backslash (T \uparrow t) : t \in T \backslash C\}.$$

This is the ultrafilter associated with C. We now consider several cases. *Case 1.* C has a maximal element t, and t has finitely many immediate successors. Then $\{t\} \in F_C$, which is thereby principal, so that $\pi\chi F = 1$. *Case 2.* C has infinitely many immediate successors. Let M be a countable set of such immediate successors, and let $X = \{T \uparrow t : t \in M\}$. Then X is dense in F_C. So $\pi\chi F_C \leq \omega$ in this case. *Case 3.* C has no maximal element, but has finitely many immediate successors. Let M be the set of all immediate successors of C, and let N be a cofinal subset of C of size $\text{cf}\,C$. Then $\{(T \uparrow t) \backslash \bigcup_{s \in M} (T \uparrow s) : t \in N\}$ is dense in F_C. Suppose that X is dense in F_C but $|X| < \text{cf}\,C$. Wlog each element $x \in X$ has the form $(T \uparrow t_x) \backslash \bigcup_{s \in P_x} (T \uparrow s)$. Choose $u \in N$ such that $t_x < u$ for all $x \in X$. Then $(T \uparrow u) \backslash \bigcup_{s \in M} (T \uparrow s) \in F_C$, and no element of X is below it, contradiction. Thus $\pi\chi F_C = \text{cf}\,C$ in this case.   □

For tree algebras we have $\pi\chi A \leq \text{Depth}A$, since $\text{Depth}A = \text{t}A$ for them. The difference can be arbitrarily large; this is an observation of Douglas Peterson. Namely, given κ, consider the tree

$$T \stackrel{\text{def}}{=} \{f : f : \alpha + 1 \to \omega \text{ for some } \alpha \leq \kappa\} \cup \{0\}$$

under $\subseteq$. Every initial chain of T has countably many immediate successors, so $\pi\chi(\mathrm{Treealg}\,T) = \omega$ by Theorem 11.12; but $\mathrm{Depth}(\mathrm{Treealg}\,T) = \kappa$.

In Dow, Monk [94] the relationship between depth and π-character for super-atomic BAs is described. There is a BA A such that $\mathrm{Depth}A = \omega$ and $\pi\chi A = \omega_1$. If $\pi\chi A \geq \omega_2$, then $\pi\chi A = \mathrm{Depth}A$. Above we showed that one can have $\pi\chi A <$ $\mathrm{Depth}A$ with any prescribed gap for A an arbitrary BA.

12. Tightness

Again we note first of all that if F is a non-principal ultrafilter in a BA A, then $tF \geq \omega$. To see this, note that for each $x \in F$ there is a $y \notin F$ such that $0 < y < x$; hence there is an ultrafilter G_x such that $x \in G_x$ but $G_x \neq F$. Let $Y = \{G_x : x \in F\}$. Thus $F \subseteq \bigcup Y$. Suppose that Z is a finite subset of Y such that $F \subseteq \bigcup Z$. But it is a very elementary exercise to show that no ultrafilter is included in a finite union of other, different, ultrafilters. So, $tF \geq \omega$, and hence $tA \geq \omega$ for every infinite BA A.

From the definition of tightness it is clear that $t(A \times B) = \max\{tA, tB\}$. Furthermore, $t(\prod_{i \in I}^{w} A_i) = \sup_{i \in I} tA_i$ for any system $\langle A_i : i \in I \rangle$ of non-trivial BAs with I infinite. By the topological description of weak products, to prove this it suffices to show that $tF = \omega$ for the "new" ultrafilter $F \overset{\text{def}}{=} \{x \in \prod_{i \in I}^{w} A_i :$ there is a finite subset G of I such that $x_i = 1$ for all $i \in I \backslash G\}$. To see this, first note that if $G \in \text{Ult}(\prod_{i \in I} A_i)$ and $G \neq F$, then there is an $i_G \in I$ and an ultrafilter K_G on A_i such that $G = \{x \in \prod_{i \in I} A_i : x_{i_G} \in K_G\}$. Next, for H a finite subset of I let

$$x_H i = \begin{cases} 1 & \text{if } i \in I \backslash H \\ 0 & \text{if } i \in H. \end{cases}$$

Now suppose that $F \subseteq \bigcup Y$ with $Y \subseteq \text{Ult}(\prod_{i \in I} A_i)$. The case $F \in Y$ is easy, so suppose that $F \notin Y$. Now $H \overset{\text{def}}{=} \{i_G : G \in Y\}$ is infinite; otherwise $x_H \in F$ gives a contradiction. Let Z be a countable subset of Y such that $\{i_G : G \in Z\}$ is infinite. Suppose that $x \in F$. Say $x_i = 1$ for all $i \in I \backslash L$, L finite. Choose $G \in Z$ such that $i_G \notin L$. Then $x \in G$, as desired.

Note that this argument again shows that tightness is attained in $\prod_{i \in I}^{w} A_i$ iff there is an $i \in I$ such that $t(\prod_{i \in I}^{w} A_i) = tA_i$ and tightness is attained in A_i (for infinite A_i's). From this, the attainment property of tightness follows: for each limit cardinal $\kappa > \omega$ there is a BA A with tightness κ not attained: take the weak product of $\langle A_\lambda : \omega < \lambda < \kappa, \lambda$ a cardinal$\rangle$, where A_λ is the free BA of size λ.

For the free sequence equivalents of tightness see Chapters 4 and 11. The free sequence characterization shows that if A is a subalgebra or homomorphic image of B, then $tA \leq tB$. Clearly the difference can be arbitrarily large.

Concerning attainment in the free sequence sense, we first show

Theorem 12.1. *If κ is an infinite cardinal with $\text{cf}\kappa > \omega$ and $\langle A_i : i \in I \rangle$ is a system of BAs none of which has a free sequence of type κ, then also $\prod_{i \in I}^{w} A_i$ does not have a free sequence of type κ.*

Proof. Suppose that $\langle F_\alpha : \alpha < \kappa \rangle$ is a free sequence in $\text{Ult}(\prod_{i \in I}^{w} A_i)$. We think of $\text{Ult}(\prod_{i \in I}^{w} A_i)$ as the one-point compactification of the disjoint union of all of the spaces $\text{Ult}A_i$. We may assume that the "new" ultrafilter G is not among the F_α's. For each $\alpha < \kappa$ let iF_α be the unique $i \in I$ such that $F_\alpha \in \text{Ult}A_i$. Set $J = \{iF_\alpha : \alpha < \kappa\}$. Then $|J| \geq \text{cf}\kappa$, since $\kappa = \bigcup_{j \in J}\{\eta < \kappa : iF_\eta = j\}$. Now $J = \bigcup_{\xi < \kappa}\{iF_\eta : \eta < \xi\}$, so it follows from $\text{cf}\kappa > \omega$ that there is a $\xi < \kappa$ such that

$\{iF_\eta : \eta < \xi\}$ is infinite. Clearly $|\{iF_\eta : \xi \le \eta\}| \ge \mathrm{cf}\kappa$ by the above argument, so it follows that

$$G \in \overline{\{F_\eta : \eta < \xi\}} \cap \overline{\{F_\eta : \xi \le \eta < \kappa\}},$$

which contradicts the free sequence property. □

It follows from Theorem 12.1 that for every κ with $\mathrm{cf}\kappa > \omega$ there is a BA with tightness κ not attained in the free sequence sense.

Now we turn to the case of cofinality ω:

Theorem 12.2. *Let* $\mathrm{t}A = \kappa$, *where* κ *is a singular cardinal of cofinality* ω. *Then* A *has a free sequence of length* κ.

Proof. This will be a modification of the proof of 4.2; see also Theorem 4.21. An element $a \in A$ is called a μ-*element* if for some ideal I of $A \restriction a$, the algebra $(A \restriction a)/I$ has a strictly increasing sequence of type μ. Let $\langle \lambda_i : i < \omega \rangle$ be a strictly increasing sequence of infinite regular cardinals with supremum κ. We call an element $a \in A$ an ∞-*element* if it is a λ_i-element for all $i < \omega$.

(1) If a is an ∞-element and $a = b + c$ with $b \cdot c = 0$, then b is an ∞-element or c is an ∞-element.

For, it is enough to show that for every $i < \omega$, either b is a λ_i-element or c is a λ_i-element. Suppose that for some $i < \omega$, neither b nor c is a λ_i-element. Let I be an ideal in $A \restriction a$ and $\langle [x_\alpha] : \alpha < \lambda_i \rangle$ a strictly increasing sequence of elements in $(A \restriction a)/I$. Now if $\alpha < \beta < \lambda_i$, then

$$x_\alpha \cdot b \cdot -(x_\beta \cdot b) = x_\alpha \cdot -x_\beta \cdot b \in I \cap (A \restriction b),$$

and hence in $A \restriction b$ we have $[x_\alpha \cdot b] \le [x_\beta \cdot b]$. Hence there is an $\alpha < \lambda_i$ such that if $\alpha < \beta < \gamma < \lambda_i$ then $x_\gamma \cdot -x_\beta \cdot b \in I$. Similarly for c: there is an $\alpha' < \lambda_i$ such that if $\alpha' < \beta < \gamma < \lambda_i$, then $x_\gamma \cdot -x_\beta \cdot c \in I$. But then if $\max(\alpha, \alpha') < \beta < \gamma < \lambda_i$ we get $x_\gamma \cdot -x_\beta \in I$, contradiction. This proves (1).

Now we construct disjoint elements $a_0, a_1, \ldots$ such that a_i is a λ_i-element for all $i < \omega$. Suppose that a_i has been constructed for all $i < n$ so that $\prod_{i<n} -a_i$ is an ∞-element. Now there exists an ideal I in $A \restriction \prod_{i<n} -a_i$ with a sequence $\langle [x_\alpha] : \alpha < \lambda_{n+1} \rangle$ strictly increasing in $(A \restriction \prod_{i<n} -a_i)/I$. Then clearly

(2) x_{λ_n} is a λ_n-element.

Now by (1) and (2) there is a λ_n-element a_n such that $\prod_{i \le n} -a_i$ is an ∞-element.

Now for each $i < \omega$ choose an ideal I_i in $A \restriction a_i$ such that $(A \restriction a_i)/I_i$ has a chain of type λ_i. Let $J = \langle \bigcup_{i<\omega} I_i \rangle^{Id}$. Then $J \cap (A \restriction a_i) = I_i$ for each $i < \omega$, and hence A/J has a chain of type λ_i for all $i < \omega$. Hence as in the proof of 4.2, A/J has a chain of type κ, as desired (see the proof of 4.21). □

We also recall from Theorem 11.8 that $\mathrm{t}A = \pi\chi_{\mathrm{H}+}A = \pi\chi_{\mathrm{h}+}A$. And, as mentioned after the proof of Theorem 11.8, $\pi\chi_{\mathrm{H}+}$ and $\pi\chi_{\mathrm{h}+}$ have the same attainment properties, while $\pi\chi_{\mathrm{H}+}$ attained implies that t is attained in the free sequence sense. Another of the attainment problems is answered by the following theorem.

Theorem 12.3. *Suppose that κ is a singular cardinal. Then tightness is not attained in* Intalg κ.

Proof. Let $\langle \lambda_\alpha : \alpha < \mathrm{cf}\kappa \rangle$ be a strictly increasing continuous sequence of cardinals with supremum κ. By the Handbook, each ultrafilter on Intalg κ is determined by an end segment of κ not containing 0 (this last restriction is not found in the Handbook, but it is clearly necessary). If C is such an end segment of κ, then its associated ultrafilter F_C is generated by

$$\{[0, c) : c \in C\} \cup \{[c, \infty) : c \in \kappa \backslash C\}.$$

So, take any end segment C; we want to show that $tF_C < \kappa$. *Case 1.* $C = 0$. In this case we claim that $tF_C \leq \mathrm{cf}\kappa$. In fact, suppose that $F_C \subseteq \bigcup Y$, where $Y \subseteq \mathrm{Ult}(\mathrm{Intalg}\,\kappa)$. For each $\alpha < \mathrm{cf}\kappa$ we have $[\lambda_\alpha, \infty) \in F_C$, so we can choose $G_\alpha \in Y$ such that $[\lambda_\alpha, \infty) \in G_\alpha$. We claim that $F_C \subseteq \bigcup\{G_\alpha : \alpha < \mathrm{cf}\kappa\}$ (as desired). In fact, let $x \in F_C$. Without loss of generality x has the form $[c, \infty)$ for some $c \in \kappa$. Choose $\alpha < \mathrm{cf}\kappa$ such that $c < \lambda_\alpha$. Then $[c, \infty) \supseteq [\lambda_\alpha, \infty) \in G_\alpha$, as desired. *Case 2.* $C \neq 0$. Let c be the least element of C. Then we claim that $tF_C \leq \max(\omega, |c|)$. For, again suppose that $F_C \subseteq \bigcup Y$, where $Y \subseteq \mathrm{Ult}(\mathrm{Intalg}\,\kappa)$. For each $d < c$ we have $[d, c) \in F_C$, and so we can choose $G_d \in Y$ such that $[d, c) \in G_d$. Now we claim that $F_C \subseteq \bigcup\{G_d : d < c\}$, as desired. For, let $x \in F_C$. Wlog x has the form $[d, e)$ with $d \in \kappa \backslash C$ and $e \in C$. Then $x \in G_d$, as desired. $\square$

Corollary 12.4. *For every singular cardinal κ there is a BA A such that $tA = \kappa$ not attained but A has a free sequence of type κ.* $\square$

This corollary answers Problem 29 of Monk [90]. But recall from the proof of Theorem 4.20 that if tA is regular, then attainment in the free sequence sense implies attainment in the defined sense.

The description of $\pi\chi$ for interval algebras given at the end of Chapter 11 shows that if κ is singular, then $\pi\chi(\mathrm{Intalg}\,\kappa) = \kappa$ not attained. Thus attainment in the free sequence sense does not imply attainment in the $\pi\chi_{\mathrm{H}+}$ sense, answering Problem 30 in Monk [90] negatively. But again if tA is regular and it is attained in the free sequence sense then it is attained in the $\pi\chi_{\mathrm{h}+}$ sense. The argument here is a little lengthy, but will be useful in discussing character too. Let $tA = \kappa$, κ regular, and suppose that $\langle F_\alpha : \alpha < \kappa \rangle$. is a free sequence in Ult$A$. For each $\xi < \kappa$ choose $a_\xi \in A$ such that $\{F_\alpha : \alpha < \xi\} \subseteq Sa_\xi$ and $Sa_\xi \cap \{F_\alpha : \xi \leq \alpha < \kappa\} = 0$. Then

$$\{-a_\xi : \xi < \kappa\} \cup \{x \in A : \{F_\alpha : \alpha < \kappa\} \subseteq Sx\}$$

has the finite intersection property. In fact, otherwise we would get $-a_{\xi_1} \cdot \ldots \cdot$
$-a_{\xi_n} \cdot x = 0$, where $\{F_\alpha : \alpha < \kappa\} \subseteq Sx$. Choose $\alpha < \kappa$ with $\xi_i < \alpha$ for all $i = 1, \ldots, n$. Then $x \in F_\alpha$, so $a_{\xi_i} \in F_\alpha$ for some i, contradiction. So, let G be an ultrafilter containing the given set. Let $Y = \{F_\alpha : \alpha < \kappa\} \cup \{G\}$. We claim that $\pi\chi(G, Y) = \kappa$. For, suppose that $M \in [A]^{<\kappa}$ and $\{Sx \cap Y : x \in M\}$ is a π-base for G, where $Sx \cap Y \neq 0$ for all $x \in M$. Then by the regularity of κ, there is

an $x \in M$ and a $\Gamma \in [\kappa]^{\kappa}$ such that $\mathcal{S}x \cap Y \subseteq \mathcal{S}(-a_{\xi}) \cap Y$ for all $\xi \in \Gamma$. Then $\{F_{\alpha} : \alpha < \kappa\} \subseteq \mathcal{S}(-x)$. In fact, let $\alpha < \kappa$. Choose $\xi \in \Gamma$ such that $\alpha < \xi$. Then $F_{\alpha} \in \mathcal{S}a_{\xi}$, so $F_{\alpha} \notin \mathcal{S}x$, hence $F_{\alpha} \in \mathcal{S}(-x)$, proving that $\{F_{\alpha} : \alpha < \kappa\} \subseteq \mathcal{S}(-x)$. It follows that $-x \in G$ too. So $\mathcal{S}x \cap Y = 0$, contradiction.

Three problems about attainment remain; the first one is Problem 28 in Monk [90].

Problem 41. *Does attainment of tightness imply attainment in the free sequence sense?*

Problem 42. *Does attainment of tightness imply attainment in the $\pi\chi_{H+}$ sense?*

Note that "yes" on Problem 42 implies "yes" on Problem 41.

Problem 43. *Does attainment of tightness in the $\pi\chi_{H+}$ sense imply attainment in the sense of the definition?*

We return to the discussion of products.

Theorem 12.5. *If $\langle A_i : i \in I \rangle$ is a system of non-trivial BAs, with I infinite, then $\mathrm{t}(\prod_{i \in I} A_i) \geq \max(2^{|I|}, \sup_{i \in I} \mathrm{t}A_i)$.*

Proof. If $j \in I$, then A_j is isomorphic to a subalgebra of $\prod_{i \in I} A_i$; so $\mathrm{t}A_i \leq \mathrm{t}(\prod_{i \in I} A_i)$. Since independence is less than or equal to tightness, it also follows that $2^{|I|} \leq \mathrm{t}(\prod_{i \in I} A_i)$. □

Theorem 10.5 implies that tightness can jump in a product: apply it to $A = \mathrm{Finco}\ \kappa$ and use the discussion of free products below. A similar remark holds for ultraproducts.

Note that there can be superatomic interval algebras with high tightness; this is clear from the fact that $\mathrm{Depth} \leq \mathrm{t}$.

Now we consider ultraproducts, giving some results of Douglas Peterson. Recall from the introduction that tightness is an order-independence function. For such functions we have the following theorem, which uses the notion of *depth* of a linear ordering, which is the supremum of cardinalities of well-ordered subsets of the ordering.

Theorem 12.6. *Suppose that k is an order-independence function, $\langle A_i : i \in I \rangle$ is a sequence of infinite BAs, with I infinite, F is an ultrafilter on I, and $\langle \kappa_i : i \in I \rangle$ is a sequence of cardinals such that $\kappa_i < k'A_i$ for all $i \in I$. Then $k\left(\prod_{i \in I} A_i/F\right) \geq \mathrm{Depth}\left(\prod_{i \in I} \kappa_i/F\right)$.*

Proof. For each $i \in I$ let $\langle a_{\alpha}^i : \alpha < \kappa_i \rangle$ be a sequence of elements of A_i such that for all finite $G, H \subseteq \kappa_i$, if $\langle \kappa_i, <, G, H \rangle \models \varphi$ then $\prod_{\alpha \in F} a_{\alpha}^i \cdot \prod_{\alpha \in H} -a_{\alpha}^i \neq 0$. Let $\lambda = \mathrm{Depth}\left(\prod_{i \in I} \kappa_i/F\right)$. We consider two cases. *Case 1.* λ is a successor cardinal. Let $\langle f_{\alpha}/F : \alpha < \lambda \rangle$ be a sequence of elements of $\prod_{i \in I} \kappa_i/F$ such that $f_{\alpha}/F < f_{\beta}/F$ if $\alpha < \beta$. Define $g_{\alpha}i = a_{f_{\alpha}i}^i$ for all $\alpha < \lambda$ and $i \in I$. Now suppose that G and H are finite subsets of λ such that $\langle \lambda, <, G, H \rangle \models \varphi$. Let

$$K = \{i \in I : \forall \alpha, \beta \in G \cup H(\alpha < \beta \Rightarrow f_{\alpha}i < f_{\beta}i)\}.$$

Then $K \in F$. By (2) in the definition of order-independence function we have
$(\kappa_i, <, \{f_\alpha i : \alpha \in G\}, \{f_\alpha i : \alpha \in H\}) \models \varphi$ for each $i \in K$, and hence $\prod_{\alpha \in G} a^i_{f_\alpha i} \cdot$
$\prod_{\alpha \in H} -a^i_{f_\alpha i} \neq 0$. Therefore $\prod_{\alpha \in G} f_\alpha / F \cdot \prod_{\alpha \in H} -g_\alpha / F \neq 0$, as desired.

$Case\ 2.$ λ is a limit cardinal. Then $k\left(\prod_{i \in I} A_i / F\right) \geq \kappa$ for each successor
$\kappa < \lambda$, by the above argument; hence $k\left(\prod_{i \in I} A_i / F\right) \geq \lambda$. $\qquad\square$

Theorem 12.7. (GCH) *Suppose that $\langle A_i : i \in I \rangle$ is a system of infinite BAs, with
I infinite, and F is a regular ultrafilter on I. Then* $t\left(\prod_{i \in I} A_i / F\right) \geq \left|\prod_{i \in I} tA_i / F\right|$.

Proof. Let $\kappa = \text{ess.sup}^F_{i \in I} tA_i$. $Case\ 1.$ $\kappa \leq |I|$. The ultraproduct has an independent subset of size $2^{|I|}$, and the desired result follows. $Case\ 2.$ $\text{cf}\kappa > |I|$. Then if
κ is a successor cardinal, we may assume that $t'A_i = \kappa^+$ for all $i \in I$, and hence
by Theorem 12.6 we have $t\left(\prod_{i \in I} A_i / F\right) \geq \text{Depth}\left(\prod_{i \in I} \kappa / F\right) = \kappa$. The limit case
clearly follows from this case. $Case\ 3.$ $\text{cf}\kappa \leq |I| < \kappa$. Using Lemma 3.12 or Lemma
3.13, we obtain a system $\langle \lambda_i : i \in I \rangle$ of infinite cardinals such that $\lambda_i < tA_i$ for
each $i \in I$, and $\text{ess.sup}^F_{i \in I} \lambda_i = \kappa$. Hence by Theorem 12.6 again, and by the proof
of Theorem 4.13, $t\left(\prod_{i \in I} A_i / F\right) \geq \text{Depth}\left(\prod_{i \in I} \lambda_i / F\right) \geq \kappa^+$. $\qquad\square$

As usual, Donder's theorem then says that $\geq$ holds for any uniform ultrafilter,
assuming $V = L$. The inequality can be strict, from the discussion of independence.
A consistent example exists for the other direction by Magidor, Shelah [91]; see
also Rosłanowski, Shelah [94].

The tightness of free products is described by a theorem of Malyhin [72]; we give
the result here. The proof we give is due to Todorčević (private communication);
he uses the idea of this proof to strengthen Malyhin's result.

Theorem 12.8. $t(A \oplus B) = \max(tA, tB)$.

Proof. The inequality $\geq$ is clear. For the other inequality it suffices to show that
if $\langle c_\alpha : \alpha < \theta \rangle$ is a free sequence in $A \oplus B$ with θ regular and uncountable, then
either A or B has a free sequence of that length too. We use *free sequence* here in
the algebraic sense described in Chapter 11. First we claim:

(1) We may assume that each c_α has the form $a_\alpha \cdot b_\alpha$ with $a_\alpha \in A$ and $b_\alpha \in B$.

To see this, first write $c_\alpha = \sum_{i < m_\alpha} a_{\alpha i} \cdot b_{\alpha i}$ with each $a_{\alpha i} \in A$ and each $b_{\alpha i} \in B$.
Since θ is regular and uncountable, we may assume that $m_\alpha = m$ does not depend
on α. Now for each $\alpha < \theta$ let F_α be an ultrafilter on $A \oplus B$ such that $\{c_\xi : \xi \leq$
$\alpha\} \cup \{-c_\xi : \alpha < \xi < \theta\} \subseteq F_\alpha$. Then by the first part of the proof of Theorem 4.20
we get an ultrafilter G on $A \oplus B$ such that

(2) $|\{\alpha < \theta : a \in F_\alpha\}| = \theta$ for all $a \in G$.

Then $c_\alpha \in G$ for all $\alpha < \theta$; for if $-c_\alpha \in G$ we would get $-c_\alpha \in F_\beta$ for some $\beta \geq \alpha$
by (2), and this is impossible. It follows that for all $\alpha < \theta$ there is an $i < m$ such
that $a_{\alpha i} \cdot b_{\alpha i} \in G$. Hence there exist an $i < m$ and a $\Gamma \in [\theta]^\theta$ such that $a_{\alpha i} \cdot b_{\alpha i} \in G$
for all $\alpha \in \Gamma$. Now let

$$K = \{\delta \in \Gamma : \forall H \in [\Gamma \cap \delta]^{<\omega} \exists \alpha \in (\max H, \delta) \forall \xi \in H(a_{\xi i} \cdot b_{\xi i} \in F_\alpha)\}.$$

We claim that K is unbounded in θ. For, let $\delta_0 < \theta$. For every finite $H \subseteq \Gamma \cap \delta_0$ we have $\prod_{\xi \in H} a_{\xi i} \cdot b_{\xi i} \in G$, and hence by (2) there is an $\alpha_H > \max H$ such that $\prod_{\xi \in H} a_{\xi i} \cdot b_{\xi i} \in F_{\alpha_H}$. Choose $\delta_1 \in \Gamma$ greater than δ_0 and all ordinals α_H for $H \in [\Gamma \cap \delta_0]^{<\omega}$. Then repeat the construction for δ_1, obtaining $\delta_2 \in \Gamma$, etc. Finally, let δ_ω be the least member of Γ greater than all δ_i, $i < \omega$. Clearly $\delta_\omega \in K$, proving the claim about K.

Let $\langle \delta_\xi : \xi < \theta \rangle$ enumerate K in increasing order. We claim, then, that $\langle a_{\delta_\xi i} \cdot b_{\delta_\xi i} : \xi < \theta \rangle$ is a free sequence in $A \oplus B$; this will prove the claim (1). To prove this, let M and N be finite subsets of θ such that each member of M is less than each member of N. We may assume that N is nonempty. Let ξ be the least member of N. We then apply the definition of K to its member δ_ξ to get an $\alpha \in (\max\{\delta_\eta : \eta \in M\}, \delta_\xi)$ such that $a_{\delta_\eta i} \cdot b_{\delta_\eta i} \in F_\alpha$ for all $\eta \in M$. Note that we also have $-c_{\delta_\eta} \in F_\alpha$ for all $\eta \in N$. Now

$$\prod_{\eta \in M} a_{\delta_\eta i} \cdot b_{\delta_\eta i} \cdot \prod_{\eta \in N} -c_{\delta_\eta} \leq \prod_{\eta \in M} a_{\delta_\eta i} \cdot b_{\delta_\eta i} \cdot \prod_{\eta \in N} -(a_{\delta_\eta i} \cdot b_{\delta_\eta i}),$$

and the left side is in F_α and hence is nonzero, so the right side is nonzero too, and this proves that $\langle a_{\delta_\xi i} \cdot b_{\delta_\xi i} : \xi < \theta \rangle$ is a free sequence in $A \oplus B$.

So now we assume (1). We consider two cases. *Case 1.* $\forall \alpha < \theta \exists \beta \geq \alpha \forall K \in [\alpha]^{<\omega} \forall L \in [\theta \backslash \beta]^{<\omega} (a_{KL} \overset{\text{def}}{=} \prod_{\xi \in K} a_\xi \cdot \prod_{\xi \in L} -a_\xi \neq 0)$. Define $\langle \alpha_\xi : \xi < \theta \rangle$ as follows. If α_η has been defined for all $\eta < \xi$, let $\beta_\xi = \sup_{\eta < \xi} \alpha_\eta$ and choose $\alpha_\xi > \beta_\xi$ such that $\forall K \in [\beta_\xi]^{<\omega} \forall L \in [\theta \backslash \alpha_\xi]^{<\omega} (a_{KL} \neq 0)$. Then $\langle a_{\alpha_\xi} : X < \theta \rangle$ is a free sequence in A. For, assume that M and N are finite subsets of θ, each member of M less than each member of N. Let $\xi = \sup_{\eta \in M}(\eta + 1)$ ($\xi = 0$ if $M = 0$). Then $\{\alpha_\eta : \eta \in M\} \in [\beta_\xi]^{<\omega}$ and $\{\alpha_\eta : \eta \in N\} \in [\theta \backslash \alpha_\xi]^{<\omega}$, so $\prod_{\eta \in M} a_{\alpha_\eta} \cdot \prod_{\eta \in N} -a_{\alpha_\eta} \neq 0$, as desired.

Case 2. Case 1 fails: $\exists \alpha_0 < \theta \forall \beta \geq \alpha_0 \exists K_\beta \in [\alpha_0]^{<\omega} \exists L \in [\theta \backslash \beta]^{<\omega} (a_{K_\beta L} = 0)$. So $\exists K \in [\alpha_0]^{<\omega} \exists \Gamma \in [\theta \backslash \alpha_0]^{\theta} \forall \beta \in \Gamma \exists L \in [\theta \backslash \beta]^{<\omega} (a_{KL} = 0)$. Hence we get $\langle L_\alpha : \alpha < \theta \rangle$ such that $(\alpha < \beta < \theta \Rightarrow \forall \xi \in L_\alpha \forall \eta \in L_\beta (\xi < \eta))$, $(\alpha < \theta \Rightarrow \forall \xi \in K \forall \eta \in L_\alpha (\xi < \eta))$, and $a_{KL_\alpha} = 0$ for all $\alpha < \theta$. Let $\overline{b}_\alpha = \prod_{\xi \in L_\alpha} b_\xi$ for all $\alpha < \theta$. Then $\langle \overline{b}_\alpha : \alpha < \theta \rangle$ is a free sequence in B. For, suppose that M and N are finite subsets of θ, each member of M less than each member of N. Then, with $P = \bigcup_{\alpha \in N} L_\alpha$,

$$0 \neq \prod_{\xi \in K} a_\xi \cdot b_\xi \cdot \prod_{\alpha \in M, \xi \in L_\alpha} a_\xi \cdot b_\xi \cdot \prod_{\alpha \in N, \xi \in L_\alpha} -(a_\xi \cdot b_\xi)$$

$$= \prod_{\xi \in K} a_\xi \cdot b_\xi \cdot \prod_{\alpha \in M, \xi \in L_\alpha} a_\xi \cdot b_\xi \cdot \sum_{\Gamma \subseteq P} \left(\prod_{\xi \in \Gamma} -a_\xi \cdot \prod_{\xi \in P \backslash \Gamma} -b_\xi \right),$$

so choose $\Gamma \subseteq P$ so that

$$0 \neq \prod_{\xi \in K} a_\xi \cdot b_\xi \cdot \prod_{\alpha \in M, \xi \in L_\alpha} a_\xi \cdot b_\xi \cdot \prod_{\xi \in \Gamma} -a_\xi \cdot \prod_{\xi \in P \backslash \Gamma} -b_\xi.$$

Now if $L_\alpha \subseteq \Gamma$ for some $\alpha \in N$, then

$$\prod_{\xi \in K} a_\xi \cdot b_\xi \cdot \prod_{\alpha \in M, \xi \in L_\alpha} a_\xi \cdot b_\xi \cdot \prod_{\xi \in \Gamma} -a_\xi \cdot \prod_{\xi \in P \backslash \Gamma} -b_\xi \leq a_{KL_\alpha} = 0,$$

contradiction. So for all $\alpha \in N$ there is a $\xi \in L_\alpha \backslash \Gamma$. Thus

$$0 \neq \prod_{\xi \in K} b_\xi \cdot \prod_{\alpha \in M, \xi \in L_\alpha} b_\xi \cdot \prod_{\xi \in P \backslash \Gamma} -b_\xi \leq \prod_{\alpha \in M} \bar{b}_\alpha \cdot \prod_{\alpha \in N} \sum_{\xi \in L_\alpha} -b_\xi$$

$$= \prod_{\alpha \in M} \bar{b}_\alpha \cdot \prod_{\alpha \in N} -\bar{b}_\alpha,$$

as desired.   □

Theorem 12.9. *If $\langle A_i : i \in I \rangle$ is a system of BAs each with at least four elements, then $\mathrm{t}(\oplus_{i \in I} A_i) = \max(|I|, \sup_{i \in I} \mathrm{t} A_i)$.*

Proof. . Obviously $\mathrm{t} A_j \leq \mathrm{t}(\oplus_{i \in I} A_i)$ for each $j \in I$; and $|I| \leq \oplus_{i \in I} A_i$ since $\mathrm{Ind} \leq \mathrm{t}$. Thus $\geq$ holds. To prove $\leq$, let $\kappa = \max(|I|, \sup_{i \in I} \mathrm{t} A_i)$, and suppose that $\langle c_\alpha : \alpha < \kappa^+ \rangle$ is a free sequence in $\oplus_{i \in I} A_i$; we shall get a contradiction. For each $\alpha < \kappa^+$ there is a finite $S_\alpha \subseteq I$ such that $c_\alpha \in \oplus_{i \in S_\alpha} A_i$. We may assume that $S = S_\alpha$ does not depend on α. But then $\kappa^+ \leq \sup_{i \in S} \mathrm{t} A_i$ by Theorem 12.8, contradiction.   □

The behaviour of tightness in the free sequence sense under unions of chains of BAs is similar to the case of cellularity (Theorem 3.11). The definition of ordinary sup-function does not quite fit, but essentially the same proof can be used:

Theorem 12.10. *Let κ and λ be infinite cardinals, with λ regular. Then the following conditions are equivalent:*

(i) $\mathrm{cf} \kappa = \lambda$.

(ii) There is a strictly increasing sequence $\langle A_\alpha : \alpha < \lambda \rangle$ of infinite Boolean algebras each with no free sequence of type κ such that $\bigcup_{\alpha < \lambda} A_\alpha$ has a free sequence of type κ.   □

In view of the equivalence of tightness with its free sequence variant, 12.10 also applies to tightness when κ is a successor cardinal. And actually 12.10 extends in the following form to tightness itself; this answers, negatively, Problem 31 in Monk [90].

Theorem 12.11. *Let κ and λ be infinite cardinals, with λ regular. Then the following conditions are equivalent:*

(i) $\mathrm{cf} \kappa = \lambda$.

(ii) There is a strictly increasing sequence $\langle A_\alpha : \alpha < \lambda \rangle$ of Boolean algebras each with tightness less than κ such that $\bigcup_{\alpha < \lambda} A_\alpha$ has tightness κ.

Proof. By the comment before the theorem, we assume that κ is a limit cardinal. Let $B = \bigcup_{\alpha < \lambda} A_\alpha$. (i)$\Rightarrow$(ii): Take a free BA of size κ and write it as an increasing union of smaller algebras in the obvious way.

(ii)$\Rightarrow$(i): Assume that (ii) holds and (i) fails. Let $\mu = \sup_{\alpha < \lambda} tA_\alpha$; the first part of the proof will consist in showing that $\mu = \kappa$; to this end, suppose that $\mu < \kappa$. Fix ν such that $\mu < \nu < \kappa$. Let $\langle a_\alpha : \alpha < \nu \rangle$ be a free sequence in B. For each $\beta < \lambda$ let $S_\beta^\nu = \{\alpha < \nu : a_\alpha \in A_\beta\}$. Thus $S_\beta^\nu \subseteq S_\gamma^\nu$ for $\beta < \gamma < \lambda$, and $\nu = \bigcup_{\beta < \lambda} S_\beta^\nu$. If $\exists \beta < \lambda \forall \gamma \in (\beta, \lambda)[S_\beta^\nu = S_\gamma^\nu]$, then $\nu = S_\beta^\nu$ and so $\{a_\alpha : \alpha < \nu\} \subseteq A_\beta$, hence $tA_\beta \geq \nu$, contradiction. Thus $\forall \beta < \lambda \exists \gamma \in (\beta, \lambda)[S_\beta^\nu \subset S_\gamma^\nu]$. Applying this to $\nu = \mu^+$ we get $\lambda \leq \mu^+$; then applying it to $\nu = \mu^{++}$ we get $\mu^{++} = \bigcup_{\beta < \lambda} S_\beta^{\mu^{++}}$, so there is a $\beta < \lambda$ such that $|S_\beta^{\mu^{++}}| = \mu^{++}$, so A_β has a free sequence of type μ^{++}, which contradicts $tA_\beta \leq \mu$. This contradiction proves that $\mu = \kappa$.

Since each tA_α is less than κ, from $\mu = \kappa$ it follows that $\mathrm{cf}\kappa \leq \lambda$; since (i) fails, we have in fact that $\mathrm{cf}\kappa < \lambda$. Now $\forall \alpha < \lambda \exists \beta \in (\alpha, \lambda)[tA_\alpha < tA_\beta]$, since otherwise we would have $\mu < \kappa$. Hence $\lambda \leq \sup_{\alpha < \lambda} tA_\alpha = \kappa$. Since λ is regular, $\lambda < \kappa$. So κ is singular. Let $\langle \nu_\alpha : \alpha < \mathrm{cf}\kappa \rangle$ be a strictly increasing sequence of cardinals with supremum κ. For each $\alpha < \mathrm{cf}\kappa$ there is a $\beta_\alpha < \lambda$ such that $tA_{\beta_\alpha} \geq \nu_\alpha$, since $\mu = \kappa$. Let $\gamma = \sup_{\alpha < \mathrm{cf}\kappa} \beta_\alpha$; then $\gamma < \lambda$ since $\mathrm{cf}\kappa < \lambda$. But then $tA_\gamma = \kappa$, contradiction. $\square$

A more natural version of Theorem 12.10 for tightness itself would be the equivalence expressed in the following problem.

Problem 44. *Is the following true? Let κ and λ be infinite cardinals, with λ regular. Then the following conditions are equivalent:*

(i) $\mathrm{cf}\kappa = \lambda$.

(ii) There is a strictly increasing sequence $\langle A_\alpha : \alpha < \lambda \rangle$ of Boolean algebras each having no ultrafilter with tightness κ such that $\bigcup_{\alpha < \lambda} A_\alpha$ has an ultrafilter with tightness κ.

We turn to derived functions for tightness. By Theorem 11.8 we have that $t_{H+} = t_{h+} = tA$. Clearly $t_{S+} = tA$, $t_{S-}A = \omega$, and $_d t_{S+} A = tA$. In the algebra of Fedorchuk [75] we have $t_{H-}A \leq tA < \mathrm{Card}_{H-}A$; so Problem 32 of Monk [90] was solved long ago. See Chapter 16 for Fedorchuk's algebra. Note that $_d t_{S-}A \neq tA$ in general; this can be seen by considering $\mathscr{P}\omega$ and its dense subalgebra consisting of the finite and cofinite subsets of ω.

Recall also our earlier results that $\mathrm{Depth}_{H+}A = tA = \pi\chi_{H+}A$; see Theorems 4.21, 11.8.

Next we mention more about the relationships between tightness and our previously introduced functions. By Theorem 4.21 we have $\mathrm{Depth}A \leq tA$ for any BA A; the difference can be big, for example in a free algebra. $\pi\chi A \leq tA$ by Theorem 11.8. We observed in Chapter 11 that one can have $\pi\chi A < \mathrm{Depth}A$ in an interval algebra with the difference arbitrarily large. This solves Problem 33 in Monk [90]. In particular, it is possible to have $\pi\chi A < tA$ with the difference arbitrarily large.

We observed at the beginning of this chapter that $\mathrm{Ind}A \leq tA$; the difference is large in some interval algebras. Obviously $tA \leq |A|$. Note that $tA \leq sA \leq \mathrm{Irr}A$ by Theorems 3.25 and 4.21. Thus Problem 34 in Monk [90] has the obvious answer "no". $tA > \pi A$ for $A = \mathscr{P}\omega$. $tA > \text{Length}A$ for A an uncountable free BA. $\text{Length}A > tA$ for A the interval algebra on the reals. $cA > tA$ for A an uncountable finite-cofinite algebra.

We also give the following result relating π with t; it is from Todorčević [90a].

Theorem 12.12. *For every infinite BA A there is a sequence $\langle a_\alpha : \alpha < \beta \rangle$ of nonzero elements of A such that $\{a_\alpha : \alpha < \beta\}$ is dense in A and for every subset Γ of β with no maximum element, the sequence $\langle a_\alpha : \alpha \in \Gamma \rangle$ is free iff $\{a_\alpha : \alpha \in \Gamma\}$ has the finite intersection property. (Since Γ has a natural order from β, the meaning of "free" in this extended sense is clear.)*

Proof. Let P be a maximal disjoint subset of A^+ such that $A \restriction b$ is π-homogeneous for every $b \in P$, that is, $\pi(A \restriction c) = \pi \restriction b)$ for every nonzero $c \leq b$. Temporarily fix $b \in P$. Let $\pi(A \restriction b) = \kappa_b$, and let $\langle c_\alpha^b : \alpha < \kappa_b \rangle$ enumerate a dense subset of $(A \restriction b)^+$ of size κ_b. Now we define $\langle a_\alpha^b : \alpha < \kappa_b \rangle$ by induction. Suppose that a_α^b has been defined for all $\alpha < \beta$. Let $\mathscr{F}$ be the collection of all nonzero elements of the form $c_\beta^b \cdot \prod_{\alpha \in F} (a_\alpha^b)^{\varepsilon\alpha}$ for F a finite subset of β and $\varepsilon \in {}^F 2$. Then $\mathscr{F}$ is not dense in $A \restriction c_\beta^b$, so there is an $a_\beta^b \in (A \restriction c_\beta^b)^+$ such that $x \not\leq a_\beta^b$ for all $x \in \mathscr{F}$. This finishes the construction.

Concatenating the so obtained sequences $\langle a_\alpha^b : \alpha < \kappa_b \rangle$ in any order, we obtain a sequence $\langle a_\alpha : \alpha < \beta \rangle$ as desired in the theorem. In fact, first we check that $\{a_\alpha : \alpha < \beta\}$ is dense in A. Suppose that $a \in A^+$. Choose $b \in P$ such that $a \cdot b \neq 0$. There is a $\gamma < \kappa_b$ such that $c_\gamma^b \leq a \cdot b$. By construction, $a_\gamma^b \leq c_\gamma^b$, as desired. Next we check that for any subset Γ of β with no maximum element, $\langle a_\alpha : \alpha \in \Gamma \rangle$ is free iff $\{a_\alpha : \alpha \in \Gamma\}$ has the finite intersection property. $\Rightarrow$: obvious. $\Leftarrow$: Assume that $\{a_\alpha : \alpha \in \Gamma\}$ has the finite intersection property. Then there is a $b \in P$ such that each of the a_α's for $\alpha \in \Gamma$ of the form a_γ^b. So without loss of generality we assume that $\{a_\alpha^b : \alpha \in \Gamma\}$ has the finite intersection property, and we want to show that $\langle a_\alpha^b : \alpha \in \Gamma \rangle$ is free. We prove

(*) If F and G are finite subsets of γ and $F < G$, then $\prod_{\alpha \in F} a_\alpha^b \cdot \prod_{\alpha \in G} -a_\alpha^b \neq 0$.

This we do by induction on $|G|$. The case $G = 0$ is given. Assume that (*) is true for G, and $G < \gamma \in \Gamma$. If $\prod_{\alpha \in F} a_\alpha^b \cdot \prod_{\alpha \in G} -a_\alpha^b \cdot c_\gamma^b = 0$, then also $\prod_{\alpha \in F} a_\alpha^b \cdot \prod_{\alpha \in G} -a_\alpha^b \cdot a_\gamma^b = 0$, and so $0 \neq \prod_{\alpha \in F} a_\alpha^b \cdot \prod_{\alpha \in G} -a_\alpha^b = \prod_{\alpha \in F} a_\alpha^b \cdot \prod_{\alpha \in G} -a_\alpha^b \cdot -a_\gamma^b$, as desired. If $\prod_{\alpha \in F} a_\alpha^b \cdot \prod_{\alpha \in G} -a_\alpha^b \cdot c_\gamma^b \neq 0$, then $\prod_{\alpha \in F} a_\alpha^b \cdot \prod_{\alpha \in G} -a_\alpha^b \cdot -a_\gamma^b \neq 0$ by construction. $\qquad \square$

There are several natural finite versions of tightness, using the free sequence equivalent. For $m, n \in \omega$, an m, n-*free sequence* is a sequence $\langle a_\alpha : \alpha < \kappa \rangle$ such that if $\Gamma, \Delta \subseteq \alpha$ with $|\Gamma| = m$, $|\Delta| = n$, and $\Gamma < \Delta$, then $\prod_{\alpha \in \Gamma} a_\alpha \cdot \prod_{\beta \in \Delta} -a_\beta \neq 0$. Then we set

$$t_{mn}A = \sup\{\kappa : \text{there is an } m, n\text{-free sequence of length } \kappa\}.$$

Similarly we get four more notions:

An m-free sequence is a sequence $\langle a_\alpha : \alpha < \kappa \rangle$ such that if $\Gamma, \Delta \subseteq \alpha$ with $|\Gamma| = m$, Δ finite and $\Gamma < \Delta$, then $\prod_{\alpha \in \Gamma} a_\alpha \cdot \prod_{\beta \in \Delta} -a_\beta \neq 0$;

$$t_m A = \sup\{\kappa : \text{there is an } m\text{-free sequence of length } \kappa\}.$$

$$\mathrm{ut}_{mn} A = \sup\{|X| : \forall Y \in [X]^m \text{ and } \forall Z \in [X]^n (Y \cap Z = 0 \Rightarrow \prod_{y \in Y} \cdot \prod_{z \in Z} -z \neq 0)\};$$

$$\mathrm{ut}_m A = \sup\{|X| : \forall Y \in [X]^m \text{ and } \forall \text{ finite } Z(Y \cap Z = 0 \Rightarrow \prod_{y \in Y} \cdot \prod_{z \in Z} -z \neq 0)\}.$$

These notions are studied in Rosłanowski, Shelah, S. [94].

Concerning tightness for special classes of algebras, note first of all that $tA = |A|$ whenever A is complete. The description of t for interval algebras is similar to that for $\pi\chi$. Since t coincides with $\mathrm{Depth}_{\mathrm{H}+}$, $tA = \mathrm{Depth} A$ for A an interval algebra, by retractiveness. But it is of some interest to describe tF for each ultrafilter F on an interval algebra. Let A be the interval algebra on a linearly ordered set L with first element. Let C be a terminal segment of L not containing 0, and let κ be the type of a shortest cofinal sequence in $L \backslash C$ and λ the type of a shortest coinitial sequence in C. Then, we claim, the tightness of the ultrafilter F_C associated with C is the maximum of κ and λ. Let $\langle a_\alpha : \alpha < \kappa \rangle$ be a strictly increasing cofinal sequence in $L \backslash C$, and let $\langle b_\alpha : \alpha < \lambda \rangle$ be a strictly decreasing coinitial sequence in C. First we show that $tF_C \leq \max\{\kappa, \lambda\}$. So, assume that $F_C \subseteq \bigcup Y$, where $Y \subseteq \mathrm{Ult} A$. For each $\alpha < \kappa$ and $\beta < \lambda$ we have $[a_\alpha, b_\beta) \in F_C$, so choose $G_{\alpha\beta} \in Y$ such that $[a_\alpha, b_\beta) \in G_{\alpha\beta}$. Clearly $F_C \subseteq \bigcup\{G_{\alpha\beta} : \alpha < \kappa, \beta < \lambda\}$, as desired. Second we show that $tF_C = \max\{\kappa, \lambda\}$. Say wlog $\kappa = \max\{\kappa, \lambda\}$. For each $\alpha < \kappa$ let G_α be an ultrafilter such that $[a_\alpha, a_{\alpha+1}) \in G_\alpha$. each $\beta < \lambda$ let H_β be an ultrafilter such that Then

$$F_C \subseteq \bigcup_{\alpha < \kappa} G_\alpha;$$

and it is clear that no subset with fewer than κ elements will work.

For tree algebras the situation is similar: $t(\mathrm{Treealg}\, T) = \mathrm{Depth}(\mathrm{Treealg}\, T)$ by retractiveness. Now take any ultrafilter F on $\mathrm{Treealg}\, T$. It corresponds to an initial chain C of T; see the Handbook. A description of tF, due to Brenner [82], is as follows:

Theorem 12.13. *Let T be a tree with a single root and F an ultrafilter on* $\mathrm{Treealg} T$. *Let $C = \{t \in T : (T \uparrow t) \in F\}$. Then one of the following holds:*

(i) C has a greatest element t, and t has only finitely many immediate successors. Then F is principal, and $tF = 1$.

(ii) C has a greatest element t, and t has infinitely many immediate successors. Then $tF = \omega$.

(iii) C has no greatest element. Then $tF = \mathrm{cf} C$.

Proof. (i) is obvious. For (ii), suppose that C has a greatest element t and t has infinitely many immediate successors. Suppose that $F \subseteq \bigcup Y$, where $Y \subseteq \mathrm{Ult}(\mathrm{Treealg}T)$. Without loss of generality $F \notin Y$. We claim

(1) $S \overset{\mathrm{def}}{=} \{s : s$ is an immediate successor of t and $(T \uparrow s) \in G$ for some $G \in Y\}$ is infinite.

For, suppose that S is finite. Now $(T \uparrow t) \backslash \bigcup_{s \in S}(T \uparrow s) \in F$, so choose $G \in Y$ such that $(T \uparrow t) \backslash \bigcup_{s \in S}(T \uparrow s) \in G$. For every immediate successor s of t we have $(T \uparrow s) \notin G$. So $F = G$, contradiction. So (1) holds.

Let $U \in [S]^\omega$. For each $u \in U$ choose $G_u \in Y$ such that $(T \uparrow u) \in G_u$. Now suppose that $x \in F$. Without loss of generality x has the form $(T \uparrow t) \backslash \bigcup_{v \in V}(T \uparrow v)$ where V is a finite set of immediate successors of t. Choose $u \in U \backslash V$. Clearly $(T \uparrow t) \backslash \bigcup_{v \in V}(T \uparrow v) \in G_u$, as desired.

In the present case it is clear that F is nonprincipal, so $\mathrm{t}F = \omega$.

For (iii), suppose that C has no greatest element. Let $\langle s_\alpha : \alpha < \mathrm{cf}C \rangle$ be a strictly increasing cofinal sequence of elements of C.

First we show that $\mathrm{t}F \geq \mathrm{cf}C$. For each $\alpha < \mathrm{cf}C$ the set

$$\{T \uparrow s_\alpha\} \cup \{T \backslash (T \uparrow u) : u \text{ is an immediate successor of } s_\alpha\}$$
$$\cup \{T \backslash (T \uparrow v) : v \text{ and } s_\alpha \text{ are incomparable}\}$$

has the fip, as is easily seen; let G_α be an ultrafilter containing this set. We claim that $F \subseteq \bigcup_{\alpha < \mathrm{cf}C} G_\alpha$. For, suppose that $x \in F$. We may assume that

(2) $x = (T \uparrow r) \backslash \bigcup_{u \in U}(T \uparrow u)$ where $r \in C$, U is a set of immediate successors of r, and $U \cap C = 0$.

Choose $\alpha < \mathrm{cf}C$ such that $r \leq s_\alpha$. Clearly $x \in G_\alpha$, as desired.

Now suppose that $\Gamma \subseteq \mathrm{cf}C$ and $|\Gamma| < \mathrm{cf}C$. We claim that $F \not\subseteq \bigcup_{\alpha \in \Gamma} G_\alpha$. For, choose $\beta < \mathrm{cf}C$ such that $\Gamma < \beta$. Clearly $(T \uparrow s_\beta) \in F$ but $(T \uparrow s_\beta) \notin \bigcup_{\alpha \in \Gamma} G_\alpha$.

So, we have shown that $\mathrm{t}F \geq \mathrm{cf}C$.

Now suppose that $F \subseteq \bigcup Y$, $Y \subseteq \mathrm{Ult}(\mathrm{Treealg}\,T)$. We want to find $Z \in [Y]^{\leq \mathrm{cf}C}$ such that $F \subseteq \bigcup Z$. We may assume that $F \notin Y$. For each $\alpha < \mathrm{cf}C$ let $y_\alpha = (T \uparrow s_\alpha)$ and let $G_\alpha \in Y$ be such that $y_\alpha \in G_\alpha$. We claim that $F \subseteq \bigcup_{\alpha < \mathrm{cf}C} G_\alpha$. Let $x \in F$. We may assume that x is as in (2). Choose $\alpha < \mathrm{cf}C$ such that $r < \alpha$. Then $y_\alpha \subseteq x$, and so $x \in G_\alpha$, as desired. $\square$

The connection between tightness and depth in superatomic BAs is not completely known. In Dow, Monk [94] it is shown that if $\kappa \to (\kappa)^{<\omega}$, then every superatomic BA with tightness at least κ^+ has depth at least κ. But the following problem, for example, is open.

Problem 45. *Is there a superatomic BA A such that* $\mathrm{t}A = (2^\omega)^+$ *and* $\mathrm{Depth}A = \omega$?

13. Spread

The following theorem gives some equivalent definitions of spread.

Theorem 13.1. *For any infinite BA A, sA is equal to each of the following cardinals:*

$\sup\{|X| : X$ *is a minimal set of generators of* $\langle X \rangle^{\mathrm{Id}}\}$;

$\sup\{|X| : X$ *is ideal-independent*$\}$;

$\sup\{|X| : X$ *is the set of all atoms in some homomorphic image of* $A\}$;

$\sup\{|\mathrm{At}B| : B$ *is an atomic homomorphic image of* $A\}$;

$\sup\{cB : B$ *is a homomorphic image of* $A\}$.

Proof. Six cardinals are mentioned in this theorem; let them be denoted by $\kappa_0, \ldots \kappa_5$ in the order that they are mentioned. In Theorem 3.24 we proved that $\kappa_0 = \kappa_2$, and in Theorem 3.25 that $\kappa_2 = \kappa_5$. It is obvious that $\kappa_1 = \kappa_2$. To show that $\kappa_3 \leq \kappa_4$, suppose that B is a homomorphic image of A with an infinite number of atoms. Let I be the ideal $\langle\{x : x \cdot a = 0$ for every atom a of $B\}\rangle^{\mathrm{Id}}$ of B. Clearly B/I is atomic with the same number of atoms as B. This shows that $\kappa_3 \leq \kappa_4$. Obviously $\kappa_4 \leq \kappa_5$ Finally, for $\kappa_5 \leq \kappa_3$, let B be a homomorphic image of A, and let D be an infinite disjoint subset of B. We show how to find an atomic homomorphic image C of B with exactly $|D|$ atoms. Let M be the subalgebra of B generated by D. Let f be an extension of the identity on D to a homomorphism of B into $\overline{D}$; the image of B under f is as desired. $\qquad\square$

From these characterizations it follows that if A is a subalgebra or homomorphic image of B, then $sA \leq sB$. Clearly the difference can be arbitrarily large.

As to attainment of spread, first note that all of the equivalents of spread given in Theorem 13.1 have the same attainment properties. We state the facts known about attainment of spread without proof: (1) Spread is always attained for singular strong limit cardinals: see Juhász [80] Theorem 4.2; (2) Spread is always attained for singular cardinals of cofinality ω; see Juhász [80], Theorem 4.3; (3) Assuming V=L, if κ is inaccessible but not weakly compact, then there is a BA A with spread κ not attained: see Juhász [71], example 6.6; (4) If sA is weakly compact, then sA is attained: see Juhász [71], remark following 3.2; (5) If 2^ω is a limit cardinal, then there is a BA A with spread 2^ω not attained; see Corollary 3.31.

An infinite BA A has an infinite disjoint subset D, which gives rise to an infinite discrete subspace of Ult$A$. So sA is always infinite.

The following theorem is obvious upon looking at its topological dual:

Theorem 13.2. *Suppose that $\langle A_i : i \in I \rangle$ is a system of BAs each with at least two elements. Then* $s\left(\prod_{i \in I}^{\mathrm{w}} A_i\right) = \max(|I|, \sup_{i \in I} sA_i)$. $\qquad\square$

Clearly $s\left(\prod_{i \in I} A_i\right) \geq \max(2^{|I|}, \sup_{i \in I} sA_i)$. Shelah and Peterson independently oberved that strict inequality is possible, thus answering Problem 35 of Monk [90]. Namely, let κ be the first limit cardinal bigger than 2^ω (thus κ has cofinality ω), let

A be the finite-cofinite algebra on κ, and consider $^\omega A$. Then for any non-principal ultrafilter on ω we have

$$\kappa^\omega = |{}^\omega A| \geq s({}^\omega A) \geq c({}^\omega A/F) = \kappa^\omega$$

by the discussion of ultraproducts for cellularity. Thus $^\omega A$ gives a product where this inequality is strict.

Turning to ultraproducts, note that spread is an ultra-sup function, so Theorems 3.15–3.17 apply; Theorem 3.17 says that $s\left(\prod_{i\in I} A_i/F\right) \geq \left|\prod_{i\in I} sA_i/F\right|$ for F regular, and Donder's theorem says that under $V = L$ the regularity assumption can be removed. The example of Laver mentioned in Chapter 4 shows also that $>$ is consistent; see Rosłanowski, Shelah [94] for another example, consistently.

Problem 46. *Can one construct an example with* $s\left(\prod_{i\in I} A_i/F\right) > \left|\prod_{i\in I} sA_i/F\right|$ *in ZFC?*

Problem 47. *Is an example with* $s\left(\prod_{i\in I} A_i/F\right) < \left|\prod_{i\in I} sA_i/F\right|$ *consistent?*

Here the methods of the paper Magidor, Shelah [91] might yield a solution; and note from Rosłanowski, Shelah [91] that in any such example the invariants sA_i are inaccessible.

Theorem 13.3. *If $\langle A_i : i \in I \rangle$ is a system of BAs each with at least 4 elements, then $s(\oplus_{i\in I} A_i) \geq \max(|I|, \sup_{i\in I} sA_i)$.* $\qquad\square$

Equality does not hold in Theorem 13.3, in general. For example, let A be the interval algebra on the reals. We observed in Corollary 3.29 that $sA = \omega$. Here is a system of 2^ω ideal independent elements in $A \oplus A$: for each real number r, let $a_r = [r, \infty) \times [-\infty, r)$ (considered as an element of $A \oplus A$). Suppose that F is a finite subset of $\mathbb{R}$, $r \in \mathbb{R}\backslash F$, and $a_r \in \langle a_s : s \in F \rangle^{\mathrm{Id}}$. Thus

$$[r, \infty) \times [-\infty, r) \cdot \prod_{s \in F} ([-\infty, s) + [s, \infty)) = 0.$$

But if $T \overset{\text{def}}{=} \{s \in F : r < s\}$ and $U \overset{\text{def}}{=} F\backslash T$, then

$$[r, \infty) \times [-\infty, r) \cdot \prod_{s \in F} ([-\infty, s) + [s, \infty)) \geq$$

$$[r, \infty) \times [-\infty, r) \cdot \prod_{s \in T} [-\infty, s) \cdot \prod_{s \in U} [s, \infty) \neq 0,$$

contradiction.

We can, however, give an upper bound for the spread of a free product, namely $\max(|I|, 2^{\sup_{i\in I} sA_i})$. This is true because $|B| \leq 2^{sB}$ for any BA B (see Theorem 13.6 below); so

$$\max(|I|, \sup_{i\in I} sA_i) \leq s(\oplus_{i\in I} A_i) \leq \max(|I|, 2^{\sup_{i\in I} sA_i}).$$

Both equalities here can be attained.

We give now the proof that $|B| \leq 2^{sB}$ for any BA B. It depends on several other results which are of interest. A *network* for a space X is a collection $\mathcal{N}$ of subsets of X such that every open set in X is a union of members of $\mathcal{N}$ (the members of $\mathcal{N}$ are not assumed to be open).

Theorem 13.4. *For any infinite BA A, $|A| = \min\{|\mathcal{N}| : \mathcal{N}$ is a network for* $\mathrm{Ult}A\}$.

Proof. Clearly $\geq$ holds. Now suppose that $\mathcal{N}$ is a network for $\mathrm{Ult}A$. Let $\mathscr{P}$ be the set of all pairs (C, D) such that $C, D \in \mathcal{N}$ and for some disjoint open sets U and V, $C \subseteq U$ and $D \subseteq V$; and for each $(C, D) \in \mathscr{P}$, choose open sets of this sort — call them U_{CD} and V_{CD}. Then let $\mathcal{W}$ be the closure of the set $\{U_{CD}, V_{CD} : (C, D) \in \mathscr{P}\}$ under $\cap$ and $\cup$. We shall now show that $\{Sa : a \in A\} \subseteq \mathcal{W}$, which will prove $\leq$. So, let $a \in A$. For each $F \in Sa$ and $G \notin Sa$ choose disjoint open sets X, Y such that $F \in X$ and $G \in Y$; then choose $C(F, G), D(F, G) \in \mathcal{N}$ such that $F \in C(F, G) \subseteq X$ and $G \in D(F, G) \subseteq Y$. Thus $(C(F, G), D(F, G)) \in \mathscr{P}$; so in particular $F \in U_{C(F,G)D(F,G)}$ and $G \in V_{C(F,G)D(F,G)}$. Now fix $G \notin Sa$. Thus by compactness of Sa we get a finite subset $\mathcal{F}$ of Sa such that $Sa \subseteq \bigcup_{F \in \mathcal{F}} U_{C(F,G)D(F,G)}$. Let $U(G) = \bigcup_{F \in \mathcal{F}} U_{C(F,G)D(F,G)}$ and $V(G) = \bigcap_{F \in \mathcal{F}} V_{C(F,G)D(F,G)}$. Thus $Sa \subseteq U(G)$ and $G \in V(G)$, and $U(G)$ and $V(G)$ are disjoint. By compactness of $\mathrm{Ult}A \backslash Sa$ there is a finite subset $\mathcal{G}$ of $\mathrm{Ult}A \backslash Sa$ such that $\mathrm{Ult}A \backslash Sa \subseteq \bigcup_{G \in \mathcal{G}} V(G)$. Since also $Sa \subseteq \bigcap_{G \in \mathcal{G}} U(G)$, and $\bigcup_{G \in \mathcal{G}} V(G)$ and $\bigcap_{G \in \mathcal{G}} U(G)$ are disjoint, we have $Sa = \bigcap_{G \in \mathcal{G}} U(G) \in \mathcal{W}$, as desired. $\square$

Lemma 13.5. *If X is a Hausdorff space and $2^\kappa < |X|$, then there is a sequence $\langle F_\alpha : \alpha < \kappa^+ \rangle$ of closed subsets of X such that $\alpha < \beta$ implies $F_\beta \subset F_\alpha$.*

Proof. For each $f \in \bigcup_{\alpha < \kappa^+} {}^\alpha 2$ we define a closed subset X_f of X. Let $X_0 = X$. For $\mathrm{dom}f$ limit, let $X_f = \bigcap_{\alpha < \mathrm{dom}f} X_{f \restriction \alpha}$. Now suppose that X_f has been constructed. If $|X_f| \leq 1$, let $X_{f \frown \langle 0 \rangle} = X_{f \frown \langle 1 \rangle} = X_f$. Otherwise, let $X_{f \frown \langle 0 \rangle}$ and $X_{f \frown \langle 1 \rangle}$ be two proper closed subsets of X_f whose union is X_f. This finishes the construction. Clearly $\bigcup_{\mathrm{dom}f = \alpha} X_f = X$ for all $\alpha < \kappa^+$. Now

(*) there is an $f \in {}^{\kappa^+} 2$ such that $|X_{f \restriction \alpha}| \geq 2$ for all $\alpha < \kappa^+$.

For, otherwise, for all $x \in X$ there is an $f \in \bigcup_{\alpha < \kappa^+} {}^\alpha 2$ such that $X_f = \{x\}$, and so

$$|X| \leq \left| \bigcup_{\alpha < \kappa^+} {}^\alpha 2 \right| = 2^\kappa,$$

contradiction. So (*) holds, and it clearly gives the desired result. $\square$

Theorem 13.6. $|B| \leq 2^{sB}$ *for any BA B.*

Proof. To start with, we prove:

(1) $\mathrm{d}A \leq 2^{sA}$.

In fact, suppose that (1) fails. Note that for every $Y \subseteq \mathrm{Ult}A$ of power $< \mathrm{d}A$ we have $\overline{Y} \neq \mathrm{Ult}A$. Hence one can construct two sequences $\langle F_\alpha : \alpha < (2^{\mathrm{s}A})^+ \rangle$ and $\langle a_\alpha : \alpha < (2^{\mathrm{s}A})^+ \rangle$ such that $a_\alpha \in F_\alpha \in \mathrm{Ult}A$ and $\mathcal{S}a_\alpha \cap \{F_\beta : \beta < \alpha\} = 0$ for all $\alpha < (2^{\mathrm{s}A})^+$. Let $X = \{F_\alpha : \alpha < (2^{\mathrm{s}A})^+\}$. Clearly F is one-one, so $|X| > 2^{\mathrm{s}A}$. By Lemma 13.5, let $\langle K_\alpha : \alpha < (\mathrm{s}A)^+ \rangle$ be a system of closed subsets of X such that $\alpha < \beta$ implies that $K_\beta \subset K_\alpha$. Say $F_{\beta_\alpha} \in K_\alpha \backslash K_{\alpha+1}$ for all $\alpha < (\mathrm{s}A)^+$, and choose $b_\alpha \in A$ so that $F_{\beta_\alpha} \in \mathcal{S}b_\alpha \cap X \subseteq X \backslash K_{\alpha+1}$. Then

(2) $\mathcal{S}b_\alpha \cap \{F_{\beta_\gamma} : \gamma > \alpha\} = 0.$

For, suppose $\gamma > \alpha$ and $F_{\beta_\gamma} \in \mathcal{S}b_\alpha$. But $F_{\beta_\gamma} \in K_\gamma \subseteq K_{\alpha+1}$, contradiction.

Define $f : [(\mathrm{s}A)^+]^2 \to 2$ as follows: $f\{\gamma, \delta\} = 0$ iff when $\gamma < \delta$ we have $\beta_\gamma > \beta_\delta$. We now use the partition relation $\mu^+ \to (\omega, \mu^+)$. Since there is no infinite decreasing sequence of ordinals, we get a subset Γ of $(\mathrm{s}A)^+$ of size $(\mathrm{s}A)^+$ such that if $\gamma, \delta \in \Gamma$ and $\gamma < \delta$, then $\beta_\gamma < \beta_\delta$. Hence for any $\alpha \in \Gamma$ we have

$$\mathcal{S}(a_{\beta_\alpha} \cdot b_\alpha) \cap \{F_{\beta_\gamma} : \gamma \in \Gamma\} = \{F_{\beta_\alpha}\},$$

and $\{F_{\beta_\gamma} : \gamma \in \Gamma\}$ is discrete, contradiction. So, we have finally proved (1).

Let Y be a subset of $\mathrm{Ult}A$ which is dense in $\mathrm{Ult}A$ and of cardinality $\mathrm{d}A$. Let

$$\mathcal{N} = \{\overline{Z} : Z \subseteq Y, \ |Z| \leq \mathrm{t}A\}.$$

From (1) and Lemma 13.5 we see that $|\mathcal{N}| \leq 2^{\mathrm{s}A}$. So, we will be finished, by Theorem 13.4, after we show that $\mathcal{N}$ is a network for A. Let $F \in U$, with U open. Say $F \in V \subseteq \overline{V} \subseteq U$, with V open. Choose $Z \subseteq Y$ with $|Z| \leq \mathrm{t}A$ such that $F \in \overline{Z}$. Let $Z' = V \cap Z$. Then $F \in \overline{Z'} \subseteq U$ and $\overline{Z'} \in \mathcal{N}$, as desired. $\qquad \square$

By Theorem 13.1, spread can be considered to be an ordinary sup-function, and so its behaviour under unions is given by Theorem 3.11.

We turn to the derived functions for spread. The following facts are clear: $\mathrm{s_{H+}}A = \mathrm{s}A$; $\mathrm{s_{S+}}A = \mathrm{s}A$; $\mathrm{s_{S-}}A = \omega$; $\mathrm{s_{h-}}A = \omega$; $\mathrm{_dss_{S+}}A = \mathrm{s}A$. The algebra A of Fedorchuk [75] (constructed under $\lozenge$ and presented in Chapter 16) is such that $\mathrm{s_{H-}}A \leq \mathrm{s}A < \mathrm{Card_{H-}}A$. Thus Problem 36 of Monk [90] was solved long ago. It is also easy to see that $\mathrm{s_{h+}}A = \mathrm{s}A$. The status of the derived function $\mathrm{_dss_{-}}$ is not clear; note that $\mathrm{_dss_{-}}A < \mathrm{s}A$ for $A = \mathscr{P}\kappa$.

Turning to the relationships of spread to our other functions, we first list out the things already proved: $\mathrm{c_{H+}}A = \mathrm{s}A$ by Theorem 3.25; $\mathrm{Depth_{h+}}A = \mathrm{s}A$ in Theorem 4.23; $\mathrm{t}A \leq \mathrm{s}A$ in Theorem 5.11; and $|A| \leq 2^{\mathrm{s}A}$ in Theorem 13.6. Now we prove the important fact that $\pi A \leq \mathrm{s}A \cdot (\mathrm{t}A)^+$ for any infinite BA A, following Todorčević [90a]. The result he proves is somewhat stronger, and to state it we need two definitions. First, $\mathrm{dd}A$ is the least cardinality of a collection of discrete subsets of $\mathrm{Ult}A$ whose union is dense in $\mathrm{Ult}A$. Second, $\mathrm{f'}A$ is the smallest cardinal such that A does not have a free sequence of length $\mathrm{f'}A$. Thus if $\mathrm{t}A$ is attained in the free sequence sense, then $\mathrm{f'}A = (\mathrm{t}A)^+$, while $\mathrm{t}A = \mathrm{f'}A$ otherwise.

Theorem 13.7. $\mathrm{dd}A \leq \mathrm{f}'A$, and $\pi A \leq \mathrm{s}A \cdot \mathrm{f}'A$ for any infinite BA A.

Proof. Choose $\langle a_\alpha : \alpha < \beta \rangle$ in accordance with Theorem 12.12. Let $E = \{a_\alpha : \alpha < \beta\}$. Now we define $D_\gamma \subseteq \mathrm{Ult}A$ and $S_\gamma \subseteq E$ for $\gamma < \mathrm{f}'A$ by induction. Suppose that they have been defined for all $\gamma < \delta$. Let

$$S_\delta = \{x \in E : Sx \cap D_\gamma = 0 \text{ for all } \gamma < \delta\}.$$

Then we let D_δ be a maximal subset of $\bigcup_{x \in S_\delta} Sx$ having at most one element in common with each Sx for $x \in S_\delta$. This finishes the construction. Note that D_δ is discrete: if $F \in D_\delta$, choose $x \in S_\delta$ such that $F \in Sx$. Then $D_\delta \cap Sx = \{F\}$ by the defining property of D_δ

For each $F \in D_\delta$ choose $gF \in S_\delta$ such that $F \in SgF$. Then g is a one-one function, and its range is $S'_\gamma \overset{\text{def}}{=} \{x \in S_\delta : D_\delta \cap Sx \neq 0\}$. Now

(1) $Sx \cap \bigcup_{\delta < \mathrm{f}'A} D_\delta \neq 0$ for all $x \in E$.

For, suppose that (1) fails for a certain $x \in E$. Then

(2) $Sx \subseteq \bigcup_{y \in S'_\delta} Sy$ for all $\delta < \mathrm{f}'A$.

For, suppose that (2) fails for a certain $\delta < \mathrm{f}'A$. Choose $F \in Sx \backslash \bigcup_{y \in S'_\delta} Sy$. Now if $G \in D_\delta \cap Sy$ with $y \in S_\delta$, then $y \in S'_\delta$ and so $F \notin Sy$. Also, $D_\delta \cap Sx = 0$ by (1) failing. So $D_\delta \cup \{F\}$ has at most one element in common with each Sy for $y \in S_\delta$, and $F \notin D_\delta$, contradicting the maximality of D_δ. Thus (2) holds.

Now if $\gamma < \delta < \mathrm{f}'A$, then $S'_\gamma \cap S'_\delta = 0$, since if $z \in S'_\gamma \cap S'_\delta$, then $Sz \cap D_\gamma = 0$ because $z \in S'_\delta \subseteq S_\delta$, but $Sz \cap D_\gamma \neq 0$ by the definition of S'_γ, contradiction. It follows now that for any $F \in Sx$ we have $F \in Sy$ for a collection of $\mathrm{f}'A$ y's, and this contradicts the condition of Theorem 12.12. So we have proved (1).

By (1) we have $\mathrm{dd}A \leq \mathrm{f}'A$, since E is dense in A. Next,

(3) $\bigcup_{\delta < \mathrm{f}'A} S'_\delta$ is dense in A.

In fact, suppose not. Then there is an $x \in E$ such that $x \cdot y = 0$ for all $y \in \bigcup_{\delta < \mathrm{f}'A} S'_\delta$. By (1), choose $\delta < \mathrm{f}'A$ such that $D_\delta \cap Sx \neq 0$; say $F \in D_\delta \cap Sx$. Say $F \in Sz$ with $z \in S'_\delta$. So $x \cdot z \in F$, contradicting the fact that $x \cdot z = 0$. Thus (3) holds.

Now $|D_\delta| = |S'_\delta|$ for all $\delta < \mathrm{f}'A$. Hence

$$\pi A \leq \left| \bigcup_{\delta < \mathrm{f}'A} S'_\delta \right| \leq \mathrm{s}A \cdot \mathrm{f}'A,$$

as desired.   □

Note that $\mathrm{t}A$ can be much smaller than $\mathrm{s}A$, for example in the finite-cofinite algebra on an infinite cardinal κ. Also note that, obviously, $\mathrm{c}A \leq \mathrm{s}A$; and the difference is big in, e.g., free algebras. We have $\mathrm{s}A > \mathrm{Length}A$ for A a free algebra; $\mathrm{s}A < \mathrm{Length}A$ for A the interval algebra on the reals. Also, $\mathrm{s}A > \pi A$ for $A = \mathscr{P}\kappa$. The interval algebra of a Suslin line provides an example of a BA A with $sA = \omega$ and $dA > \omega$. In fact, clearly $sA = cA$ for A an interval algebra, by the retractiveness of interval algebras.

An example with $sA = \omega < dA$ cannot be given in ZFC; this follows from the following rather deep results. Juhász [71] showed that under the assumption of MA+¬CH, for every compact Hausdorff space X, if $sX = \omega$ then $hLX = \omega$. Todorčević [83] showed that it is consistent with MA+¬CH that for every regular space X, if $sX = \omega$ then $hLX = \omega$. Hence it is consistent that for every BA A, if $sA = \omega$ then $hLA = \omega = hdA$.

Bounded versions of spread can be defined as follows. For m a positive integer, a subset X of A is called *m-ideal-independent* if for all distinct $x_0, \ldots, x_m \in X$ we have $x_0 \not\leq x_1 + \cdots + x_m$. Then we let $s_m A = \sup\{|X| : X \subseteq A \text{ and } X \text{ is } m\text{-ideal-independent}\}$. For these functions see Rosłanowski, Shelah [94].

14. Character

First note that we can define χA as a sup; namely, for any ultrafilter F on A let $\chi F = \min\{|X| : X$ is a set of generators of $F\}$—then $\chi A = \sup\{\chi F : F$ is an ultrafilter on $A\}$. Clearly then, by topological duality, $\chi(A \times B) = \sup(\chi A, \chi B)$. For a weak product we have $\chi(\prod_{i\in I}^{w} A_i) = \max(|I|, \sup_{i\in I}\chi A_i)$. To show this, it suffices to show that $\chi F = |I|$ for the "new" ultrafilter F. This ultrafilter is defined as follows. For each subset M of I, let x_M be the element of $\prod_{i\in I} A_i$ such that $x_M i = 1$ if $i \in M$ and $x_M i = 0$ for $i \notin M$. Then F is the set of all $y \in \prod_{i\in I}^{w} A_i$ such that $x_M \leq y$ for some cofinite subset M of I. So, it is clear that $\chi F \leq |I|$. If X is a set of generators for F with $|X| < |I|$, then there is a $y \in X$ such that $y \subseteq x_M$ for infinitely many cofinite subsets M of I; this is clearly impossible.

As usual, weak products enable us to discuss the attainment problem. Any infinite BA has a non-principal ultrafilter, and hence if A has character ω, then it is attained. Next, if κ is a singular cardinal, then we can construct a BA A with $\chi A = \kappa$ not attained. Namely, let $\langle \mu_\xi : \xi < \mathrm{cf}\kappa \rangle$ be an increasing sequence of infinite cardinals with sup κ. For each $\xi < \mathrm{cf}\kappa$ let A_ξ be the free BA on μ_ξ free generators; thus $\chi A_\xi = \mu_\xi$. By the above remarks on weak products, $\prod_{\xi<\mathrm{cf}\kappa}^{w} A_\xi$ has character κ not attained. For regular limit cardinals, there is an old theorem of Parovichenko [67] which characterizes attainment; this solves Problem 37 of Monk [90]. We give this theorem here; it requires a simple lemma, and a definition. If X is a topological space and $x \in X$, the *character* of x is $\min\{|\mathscr{O}| : \mathscr{O}$ is a neighborhood base of $x\}$. Clearly for any BA A and any $F \in \mathrm{Ult}A$, the character of F as a point in $\mathrm{Ult}A$ coincides with χF as defined above.

Lemma 14.1. *For any Hausdorff topological space X and any $Y \subseteq X$ we have $|\overline{Y}| \leq 2^{2^{|Y|}}$.*

Proof. For any $z \in \overline{Y}$ let $fz = \{Y \cap U : U$ an open neighborhood of $z\}$. Thus f maps $\overline{Y}$ into $\mathscr{P}\mathscr{P}Y$, so it suffices to show that f is one-one. Suppose that z and w are distinct points of $\overline{Y}$, and suppose that $fz = fw$. Let U and V be disjoint open neighborhoods of z and w respectively. Then $Y \cap U \in fz = fw$, so let W be an open neighborhood of w such that $Y \cap U = Y \cap W$. Then $0 \neq Y \cap W \cap V = Y \cap U \cap V = 0$, contradiction. $\square$

Theorem 14.2. *Let κ be a limit cardinal. Then the following conditions are equivalent:*

    (i) κ is weakly compact.

    (ii) For every compact Hausdorff space X of size at least κ, X has a point of character at least κ.

    (iii) For every BA A, if $\chi A = \kappa$, then A has an ultrafilter with character κ.

Proof. (i)$\Rightarrow$(ii): Assume (i), and suppose that X is a compact Hausdorff space with $|X| \geq \kappa$ such that X has no point of character $\geq \kappa$.

(1) We may assume that $|X| = \kappa$.

For, let $Y \in [X]^\kappa$. We claim that $|\overline{Y}| = \kappa$. Since the character of a point of $\overline{Y}$ is clearly still $< \kappa$, (1) follows from the claim. By Lemma 14.1, it suffices to show that if $y \in \overline{Y}$, then there is a $Z \in [Y]^{<\kappa}$ such that $y \in \overline{Z}$, since then $\overline{Y} = \bigcup_{Z \in [Y]^{<\kappa}} \overline{Z}$, and by κ being strongly inaccessible $|\overline{Y}| = \kappa$ follows. Let $\mathscr{U}$ be an open neighborhood base for y of size $< \kappa$. For every $U \in \mathscr{U}$ choose $z_U \in U \cap Y$. Clearly $Z \stackrel{\text{def}}{=} \{z_U : u \in \mathscr{U}\}$ is as desired.

So we now assume that $|X| = \kappa$. Let $X = \{x_\alpha : \alpha < \kappa\}$. For each $\alpha < \kappa$, let $\mathscr{U}_\alpha$ be an open neighborhood base for x_α of size $< \kappa$. Set $\mathscr{F}_\alpha = \{F : X\backslash F \in \mathscr{U}_\alpha\}$. So $\mathscr{F}_\alpha$ is a collection of closed sets, $\bigcup \mathscr{F}_\alpha = X\backslash\{x_\alpha\}$, and $|\mathscr{F}_\alpha| < \kappa$. Let T be the collection of all functions f such that there is an $\alpha < \kappa$ such that $\mathrm{dmn}\, f = \alpha$, $\forall \beta < \alpha(f_\beta \in \mathscr{F}_\beta)$, and $\bigcap_{\beta < \alpha} f_\beta \neq 0$. Thus T is a tree under $\subseteq$.

(2) $\forall \alpha < \delta \exists f \in T(\mathrm{dmn}\, f = \alpha)$.

For,

$$0 \neq X\backslash\{x_\beta : \beta < \alpha\} = \bigcap_{\beta < \alpha} \bigcup \mathscr{F}_\alpha$$

$$= \bigcup_{f \in \prod_{\beta < \alpha} \mathscr{F}_\beta} \bigcap_{\beta < \alpha} f_\beta,$$

so there is an $f \in \prod_{\beta < \alpha} \mathscr{F}_\beta$ such that $\bigcap_{\beta < \alpha} f_\beta \neq 0$. Thus f is as desired in (2).

(3) Every level of T has size $< \kappa$.

This is true since κ is strongly inaccessible.

Now by the weak compactness of κ, let f with domain κ be a branch through T. By compactness, $\bigcap_{\alpha < \kappa} f_\alpha \neq 0$. But if $y \in \bigcap_{\alpha < \kappa} f_\alpha$, then $y \neq x_\alpha$ for all $\alpha < \kappa$, contradiction.

(ii)$\Rightarrow$(iii): obvious.

(iii)$\Rightarrow$(i): Assume that κ is not weakly compact; we want to find a BA A with character κ not attained. By the comments before 14.1, we may assume that κ is regular. Let L be a linear order of size κ such that neither κ nor κ^* embeds in L. By replacing points of L by ordinals less than κ we may assume that each ordinal less than κ embeds in L. More precisely, write $L = \{a_\alpha : \alpha < \kappa\}$ with no repetitions. let $M = \{(\beta, a_\alpha) : \beta \leq \alpha, \alpha < \kappa\}$, ordered anti-lexicographically. We show that M has no increasing chain of type κ. For, suppose that $\langle (\beta_\xi, x_\xi) : \xi < \kappa\rangle$ is such a chain. Since κ is regular, the set $\{x_\xi : \xi < \kappa\}$ has κ elements, and hence determines a chain in L of type κ, contradiction. Similarly, M has no decreasing chain of type κ. Let $A = \mathrm{Intalg}\, M$. Then by the description of character for interval algebras below, A has the desired properties. $\square$

To treat arbitrary direct products, note that obviously $tA \leq \chi A$; hence $\mathrm{Ind} A \leq \chi A$, and so clearly $\chi(\prod_{i \in I} A_i) \geq \max(2^{|I|}, \sup_{i \in I} \chi A_i)$. Shelah and Peterson independently observed that strict inequality is possible. This solves Problem 38 in Monk [90]. The same example used for spread works here: let κ be the first limit cardinal $> 2^\omega$, let A be the finite-cofinite algebra on κ, and consider $^\omega A$. Character does not increase when going to a homomorphic image (see below), and Theorem 6.1 can be applied.

We now discuss ultraproducts, giving some results of Douglas Peterson. Character is a sup-min function, and so Theorems 6.1–6.3 apply. Then a proof similar to that of Theorem 4.14 shows that if GCH holds then $\chi\left(\prod_{i\in I} A_i/F\right) \geq \left|\prod_{i\in I} \chi A_i/F\right|$ for F regular, and Donder's theorem says that under $V = L$ the regularity assumption can be removed. Whether there is consistently an example with $<$ is open.

Problem 48. *Is it consistent that there exist a system $\langle A_i : i \in I\rangle$ of infinite BAs with I infinite, and an ultrafilter F such that $\chi\left(\prod_{i\in I} A_i/F\right) < \left|\prod_{i\in I} \chi A_i/F\right|$?*

On the other hand, it is easy to give an example in which $>$ holds. Let κ be any infinite cardinal such that $\kappa^\omega = \kappa$. We will shortly show that the Aleksandroff duplicate A of a free BA on κ generators has character κ and cellularity 2^κ. By Theorem 11.5 this implies that $\chi(A^{*i}) = \kappa$ for all $i \in \omega\backslash 1$ while $\chi\left(\prod_{i\in\omega\backslash 1} A^{*i}/F\right) = 2^\kappa$ for any nonprincipal ultrafilter F on $\omega\backslash 1$. (See below for the character of free products.) The assumption $\kappa^\omega = \kappa$ implies that $\left|\prod_{i\in\omega\backslash 1} \chi(A^{*i})/F\right| = \kappa$.

Character can increase in going from an algebra to a subalgebra. To construct an example of this sort, first notice that if A is the finite-cofinite algebra on an infinite cardinal κ, then $\chi A = \kappa$, by our initial remarks (since $A = \prod^w_{\alpha<\kappa} 2$). The algebra that we want is the Aleksandroff duplicate of the free algebra on κ free generators, where κ is any infinite cardinal. Recall from Chapter 1 the definition of the Aleksandroff duplicate. Now let B be the free BA on κ free generators, κ any infinite cardinal. We claim that $\chi\mathrm{Dup}B = \kappa$. To see this, we describe the ultrafilters on $\mathrm{Dup}B$. Note that $\mathrm{Dup}B$ is atomic, and its atoms are all of the elements $(0, \{F\})$ for $F \in \mathrm{Ult}A$. So there is a principal ultrafilter corresponding to each of these atoms. Next, if G is an ultrafilter on B, then $G^+ \overset{\text{def}}{=} \{(a, X) : a \in G, X \subseteq \mathrm{Ult}B, Sa\triangle X \text{ finite}\}$ is an ultrafilter on $\mathrm{Dup}B$. Conversely, any nonprincipal ultrafilter on $\mathrm{Dup}B$ is easily seen to have this form. Thus it suffices to show that any ultrafilter of this form has character κ. So, let F be an arbitrary ultrafilter on B. We claim that the set X of all elements of F^+ of the form $(a, Sa\backslash\{F\})$ generates F^+. For, let (a, Y) be any element of F^+; thus $Sa\backslash Y$ is finite. For each $G \in Sa\backslash(Y \cup \{F\})$ choose $a_G \in F\backslash G$. Then let $b = a \cdot \prod_{G\in Sa\backslash(Y\cup\{F\})} a_G$. Then $(b, Sb\backslash\{F\}) \in F^+$ and $(b, Sb\backslash\{F\}) \leq (a, Y)$, as desired. So, this shows that $\chi F^+ \leq \kappa$. An easy argument shows that actually $\chi F^+ = \kappa$. Namely, if Z generates F^+ and $|Z| < \kappa$, then choose $(a, X) \in Z$ such that $(a, X) \leq (b, Sb)$ for infinitely many $b \in F$ such that b or $-b$ is one of the free generators of B; this is impossible. So, $\chi\mathrm{Dup}B = \kappa$. But the finite-cofinite algebra A on $\mathrm{Ult}B$ is isomorphic to a subalgebra of $\mathrm{Dup}B$, and by the previous remarks it has character 2^κ.

If A is a homomorphic image of B, then $\chi A \leq \chi B$ (let f be a homomorphism from B onto A; if $F \in \mathrm{Ult}A$, then $f^{-1}[F] \in \mathrm{Ult}B$, and if we choose $X \subseteq f^{-1}[F]$ with $|X| \leq \chi B$ such that X generates $f^{-1}[F]$, then $f[X]$ generates F). It is also easy to see that if $\langle A_i : i \in I \rangle$ is a system of BAs each with at least four elements, then $\chi(\oplus_{i \in I} A_i) = \max(|I|, \sup_{i \in I} \chi A_i)$. In fact, for $\geq$, first let $j \in I$ and let $F \in \mathrm{Ult}A_j$. Let G be any ultrafilter on $\oplus_{i \in I} A_i$ which includes F. Suppose that $X \subseteq G$ generates G. Without loss of generality, each member of X is a product of elements from distinct A_i's. Then it is clear that $X \cap A_j \subseteq F$ and $X \cap A_j$ generates F. So $\chi F \leq |X|$. It follows that $\chi F \leq \chi G \leq \chi(\oplus_{i \in I} A_i)$. Hence $\chi A_j \leq \chi(\oplus_{i \in I} A_i)$. It is clear that $\chi H \geq |I|$ for any ultrafilter H on $\oplus_{i \in I} A_i$. Altogether, this proves $\geq$. For $\leq$, for any ultrafilter G on $\oplus_{i \in I} A_i$, and for each $i \in I$ let $X_i \subseteq G \cap A_i$ generate $G \cap A_i$, with $|X_i| = \chi(G \cap A_i)$. Clearly the set of all finite products of elements of $\bigcup_{i \in I} X_i$ generates G, as desired.

We turn to the derived functions for character. By a remark above, we have $\chi_{\mathrm{H}+}A = \chi A$ for any infinite BA A. Under CH we have $\chi_{\mathrm{H}-}A = \mathrm{Card}_{\mathrm{H}-}A$; indeed, inequality would imply that $\chi_{\mathrm{H}-}A = \omega$, and then results from van Douwen [89] would imply that $\mathrm{Card}_{\mathrm{H}-}A = \omega$ too; see below. On the other hand, Koszmider (email message) has shown that it is consistent to have $\mathrm{Card}_{\mathrm{H}-}A = \omega_2 = 2^\omega$ while $\chi_{\mathrm{H}-}A = \omega_1$. This solves Problem 39 in Monk [90].

We also do not know the status of $\chi_{\mathrm{S}+}A$; we observed above that it can happen that $\chi_{\mathrm{S}+}A > \chi A$. Clearly $\chi_{\mathrm{S}-}A = \omega$ for any infinite BA A. The topological version of character is this: for any space X and any $x \in X$, $\chi(x, X)$ is the minimum of the cardinalities of neighborhood bases for x in X, and $\chi X = \sup\{\chi(x, X) : x \in X\}$. Clearly then $\chi_{\mathrm{h}+}A = \chi A$, and $\chi_{\mathrm{h}-}A = 1$ for any infinite BA A, since A has an infinite discrete subspace.

The function $\chi_{\inf}$ is of some interest; recall from the introduction that $\chi_{\inf}A = \inf\{\chi F : F \in \mathrm{Ult}A\}$ for any infinite BA A. It has not been investigated much, but we give the following classical result of Čech and Pospíšil concerning it:

Theorem 14.3. $2^{\chi_{\inf}A} \leq |\mathrm{Ult}A|$ *for any infinite BA A.*

Proof. For brevity set $\kappa = \chi_{\inf}A$. It clearly suffices to construct a function f mapping $^{<\kappa}2$ into A such that

(1) For each $s \in {}^{<\kappa}2$, the set $\{f(s \restriction \alpha) : \alpha \leq \mathrm{dom}\, s\}$ has the finite intersection property;

(2) $f(s^\frown 0) \cdot f(s^\frown 1) = 0$ for each $s \in {}^{<\kappa}2$.

Suppose $s \in {}^{<\kappa}2$ and $f(s \restriction \alpha)$ has been defined for all $\alpha \in \mathrm{dom}\, s$. By the induction hypothesis, $\{f(s \restriction \alpha) : \alpha \in \mathrm{dom}\, s\}$ has the finite intersection property; since this set has $< \kappa$ elements, it does not generate an ultrafilter, and hence there is a $a \in A$ such that both a and $-a$ fail to be in the filter generated by it. Hence if we set $f(s^\frown 0) = a$ and $f(s^\frown 1) = -a$ we extend our function f so that (1) and (2) will hold. This completes the proof. $\qquad \square$

We give some more results related to $\chi_{\inf}$. For any topological space X and infinite cardinal κ, we say that an infinite sequence $\langle a_\alpha : \alpha < \kappa \rangle$ of elements of X *converges* to a point $y \in X$ provided that for every open neighborhood U of y there is a $\beta < \kappa$ such that $a_\alpha \in U$ for all $\alpha \in \kappa \backslash \beta$. We introduce a cardinal function aA, the *altitude* of A, on an arbitrary infinite BA A by

$$aA = \min\{\kappa : \text{there is a one-one convergent sequence of length } \kappa \text{ in Ult}A\}.$$

It may not be completely clear that there always is an infinite one-one convergent sequence in Ult$A$. Rather than proving this directly, we give it as a consequence of the following theorem.

Theorem 14.4. *Let B be a homomorphic image of A, and G a nonprincipal ultrafilter on B. Then $aA \leq \chi G$; in particular, aA exists for any BA A.*

Proof. Let f be a homomorphism from A onto B, and let $\langle b_\alpha : \alpha < \chi G \rangle$ be an enumeration of a set of generators of G. For each $\alpha < \chi G$ choose $a_\alpha \in A$ such that $fa_\alpha = b_\alpha$. Now for each $\beta < \chi G$, the set $\{b_\alpha : \alpha < \beta\}$ does not generate G, so we can choose $c_\beta \in G$ such that for all finite $\Gamma \subseteq \beta$ we have $\prod_{\alpha \in \Gamma} b_\alpha \not\leq c_\beta$. Say $fd_\beta = c_\beta$. Then

(1) $\{a_\alpha : \alpha < \beta\} \cup \{-d_\beta\} \cup \{x \in A : fx = 1\}$ has fip.

For, otherwise there exist a finite $\Gamma \subseteq \beta$ and an $x \in A$ such that $fx = 1$ and $\prod_{\alpha \in \Gamma} a_\alpha \cdot d_\beta \cdot x = 0$. Applying f to this equation we get $\prod_{\alpha \in \Gamma} b_\alpha \cdot -c_\beta = 0$, contradiction. So (1) holds. Let F_β be an ultrafilter on A extending the set in (1). Now we claim

(2) $\langle F_\beta : \beta < \chi G \rangle$ converges to $f^{-1}[G]$.

For, let $e \in f^{-1}[G]$. Choose a finite $\Gamma \subseteq \chi G$ such that $\prod_{\alpha \in \Gamma} b_\alpha \leq fe$. Let $\beta = (\sup \Gamma) + 1$. Suppose that $\gamma \in \chi G \backslash \beta$. Then

$$\{a_\alpha : \alpha \in \Gamma\} \cup \{-\prod_{\alpha \in \Gamma} a_\alpha + e\} \subseteq F_\gamma,$$

so $e \in F_\gamma$. Thus (2) holds.

(3) $F_\beta \neq f^{-1}[G]$ for all $\beta < \chi G$.

This is clear by the definition of F_β, since $d_\beta \in f^{-1}[G]$.

From (2) and (3) we see that by taking a one-one subsequence of $\langle F_\beta : \beta < \chi G \rangle$ we get a sequence which proves that $aA \leq \chi G$. $\square$

This theorem suggests a variant of χ_{inf}:

$$\chi_{\text{npinf}} A = \min\{\chi G : G \text{ is a nonprincipal ultrafilter on } A\}.$$

Thus we get the following corollary:

Corollary 14.5. $aA \leq (\chi_{\text{npinf}})_{\text{H}} - A \leq \chi_{\text{H}} - A$ *for any infinite BA A.* $\square$

We also give a related theorem.

Theorem 14.6. *For any infinite BA A, if $\mathrm{a}A = \omega$ then $\mathrm{Card_{H-}}A = \omega$.*

Proof. By hypothesis, choose a one-one convergent sequence $\langle F_n : n < \omega \rangle$ of ultrafilters of A. We may assume that no F_n is equal to the limit G of this sequence. For each $a \in A$ let

$$fa = \mathcal{S}a \cap \{F_n : n \in \omega\}.$$

Clearly f is a homomorphism from A into the finite-cofinite algebra of subsets of $\{F_n : n \in \omega\}$. It remains only to show that the range of f is infinite. Take any $n \in \omega$; we show that $\{F_n\}$ is in the range of f. Choose $a \in F_n \backslash G$. Then choose m so that for all $p \in \omega \backslash m$ we have $-a \in F_p$. For each $p < m$ with $p \neq n$ choose $b_p \in F_n \backslash F_p$. Then $f(a \cdot \prod_{p<m, p\neq n} b_p) = \{F_n\}$, as desired.   □

Corollary 14.7. *If $\chi_{\mathrm{H-}}A = \omega$ then $\mathrm{Card_{H-}}A = \omega$.*   □

We turn to the relationship of character with our previously treated functions. Obviously $\mathrm{t}A \leq \chi A$ for any infinite BA A; the difference can be big—for example for the finite-cofinite algebra on an infinite cardinal κ. Now consider the possibility that $\chi A > \mathrm{s}A$. By the comment at the end of the last section, plus the fact that $\chi A \leq \mathrm{hL}A$ (easy, and proved in Chapter 15), it is consistent that $\mathrm{s}A = \omega$ implies $\chi A = \omega$. The Kunen line, constructed in Chapter 8, has uncountable character but countable spread; it was constructed using CH. To show that it has uncountable character, assume the notation in the construction.

(*) If $U \in \tau_{\omega_1}$ is open and $\{\xi < \omega_1 : x_\xi \in U\}$ is uncountable, then U is not compact in τ_{ω_1}.

To see this, note that for each $x_\xi \in U$ we have $x_\xi \in U \cap \mathbb{R}_{\xi+1} \in \tau_{\xi+1}$. Thus $\{U \cap \mathbb{R}_{\xi+1} : x_\xi \in U\}$ is a cover of U, and it clearly has no finite subcover, proving (*).

Now suppose that the new point y of Y has a countable base $\{\{y\} \cup U_m : m \in \omega\}$; we may assume that these are clopen. Hence $Y \backslash (\{y\} \cup U_m) = X \backslash U_m$ is open in Y and hence in X; and it is also compact. It follows from (*) that $X \backslash U_m$ is countable. Choose $x_\alpha \in \bigcap_{m \in \omega} U_m$. Then $Y \backslash \{x_\alpha\}$ is an open neighborhood of y not containing any of the basis sets $\{y\} \cup U_m$, contradiction.

Problem 49. *Can one construct in ZFC a BA A such that $\mathrm{s}A < \chi A$?*

This was Problem 40 in Monk [90]. It is equivalent to the problem whether one can construct in ZFC a BA A such that $\mathrm{s}A < \mathrm{hL}A$; see the end of Chapter 15.

An example of a BA A with $\mathrm{c}A > \chi A$ is provided by the Aleksandroff duplicate of the free algebra on κ free generators, as discussed above. The interval algebra on $\mathbb{R}$ gives an example of an algebra A with $\mathrm{Length}A > \chi A$.

Now we turn to Arhangelskiĭ's theorem that $|\mathrm{Ult}A| \leq 2^{\chi A}$ for any infinite BA A. We need some lemmas.

Lemma 14.8. *If $Y \subseteq \mathrm{Ult}A$ and $F \subseteq \bigcup Y$ has the finite intersection property, then there is an ultrafilter G such that $F \subseteq G \subseteq \bigcup Y$.*

Proof. Let G be maximal among the filters H such that $F \subseteq H \subseteq \bigcup Y$. Suppose that G is not an ultrafilter; say $a, -a \notin G$. Then $\langle G \cup \{a\} \rangle^{\mathrm{fi}} \not\subseteq \bigcup Y$. Say $b \in G$ and $b \cdot a \notin \bigcup Y$. Similarly obtain $c \in G$ such that $c \cdot -a \notin \bigcup Y$. Choose $H \in Y$ such that $b \cdot c \in H$. Then $b \cdot c \cdot a \notin H$ and $b \cdot c \cdot -a \notin H$, contradiction. $\square$

Note that for any subset Y of $\mathrm{Ult}A$, the closure of Y is $\{F \in \mathrm{Ult}A : F \subseteq \bigcup Y\}$.

Lemma 14.9. *If $Z \subseteq \mathrm{Ult}A$ is closed, then $\mathrm{Ult}A \backslash Z$ is the union of at most $\max\{\omega, |Z|, \sup_{G \in Z} \chi G\}$ clopen sets.*

Proof. For every $G \in Z$ let $\{a_\alpha^G : \alpha < \chi G\}$ be a set of generators of G, closed under multiplication. Let

$$B = \{a_{\alpha_0}^{G_0} + \cdots + a_{\alpha_{n-1}}^{G_{n-1}} : n \in \omega, \ G_0, \ldots, G_{n-1} \in Z, \ \alpha_i < \chi G_i \text{ for all } i < n\},$$

and let $C = \{y : -y \in B \cap \bigcap Z\}$. We claim that

$$\mathrm{Ult}A \backslash Z = \bigcup_{y \in C} Sy,$$

which gives the desired result. $\supseteq$ is clear. Now suppose that $F \in \mathrm{Ult}A \backslash Z$. For every $G \in Z$ choose $b_G \in F \backslash G$; say $a_{\alpha(G)}^G \leq -b_G$. Then

(*) There exist an integer $n \in \omega$ and elements $G_0, \ldots, G_{n-1} \in Z$ with the property that $a_{\alpha(G_0)}^{G_0} + \cdots + a_{\alpha(G_{n-1})}^{G_{n-1}} \in H$ for all $H \in Z$.

Otherwise, $L \overset{\mathrm{def}}{=} \{-a_{\alpha(G_0)}^{G_0} \cdot \ldots \cdot -a_{\alpha(G_{n-1})}^{G_{n-1}} : n \in \omega, \ G_0, \ldots, G_{n-1} \in Z\}$ has the finite intersection property and is contained in $\bigcup Z$. Hence by Lemma 14.8, there is an ultrafilter K such that $L \subseteq K \subseteq \bigcup Z$. Hence $K \in Z$ and $-a_{\alpha(K)}^K \in K$, contradiction.

We choose $n \in \omega$ and $G_0, \ldots, G_{n-1} \in Z$ as in (*). Let y be the element $-a_{\alpha(G_0)}^{G_0} \cdot \ldots \cdot -a_{\alpha(G_{n-1})}^{G_{n-1}}$. Then $y \in C$ and $F \in Sy$, as desired. $\square$

Lemma 14.10. *If $Y \subseteq \mathrm{Ult}A$ and $|Y| \leq \chi A$, then $|\overline{Y}| \leq 2^{\chi A}$.*

Proof. For every ultrafilter G on A let $\{a_\alpha^G : \alpha < \chi A\}$ be a set of generators of G, and set $fG = \{\{F \in Y : a_\alpha^G \in F\} : \alpha < \chi A\}$. Thus $fG \in [\mathscr{P}Y]^{\leq \chi A}$. Hence it is enough to show that $f \restriction \overline{Y}$ is one-one. Suppose that G and H are distinct ultrafilters on A such that $G, H \in \overline{Y}$. Say $a_\alpha^G \in G \backslash H$, and choose $a_\beta^H \leq -a_\alpha^G$. Suppose that $fG = fH$; then there is a $\gamma < \chi A$ such that $\{F \in Y : a_\gamma^G \in F\} = \{F \in Y : a_\beta^H \in F\}$. Then $a_\alpha^G \cdot a_\gamma^G \in G$; say then $a_\alpha^G \cdot a_\gamma^G \in F \in Y$. Then $a_\beta^H \in F$, $a_\alpha^G \in F$, and $-a_\alpha^G \in F$, contradiction. $\square$

Now we are ready for the proof of Arhangelskiĭ's theorem:

Theorem 14.11. $|\text{Ult}A| \leq 2^{\chi A}$ *for any infinite BA* A.

Proof. Suppose that $2^{\chi A} < |\text{Ult}A|$. Fix an ultrafilter F on A. For each $f \in \bigcup_{\alpha<(\chi A)^+}{}^{\alpha}(2^{\chi A})$ we define a closed set $X_f \subseteq \text{Ult}A$ and a $G_f \in \text{Ult}A$. Let $X_0 = \text{Ult}A$ and $G_0 = F$. For dom f limit let $X_f = \bigcap_{\alpha<\text{dom } f} X_{f\restriction\alpha}$, and if $X_f \neq 0$ choose $G_f \in X_f$, and otherwise let $G_f = F$. Now suppose that dom f is a successor ordinal $\alpha + 1$. Let $g = f \restriction \alpha$, and set $Y_g = \{G_{g\restriction\beta} : \beta \leq \alpha\}$. Thus $|Y_g| \leq \chi A$, so by Lemma 14.10, $|\overline{Y}_g| \leq 2^{\chi A}$, and so by Lemma 14.9 we can let $\langle a_\beta^g : \beta < 2^{\chi A}\rangle$ be such that $\text{Ult}A\setminus\overline{Y}_g = \bigcup_{\beta<2^{\chi A}} \mathcal{S}(a_\beta^g)$ and set $X_f = X_g \cap \mathcal{S}(a_{f\alpha}^g)$. Again let $G_f \in X_f$ if $X_f \neq 0$, and $G_f = F$ otherwise. This finishes the construction.

Now choose

$$H \in \text{Ult}A\setminus\bigcup\left\{\overline{\{G_{f\restriction\beta} : \beta \leq \text{dom } f\}} : f \in \bigcup_{\alpha<(\chi A)^+}{}^{\alpha}(2^{\chi A})\right\}$$

Now we define f mapping $(\chi A)^+$ into $2^{\chi A}$ by induction. Suppose that $f\beta$ has been defined for all $\beta < \alpha$. Now $H \notin \overline{\{G_{f\restriction\beta} : \beta \leq \alpha\}}$, so there is a $\gamma < 2^{\chi A}$ such that $H \in X_{(f\restriction\alpha)\cup\{(\alpha,\gamma)\}}$; set $f\alpha = \gamma$. Thus $H \in X_{f\restriction\alpha}$ for all $\alpha < (\chi A)^+$. We claim that $\langle G_{f\restriction(\beta+1)} : \beta < (\chi A)^+\rangle$ is a free sequence, which contradicts $tA \leq \chi A$. Let $\alpha < (\chi A)^+$ and suppose that

$$K \in \overline{\{G_{f\restriction(\beta+1)} : \beta < \alpha\}} \cap \overline{\{G_{f\restriction(\beta+1)} : \alpha \leq \beta < (\chi A)^+\}}$$

Then $K \in \overline{\{G_{f\restriction\beta} : \beta \leq \alpha\}}$, so $K \in \text{Ult}A\setminus X_{f\restriction(\alpha+1)}$, and this set is open, so there is a $\beta \geq \alpha$ such that $G_{f\restriction(\beta+1)} \in \text{Ult}A\setminus X_{f\restriction(\alpha+1)} \subseteq \text{Ult}A\setminus X_{f\restriction(\beta+1)}$, contradiction. $\square$

We describe character for interval algebras. Let L be an ordering, and A the interval algebra on L. As mentioned at the end of Chapter 11, the ultrafilters on A are in one-one correspondence with the terminal segments T of L such that $0 \notin L$. The *character* of such a terminal segment is the pair (κ, λ^*) such that $L\setminus T$ has cofinality κ and T has coinitiality λ. And χF is the maximum of κ and λ. χA is the supremum of all χF. From this description it is clear that $\text{Depth}A = \chi A$ (and hence both are equal to tA), for any interval algebra A.

For tree algebras we have the following theorem (Brenner [82]):

Theorem 14.12. *Let* T *be a tree, and set* $A = \text{Treealg } T$. *Then* $\chi A = \sup\{|\{x : x$ *is an immediate successor of* $C\}|, \text{cf } C : C$ *an initial chain of* $T\}$.

Proof. Let $\kappa = \sup\{|\{x : x$ is an immediate successor of $C\}|, \text{cf } C : C$ an initial chain of $T\}$. Let F be an ultrafilter of A and let C be the associated initial chain. We shall show that F has a set of generators of size at most κ; this will prove $\leq$.

If C has a maximal element x, then $\{(T \uparrow x)\setminus\bigcup_{y\in F}(T \uparrow y) : F$ a finite set of immediate successors of $x\}$ generates F, as desired.

Suppose that C has no maximal element. Let $\langle x_\alpha : \alpha < \operatorname{cf} C \rangle$ be an increasing cofinal sequence in C. Then the set

$$\{(T \uparrow x_\alpha) \backslash \bigcup_{y \in F} (T \uparrow y) : \alpha < \operatorname{cf} C, \ F \text{ a finite set of immediate successors of } C\}$$

generates F, as desired.

Conversely, Let C be an initial chain of T and F the associated ultrafilter. Suppose X generates F and $|X| < \max(|\{x : x \text{ is an immediate successor of } C\}|, \operatorname{cf} C)$. Without loss of generality each element $x \in X$ has the form $(T \uparrow t_x) \backslash \bigcup_{y \in F_x} (T \uparrow y)$, where F_x is a finite set of successors of t_x. If C has a maximal element z, then $|X| < \{u : u \text{ is an immediate successor of } z\}$, so there is an immediate successor u of z such that $z \notin \bigcup_{x \in X} F_x$. Then $(T \uparrow z) \backslash (T \uparrow u) \in F$, but no element of X is $\leq$ it, contradiction. Suppose that C has no maximal element. If $|X| < \operatorname{cf} C$, choose $z \in C$ such that $t_x < z$ for all $x \in X$; then $T \uparrow z$ is in F but has no element of X less than it. Finally, if $\operatorname{cf} C \leq |X|$, then $|X| < \{u : u \text{ is an immediate successor of } C\}$, and we get a contradiction as above. $\qquad\square$

Concerning superatomic algebras we have the following result (due to Monk):

Theorem 14.13. $\chi A = |A|$ *for A superatomic.*

Proof. Let λ be a regular cardinal $\leq |A|$; we want to show that $\chi A \geq \lambda$. Let R be a complete system of representatives of atoms (of all levels) of A. We may assume that the top atoms of R form a finite partition of unity. Recall the notion of *rank* of an element of A: this is the least α such that $a \in I_{\alpha+1}$. An element $a \in A$ is *big* if $|\{x \in R : x \leq^* a\}| \geq \lambda$. ($u \leq^* v$ means that if β is the rank of u then $u/I_\beta \leq v/I_\beta$). Note that at least one of the top atoms of R is big. Let α be minimum such that there is a big $a \in R$ of rank α, and fix such an a, and the ultrafilter F associated with it. Note that if x is any element of rank less than α, then x is small. In fact, otherwise suppose that x is of smallest rank $\beta < \alpha$ such that x is big. Then $x/I_\beta = c_1/I_\beta + \cdots + c_m/I_\beta$ for certain $c_1, \ldots, c_m \in R$ such that each c_i/I_β is an atom. Thus $x \cdot -c_1 \cdot \ldots \cdot -c_m \in I_\beta$, and hence it is small. If $y \in R$ and $y \leq^* x$, say y has rank γ. Then

$$y/I_\gamma \leq x/I_\gamma = (x \cdot c_1)/I_\gamma + \cdots + (x \cdot c_m)/I_\gamma + (x \cdot -c_1 \cdot \ldots \cdot -c_m)/I_\gamma$$
$$\leq c_1/I_\gamma + \cdots + c_m/I_\gamma + (x \cdot -c_1 \cdot \ldots \cdot -c_m)/I_\gamma,$$

and so y/I_γ is $\leq$ one of these last summands. This means that one of $c_1, \ldots, c_m, x \cdot -c_1 \cdot \ldots \cdot -c_m$ is big, contradiction.

We claim that $\chi F \geq \lambda$ (as desired). In fact, suppose that $X \subseteq F$ generates F, where $|X| < \lambda$. If $b \in R$, $b \leq^* a$, and b has rank less than α, then $-b \in F$, and hence there is an $x_b \in X$ such that $x_b \leq -b$. Since λ is regular there is an $S \subseteq R$ with $|S| \geq \lambda$ and an $x \in X$ such that each member of S has rank less than α and $x \leq -b$ for each $b \in S$. Thus $b \leq -x$, and $b \leq^* a \cdot -x$. So, $a \cdot -x$ is big.

Case 1. $a/I_\alpha \leq x/I_\alpha$. Then $a \cdot -x \in I_\alpha$, so $a \cdot -x$ is small, contradiction.

Case 2. $a/I_\alpha \cdot x/I_\alpha = 0$. Then $a \cdot x \in I_\alpha$, and hence $-a + -x \in F$ and $-x \in F$, contradiction. $\qquad\square$

15. Hereditary Lindelöf degree

We begin with some equivalent definitions. Two of them involve new notions. A sequence $\langle x_\xi : \xi < \kappa \rangle$ of distinct elements of a topological space X is *right-separated* provided that for every $\xi < \kappa$ the set $\{x_\eta : \eta \leq \xi\}$ is open in $\{x_\xi : \xi < \kappa\}$. A sequence $\langle a_\alpha : \alpha < \kappa \rangle$ of elements of a BA A is *right separated* provided that if Γ is a finite subset of κ and $\alpha < \kappa$ with $\beta < \alpha$ for all $\beta \in \Gamma$, then $a_\alpha \cdot \prod_{\beta \in \Gamma} -a_\beta \neq 0$.

Theorem 15.1. *For any infinite BA A, hLA is equal to each of the following cardinals:*

> $\sup\{\kappa$:*there is an ideal not generated by less than κ elements*$\}$;
> $\sup\{\kappa$:*there is a strictly increasing sequence of ideals of length κ*$\}$;
> $\sup\{\kappa$:*there is a strictly increasing sequence of filters of length κ*$\}$;
> $\sup\{\kappa$:*there is a strictly increasing sequence of open sets of length κ*$\}$;
> $\sup\{\kappa$:*there is a strictly decreasing sequence of closed sets of length κ*$\}$;
> $\sup\{\kappa$:*there is a right-separated sequence in* Ult$A$ *of length κ*$\}$;
> $\sup\{\kappa$:*there is a right-separated sequence in A of length κ*;
> $\min\{\kappa$:*every open cover of a subspace of* Ult$A$ *has a subcover of size $\leq \kappa$*$\}$.

Proof. Nine cardinals are mentioned; let them be denoted by $\kappa_0, \ldots, \kappa_8$ in their order of mention (starting with hL itself). First we take care of easy relations: $\kappa_2 = \kappa_3$ since, if I is an ideal then $I^f \overset{\text{def}}{=} \{a \in A : -a \in I\}$ is a filter, and $I \subset J$ iff $I^f \subset J^f$; similarly, going from filters to ideals. So $\kappa_2 = \kappa_3$ follows. Next, $\kappa_2 \leq \kappa_4$. For, if I is an ideal, let $I^u = \bigcup_{a \in I} Sa$. Then I^u is open, and $I \subset J$ implies $I^u \subset J^u$. (If $a \in J \backslash I$, then $Sa \subseteq J^u$, of course, but $Sa \not\subseteq I^u$, since otherwise compactness of Sa would easily yield $a \in I$.) This shows $\kappa_2 \leq \kappa_4$. It is clear that $\kappa_4 = \kappa_5$, by taking complements. $\kappa_4 \leq \kappa_6$: If $\langle U_\alpha : \alpha < \kappa \rangle$ is a strictly increasing sequence of open sets, for every $\alpha < \kappa$ choose $x_\alpha \in U_{\alpha+1} \backslash U_\alpha$. Clearly $\langle x_\alpha : \alpha < \kappa \rangle$ is right-separated. $\kappa_6 = \kappa_7$: First suppose that $\langle F_\alpha : \alpha < \kappa \rangle$ is right-separated in Ult A. For all $\alpha < \kappa$ choose $a_\alpha \in A$ such that $F_\alpha \in Sa_\alpha \cap \{F_\beta : \beta < \kappa\} \subseteq \{F_\beta : \beta \leq \alpha\}$. We claim that $\langle a_\alpha : \alpha < \kappa \rangle$ is right-separated in A. To see this, suppose that Γ is a finite subset of κ, $\alpha < \kappa$, and $\beta < \alpha$ for all $\beta \in \Gamma$. Thus $a_\alpha \in F_\alpha$. If $\beta \in \Gamma$, then $a_\beta \notin F_\alpha$ by the choice of a_β, and hence $-a_\beta \in F_\alpha$. Therefore the element $a_\alpha \cdot \prod_{\beta \in \Gamma} -a_\beta$ is in F_α, and hence it must be non-zero, as desired. Second, suppose that $\langle a_\alpha : \alpha < \kappa \rangle$ is right-separated in A. Then for each $\alpha < \kappa$ the set $\{a_\alpha\} \cup \{-a_\beta : \beta < \alpha\}$ has the finite intersection property, and hence is included in an ultrafilter F_α. It is easy to check that $\langle F_\alpha : \alpha < \kappa \rangle$ is right-separated in Ult A, as desired. $\kappa_8 \leq \kappa_0$: for any subspace X of Ult$A$, any cover of X has a subcover of power $\leq$ L$X \leq \kappa_0$, so $\kappa_8 \leq \kappa_0$.

It remains only to prove that $\kappa_0 \leq \kappa_1$, $\kappa_1 \leq \kappa_2$, and $\kappa_7 \leq \kappa_8$. For the first one, suppose that $X \subseteq$ Ult$A$ and $\mathcal{O}$ is an open cover of X with no subcover of power λ; we construct an ideal not generated by λ or fewer elements. Let

$$I = \langle \{a \in A : \exists U \in \mathcal{O}(Sa \cap X \subseteq U)\} \rangle^{\text{Id}}.$$

Suppose that I is generated by J, where $|J| \leq \lambda$. For every $a \in J$ there is a finite subset $\mathscr{P}_a$ of $\mathcal{O}$ such that $Sa \cap X \subseteq \bigcup \mathscr{P}_a$. Let $\mathcal{O}' = \bigcup_{a \in J} \mathscr{P}_a$. We claim that $\mathcal{O}'$ covers X, which is the desired contradiction. Indeed, let $x \in X$. Say $x \in U \in \mathcal{O}$. Say $x \in Sa \cap X \subseteq U$. Choose a finite subset F of J such that $a \leq \sum F$. Then $x \in Sa \cap X \subseteq \bigcup_{b \in F} \bigcup \mathscr{P}b$, as desired.

Next, $\kappa_1 \leq \kappa_2$: suppose that I is an ideal not generated by fewer than λ elements. Then it is easy to construct a sequence $\langle a_\alpha : \alpha < \lambda \rangle$ of elements of I such that $a_\alpha \notin \langle \{a_\beta : \beta < \alpha\} \rangle^{\mathrm{Id}}$ for all $\alpha < \lambda$. Thus $\langle \langle \{a_\beta : \beta < \alpha\} \rangle^{\mathrm{Id}} : \alpha < \lambda \rangle$ is a strictly increasing sequence of ideals, as desired.

For $\kappa_7 \leq \kappa_8$, suppose that λ is a regular cardinal $\leq \kappa_7$ and $\langle x_\alpha : \alpha < \lambda \rangle$ is right separated. Thus for each $\alpha < \lambda$ we can choose an open set U_α such that $U_\alpha \cap \{x_\xi : \xi < \lambda\} = \{x_\xi : \xi < \alpha + 1\}$. Then $\{U_\alpha : \alpha < \lambda\}$ is a cover of $\{x_\xi : \xi < \lambda\}$ with no subcover of size $< \lambda$. Hence $\lambda < \kappa_8$, and this shows that $\kappa_7 \leq \kappa_8$. □

In Theorem 15.1, eight of the nine equivalents involve sups, and thus give rise to attainment problems. The proof of the theorem shows the following: attainment is the same for κ_2 and κ_3, for κ_4 and κ_5, and for κ_6 and κ_7; moreover, attainment in the sense κ_2 implies attainment in the sense κ_4, attainment in the sense κ_4 implies attainment in the sense κ_6, and attainment in the sense κ_1 implies attainment in the sense κ_2. It is also easy to see that attainment in the sense κ_4 implies attainment in the sense κ_2. In fact, if $\langle U_\alpha : \alpha < \kappa_3 \rangle$ is an increasing sequence of open sets, for each $\alpha < \kappa_3$ let $I_\alpha = \{a : Sa \subseteq U_\alpha\}$. Clearly I_α is an ideal. To show properness, pick $F \in U_{\alpha+1} \backslash U_\alpha$. Say $F \in Sa \subseteq U_{\alpha+1}$. Thus $a \in I_{\alpha+1} \backslash I_\alpha$. And attainment in the sense κ_6 implies attainment in the sense κ_4. In fact, suppose that $\langle F_\alpha : \alpha < \kappa \rangle$ is right separated. For each $\alpha < \kappa$ choose $a_\alpha \in F_\alpha$ such that $Sa_\alpha \cap \{F_\beta : \beta > \alpha\} = 0$, and let $U_\alpha = \bigcup_{\beta < \alpha} Sa_\beta$. Note that $F_\alpha \in U_{\alpha+1} \backslash U_\alpha$, as desired. Also note that if hLA is regular, then attainment in the right-separated sense implies attainment in the defined sense.

Thus we have seen that there are only three versions of the definition of hL that might lead to different attainment properties: hL as defined, in the ideal-generated sense, and in the right-separated sense, where we know only that attainment in the ideal-generated sense implies attainment in the right-separated sense. So we have the following problem (Problems 41–43 in Monk [90]):

Problem 50. *Describe the implications between attainment of* hL *as defined, in the ideal-generation sense, and in the right-separated sense.*

It is known that hL is attained in the right-separated sense for cardinals of cofinality ω, and for strong limit singular cardinals; see Juhász [80].

We turn to algebraic operations. If A is a subalgebra or homomorphic image of B, then hL$A \leq$ hLB. Furthermore, looking at the right-separated equivalent and the topological dual it is clear that

$$\mathrm{hL}\left(\prod_{i \in I}^{\mathrm{w}} A_i\right) = \max(|I|, \sup_{i \in I} \mathrm{hL}A).$$

Note that $\mathrm{Ind}A \leq \mathrm{hL}A$, using the equivalent concerning ideals, for example. Hence it is clear that $\mathrm{hL}(\prod_{i \in I} A_i) \geq \max(2^{|I|}, \sup_{i \in I} \mathrm{hL}A_i)$. Strict inequality is possible, as was noticed by Shelah and Peterson independently, solving Problem 44 of Monk [90]. Again the example used for spread applies here.

Concerning ultraproducts, note that hL is an order-independence function, and hence Theorem 12.6 holds. By the proof of Theorem 12.7 it follows that under GCH we have $\mathrm{hL}\left(\prod_{i \in I} A_i/F\right) \geq \left|\prod_{i \in I} \mathrm{hL}A_i/F\right|$ for F regular, and Donder's theorem says that under $V = L$ the regularity assumption can be removed. The example of Laver for depth works for hL also: it is consistent to have a situation where $\mathrm{hL}\left(\prod_{i \in I} A_i/F\right) > \left|\prod_{i \in I} \mathrm{hL}A_i/F\right|$ for F regular; see also Rosłanowski, Shelah [94] for another consistent example.

Problem 51. *Can one construct in ZFC an example with* $\mathrm{hL}\left(\prod_{i \in I} A_i/F\right) > \left|\prod_{i \in I} \mathrm{hL}A_i/F\right|$*?* We do not know whether $<$ is possible:

Problem 52. *Can one have* $\mathrm{hL}\left(\prod_{i \in I} A_i/F\right) < \left|\prod_{i \in I} \mathrm{hL}A_i/F\right|$ *for some system of BAs (consistently)?*

As usual, it may be that Magidor, Shelah [91] essentially answers this problem, and an example of this sort implies that the invariants $\mathrm{hL}A_i$ are (for most i) inaccessible.

Next come free products:

Theorem 15.2. *If* $\langle A_i : i \in I \rangle$ *is a system of BAs each with at least 4 elements, then*

$$\max(|I|, \sup_{i \in I} \mathrm{hL}A_i) \leq \mathrm{hL}(\oplus_{i \in I} A_i) \leq \max\left(|I|, 2^{\sup_{i \in I} \mathrm{S}A_i}\right).$$

Proof. The first inequality is easy. For the second,

$$\mathrm{hL}(\oplus_{i \in I} A_i) \leq |\oplus_{i \in I} A_i| = |I| \cdot \sup_{i \in I} |A_i|$$
$$\leq |I| \cdot \sup_{i \in I} 2^{\mathrm{S}A_i}$$
$$\leq \max\left(|I|, 2^{\sup_{i \in I} \mathrm{S}A_i}\right). \qquad \square$$

The inequalities in Theorem 15.2 are sharp, in the sense that all possibilities can occur. Thus both are equalities if $I = \omega_1$ and each A_i is a four-element algebra. The first is an equality and the second not for $I = \omega_1$ and each A_i the free BA on ω_1 free generators. The first is a strict inequality and the second an equality for $A \oplus A$, where $A = \mathrm{Intalg}\,\mathbb{R}$. Finally, both inequalities are strict for $A \oplus A$, where A is the tree algebra on a Suslin tree in which each element has infinitely many immediate successors, and $\neg$CH holds. It is clear that $\neg$CH is needed to get an example where both inequalities are strict.

Concerning derived functions of hL, we mention these obvious facts:

$$\mathrm{hL}A = \mathrm{hL}_{\mathrm{H}+}A = \mathrm{hL}_{\mathrm{S}+}A = \mathrm{hL}_{\mathrm{h}+}A = {}_\mathrm{d}\mathrm{hL}_{\mathrm{S}+}A;$$

and $\mathrm{hL}_{\mathrm{S}-}A = \mathrm{hL}_{\mathrm{h}-}A = \omega$. The following theorem is a corollary of 14.5 and 14.6, using the fact given below that $\chi A \leq \mathrm{hL}A$ for any infinite BA.

Theorem 15.3. *If* $\mathrm{hL}A = \omega$, *then* $\mathrm{Card}_{\mathrm{H}-}A = \omega$.   □

On the relationship of hL with the previously defined functions: obviously $sA \leq \mathrm{hL}A$ for any infinite BA A. Next, $\chi A \leq \mathrm{hL}A$. In fact, suppose that F is any ultrafilter on A; we want to find a subset X of F which generates F and has at most $\mathrm{hL}A$ elements. The set $\{Sa : -a \in F\}$ covers $\mathrm{Ult}A \backslash \{F\}$. Hence there is a subset X of F such that $\{Sa : -a \in X\}$ also covers $\mathrm{Ult}A \backslash \{F\}$, and $|X| \leq \mathrm{hL}A$. We claim that X generates F. For suppose that $a \in F$. Then $X \cup \{-a\}$ does not have the finite intersection property; otherwise, there would exist an ultrafilter G containing this set—then $G \neq F$, so $b \in G$ for some b such that $-b \in X$, contradiction. But $X \cup \{-a\}$ not having the finite intersection property means that a is in the filter generated by X, as desired.

The following theorem is due to Todorčević [90]:

Theorem 15.4. $|A| \leq \mathrm{Irr}A \cdot (\mathrm{hL}A)^+$ *for any infinite BA* A.

Proof. Let $\theta = \mathrm{Irr}A \cdot (\mathrm{hL}A)^+$ and $\kappa = (\mathrm{hL}A)^+$. Assume that $|A| > \theta$, in order to work for a contradiction. Wlog $|A| = \theta^+$. Write A as a strictly increasing sequence $\langle A_\xi : \xi < \theta^+ \rangle$ of subalgebras of size $\leq \theta$. Let $S_0 = \{\delta < \theta^+ : \mathrm{cf}\,\delta = \kappa\}$. So S_0 is stationary in θ^+. For each $\delta \in S_0$ choose $a_\delta \in A_{\delta+1} \backslash A_\delta$. Define

$$I_\delta = \{b \in A_\delta : b \cdot a_\delta = 0\}; \quad J_\delta = \{b \in A_\delta : b \cdot -a_\delta = 0\}.$$

Note that I_δ and J_δ are ideals in A_δ. Let I'_δ be the ideal of A generated by I_δ. Thus $I'_\delta = \{a \in A : a \leq b \text{ for some } b \in I_\delta\}$. Now I'_δ has a generating set of size $\leq \mathrm{hL}A$; so I_δ itself has such a generating set. Since κ is a regular cardinal $> \mathrm{hL}A$, there is an $f\delta < \delta$ such that a generating set for I_δ is a subset of $A_{f\delta}$. So by Fodor's theorem there is a $\xi_0 < \theta^+$ and a stationary subset S_1 of S_0 such that $f\delta = \xi_0$ for all $\delta \in S_1$. Similarly, we can get a stationary subset S_2 of S_1 and a $\xi_1 < \theta^+$ such that every ideal J_δ for $\delta \in S_2$ has a generating set in A_{ξ_1}. Let ξ_2 be the maximum of ξ_0, ξ_1. We now claim that $\langle a_\delta : \delta \in S_2 \rangle$ is irredundant, which, of course, is a contradiction. To prove this claim we first show

(*) Suppose $\xi_2 < \delta < \delta_0 < \cdots \delta_n$ are elements of S_2 and $\varepsilon \in {}^{n+1}2$. Suppose that $c \in A_\delta$ and

$$c \cdot \prod_{i \leq n} a_{\delta_i}^{\varepsilon i} \leq a_\delta.$$

Then there exist $b_0, \ldots b_n \in A_{\xi_2}$ such that

$$c \cdot \prod_{i \leq n} a_{\delta_i}^{\varepsilon i} \leq c \cdot \prod_{i \leq n} b_i \leq a_\delta.$$

We prove (*) by induction on n; the following argument will work when $n = 0$ and also for the inductive step. Let $d = \left(c \cdot \prod_{i < n} a_{\delta_i}^{\varepsilon i}\right) \cdot -a_\delta$. Then $d \in A_{\delta_n}$ and

$d \leq -a_{\delta_n}^{\varepsilon n}$. *Case 1.* $\varepsilon n = 1$. Then $d \in I_{\delta_n}$. Hence there is an $x \in A_{\xi_0} \cap I_{\delta_n}$ such that $d \leq x$. So $d \leq x \leq -a_{\delta_n}^{\varepsilon n}$. *Case 2.* $\varepsilon n = 0$. Then $d \in J_{\delta_n}$, so there is an $x \in A_{\xi_1} \cap J_{\delta_n}$ such that $d \leq x$. So again $d \leq x \leq -a_{\delta_n}^{\varepsilon n}$.

So, in either case we get an $x \in A_{\xi_2}$ such that $d \leq x \leq -a_{\delta_n}^{\varepsilon n}$. It follows that

$$c \cdot \prod_{i \leq n} a_{\delta_i}^{\varepsilon i} \leq c \cdot -x \cdot \prod_{i < n} a_{\delta_i}^{\varepsilon i} \leq a_\delta,$$

so we have started the induction if $n = 0$, and continued the induction otherwise.

Now suppose that $\langle a_\delta : \xi_2 < \delta \in S_2 \rangle$ is redundant. So we can find $\delta < \delta_0 < \cdots < \delta_n$ with $\xi_2 < \delta$ such that a_δ is generated by $A_\delta \cup \{a_{\delta_0}, \ldots, a_{\delta_n}\}$. Therefore a_δ is a finite union of elements of the form

$$c \cdot \prod_{i \leq n} a_{\delta_i}^{\varepsilon i},$$

where $c \in A_\delta$. By (*), every such intersection can be replaced by one of the form

$$c \cdot \prod_{i \leq n} b_i$$

for some $b_0, \ldots, b_n \in A_{\xi_k}$. It follows that $a_\delta \in A_\delta$, contradiction. □

The BA of the Kunen line constructed in Chapter 8 (assuming CH) has character ω_1 (see Chapter 14), hence hereditary Lindelöf degree ω_1, and countable spread.

If one can construct in ZFC a BA A such that $sA < hLA$, then one can also construct in ZFC a BA B such that $sB < \chi B$ (see problem 49). For, let I be an ideal of A such that I is not generated by fewer than $(sA)^+$ elements, and let $B = I \cup -I$.

An example where $\chi A < hLA$ is provided by the Aleksandroff duplicate of a free algebra; see Chapter 14. An example with $hLA < dA$ is provided by the interval algebra on a complete Suslin line, using the argument of Lemma 3.28; on the other hand, in the first edition of Juhász's book, it is shown that MA+¬CH implies that $hLA = \omega$ implies $hdA = \omega$.

These observations leave the following question open; this is Problem 46 in Monk [90]:

Problem 53. *Is there an example in ZFC of a BA A such that $hLA < dA$?*

This problem is equivalent to the problem of constructing in ZFC a BA A such that $hLA < hdA$; see the end of Chapter 16.

Bounded versions of hL can be defined as follows. For m a positive integer, a sequence $\langle x_\alpha : \alpha < \kappa \rangle$ of elements of A is said to be *m-right-separated* provided that if $\Gamma \in [\kappa]^m$, $\alpha < \kappa$, and $\beta < \alpha$ for all $\beta \in \Gamma$, then $a_\alpha \cdot \prod_{\beta \in \Gamma} -a_\beta \neq 0$. Then we define

$$hL_m A = \sup\{\kappa : \text{there is an } m\text{-right-separated sequence in } A\}.$$

For this notion see Rosłanowski, Shelah [94].

For an interval algebra A we have $\mathrm{hL}A = \mathrm{c}A$. In fact, suppose that A is the interval algebra on L and I is an ideal of A. Define $a \equiv b$ iff $a, b \in L$ and either $a = b$ or else if, say, $a < b$, then $[a, b) \in I$. Then $\equiv$ is a convex equivalence relation on L. For each $\equiv$-class k having more than one element, let $\langle a_\alpha^k : \alpha < \lambda_k \rangle$ be a strictly decreasing coinitial sequence in k (with $\lambda_k = 1$ if k has a first element), and let $\langle b_\alpha^k : \alpha < \mu_k \rangle$ be a strictly increasing cofinal sequence in k (with $\mu_k = 1$ if k has a greatest element), and with $a_0^k < b_0^k$. Note that there are at most $\mathrm{c}A$ $\equiv$-classes with more than one element, and always $\lambda_k, \mu_k < (\mathrm{c}A)^+$. Hence

$$\{[a_\alpha^k, b_\beta^k) : k \text{ an } \equiv\text{-class with more than one element, } \alpha < \lambda_k, \ \beta < \mu_k\}$$

is a collection of at most $\mathrm{c}A$ elements which generates I; so $\mathrm{hL}A \leq \mathrm{c}A$ by Theorem 15.1.

For any tree algebra B on an infinite tree T we also have $\mathrm{c}B = \mathrm{hL}B$. For, Treealg T embeds in an interval algebra A, and we may assume that Treealg T is dense in A (extend the identity from Treealg T onto itself to a homomorphism from A into the completion of Treealg T, and then take the image of A). Hence

$$\mathrm{hL}A \geq \mathrm{hL}(\text{Treealg}\, T) \geq \mathrm{c}(\text{Treealg}\, T) = \mathrm{c}A = \mathrm{hL}A.$$

16. Hereditary density

We begin again with some equivalent definitions, which are similar to the case of hereditary Lindelöf degree. Recall the definition of left-separated sequence from Chapter 6, before Theorem 6.7.

Theorem 16.1. *For any infinite BA A, hd$A$ is equal to each of these cardinals:*

$\sup\{\kappa$:*there is a strictly decreasing sequence of ideals of length $\kappa\}$;*
$\sup\{\kappa$:*there is a strictly decreasing sequence of filters of length $\kappa\}$;*
$\sup\{\kappa$:*there is a strictly decreasing sequence of open sets of length $\kappa\}$;*
$\sup\{\kappa$:*there is a strictly increasing sequence of closed sets of length $\kappa\}$;*
$\sup\{\kappa$:*there is a left-separated sequence of length $\kappa\}$;*
$\min\{\kappa$:*every subspace S of Ult$A$ has a dense subset of power $\leq \kappa\}$;*
$\sup\{\pi B : B$ *is a homomorphic image of $A\}$;*
$\sup\{dB : B$ *is a homomorphic image of $A\}$.*

(Note that *left-separated* can be taken in the topological or algebraic sense.)

Proof. This time there are nine cardinals, named $\kappa_0, \ldots, \kappa_8$ in their order of mention, starting with hd itself. The following relationships are easy, following the pattern of the proof of Theorem 15.1: $\kappa_1 = \kappa_2$; $\kappa_1 \leq \kappa_3$; $\kappa_3 = \kappa_4$; $\kappa_3 \leq \kappa_5$; and $\kappa_0 = \kappa_6$. Furthermore, $\kappa_8 = \kappa_0$ by Theorem 5.13, $\kappa_0 = \kappa_5$ by Theorem 6.7, and $\kappa_0 = \kappa_7$ by Theorem 6.10. Hence only two inequalities remain.

$\kappa_6 \leq \kappa_2$: Suppose that X is a subspace of Ult$A$, and $dX = \kappa$; we construct a strictly decreasing sequence of filters of type κ. By induction let

$$F_\alpha \in X \backslash \overline{\{F_\beta : \beta < \alpha\}}$$

for each $\alpha < \kappa$. Then set $C_\alpha = \bigcap_{\beta \leq \alpha} F_\beta$. Thus $\langle C_\alpha : \alpha < \kappa \rangle$ is a decreasing sequence of filters. It is strictly decreasing, since if $\alpha < \kappa$ we can choose $a \in F_{\alpha+1}$ such that $Sa \cap \{F_\beta : \beta \leq \alpha\} = 0$, so that $-a \in C_\alpha \backslash C_{\alpha+1}$.

$\kappa_5 \leq \kappa_6$: Suppose $\langle x_\alpha : \alpha < \kappa \rangle$ is left separated, where κ is regular. Clearly then $\{x_\alpha : \alpha < \kappa\}$ has no dense subset of power $< \kappa$. $\qquad \square$

The equivalents in Theorem 16.1 give rise to eight possible attainment problems, on the face of it. However, proofs of previous results set some limits:

Proof of Theorem 5.13:

attainment in the κ_8 sense implies attainment in the κ_0 sense;

Proof of Theorem 6.7:

attainment in the κ_0 sense implies attainment in the κ_5 sense;

for hd$A$ regular, attainment in the κ_5 sense implies attainment in the κ_0 sense;

Proof of Theorem 6.10:

attainment in the κ_7 sense implies attainment in the κ_5 sense;

attainment in the κ_0 sense implies attainment in the κ_7 sense;

Proof of Theorem 16.1:

attainment for $\kappa_1, \kappa_2, \kappa_3, \kappa_4$ are equivalent;

attainment in the κ_0 sense implies attainment in the κ_2 sense;

Now we note two other implications.

κ_1 *attained implies* κ_5 *attained.* Suppose that $\langle I_\alpha : \alpha < \kappa_1 \rangle$ is a strictly decreasing sequence of ideals. For each $\alpha < \kappa_1$ choose $a_\alpha \in I_\alpha \backslash I_{\alpha+1}$. Then $\langle a_\alpha : \alpha < \kappa_1 \rangle$ is left-separated.

κ_5 *attained implies* κ_1 *attained.* Similarly.

Thus we are left with four possible attainment questions, represented by κ_0, κ_5, κ_7, and κ_8, where attainment implications $\kappa_8 \Rightarrow \kappa_0 \Rightarrow \kappa_7 \Rightarrow \kappa_5$ hold.

Problem 54. *Describe completely the attainment relations for the equivalent definitions of* hd.

This extends Problems 47 and 48 from Monk [90].

Like for hL, it is known that hd in the sense of left-separation is attained for singular cardinals of cofinality ω and for strong limit singular cardinals.

If A is a subalgebra or homomorphic image of B, then hd$A \leq$ hd$B$. It is also clear that

$$\text{hd} \left(\prod_{i \in I}^{\text{w}} A_i \right) = \max(|I|, \sup_{i \in I} \text{hd} A_i).$$

Obviously s$A \leq$ hd$A$, and hence Ind$A \leq$ hd$A$. It follows that for arbitrary products we have, as usual , hd$\left(\prod_{i \in I} A_i \right) \geq \max(2^{|I|}, \sup_{i \in I} \text{hd} A_i)$. Shelah and Peterson independently noticed that strict inequality is possible; this answers Problem 49 of Monk [90]. The example for spread can be used here.

The situation for ultraproducts is like for hL. hd is an order-independence function, and hence Theorem 12.6 holds. By the proof of Theorem 12.7 it follows that under GCH we have hd $\left(\prod_{i \in I} A_i / F \right) \geq \left| \prod_{i \in I} \text{hd} A_i / F \right|$ for F regular, and Donder's theorem says that under $V = L$ the regularity assumption can be removed. A consistent example with $>$ is due to Laver, as before, and other consistent examples can be found in Rosłanowski, Shelah [94].

Problem 55. *Can one get an example with* hd $\left(\prod_{i \in I} A_i / F \right) > \left| \prod_{i \in I} \text{hd} A_i / F \right|$ *in* ZFC?

Problem 56. *Is an example with* hd $\left(\prod_{i \in I} A_i / F \right) < \left| \prod_{i \in I} \text{hd} A_i / F \right|$ *consistent?*

For this problem see Magidor, Shelah [91] and Rosłanowski, Shelah [94] for a possible solution, as usual.

For free products, the analog of Theorem 15.2 holds, with essentially the same proof:

Theorem 16.2. *If $\langle A_i : i \in I \rangle$ is a system of non-trivial BAs, for brevity let $\lambda = \sup_{i \in I} \mathrm{hd} A_i$; then*

$$\max(|I|, \sup_{i \in I} \mathrm{hd} A_i) \le \mathrm{hd}(\oplus_{i \in I} A_i) \le \max\left(|I|, 2^{\sup_{i \in I} SA_i}\right). \qquad \square$$

The inequalities in Theorem 16.2 are sharp. This is seen as for Theorem 15.2, except for both $<$; for this case one can take S such that $\mathbb{Q} \subseteq S \subseteq \mathbb{R}$, $|S| = \aleph_1$, let $A = \mathrm{Intalg} S$, and consider $A \oplus A$, assuming $\neg$CH.

Another important fact about free products is given in the following theorem.

Theorem 16.3. *For infinite BAs A and B we have $\mathrm{s}(A \oplus B) \ge \min(\mathrm{hL} A, \mathrm{hd} B)$.*

Proof. Let $\kappa = \min(\mathrm{hL} A, \mathrm{hd} B)$, and let $\lambda^+ \le \kappa$. Let $\langle a_\alpha : \alpha < \lambda^+ \rangle$ be right-separated in A, and let $\langle b_\alpha : \alpha < \lambda^+ \rangle$ be left-separated in B. We claim that $\langle a_\alpha \cdot b_\alpha : \alpha < \lambda^+ \rangle$ is ideal independent. For, suppose that $\Gamma \in [\lambda^+]^{<\omega}$ and $\alpha \in \lambda^+ \backslash \Gamma$. Let $\Delta = \{\beta \in \Gamma : \beta < \alpha\}$. then

$$a_\alpha \cdot b_\alpha \cdot \prod_{\beta \in \Gamma} -(a_\beta \cdot b_\beta) \ge \left(a_\alpha \cdot \prod_{\beta \in \Delta} -a_\beta \right) \cdot \left(b_\alpha \cdot \prod_{\beta \in \Gamma \backslash \Delta} -b_\beta \right) \ne 0,$$

as desired. $\square$

Now we want to give some results concerning the exponential due to Malyhin [72].

Lemma 16.4. *If X is a topological space and $X \in \overline{Z}$ in $\mathrm{Exp} X$, then $\bigcup Z$ is dense in X.*

Proof. Let $0 \ne U$ be open in X. Then $X \in \mathscr{V}(X, U)$, so there is an $F \in Z \cap \mathscr{V}(X, U)$. Thus $F \cap U \ne 0$, so $\bigcup Z \cap U \ne 0$. $\square$

Lemma 16.5. *For any Hausdorff space X we have $\mathrm{d} X \le \mathrm{t}(\mathrm{Exp}\, X)$.*

Proof. Let Z be the collection of all finite non-empty subsets of X. Then Z is a subset of $\mathrm{Exp}\, X$. Moreover, $X \in \overline{Z}$, since if $\mathscr{V}(U_0, \ldots, U_{m-1})$ is any neighborhood of X, choose $a_i \in U_i$ for all $i < m$; then $\{a_i : i < m\} \in Z \cap \mathscr{V}(U_0, \ldots, U_{m-1})$. Now choose a subset Y of Z of size $\le \mathrm{t}(\mathrm{Exp}\, X)$ such that $X \in \overline{Y}$. Then by Lemma 16.4, $\bigcup Y$ is dense in X. Clearly $|\bigcup Y| \le \mathrm{t}(\mathrm{Exp}\, X)$. $\square$

Theorem 16.6. $\mathrm{hd} A \le \mathrm{t}(\mathrm{Exp}\, A)$.

Proof. We use the fact that $\mathrm{hd} A = \sup\{\mathrm{d} B : B$ a homomorphic image of $A\}$, given in Theorem 16.1. Let B be any homomorphic image of A. Then by Proposition 2.7, $\mathrm{Exp}\, B$ is a homomorpic image of $\mathrm{Exp}\, A$, so $\mathrm{d} B \le \mathrm{t}(\mathrm{Exp}\, B) \le \mathrm{t}(\mathrm{Exp}\, A)$. $\square$

Lemma 16.7. $\chi A \le \mathrm{t}(\mathrm{Exp}\, A)$; *in fact, every closed set in $\mathrm{Ult} A$ has a neighborhood basis with at most $\mathrm{t}(\mathrm{Exp}\, A)$ elements.*

Proof. Let F be a closed subset of $\mathrm{Ult}A$. Then

$$F \in \overline{\{U : U \text{ clopen}, F \subseteq U\}}.$$

For, suppose that $F \in \mathscr{V}(U_0, \ldots, U_{m-1})$ with each U_i clopen. Then $U_0 \cup \ldots \cup U_{m-1} \in \mathscr{V}(U_0, \ldots, U_{m-1})$, as desired.

It follows that there is a subset $\mathscr{O}$ of $\{U : U \text{ clopen}, F \subseteq U\}$ such that $|\mathscr{O}| \leq \mathrm{t}(\mathrm{Exp}\, A)$ and $F \in \overline{\mathscr{O}}$. Then $\mathscr{O}$ is the desired neighborhood base for F. For, suppose $F \subseteq W$ with W clopen. Then $F \in \mathscr{V}(W)$, so there is a $U \in \mathscr{O}$ such that $U \in \mathscr{V}(W)$. So $U \subseteq W$, as desired. $\qquad\square$

Theorem 16.8. $\chi(\mathrm{Exp}\, A) = \mathrm{t}(\mathrm{Exp}\, A)$.

Proof. Since $\mathrm{t} \leq \chi$ in general, it suffices to show $\chi(\mathrm{Exp}\, A) = \mathrm{t}(\mathrm{Exp}\, A)$. Let F be a nonempty closed subset of $\mathrm{Ult}A$. By Lemma 16.7, let $\mathscr{O}$ be a collection of clopen subsets of $\mathrm{Ult}A$ which forms a neighborhood base for F, with $|\mathscr{O}| \leq \mathrm{t}(\mathrm{Exp}\, A)$. And by Theorem 16.6, let Y be a dense subset of F of size at most $\mathrm{t}(\mathrm{Exp}\, A)$. For each $y \in Y$, let $\mathscr{P}_y$ be a collection of clopen subsets of $\mathrm{Ult}A$ which forms a neighborhood base for y, with $|\mathscr{P}_y| \leq \mathrm{t}(\mathrm{Exp}\, A)$, again using Lemma 16.7. Now let $\mathscr{Q}$ be the collection of all open sets in $\mathrm{Exp}\,(\mathrm{Ult}A)$ of the form

$$\mathscr{V}(W_0, \ldots, W_{m-1}, S),$$

where $S \in \mathscr{O}$ and for each $i < m$ there is a $y_i \in Y$ such that $W_i \in \mathscr{P}_{y_i}$. We claim that $\mathscr{Q}$ forms a neighborhood base for F in $\mathrm{Exp}\,(\mathrm{Ult}A)$. Clearly F is a member of each member of $\mathscr{Q}$. Now suppose that $F \in \mathscr{V}(U_0, \ldots, U_{m-1})$, where each U_i is open in $\mathrm{Ult}A$. Choose $S \in \mathscr{O}$ such that $S \subseteq U_0 \cup \ldots \cup U_{m-1}$. For each $i < m$ we have $F \cap U_i \neq 0$, and so we can choose $y_i \in Y \cap U_i$. Then choose $W_i \in \mathscr{P}_{y_i}$ such that $y_i \in W_i \subseteq U_i$. Clearly then

$$F \in \mathscr{V}(W_0, \ldots, W_{m-1}, S) \subseteq \mathscr{V}(U_0, \ldots, U_{m-1}),$$

as desired. $\qquad\square$

Theorem 16.9. $\mathrm{hL}A \leq \mathrm{s}(\mathrm{Exp}\, A)$.

Proof. By Theorem 15.1, let $\langle F_\alpha : \alpha < \kappa \rangle$ be a strictly decreasing sequence of closed subsets of $\mathrm{Ult}A$, where $\kappa = \mathrm{hL}A$. We claim that $D \overset{\mathrm{def}}{=} \{F_{\alpha+1} : \alpha < \lambda\}$ is a discrete set of points of $\mathrm{Exp}\, A$. For, let $\alpha < \kappa$. Choose $x \in F_\alpha \backslash F_{\alpha+1}$ and $y \in F_{\alpha+1} \backslash F_{\alpha+2}$. Then there is a clopen subset S of $\mathrm{Ult}A$ such that $y \in S$, $F_{\alpha+2} \cap S = 0$, and $x \notin S$. And there is a clopen U such that $F_{\alpha+1} \subseteq U$ and $x \notin U$. Now $\mathscr{V}(U, X) \cap D = \{F_{\alpha+1}\}$. For, obviously $F_{\alpha+1} \in \mathscr{V}(U,,X)$. Suppose that $\alpha < \beta$ and $F_{\beta+1} \in \mathscr{V}(U, S)$. Then $F_{\beta+1} \subseteq F_{\alpha+2}$ and $F_{\beta+1} \cap S \neq 0$, contradicting $F_{\alpha+2} \cap S = 0$. Suppose that $\beta < \alpha$ and $F_{\beta+1} \in \mathscr{V}(U, S)$. Then $F_\alpha \subseteq F_{\beta+1}$, so $x \in F_{\beta+1}$. But $x \notin U$ and $x \notin S$, contradiction. $\qquad\square$

We also want to give an important result from Bell, Ginsburg, Todorčević, S. [82]. We need a well-known lemma first.

Lemma 16.10. *For any infinite BA A and any $X \subseteq A$ the following are equivalent:*
  (i) *X generates A.*
  (ii) *$\{\mathcal{S}x : x \in X\}$ separates points in $\mathrm{Ult}A$.*

Proof. $(i) \Rightarrow (ii)$: Suppose that $F, G \in \mathrm{Ult}A$, $F \neq G$. If $\forall x \in X (x \in F$ iff $x \in G)$, then $\forall x \in \langle X \rangle (x \in F$ iff $x \in G)$, by an easy argument.
   $(ii) \Rightarrow (i)$: Suppose that $a \in A \backslash \langle X \rangle$. Let $\mathscr{A} = \{x_0^{\varepsilon_0} \cdot \ldots \cdot x_{m-1}^{\varepsilon_{m-1}} :$ each $x_i \in X$ and $x_0^{\varepsilon_0} \cdot \ldots \cdot x_{m-1}^{\varepsilon_{m-1}} \leq a\}$. Then $\bigcup_{b \in \mathscr{A}} \mathcal{S}b \subset \mathcal{S}a$ by compactness, since $a \notin \langle X \rangle$, so choose $F \in \mathcal{S}a \backslash \bigcup_{b \in \mathscr{A}} \mathcal{S}b$. Now $(F \cap \langle X \rangle) \cup \{-a\}$ has fip, and so is contained in an ultrafilter G. But then F and G are distinct ultrafilters which cannot be separated by $\{\mathcal{S}x : x \in X\}$. $\square$

Theorem 16.11. *For any infinite BA A we have $\mathrm{hd}(\mathrm{Exp}\, A) = \mathrm{s}(\mathrm{Exp}\, A)$.*

Proof. Since $\mathrm{s} \leq \mathrm{hd}$ in general, we assume that $\kappa \stackrel{\mathrm{def}}{=} \mathrm{s}(\mathrm{Exp}\, A) < \mathrm{hd}(\mathrm{Exp}\, A)$ and try to get a contradiction. Let $\langle C_\alpha : \alpha < \kappa^+ \rangle$ be left separated in the space $\mathrm{Exp}\, A$. Set

$$I = \{a \in A : |A \upharpoonright a| \leq \kappa\} \quad \text{and} \quad W = \bigcup_{a \in I} \mathcal{S}a.$$

Thus I is an ideal of A. Let $F = \mathrm{Ult}A \backslash W$. By 16.6, $\mathrm{hd}A \leq \kappa$, so we can choose a dense subset D of F with $|D| \leq \kappa$.

(1) $|\{\alpha < \kappa^+ : d \notin C_\alpha\}| \leq \kappa$ for all $d \in D$.

For, suppose not: this gives us $d \in D$ such that $\Gamma \stackrel{\mathrm{def}}{=} \{\alpha < \kappa^+ : d \notin C_\alpha\}| \geq \kappa^+$. By 16.7 we have $\chi A \leq \kappa$, so we can choose a clopen neighborhood base $\mathscr{B}$ for d such that $|\mathscr{B}| \leq \kappa$. For every $\alpha \in \Gamma$ choose $U_\alpha \in \mathscr{B}$ such that $U_\alpha \cap C_\alpha = 0$. Then there exist a $\Delta \in [\Gamma]^{\kappa^+}$ and a $b \in A$ such that $\mathcal{S}b \cap C_\alpha = 0$ for all $\alpha \in \Delta$ and $\mathcal{S}b \in \mathscr{B}$. Now $d \in F \cap \mathcal{S}b$, so $b \notin I$. Thus $|A \upharpoonright b| \geq \kappa^+$.

(2) $\mathrm{hL}(\mathrm{Exp}\,(A \upharpoonright b)) \geq \kappa^+$.

For, suppose that $\mathrm{hL}(\mathrm{Exp}\,(A \upharpoonright b)) \leq \kappa$. Let $Y = \{X \in \mathrm{Exp}\,(A \upharpoonright b) : |X| = 2\}$. For all $X \in Y$ choose U_X clopen in $\mathrm{Ult}(A \upharpoonright b)$ such that $|U_X \cap X| = 1$. Then $\{\mathscr{V}(U_X, -U_X) : X \in Y\}$ covers Y. Let $\{\mathscr{V}(U_X, -U_X) : X \in Y'\}$ be a subcover with $|Y'| \leq \kappa$. Now $\{U_X : X \in Y'\}$ separates the points of $\mathrm{Ult}(A \upharpoonright b)$. For, suppose that $u, v \in \mathrm{Ult}(A \upharpoonright b)$, $u \neq v$. Say $\{u, v\} \in \mathscr{V}(U_X, -U_X)$ with $X \in Y'$. Clearly U_X separates u and v. Now by Lemma 16.10 we get $|A \upharpoonright b| \leq \kappa$, contradiction. So (2) holds.
   Now $C_\alpha \subseteq \mathcal{S}(-b)$ for all $\alpha \in \Delta$. For each $\alpha \in \Delta$ let $C'_\alpha = \{F \cap (A \upharpoonright -b) : F \in C_\alpha\}$. Then it is easy to check that $\langle C'_\alpha : \alpha \in \Delta \rangle$ is a left-separated sequence in $\mathrm{Ult}(\mathrm{Exp}\,(A \upharpoonright -b))$. So $\mathrm{hd}(\mathrm{Exp}\,(A \upharpoonright -b)) \geq \kappa^+$. Now by Theorem 16.3 it follows that $\mathrm{s}(\mathrm{Exp}\,(A \upharpoonright b) \oplus \mathrm{Exp}\,(A \upharpoonright -b)) \geq \kappa^+$. Then 1.14 implies that $\mathrm{s}(\mathrm{Exp}\, A) \geq \kappa^+$, contradiction. This proves (1).
   By (1) we may assume that $d \in C_\alpha$ for all $d \in D$ and all $\alpha < \kappa^+$, i.e., $D \subseteq C_\alpha$ for all $\alpha < \kappa^+$. Hence $F \subseteq C_\alpha$ for all $\alpha < \kappa^+$.

Now let $B = I \cup -I$. We claim that $|B| \leq \kappa$. In fact, $\mathrm{hL}B \leq \kappa$ by 16.9, so I is generated by $\leq \kappa$ elements, so the claim follows. As a consequence, $|\mathrm{Exp}\,B| \leq \kappa$ too. But now we show that $\mathrm{hd}B \geq \kappa^+$, which is the final contradiction. We can write

$$C_\alpha \in \mathscr{V}(\mathcal{S}c_\alpha^0, \ldots, \mathcal{S}c_\alpha^{m_\alpha-1}, \ldots, \mathcal{S}c_\alpha^{n_\alpha-1}),$$
$$\mathscr{V}(\mathcal{S}c_\alpha^0, \ldots, \mathcal{S}c_\alpha^{m_\alpha-1}, \ldots, \mathcal{S}c_\alpha^{n_\alpha-1}) \cap \{C_\beta : \beta < \kappa^+\} \subseteq \{C_\beta : \alpha \leq \beta\},$$
$$F \cap \mathcal{S}c_\alpha^i \neq 0 \text{ iff } i < m_\alpha.$$

For all $\alpha < \kappa^+$ let $d_\alpha = c_\alpha^0 + \cdots + c_\alpha^{m_\alpha-1}$. Thus $F \subseteq \mathcal{S}d_\alpha$, and so $d_\alpha \in I \subseteq B$. Also, $c_\alpha^{m_\alpha}, \ldots, c_\alpha^{n_\alpha-1} \in B$. Let $C'_\alpha = \{u \cap B : u \in C_\alpha\}$ for each $\alpha < \kappa^+$. Clearly each C'_α is nonempty and closed in $\mathrm{Ult}B$. We claim that $\langle C'_\alpha : \alpha < \kappa^+ \rangle$ is left-separated in $\mathrm{Ult}B$. In fact, it is easy to check that

$$C'_\alpha \in \mathscr{V}^B(\mathcal{S}^B d_\alpha, \mathcal{S}^B c_\alpha^{m_\alpha}, \ldots, \mathcal{S}^B c_\alpha^{n_\alpha-1})$$
$$\text{and } \mathscr{V}^B(\mathcal{S}^B d_\alpha, \mathcal{S}^B c_\alpha^{m_\alpha}, \ldots, \mathcal{S}^B c_\alpha^{n_\alpha-1}) \cap \{C'_\beta : \beta < \kappa^+\} \subseteq \{C'_\beta : \alpha \leq \beta\},$$

as desired. $\square$

Concerning derived functions, we have the following obvious facts:

$$\mathrm{hd}A = \mathrm{hd}_{\mathrm{H}+}A = \mathrm{hd}_{\mathrm{S}+}A = \mathrm{hd}_{\mathrm{h}+}A = {}_{\mathrm{d}}\mathrm{hd}_{\mathrm{S}+}A;$$

and $\mathrm{hd}_{\mathrm{S}-}A = \mathrm{hd}_{\mathrm{h}-}A = \omega$.

Now we want to go into a result of Fedorchuk [75], which provides an example for several of the questions in Monk [90]: assuming $\diamondsuit$, there is a BA A with $\mathrm{hd}A = \omega$ and $\mathrm{Card}_{\mathrm{H}-}A = \omega_1$. This is a weakened form of his main theorem. We give two constructions: a fairly short one of Kunen [75] (quite different from that of Fedorchuk but done upon looking at that article and noticing some problems with the construction), and a longer one which follows Fedorchuk rather closely, except for using the ideas of Kunen at crucial places which were unclear in the Fedorchuk construction.

Special $\diamondsuit$-sequences. This material is from Kunen [75] (except for the name *special* and the proof of the lemma). If $f \in {}^{\omega_1}({}^{\omega_1}\omega_1)$, let $f\lceil\lceil\alpha = \langle f_\xi \restriction \alpha : \xi < \alpha \rangle$. A *special $\diamondsuit$-sequence* is a sequence $\langle f^\alpha : \alpha < \omega_1 \rangle$ such that each $f^\alpha \in {}^\alpha({}^\alpha\omega_1)$ and for all $f \in {}^{\omega_1}({}^{\omega_1}\omega_1)$ the set $\{\alpha < \omega_1 : f\lceil\lceil\alpha = f^\alpha\}$ is stationary.

Lemma 16.12. $\diamondsuit$ *implies that there is a special $\diamondsuit$-sequence.*

Proof. Let H be a one-one function from ω_1 onto $\omega_1 \times \omega_1$. If $x \in \omega_1 \times \omega_1$, we write $x = (x_0, x_1)$. For $f \in {}^{\omega_1}({}^{\omega_1}\omega_1)$ we define $\tilde{f} \in {}^{\omega_1}\omega_1$ by $\tilde{f}\alpha = f_{(H\alpha)_0}(H\alpha)_1$. Define $G : {}^{\omega_1}\omega_1 \to {}^{\omega_1}2$ by

$$G_h\beta = \begin{cases} 1 & \text{if } h(H\beta)_0 = (H\beta)_1, \\ 0 & \text{otherwise.} \end{cases}$$

Define $F : {}^{\omega_1}2 \to {}^{\omega_1}\omega_1$ by

$$(Fk)\beta = \begin{cases} \gamma & \text{if } kH^{-1}(\beta, \gamma) = 1 \text{ and } kH^{-1}(\beta, \delta) = 0 \text{ for all } \delta \neq \gamma, \\ 0 & \text{if there is no such } \gamma. \end{cases}$$

Let $\chi : \mathscr{P}\omega_1 \to {}^{\omega_1}2$ be the natural bijection. Let $C_0 = \{\alpha < \omega_1 : H[\alpha] = \alpha \times \alpha\}$. So, C_0 is club. Let $\langle A_\alpha : \alpha < \omega_1 \rangle$ be a $\diamondsuit$-sequence. For $\alpha \in C_0$ let $f^\alpha \in {}^\alpha({}^\alpha\omega_1)$ be defined by

$$f^\alpha_\beta \gamma = (F\chi_{A_\alpha})H^{-1}(\beta, \gamma).$$

Let $f^\alpha \in {}^\alpha({}^\alpha\omega_1)$ be arbitrary if $\alpha \notin C_0$.

Now suppose that $f \in {}^{\omega_1}({}^{\omega_1}\omega_1)$. Let $B = \chi^{-1}G_{\tilde{f}}$. Now

$$C_1 \overset{\text{def}}{=} \{\alpha : \forall \xi < \alpha(f_\xi \upharpoonright \alpha \in {}^\alpha\alpha)\}$$

is club. Hence $C_0 \cap C_1 \cap \{\alpha < \omega_1 : \alpha \cap B = A_\alpha\}$ is stationary. We claim that if α is in this set, then $f \upharpoonright \alpha = f^\alpha$; this will finish the proof. Suppose $\xi < \alpha$. We want to show that $f_\xi \upharpoonright \alpha = f^\alpha_\xi$. Let $\gamma < \alpha$. Then

$$\tilde{f}H^{-1}(\xi, \gamma) = f_\xi \gamma;$$
$$G_{\tilde{f}}H^{-1}(H^{-1}(\xi, \gamma), f_\xi \gamma)) = 1;$$
$$H^{-1}(H^{-1}(\xi, \gamma), f_\xi \gamma) \in \alpha \cap B;$$
$$H^{-1}(H^{-1}(\xi, \gamma), f_\xi \gamma) \in A_\alpha;$$
$$\chi_{A_\alpha}H^{-1}(H^{-1}(\xi, \gamma), f_\xi \gamma) = 1.$$

It is easily checked that if $\delta \neq f_\xi \gamma$ then $\chi_{A_\alpha}H^{-1}(H^{-1}(\xi, \gamma), \delta) = 0$. Therefore $(F\chi_{A_\alpha})H^{-1}(\xi, \gamma) = f_\xi \gamma$. It follows that $f^\alpha_\xi \gamma = f_\xi \gamma$. $\qquad\square$

Kunen's construction. We assume $\diamondsuit$; so CH is available also. Fix a special $\diamondsuit$-sequence $\langle f^\alpha : \alpha < \omega_1 \rangle$. For any space Y, a point $y \in Y$ is a *strong limit point* of $\mathscr{H} \subseteq \mathscr{P}Y$ if for all neighborhoods V of y there is an $H \in \mathscr{H}$ such that $y \notin H$ and $H \subseteq V$. For $\alpha \leq \beta \leq \omega_1$ define $\pi^\beta_\alpha : {}^\beta2 \to {}^\alpha2$ by $\pi^\beta_\alpha g = g \upharpoonright \alpha$; thus π^β_α is continuous. We claim

(1) there is an enumeration $\langle q_\alpha : \alpha < \omega_1 \rangle$ of $\bigcup_{\sigma < \omega_1} {}^\sigma2$ such that every element is repeated ω_1 times and $q_\alpha \in {}^{\sigma_\alpha}2$ with $\sigma_\alpha \leq \alpha$.

To prove (1), for each $\sigma < \omega_1$ let ${}^\sigma2 = \{h_{\sigma\xi} : \xi < \omega_1\}$ with each element repeated ω_1 times. Let H enumerate $\omega_1 \times \omega_1$ under its natural order $((\alpha, \beta) < (\gamma, \delta)$ iff $[\max(\alpha, \beta) < \max(\gamma, \delta)$ or $(\max(\alpha, \beta) = \max(\gamma, \delta)$ and $\alpha < \gamma)$ or $(\max(\alpha, \beta) = \max(\gamma, \delta)$ and $\alpha = \gamma$ and $\beta < \delta)])$. By induction on α one can show that $(H\alpha)_0, (H\alpha)_1 \leq \alpha$ for all $\alpha < \omega_1$. Let $q_\alpha = h_{(H\alpha)_0(H\alpha)_1}$ for all $\alpha < \omega_1$. Then (1) is clear.

Our space will be a closed subspace of ${}^{\omega_1}2$. By induction on $\alpha \leq \omega_1$ we will define $X_\alpha \subseteq {}^\alpha2$ and $p_\alpha \in X_\alpha$ so that:

(2) X_α is closed and nonempty.

(3) If $\alpha \leq \beta$ then $\pi_\alpha^\beta X_\beta = X_\alpha$.

(4) If $q_\alpha \in X_{\sigma_\alpha}$, then $p_\alpha \restriction \sigma_\alpha = q_\alpha$.

(5) $p_\alpha{}^\frown 0, p_\alpha{}^\frown 1 \in X_{\alpha+1}$.

(6) If $\alpha \leq \beta$, $\{f_\xi^\alpha : \xi < \alpha\} \subseteq X_\alpha$, and $h \in X_\alpha$ is an accumulation point of $\{f_\xi^\alpha : \xi < \alpha\}$, then every point k in $X_\beta \cap (\pi_\alpha^\beta)^{-1}[\{h\}]$ is a strong limit point of $\{X_\beta \cap (\pi_\alpha^\beta)^{-1}[\{f_\xi^\alpha\}] : \xi < \alpha\}$.

Before actually making this construction we check that (2)–(6) yield the desired properties of $X \overset{\text{def}}{=} X_{\omega_1}$.

X is hereditarily separable: If not, let $f = \langle f_\xi : \xi < \omega_1 \rangle$ be a left-separated sequence in X. Let

$$C = \{\alpha < \omega_1 : \text{for all clopen } N \subseteq X_\alpha$$
$$(N \cap \{f_\xi \restriction \alpha : \xi < \alpha\} = 0 \text{ iff } (\pi_\alpha^{\omega_1})^{-1}[N] \cap \{f_\xi : \xi < \omega_1\} = 0) \text{ and}$$
$$(|N \cap \{f_\xi \restriction \alpha : \xi < \alpha\}| = 1 \text{ iff } |(\pi_\alpha^{\omega_1})^{-1}[N] \cap \{f_\xi : \xi < \omega_1\}| = 1)\}.$$

Then C is club. Since the argument for this is more complicated than usual for club arguments, we sketch it. First note, obviously:

(7) $N \cap \{f_\beta \restriction \alpha : \beta < \alpha\} \neq 0$ implies that $(\pi_\alpha^{\omega_1})^{-1}[N] \cap \{f_\beta : \beta < \omega_1\} \neq 0$, if $\alpha < \omega_1$ and N is a clopen subset of X_α.

(8) $|N \cap \{f_\beta \restriction \alpha : \beta < \alpha\}| \geq 2$ implies that $|(\pi_\alpha^{\omega_1})^{-1}[N] \cap \{f_\beta : \beta < \omega_1\}| \geq 2$, if $\alpha < \omega_1$ and N is a clopen subset of X_α.

Now to prove that C is closed, suppose that $\alpha < \omega_1$ is a limit ordinal and $\alpha \cap C$ is unbounded in α. Suppose that N is clopen in X_α and $(\pi_\alpha^{\omega_1})^{-1}[N] \cap \{f_\xi : \xi < \omega_1\} \neq 0$. Say $\xi < \omega_1$ and $f_\xi \restriction \alpha \in N$. Write $N = U_g^\alpha \overset{\text{def}}{=} \{h \in X_\alpha : g \subseteq h\}$, where $g \in {}^F 2$ for some finite $F \subseteq \alpha$. Say $F \subseteq \beta < \alpha$, $\beta \in C$. Then $f_\xi \restriction \beta \in U_g^\beta$, i.e., $(\pi_\beta^{\omega_1})^{-1}[U_g^\beta] \cap \{f_\eta : \eta < \omega_1\} \neq 0$ so, since $\beta \in C$, we get $U_g^\beta \cap \{f_\eta \restriction \beta : \eta < \beta\} \neq 0$. Hence choose $\eta < \beta$ with $f_\eta \restriction \beta \in U_g^\beta$. Hence $f_\eta \restriction \alpha \in U_g^\alpha = N$, and $N \cap \{f_\eta \restriction \alpha : \eta < \alpha\} \neq 0$, as desired. The other part of C is treated similarly. This proves that C is closed.

To prove that C is unbounded, suppose that $\alpha_0 < \omega_1$. Choose α_1 such that $\alpha_0 < \alpha_1$ and for all clopen $N \subseteq X_{\alpha_0}$ we have

$$(\pi_{\alpha_0}^{\omega_1})^{-1}[N] \cap \{f_\xi : \xi < \omega_1\} \neq 0 \Rightarrow \exists \xi < \alpha_1 (f_\xi \restriction \alpha_0 \in N) \text{ and}$$
$$|(\pi_{\alpha_0}^{\omega_1})^{-1}[N] \cap \{f_\xi : \xi < \omega_1\}| \geq 2 \Rightarrow |N \cap \{f_\xi \restriction \alpha_0 : \xi < \alpha_1\}| \geq 2.$$

This is possible since there are only countably many clopen sets $N \subseteq X_{\alpha_0}$. Continuing in this fashion with $\alpha_2, \alpha_3, \ldots$, we see that $\alpha_\omega = \sup_{n<\omega} \alpha_n$ is the desired member of C.

Fix $\alpha \in C$ such that $f \restriction \restriction \alpha = f^\alpha$.

(9) $f_\eta \restriction \alpha \in \overline{\{f_\beta^\alpha : \beta < \alpha\}}$ for all $\eta < \omega_1$.

For, if $f_\eta \restriction \alpha \in N$ with N clopen, then $(\pi_\alpha^{\omega_1})^{-1}[N] \cap \{f_\nu : \nu < \omega_1\} \neq 0$ so, since $\alpha \in C$, $N \cap \{f_\beta \restriction \alpha : \beta < \alpha\} \neq 0$, as desired.

(10) For all $\eta < \omega_1$, if $f_\eta \restriction \alpha$ is an isolated point of $\{f_\beta \restriction \alpha : \beta < \alpha\}$, then $\eta < \alpha$.

For, let N be clopen such that $N \cap \{f_\beta \restriction \alpha : \beta < \alpha\} = \{f_\eta \restriction \alpha\}$. Thus since $\alpha \in C$ we get $|(\pi_\alpha^{\omega_1})^{-1}[N] \cap \{f_\beta : \beta < \omega_1\}| = 1$. Say $f_\eta \restriction \alpha = f_\gamma \restriction \alpha$ with $\gamma < \alpha$. Since f_η and f_γ are both in $(\pi_\alpha^{\omega_1})^{-1}[N] \cap \{f_\beta : \beta < \omega_1\}$, it follows that $\eta = \gamma$, as desired.

From (9) and (10) it follows that $f_\alpha \restriction \alpha$ is an accumulation point of $\{f_\beta \restriction \alpha : \beta < \alpha\} = \{f_\beta^\alpha : \beta < \alpha\}$. Hence by (6) f_α is a strong limit point of $\{X_{\omega_1} \cap (\pi_\alpha^{\omega_1})^{-1}[\{f_\xi^\alpha\}] : \xi < \alpha\}$. So f_α is a limit point of $\{f_\xi : \xi < \alpha\}$, which contradicts the left-separatedness.

Now we show that $\chi_{H-}A \geq \omega_1$, where $A = \mathrm{Clop}X_{\omega_1}$. By Corollary 14.5 it suffices to show that X_{ω_1} has no one-one convergent sequences. So, assume that $\lim_{n\to\infty} g_n = h$ with all g_n distinct and different from h. Choose $f \in {}^{\omega_1}({}^{\omega_1}2)$ so that $\{f_\alpha : \alpha < \omega_1\} = \{g_n : n \in \omega\}$. Choose α so that $f\restriction\!\restriction\alpha = f^\alpha$, all of the functions $g_n \restriction \alpha$, $h \restriction \alpha$ are distinct, and $\{f_\beta : \beta < \alpha\} = \{g_n : n \in \omega\}$. Then $h \restriction \alpha$ is an accumulation point of $\{f_\xi^\alpha : \xi < \alpha\}$. Say $h \restriction \alpha = q_\beta$ with $\alpha \leq \beta$. Then $p_\beta{}^\frown 0$ and $p_\beta{}^\frown 1$ are both in $X_{\beta+1}$ and extend $h \restriction \alpha$. This gives by (6) two distinct points h, l which are strong limit points of $X_{\omega_1} \cap \{(\pi_\alpha^{\omega_1})^{-1}[\{f_\xi^\alpha\}] : \xi < \alpha\}$. Thus both are limit points of $\{g_n : n \in \omega\}$, contradiction.

Next we do the construction to yield (2)–(6). As soon as a space X_α is constructed, fix $p_\alpha \in X_\alpha$ such that (4) holds. Let X_0 be the one-point space. For δ limit, let $X_\delta = \{g \in {}^\delta 2 : \forall \alpha < \delta (g \restriction \alpha \in X_\alpha)\}$. It is straightforward to check (2)–(6) then. Now we do the crucial step from X_δ to $X_{\delta+1}$. We now define a nested clopen basis $\langle K_n : n \in \omega \rangle$ of p_δ. First let $\langle K_n' : n \in \omega \rangle$ be any such basis, with $K_0' = X_\delta$. If there is no $\alpha \leq \delta$ such that $\{f_\xi^\alpha : \xi < \alpha\} \subseteq X_\alpha$ and $p_\delta \restriction \alpha$ is an accumulation point of $\{f_\xi^\alpha : \xi < \alpha\}$, let $K_n = K_n'$ for all $n \in \omega$. Otherwise, let $\{\alpha_n : n \in \omega\}$ enumerate all $\alpha \leq \delta$ such that $\{f_\xi^\alpha : \xi < \alpha\} \subseteq X_\alpha$ and $p_\delta \restriction \alpha$ is an accumulation point of $\{f_\xi^\alpha : \xi < \alpha\}$, each one enumerated infinitely many times by both even and odd integers. Now define K_n by induction as follows. $K_0 = X_\delta$. If K_n has been defined, by (6) for $\beta = \delta$, p_δ is a strong limit point of $\{X_\delta \cap (\pi_{\alpha_n}^\delta)^{-1}[\{f_\xi^{\alpha_n}\}] : \xi < \alpha_n\}$, so there is a $\xi < \alpha_n$ such that $p_\delta \notin X_\delta \cap (\pi_{\alpha_n}^\delta)^{-1}[\{f_\xi^{\alpha_n}\}] \subseteq K_n$. We let $p_\delta \in K_{n+1} \subseteq K_n' \cap (K_n \backslash (\pi_{\alpha_n}^\delta)^{-1}[\{f_\xi^{\alpha_n}\}])$, K_{n+1} clopen. Thus the following condition holds:

(11) $\{K_n : n \in \omega\}$ is a nested clopen base for p_δ, and if $\alpha \leq \delta$ is such that $\{f_\xi^\alpha : \xi < \alpha\} \subseteq X_\alpha$ and $p_\delta \restriction \alpha$ is an accumulation point of $\{f_\xi^\alpha : \xi < \alpha\}$, then there are infinitely many even and infinitely many odd n such that $\exists \xi (p_\delta \notin X_\delta \cap (\pi_\alpha^\delta)^{-1}[\{f_\xi^\alpha\}] \subseteq K_n \backslash K_{n+1})$.

Finally, we set

$$X_{\delta+1} = \{g \in {}^{\delta+1}2 : g \restriction \delta = p_\delta\} \cup$$
$$\bigcup_{n \text{ even}} \{g \in {}^{\delta+1}2 : g \restriction \delta \in K_n \backslash K_{n+1}, \ g\delta = 0\} \cup$$
$$\bigcup_{n \text{ odd}} \{g \in {}^{\delta+1}2 : g \restriction \delta \in K_n \backslash K_{n+1}, \ g\delta = 1\}.$$

It remains only to check (2)–(6) for $\delta + 1$. All except (6) are easy, and (6) is obvious if $\alpha = \delta + 1$. So assume that $\alpha \leq \delta$, $h \in X_\alpha$ is an accumulation point of $\{f_\xi^\alpha : \xi < \alpha\}$, $k \in X_{\delta+1} \cap (\pi_\alpha^{\delta+1})^{-1}[\{h\}]$. To show that k is a strong limit point of $\{X_{\delta+1} \cap (\pi_\alpha^{\delta+1})^{-1}[\{f_\xi^\alpha\}] : \xi < \alpha\}$, let $k \in U_g^{\delta+1}$ with $g \in {}^F2$, $F \subseteq \delta + 1$, F finite. Without loss of generality, $\delta \in F$.

Case 1. $k \restriction \delta \neq p_\delta$. Say $k \restriction \delta \in K_n \backslash K_{n+1}$ with n even. Thus $k\delta = 0 = g\delta$. Thus $k \restriction \delta \in U_{g \restriction \delta}^\delta \cap (K_n \backslash K_{n+1})$, so by (6) for δ choose $\xi < \alpha$ such that

$$k \restriction \delta \notin (\pi_\alpha^\delta)^{-1}[\{f_\xi^\alpha\}] \subseteq U_{g \restriction \delta}^\delta \cap (K_n \backslash K_{n+1}).$$

It follows that $k \notin (\pi_\alpha^{\delta+1})^{-1}[\{f_\xi^\alpha\}] \subseteq U_g^{\delta+1}$, since if $l \in (\pi_\alpha^{\delta+1})^{-1}[\{f_\xi^\alpha\}]$ then $l \restriction \delta \in K_n \backslash K_{n+1}$ and n is even, so $l\delta = 0 = g\delta$.

Case 2. $k \restriction \delta = p_\delta$. Say $g\delta = 0$. Choose m even such that $K_m \subseteq U_{g \restriction \delta}^\delta$ and $p_\delta \notin X_\delta \cap (\pi_\alpha^\delta)^{-1}[\{f_\xi^\alpha\}] \subseteq K_m \backslash K_{m+1}$ for some $\xi < \alpha$. Then $k \notin X_{\delta+1} \cap (\pi_\alpha^{\delta+1})^{-1}[\{f_\xi^\alpha\}] \subseteq U_g^{\delta+1}$, as desired.

This completes Kunen's construction.

Fedorchuk's construction, as modified here, uses the special $\diamondsuit$-sequence introduced by Kunen, and also a general expansion construction, to which we now turn.

An expansion construction. No special set-theoretical assumptions are needed in this construction. Let X be a space, and suppose that we have associated with every $x \in X$ another space Y_x and a continuous function $f_x : X \backslash \{x\} \to Y_x$. We also assume that the spaces Y_x are pairwise disjoint. Then we set $Z = \bigcup_{x \in X} Y_x$, and we let π be the natural mapping from Z onto X: πz is the unique x such that $z \in Y_x$. We claim that the collection of all subsets of the following form constitutes a base for a topology on Z:

(1) $W \cup \pi^{-1}[U \cap f_x^{-1}[W]]$ with $x \in X$, W open in Y_x, U an open neighborhood of x in X.

To show that this collection forms a base, note first that Z can be written in the given form: $Z = Y_x \cup \pi^{-1}[X \cap f_x^{-1}[Y_x]]$ for any $x \in X$. Now suppose that we have two sets of the form (1), V and V'. Say V is exactly as in (1), and V' is similar with primes on everything. We want to show that $V \cap V'$ is a union of elements of the form (1).

Case 1. $x \in U' \cap f_{x'}^{-1}[W']$ and $x' \in U \cap f_x^{-1}[W]$. Then

$$V \cap V' = W \cup \pi^{-1}[U \cap U' \cap f_{x'}^{-1}[W'] \cap f_x^{-1}[W]]$$
$$\cup W' \cup \pi^{-1}[U \cap U' \cap f_x^{-1}[W] \cap f_{x'}^{-1}[W']].$$

Case 2. $x \in U' \cap f_{x'}^{-1}[W']$ and $x' \notin U \cap f_x^{-1}[W]$. Then

$$V \cap V' = W \cup \pi^{-1}[U \cap U' \cap f_{x'}^{-1}[W'] \cap f_x^{-1}[W]].$$

Case 3. $x \notin U' \cap f_{x'}^{-1}[W']$ and $x' \in U \cap f_x^{-1}[W]$. Similarly.
Case 4. $x \notin U' \cap f_{x'}^{-1}[W']$, $x' \notin U \cap f_x^{-1}[W]$, $x = x'$. Then

$$V \cap V' = (W \cap W') \cup \pi^{-1}[U \cap U' \cap f_x^{-1}[W \cap W']].$$

Case 5. $x \notin U' \cap f_{x'}^{-1}[W']$, $x' \notin U \cap f_x^{-1}[W]$, $x \neq x'$. Then

$$V \cap V' = \pi^{-1}[U \cap U' \cap f_x^{-1}[W] \cap f_{x'}^{-1}[W']]$$
$$= \bigcup\{Y_{x''} \cup \pi^{-1}[U \cap U' \cap f_x^{-1}[W] \cap f_{x'}^{-1}[W'] \cap f_{x''}^{-1}[Y_{x''}]] :$$
$$x'' \in \pi^{-1}[U \cap U' \cap f_x^{-1}[W] \cap f_{x'}^{-1}[W']]\}.$$

Our main aim in the next portion of the text is to show that if X and all of the spaces Y_x are Boolean, then so is Z. We do this step by step.

(2) If X and all spaces Y_x are Hausdorff, then so is Z.

In fact, let u, v be distinct members of Z. We want to find disjoint neighborhoods of them. Say $u \in Y_x$ and $v \in Y_{x'}$. If $x = x'$, let W and W' be disjoint neighborhoods of u and v respectively in Y_x. Then desired disjoint neighborhoods in Z are $W \cup \pi^{-1}[X \cap f_x^{-1}[W]]$ and $W' \cup \pi^{-1}[X \cap f_x^{-1}[W']]$. Suppose that $x \neq x'$. Let U and V be disjoint neighborhoods of x and x' respectively. Then desired disjoint neighborhoods in Z are $Y_x \cup \pi^{-1}[U \cap f_x^{-1}[Y_x]]$ and $Y_{x'} \cup \pi^{-1}[V \cap f_{x'}^{-1}[Y_{x'}]]$.

(3) If X and all spaces Y_x are compact Hausdorff, then so is Z.

To show this, let $\mathcal{O}$ be a cover of Z by basic open sets. For each $V \in \mathcal{O}$ let

$$V = W_V \cup \pi^{-1}[U_V \cap f_{x_V}^{-1}[W_V]],$$

where W_V is open in Y_{x_V} and U_V is an open neighborhood of x_V. Let $C = \{x \in X :$ for all $V \in \mathcal{O}$, $x \notin U_V \cap f_{x_V}^{-1}[W_V]\}$. If $x \in C$, then $\{W_V : x_V = x, V \in \mathcal{O}\}$ covers Y_x; let $\mathcal{O}_x$ be a finite subcover. Thus

$$\{U_V \cap f_{x_V}^{-1}[W_V] : V \in \mathcal{O}\} \cup \{\bigcap_{V \in \mathcal{O}_x} U_V : x \in C\}$$

covers X. Let $\mathscr{O}' \in [\mathscr{O}]^{<\omega}$ and $C' \in [C]^{<\omega}$ be such that

$$\{U_V \cap f_{xV}^{-1}[W_V] : V \in \mathscr{O}'\} \cup \{\bigcap_{V \in \mathscr{O}_x} U_V : x \in C'\}$$

covers X. We claim that $\mathscr{O}' \cup \bigcup_{x \in C'} \mathscr{O}_x$ covers Z (as desired). For, let $z \in Z$; say $z \in Y_x$. If $x \in U_V \cap f^{-1}[W_V]$ for some $V \in \mathscr{O}'$, then $z \in V$, as desired. Otherwise, choose $y \in C'$ such that $x \in \bigcap_{V \in \mathscr{O}_y} U_V$. Case 1. $x = y$. Then $z \in W_V \subseteq V$ for some $V \in \mathscr{O}y$, as desired. Case 2. $x \neq y$. Now $f_y x \in Y_y$, so choose $V \in \mathscr{O}_y$ such that $f_y x \in W_V$. Then $x \in U_V \cap f^{-1}[W_V]$, so $z \in V$, as desired.

(4) If X and all spaces Y_x are Boolean, then so is Z.

We first note that if W is clopen in Y_x and U is a clopen neighborhood of x, then $V \stackrel{\text{def}}{=} W \cup \pi^{-1}[U \cap f_x^{-1}[W]]$ is clopen in Z:

$$Z \backslash V = (Y_x \backslash W) \cup \pi^{-1}[U \cap f_x^{-1}[Y_x \backslash W]]$$
$$\cup \bigcup_{z \in X \backslash U} (Y_z \cup \pi^{-1}[(X \backslash U) \cap f_z^{-1}[Y_z]]).$$

We also note that if U is open in X, then $\pi^{-1}[U]$ is open in Z: if $x \in U$, then

$$\pi^{-1}[U] = Y_x \cup \pi^{-1}[U \cap f_x^{-1}[Y_x]].$$

Now suppose that $y \in V = W \cup \pi^{-1}[U \cap f_x^{-1}[W]]$ with assumptions as in (1); we want to find a clopen V' such that $y \in V' \subseteq V$.

Case 1. $y \in W$. Choose W' clopen so that $y \in W' \subseteq W$, and choose U' clopen so that $x \in U' \subseteq U$. Then $V' = W' \cup \pi^{-1}[U' \cap f_x^{-1}[W']]$ is as desired.

Case 2. $y \notin W$. Thus $\pi y \in U \cap f_x^{-1}[W]$. Let U' be clopen such that $\pi y \in U' \subseteq U \cap f_x^{-1}[W]$. Then $V' = \pi^{-1}[U']$ is as desired.

Note the following fact which was established in the course of proving (4) (true without special assumptions on the space):

(5) π is continuous.

Lemma 16.13. *Let X be a first-countable Boolean space and for each $x \in X$ let Y_x be homeomorphic to the Cantor set, the Y_x's pairwise disjoint. Let $\langle C_i^x : i < \omega \rangle$ be a system of sets such that*
  (i) $C_i^x \subseteq X \backslash \{x\}$;
  (ii) $x \in \overline{C_i^x}$.
Then there exist continuous functions $f_x : X \backslash \{x\} \to Y_x$ for $x \in X$ such that if Z is obtained from X and the functions f_x by the expansion construction, and if π is the natural mapping from Z onto X, then:
  (iii) If $D \subseteq Z$ and $\pi[D] = C_i^x$, then $Y_x \subseteq \overline{D}$.
  (iv) if W is a non-empty open set in Y_x and $i < \omega$, then $x \in \overline{C_i^x \cap f_x^{-1}[W]}$.
  (v) Z is first-countable.

Proof. Fix $x \in X$. For each $i < \omega$ let $\langle a_j^i : j < \omega \rangle$ be a sequence of distinct members of C_i^x converging to x.

(1) There are infinite subsets $A_i \subseteq \{a_j^i : j < \omega\}$ for $i < \omega$ such that $A_i \cap A_k = 0$ for distinct $i, k < \omega$.

For, we define $j(i, k) < \omega$ for $i \leq k < \omega$ by induction on k, and within that, by induction on i:

$$j(0,0) = 0;$$
$$j(i, k + 1) = \text{least } l < \omega \text{ such that } a_l^i \in \{a_j^i : j < \omega\} \backslash$$
$$(\{a_{j(s,m)}^s : s \leq m \leq k\} \cup \{a_{j(s,k+1)}^s : s < i\}).$$

Then let $A_i = \{a_{j(i,k)}^i : i \leq k < \omega\}$. Clearly A_i is an infinite subset of $\{a_j^i : j < \omega\}$. Suppose that $i \neq i'$ and $u \in A_i \cap A_{i'}$. Write $u = a_{j(i,k)}^i = a_{j(i',k')}^{i'}$. Without loss of generality, $k' \leq k$. *Case 1.* $k = 0$. Then $i = i' = 0$, contradiction. *Case 2.* $k' < k$. Then $a_{j(i,k)}^i = a_{j(i',k')}^{i'}$ contradicts the definition, for the first omitted set. *Case 3.* $0 < k = k'$. Say $i' < i$. Then $a_{j(i,k)}^i = a_{j(i',k)}^{i'}$ contradicts the definition, for the second omitted set. So (1) holds.

Now decompose each A_i into infinite subsets: $A_i = \bigcup_{j < \omega} B_j^i$, $B_j^i \cap B_k^i = 0$ for $j \neq k$. Let $\langle V_m : m < \omega \rangle$ be a decreasing sequence of open sets forming a neighborhood base for x. Then

(2) $\forall i \forall m \exists k \forall j \geq k (B_j^i \subseteq V_m)$.

For, suppose that (2) fails. So we get i, m so that for all k there is a $j \geq k$ such that $B_j^i \not\subseteq V_m$. Thus we can find an increasing sequence $k_0, k_1, \ldots$ of integers and elements $d_n \in B_{k_n}^i \backslash V_m$. So $\langle d_n : n < \omega \rangle$ is a sequence of distinct elements of A_i all outside V_m, which contradicts the fact that $x \in \overline{A_i}$. Hence (2) holds.

By (2), for each i and m choose $k(m, i)$ so that $B_j^i \subseteq V_m$ for all $j \geq k(m, i)$. Without loss of generality we may assume that $k(0, i) < k(1, i) < \cdots$. Set

$$D_m = B_{k(m,0)}^0 \cup \ldots \cup B_{k(m,m)}^m.$$

Note that $D_m \cap D_n = 0$ for $m \neq n$, $x \in \overline{D_m}$, D_m is closed in $X \backslash \{x\}$, and $D_m \subseteq V_m$.

(3) $\bigcup_{j \neq i} D_j$ is closed in $X \backslash \{x\}$ for every $i < \omega$.

In fact suppose that $z \in \overline{\bigcup_{j \neq i} D_j} \backslash \bigcup_{j \neq i} D_j$. There is then a sequence $\langle d_j : j < \omega \rangle$ of distinct elements of $\bigcup_{j \neq i} D_j$ converging to z. Say $d_j \in D_{k_j}$ for all $j < \omega$. Since $z \notin \bigcup_{j \neq i} D_j$, we may assume that the k_j's are all distinct, and in fact that $\langle k_j : j < \omega \rangle$ forms an increasing sequence of integers. Let U, an open neighborhood of z, and $m < \omega$ be such that $U \cap V_m = 0$. Now for all $t \geq m$ we have $k_t \geq t \geq m$, and so $D_{k_t} \subseteq V_m$, which is impossible. So, (3) holds.

Let $\Gamma_i = (X \backslash \{x\}) \backslash \bigcup_{j \neq i} D_j$. So Γ_i is open in $X \backslash \{x\}$, hence in X itself, $\Gamma_i \cap D_j = 0$ for $i \neq j$, and $D_i \subseteq \Gamma_i$. Write

$$\Gamma_i = \Delta_0^i \cup \Delta_1^i \cup \ldots$$

with each Δ_j^i clopen. Define $(i, j) <' (m, n)$ iff (1) $\max(i, j) < \max(m, n)$, or (2) $\max(i, j) = \max(m, n)$ and $i < m$, or (3) $\max(i, j) = \max(m, n)$ and $i = m$ and $j < n$. Then let

$$E_n^m = \Delta_n^m \backslash \bigcup \{\Delta_j^i : (i, j) <' (m, n), \ i \neq m\}$$

Note that if $i \neq m$ then $D_m \cap \Delta_j^i \subseteq D_m \cap \Gamma_i = 0$. Thus $U_m \overset{\text{def}}{=} \bigcup_{n < \omega} E_n^m \supseteq D_m$. Also, $E_n^m \cap E_q^p = 0$ for $m \neq p$: for if say $(p, q) <' (m, n)$, then

$$E_n^m \cap E_q^p \subseteq E_n^m \cap \Delta_q^p = 0.$$

It follows that $U_m \cap U_p = 0$ for $m \neq p$. Furthermore, $\bigcup_{m < \omega} U_m = X \backslash \{x\}$. For, let $y \in X \backslash \{x\}$. Then $y \in \bigcup_{i < \omega} \Gamma_i$, so there exist i, j such that $y \in \Delta_j^i$. Choose (m, n) minimum under $<'$ such that $y \in \Delta_n^m$. Then $y \in E_n^m \subseteq U_m$.

Let $\langle z_i : i < \omega \rangle$ be a sequence without repetitions of members of Y_x such that $\{z_i : i < \omega\}$ is dense in Y_x. Define $f_x : X \backslash \{x\} \to Y_x$ by setting $f[U_m] = \{z_m\}$ for all $m < \omega$. Clearly f_x is continuous.

To prove (iii), assume that $x \in X$, $i \in \omega$, $D \subseteq Z$, and $\pi[D] = C_i^x$. We want to show that $Y_x \subseteq \overline{D}$. To this end, assume that $y \in Y_x$ and V is a neighborhood of y. Without loss of generality we may assume that

$$V = W \cup \pi^{-1}[U \cap f_x^{-1}[W]],$$

with obvious assumptions. Choose $z_m \in W$ with $i \leq m$. So $B_{k(m,i)}^i \subseteq D_m \subseteq U_m \subseteq f^{-1}[W]$. Now $x \in \overline{B_{k(m,i)}^i}$, so choose $u \in U \cap B_{k(m,i)}^i \subseteq U \cap f_x^{-1}[W]$. Now $B_{k(m,i)}^i \subseteq C_i^x$, so $u \in C_i^x$. Choose $v \in D$ with $\pi v = u$. Thus $v \in V \cap D$, as desired.

For (iv), pick m such that $z_m \in W$ and $m \geq i$. Then

$$B_{k(m,i)}^i \subseteq D_m \subseteq U_m = f_x^{-1}[\{z_m\}] \subseteq f_x^{-1}[W],$$

and $B_{k(m,i)}^i \subseteq A_i \subseteq C_i^x$, so $x \in \overline{B_{k(m,i)}^i} \subseteq \overline{C_i^x \cap f_x^{-1}[W]}$, as desired.

For (v), let $z \in Z$; say $z \in Y_x$. Let $\langle U_n^x : n < \omega \rangle$ be a nested open neighborhood base for x in X, and let $\langle W_n^z : n < \omega \rangle$ be one for z in Y_x. Define

$$S_n = W_n^z \cup \pi^{-1}[U_n^x \cap f_x^{-1}[W_n^z]]$$

for all $n < \omega$. We claim that this gives a neighborhood base for z; the simple proof will be omitted. $\square$

Fedorchuk's construction. Assume $\Diamond$, and hence CH. We start with some definitions:

- $\mathfrak{F}$ is the free BA on ω free generators.
- for each $\alpha < \omega_1$, $\langle Q_\xi : \xi \in {}^\alpha\omega_1 \rangle$ is a system of pairwise disjoint spaces each homeomorphic to Ult $\mathfrak{F}$, and p_ξ is a homeomorphism from Ult $\mathfrak{F}$ onto Q_ξ.
- s is a one-one function from ω_1 onto Ult $\mathfrak{F}$.
- For each limit ordinal $\alpha \leq \omega_1$, $\langle u_\xi : \xi \in {}^\alpha\omega_1 \rangle$ is a system of distinct objects not in any of the spaces Q_ξ.
- For each non-zero $\alpha \leq \omega_1$ and each $\xi \in {}^\alpha\omega_1$ we set

$$x_\xi = \begin{cases} p_{\xi\restriction\beta}s\xi\beta & \text{if } \alpha = \beta + 1, \\ u_\xi & \text{if } \alpha \text{ is a limit ordinal.} \end{cases}$$

- For any non-zero $\alpha \leq \omega_1$ we let $X_\alpha = \{x_\xi : \xi \in {}^\alpha\omega_1\}$.
- For $0 < \alpha \leq \beta \leq \omega_1$ we define $\pi_\alpha^\beta : X_\beta \to X_\alpha$ by $\pi_\alpha^\beta x_\xi = x_{\xi\restriction\alpha}$ for any $\xi \in {}^\beta\omega_1$.

Now we begin the main part of the construction. For $0 < \alpha \leq \omega_1$ we shall construct A_α, ψ_α, and a topology on X_α so that the following conditions hold:

(1) A_α is a BA;
(2) if $\beta < \alpha$, then A_β is a subalgebra of A_α;
(3) if α is a limit ordinal, then $A_\alpha = \bigcup_{\beta<\alpha} A_\beta$.
(4) ψ_α is a homeomorphism from UltA_α onto X_α.
(5) if $\beta < \alpha$, then π_β^α is a continuous function from X_α onto X_β.
(6) if $\beta < \alpha \leq \omega_1$, then the following diagram commutes:

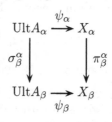

Here σ_β^α is the natural continuous mapping which is the dual of the inclusion of A_β in A_α. (Actually, $\sigma_\beta^\alpha F = F \cap A_\beta$ for any $F \in $ Ult A_α.)

(7) X_α is first-countable.

We define $A_1 = \mathfrak{F}$. Note that $X_1 = Q_0$; we take the natural topology on X_1. Let $\psi_1 = p_0$. Clearly (1)–(7) hold.

Having defined A_β and a topology on X_β, we now define $A_{\beta+1}$ and a topology on $X_{\beta+1}$. We defer until later the construction of functions f_ξ for $\xi \in {}^\beta\omega_1$; we will make this construction so that $f_\xi : X_\beta \backslash \{x_\xi\} \to Q_\xi$ is continuous. So, we can put the topology on $X_{\beta+1}$ determined by all of these functions f_ξ by the method described previously. Therefore $X_{\beta+1}$ is a Boolean space; since the natural mapping $\pi_\beta^{\beta+1}$ from $X_{\beta+1}$ onto X_β is a continuous function from $X_{\beta+1}$ onto X_β, which is homeomorphic via ψ_β to UltA_β, we hence get an extension $A_{\beta+1}$ and a homeomorphism $\psi_{\beta+1}$ so that the conditions (1)–(7) continue to hold.

Now for the limit step, assume that α is a limit ordinal and the construction has been done for all $\beta < \alpha$. We define A_α by (3). Then

(8) If $F \in \text{Ult}A_\alpha$, then there is a unique $\xi \in {}^\alpha\omega_1$ such that $\psi_\beta(F \cap A_\beta) = x_{\xi\restriction\beta}$ for all $\beta < \alpha$.

For, suppose that $\beta < \alpha$. Thus $\psi_\beta(F \cap A_\beta) \in X_\beta$, so there is a unique $\eta_\beta \in {}^\beta\omega_1$ such that $\psi_\beta(F \cap A_\beta) = x_{\eta_\beta}$. We claim that if $\beta < \gamma < \alpha$ then $\eta_\beta \subseteq \eta_\gamma$. In fact,

$$\psi_\beta(F \cap A_\beta) = \psi_\beta \sigma_\beta^\gamma(F \cap A_\gamma)$$
$$= \pi_\beta^\gamma \psi_\gamma(F \cap A_\gamma)$$
$$= \pi_\beta^\gamma x_{\eta_\gamma}$$
$$= x_{\eta_\gamma \restriction \beta},$$

so the claim follows. Hence (8) holds.

For each $F \in \text{Ult}A_\alpha$ let $\psi_\alpha F = x_\xi$, with ξ as in (10). We claim that ψ_α is one-one. For, if $F \neq G$, say $F \cap A_\beta \neq G \cap A_\beta$ for some $\beta < \alpha$. Then with $\psi_\alpha F = x_\xi$ and $\psi_\alpha G = x_\eta$ we have $x_{\xi\restriction\beta} = \psi_\beta(F \cap A_\beta) \neq \psi_\beta(G \cap A_\beta) = x_{\eta\restriction\beta}$, so $\xi \restriction \beta \neq \eta \restriction \beta$ and $\xi \neq \eta$.

Also, ψ_α is onto. For, let $\xi \in {}^\alpha\omega_1$. For all $\beta < \alpha$ let $F_\beta = \psi_\beta^{-1} x_{\xi\restriction\beta}$. Then $\beta < \gamma < \alpha$ implies that $F_\beta \subseteq F_\gamma$, since

$$\sigma_\beta^\gamma F_\gamma = \sigma_\beta^\gamma \psi_\gamma^{-1} x_{\xi\restriction\gamma}$$
$$= \psi_\beta^{-1} \pi_\beta^\gamma x_{\xi\restriction\gamma}$$
$$= \psi_\beta^{-1} x_{\xi\restriction\beta}$$
$$= F_\beta.$$

Let $G = \bigcup_{\beta<\alpha} F_\beta$. Clearly $\psi_\alpha G = x_\xi$, as desired.

Put a topology on X_α so that ψ_α is a homeomorphism. Then if $\beta < \alpha$ and $F \in \text{Ult}A_\alpha$ we have (with $\psi_\alpha F = x_\xi$)

$$\psi_\beta \sigma_\beta^\alpha F = \psi_\beta(F \cap A_\beta) = x_{\xi\restriction\beta}$$
$$= \pi_\beta^\alpha x_\xi = \pi_\beta^\alpha \psi_\alpha F.$$

Thus (6) holds. For (5), let $\beta < \alpha$ and $a \in A_\beta$; we show that $(\pi_\beta^\alpha)^{-1}[\psi_\beta[\mathscr{S}a]]$ is open in X_α. In fact, for any $F \in \text{Ult}A_\alpha$,

$$F \in \psi_\alpha^{-1}[(\pi_\beta^\alpha)^{-1}[\psi_\beta[\mathscr{S}a]]] \text{ iff } \pi_\beta^\alpha \psi_\alpha F \in \psi_\beta[\mathscr{S}a]$$
$$\text{iff } \psi_\beta \sigma_\beta^\alpha F \in \psi_\beta[\mathscr{S}a]$$
$$\text{iff } \sigma_\beta^\alpha F \in \mathscr{S}a$$
$$\text{iff } F \cap A_\beta \in \mathscr{S}a$$
$$\text{iff } a \in F,$$

so $(\pi_\beta^\alpha)^{-1}[\psi_\beta[\mathscr{S}a]] = \psi_\alpha[\mathscr{S}a]$, as desired.

(7) holds by considering duality: each ultrafilter on A_α clearly has a countable set of generators.

This finishes the construction, except for the crucial definition of the functions f_ξ, to which we now turn. So, suppose that $\xi \in {}^\beta\omega_1$, where A_β, ψ_β, and a topology on X_β have been defined so that (1)–(7) hold. We shall apply the lemma to construct f_ξ; thus we want to define a countable subset $\mathscr{C}_\xi$ of $\mathscr{P}(X_\beta \backslash \{x_\xi\})$ such that $x_\xi \in \overline{C}$ for each $C \in \mathscr{C}_\xi$. We will define $\mathscr{C}_\xi$ as $\mathscr{C}_\xi^0 \cup \mathscr{C}_\xi^1 \cup \mathscr{C}_\xi^2$.

Defining $\mathscr{C}_\xi^0$. First suppose that β is limit. For each $\alpha < \beta$ choose $\eta_\alpha \in {}^\beta\omega_1$ such that $\eta_\alpha \neq \xi$ but $\eta_\alpha \upharpoonright \alpha = \xi \upharpoonright \alpha$. Let $\mathscr{C}_\xi^0 = \{\{x_{\eta_\alpha} : \gamma < \alpha < \beta\} : \gamma < \beta\}$. We note that $x_\xi \in \overline{D}$ for each $D \in \mathscr{C}_\xi^0$. In fact, suppose $\psi_\beta^{-1} x_\xi \in \mathscr{S}a$. Say $a \in A_\alpha$, where $\alpha < \beta$. We claim that $x_{\eta_\gamma} \in \mathscr{S}a$ for all $\gamma > \alpha$. For,

$$\psi_\beta^{-1} x_\xi \cap A_\alpha = \sigma_\alpha^\beta \psi_\beta^{-1} x_\xi$$
$$= \psi_\alpha^{-1} \pi_\alpha^\beta x_\xi$$
$$= \psi_\alpha^{-1} x_{\xi \upharpoonright \alpha}$$
$$= \psi_\alpha^{-1} \pi_\alpha^\beta x_{\eta_\gamma}$$
$$= \sigma_\alpha^\beta \psi_\beta^{-1} x_{\eta_\gamma}$$
$$= \psi_\beta^{-1} x_{\eta_\gamma} \cap A_\alpha,$$

which proves the claim.

Second, assume that $\beta = \gamma + 1$ for some γ. Then $x_\xi \in Q_{\xi \upharpoonright \gamma}$, and we let Y be the range of a sequence of distinct elements of $Q_{\xi \upharpoonright \gamma} \backslash \{x_\xi\}$ which converges to x_ξ, and $\mathscr{C}_\xi^0 = \{Y\}$.

Note the following property of $\mathscr{C}_\xi^0$, true whether β is limit or not:

(9) For all $\alpha < \beta$ there is a $Y \in \mathscr{C}_\xi^0$ such that for all $x_\eta \in Y$ we have $\xi \upharpoonright \alpha \subseteq \eta$.

Defining $\mathscr{C}_\xi^1$. Let $\langle f^\alpha : \alpha < \omega_1 \rangle$ be a special $\Diamond$-sequence. We define

$$\mathscr{C}_\xi^1 = \begin{cases} \{x_{f_\gamma^\beta} : \gamma < \beta\} \backslash \{x_\xi\} & \text{if } x_\xi \text{ is an accumulation point of } \{x_{f_\gamma^\beta} : \gamma < \beta\}, \\ 0 & \text{otherwise.} \end{cases}$$

Defining $\mathscr{C}_\xi^2$. Let $g : \omega_1 \to X_\beta$ be a bijection and let $\langle J_\alpha : \alpha < \omega_1 \rangle$ be a $\Diamond$-sequence. We set

$$\mathscr{C}_\xi^2 = \{g[J_\gamma] : \gamma \leq g^{-1} x_\xi, x_\xi \in \overline{g[J_\gamma]}\}.$$

Now we prove the essential properties of the construction.

(10) Suppose that $\beta < \alpha \leq \omega_1$, β limit, $u_\eta \in X_\beta$, $E \subseteq X_{\omega_1}$, $Q_\eta \subseteq \overline{\pi_{\beta+1}^{\omega_1}[E]}$, $x_\xi \in X_\alpha$, and $\pi_\beta^\alpha x_\xi = u_\eta$. Then $x_\xi \in \overline{\pi_\alpha^{\omega_1}[E]}$.

We prove (10) by induction on α. The case $\alpha = \beta + 1$ is given. Now assume (10) for α, where $\beta + 1 \leq \alpha$. Assume that $x_\xi \in X_{\alpha+1}$ and $\pi_\beta^{\alpha+1} x_\xi = u_\eta$; we want to show that $x_\xi \in \overline{\pi_{\alpha+1}^{\omega_1}[E]}$. Suppose that $x_\xi \in V$ where V is open; without loss of generality say

$$V = W \cup (\pi_\alpha^{\alpha+1})^{-1}[U \cap f_{\xi\restriction\alpha}^{-1}[W]],$$

where W is a non-empty open subset of $Q_{\xi\restriction\alpha}$ and U is an open neighborhood of $x_{\xi\restriction\alpha}$. By (9), choose $Y \in \mathscr{C}_{\xi\restriction\alpha}^0$ so that for all $x_\rho \in Y$ we have $\xi \restriction \beta \subseteq \rho$. Now $x_{\xi\restriction\alpha} \in \overline{Y}$, so by the Lemma (iv) we get $x_{\xi\restriction\alpha} \in \overline{Y \cap f_{\xi\restriction\alpha}^{-1}[W]}$; hence choose $x_\rho \in U \cap Y \cap f_{\xi\restriction\alpha}^{-1}[W]$. Thus $\rho \in {}^\alpha\omega_1$, so by the induction hypothesis $x_\rho \in \overline{\pi_\alpha^{\omega_1}[E]}$. Say $x_\theta \in U \cap f_{\xi\restriction\alpha}^{-1}[W] \cap \pi_\alpha^{\omega_1}[E]$. Say $\pi_\alpha^{\omega_1} e = x_\theta$. Then $\pi_{\alpha+1}^{\omega_1} e \in V \cap \pi_{\alpha+1}^{\omega_1}[E]$, as desired.

Finally, suppose that γ is limit and (10) holds for each $\alpha < \gamma$. Suppose that $\pi_\beta^\gamma z = x_\eta$. Take any neighborhood V of z; without loss of generality we may assume that $V = \psi_\gamma[\mathscr{S}a]$, where $a \in A_\gamma$. Say $a \in A_\alpha$, where $\beta < \alpha < \gamma$. Thus $\psi_\gamma^{-1} z \in \mathscr{S}a$, so $\sigma_\alpha^\gamma \psi_\gamma^{-1} z \in \mathscr{S}a$, so $\psi_\alpha^{-1} \pi_\alpha^\gamma z \in \mathscr{S}a$, and finally $\pi_\alpha^\gamma z \in \psi_\alpha[\mathscr{S}a]$. Now by the induction hypothesis $\pi_\alpha^\gamma z \in \overline{\pi_\alpha^{\omega_1}[E]}$, so $\psi_\alpha[\mathscr{S}a] \cap \pi_\alpha^{\omega_1}[E] \neq 0$. This easily yields $\psi_\gamma[\mathscr{S}a] \cap \pi_\gamma^{\omega_1}[E] \neq 0$, as desired. This finishes the proof of (10).

(11) Every infinite closed subset of X_{ω_1} has cardinality 2^{ω_1}.

In fact, let C be an infinite closed subset of X_{ω_1}, and let E be a countably infinite subset of C. Choose $\gamma < \omega_1$ such that $F \cap A_\gamma \neq G \cap A_\gamma$ for all distinct $F, G \in \psi_{\omega_1}^{-1}[E]$. Choose $f \in {}^{\omega_1}({}^{\omega_1}\omega_1)$ such that $E = \{x_{f\alpha} : \alpha < \omega_1\}$. Now $D \overset{\text{def}}{=} \{\alpha < \omega_1 : f\restriction\alpha = f^\alpha\}$ is stationary, so there is a limit α with $\gamma < \alpha < \omega_1$, $f\restriction\alpha = f^\alpha$, and $E = \{x_{f\delta} : \delta < \alpha\}$. Now $\pi_\alpha^{\omega_1}[E]$ is infinite, and so it has an accumulation point x_ξ. Note that $\{x_{f_\delta^\alpha} : \delta < \alpha\} = \pi_\alpha^{\omega_1}[E]$. Let $Y = \{x_{f\delta} : \delta < \alpha\}\backslash\{x_\xi\}$. Thus $Y \in \mathscr{C}_\xi^1$, and $x_\xi \in \overline{Y}$. Hence by the Lemma (iii), $Q_\xi \subseteq \overline{\pi_{\alpha+1}^{\omega_1}[Y]}$ and so $Q_\xi \subseteq \overline{\pi_{\alpha+1}^{\omega_1}[E]}$. It now follows by (10) that $|C| = 2^{\omega_1}$.

(12) Assume that X_β is hereditarily separable. If C is a closed subset of $X_{\beta+1}$, then $B \overset{\text{def}}{=} \{x_\xi : \xi \in {}^\beta\omega_1, \, Q_\xi \cap C \neq 0 \neq Q_\xi\backslash C\}$ is countable.

For, suppose that B is uncountable. Since $Q_\xi \cap C \neq 0$ for each $x_\xi \in B$, we have $B \subseteq \pi_\beta^{\beta+1}[C]$. We use the notation for defining $\mathscr{C}_\xi^2$. Let D be a countable dense subset of B. Choose $\gamma < \omega_1$ such that $D \subseteq g[\gamma]$. Then choose δ such that $\gamma \leq \delta < \omega_1$ and $\delta \cap g^{-1}[B] = J_\delta$. Thus $B \subseteq \overline{g[\delta] \cap B} = \overline{g[J_\delta]}$, so we can choose $x_\xi \in B$ such that $\delta \leq g^{-1}x_\xi$ and $x_\xi \in \overline{g[J_\delta]}$. Now let E be a subset of C such that $\pi_\beta^{\beta+1}[E] = g[\gamma] \cap B$. Then by the choice of $\mathscr{C}_\xi^2$ and the lemma we get $Q_\xi \subseteq \overline{E} \subseteq C$, contradicting $x_\xi \in B$. Thus (12) holds.

(13) Assume that X_β is hereditarily separable. If C is a closed subset of $X_{\beta+1}$, then $B' \overset{\text{def}}{=} \{x_\xi : \xi \in {}^\beta\omega_1, \, Q_\xi \subseteq C$, and there are U, W with U an open neighborhood of x_ξ and W a non-empty subset of Q_ξ such that $(\pi_\beta^{\beta+1})^{-1}[U \cap f_\xi^{-1}[W]] \cap C = 0\}$ is countable.

The proof is similar to that of (12). Suppose that B' is uncountable. Let D be a countable dense subset of B', and choose $\gamma < \omega_1$ such that $D \subseteq g[\gamma]$. Then choose δ such that $\gamma \leq \delta < \omega_1$ and $\delta \cap g^{-1}[B'] = J_\delta$. Then $B' \subseteq \overline{g[\delta] \cap B'} = \overline{g[J_\delta]}$, so there is a $x_\xi \in B'$ such that $\delta \leq g^{-1}x_\xi$ and $x_\xi \in \overline{g[J_\delta]}$. So by the choice of $\mathscr{C}_\xi^2$ and the Lemma (iv) we get $x_\xi \in \overline{g[J_\delta] \cap f_\xi^{-1}[W]}$. Hence $U \cap g[J_\delta] \cap f_\xi^{-1}[W] \neq 0$. But $g[J_\delta] = g[\delta] \cap B' \subseteq \pi_\beta^{\beta+1}[C]$, so this contradicts $x_\xi \in B'$. So (13) holds.

(14) Each space X_β is hereditarily separable for $\beta < \omega_1$.

We prove this by induction on β. It is true for $\beta = 1$. Assume that it is true for β. Suppose that $C \subseteq X_{\beta+1}$. Since $X_{\beta+1}$ is first countable, we may assume that C is closed, as is easily seen. Let B and B' be as in (12) and (13), and let D be a countable dense subset of $\pi_\beta^{\beta+1}[C]$. For each $x_\xi \in B$ let E_ξ be a countable dense subset of $Q_\xi \cap C$, and for each $x_\xi \in B'$ let E_ξ be a countable dense subset of Q_ξ. For each $x_\xi \in D$ choose $y_\xi \in Q_\xi \cap C$. We claim that the countable set

$$\{y_\xi : x_\xi \in D\} \cup \bigcup_{x_\xi \in B \cup B'} E_\xi$$

is dense in C. To prove this, suppose that V is an open set such that $V \cap C \neq 0$. We may assume that V has the form $W \cup (\pi_\beta^{\beta+1})^{-1}[U \cap f_\xi^{-1}[W]]$, with obvious assumptions. First suppose that $(\pi_\beta^{\beta+1})^{-1}[U \cap f_\xi^{-1}[W]] \cap C \neq 0$. If c is an element of this set, then $\pi_\beta^{\beta+1}c \in U \cap f_\xi^{-1}[W]$, so we can choose $x_\eta \in D \cap U \cap f_\xi^{-1}[W]$. Then $y_\eta \in V \cap C$, as desired. Second, suppose that $W \cap C \neq 0$ and $x_\xi \in B$. Then there is some member of E_ξ which is in $C \cap V$, as desired. The only remaining case is that $W \cap X \neq 0$ and $x_\xi \in B'$, which again yields the desired conclusion.

Now suppose that β is a limit ordinal and we know that X_α is hereditarily separable for all $\alpha < \beta$. Let C be a subset of X_β. For each $\alpha < \beta$ let D_α be a countable subset of C such that $\pi_\alpha^\beta[D_\alpha]$ is dense in $\pi_\alpha^\beta[C]$. Set $E = \bigcup_{\alpha<\beta} D_\alpha$; we claim that E is dense in C. To see this, suppose that $a \in A_\beta$ and $\psi_\beta[\mathscr{S}a] \cap C \neq 0$; we want to show that $\psi_\beta[\mathscr{S}a] \cap E \neq 0$. Choose $\alpha < \beta$ so that $a \in A_\alpha$. Now

$$0 \neq \pi_\alpha^\beta[\psi_\beta[\mathscr{S}a] \cap C] \subseteq \pi_\alpha^\beta[\psi_\beta[\mathscr{S}a]] \cap \pi_\alpha^\beta[C]$$
$$= \psi_\alpha[\mathscr{S}a] \cap \pi_\alpha^\beta[C].$$

Hence there is a $z \in D_\alpha$ such that $\pi_\alpha^\beta z \in \psi_\alpha[\mathscr{S}a]$. Thus $\psi_\alpha^{-1}\pi_\alpha^\beta z \in \mathscr{S}a$, so $\sigma_\alpha^\beta \psi_\beta^{-1} z \in \mathscr{S}a$, so $z \in \psi_\beta[\mathscr{S}a]$, as desired.

(15) Suppose that C is an uncountable discrete subset of X_{ω_1}. Then $\pi_\beta^{\omega_1}[C]$ is countable for all $\beta < \omega_1$.

For, suppose not; choose $\beta < \omega_1$ with $\pi_\beta^{\omega_1}[C]$ uncountable. Let D be a countable subset of C such that $\pi_\beta^{\omega_1}[D]$ is dense in $\overline{\pi_\beta^{\omega_1}[C]}$. We again use the notation for defining $\mathscr{C}_\xi^2$. There is a $\gamma < \omega_1$ such that $\pi_\beta^{\omega_1}[D] \subseteq g[\gamma]$. Choose δ such that

$\gamma \leq \delta < \omega_1$ and $\delta \cap g^{-1}[\pi_\beta^{\omega_1}[C]] = J_\delta$. Hence $\pi_\beta^{\omega_1}[C] \subseteq \overline{g[J_\delta]}$. Hence there is a ξ such that $x_\xi \in \pi_\beta^{\omega_1}[C]$, $\delta \leq g^{-1}x_\xi$, and $x_\xi \in \overline{g[J_\delta]}$. Now $g[J_\delta] \in \mathscr{C}_\xi^2$, so by the Lemma (iii), $Q_\xi \subseteq \overline{g[J_\delta]}$. Let E be a subset of C such that $\pi_\beta^{\omega_1}[E] = g[J_\delta]$. Say $\pi_\beta^{\omega_1}c = x_\xi$, with $c \in C$. Then by (10) we have $c \in \overline{E}$. Since $c \notin E$, this contradicts the assumption that C is discrete.

(16) X_{ω_1} is hereditarily separable.

For, suppose not. Let $\langle x_{\xi_\alpha} : \alpha < \omega_1 \rangle$ be a left-separated sequence in X_{ω_1}. The following two statements are obvious.

(17) If $\alpha < \omega_1$, N is a clopen subset of X_α, and $N \cap \{x_{\xi_\beta \upharpoonright \alpha} : \beta < \alpha\} \neq 0$, then $(\pi_\alpha^{\omega_1})^{-1}[N] \cap \{x_{\xi_\beta} : \beta < \omega_1\} \neq 0$.

(18) If $\alpha < \omega_1$, N is a clopen subset of X_α, and $|N \cap \{x_{\xi_\beta \alpha} : \beta < \alpha\}| \geq 2$, then $|(\pi_\alpha^{\omega_1})^{-1}[N] \cap \{x_{\xi_\beta} : \beta < \omega_1\}| \geq 2$.

Now let

$$C = \{\alpha < \omega_1 : \text{for every clopen } N \subseteq X_\alpha (N \cap \{x_{\xi_\beta \upharpoonright \alpha} : \beta < \alpha\} = 0 \text{ iff}$$
$$(\pi_\alpha^{\omega_1})^{-1}[N] \cap \{x_{\xi_\beta} : \beta < \omega_1\} = 0) \text{and} (|N \cap \{x_{\xi_\beta \upharpoonright \alpha} : \beta < \alpha\}| = 1 \text{ iff}$$
$$|(\pi_\alpha^{\omega_1})^{-1}[N] \cap \{x_{\xi_\beta} : \beta < \omega_1\}| = 1).$$

We claim

(19) C is club in ω_1.

We shall prove this in detail, since it is a little trickier than your usual club arguments. Actually the "closed" part is straightforward, but for completeness we do that too.

First we show that C is closed. Suppose that α is a limit ordinal less than ω_1 and $\alpha \cap C$ is unbounded in α; we want to show that $\alpha \in C$. Suppose that $N \subseteq X_\alpha$ is clopen and $(\pi_\alpha^{\omega_1})^{-1}[N] \cap \{x_{\xi_\beta} : \beta < \omega_1\} \neq 0$. Say $\beta < \omega_1$ and $\pi_\alpha^{\omega_1}x_{\xi_\beta} \in N$. Thus $x_{\xi_\beta \upharpoonright \alpha} \in N$. Write $N = \psi_\alpha[\mathscr{S}a]$, $a \in A_\alpha$. Then there is a $\gamma < \alpha$ such that $\gamma \in C$ and $a \in A_\gamma$. Thus $x_{\xi_\beta \upharpoonright \alpha} \in \psi_\alpha[\mathscr{S}a]$, so by the commutative diagram we easily get $x_{\xi_\beta \upharpoonright \gamma} \in \psi_\gamma[\mathscr{S}a]$ so, since $\gamma \in C$, $x_{\xi_\delta \upharpoonright \gamma} \in \psi_\gamma[\mathscr{S}a]$ for some $\delta < \gamma$. The commutative diagram then gives $x_{\xi_\delta \upharpoonright \alpha} \in N$, as desired. Similarly $|(\pi_\alpha^{\omega_1})^{-1}[N] \cap \{x_{\xi_\beta} : \beta < \omega_1\}| \geq 2$ implies that X_α, and $|N \cap \{x_{\xi_\beta \upharpoonright \alpha} : \beta < \alpha\}| \geq 2$. Hence C is closed.

C is unbounded: Let $\alpha_0 < \omega_1$ be given; we want to find a member of C which is greater than α_0. By (15) let Γ be a countable subset of ω_1 such that $\{\xi_\beta \upharpoonright \alpha_0 : \beta < \omega_1\} = \{\xi_\beta \upharpoonright \alpha_0 : \beta \in \Gamma\}$. For each $\beta \in \Gamma$ let $\mathscr{U}_\beta$ be a countable clopen neighborhood base for $x_{\xi_\beta \upharpoonright \alpha_0}$. Let $\mathscr{V} = \bigcup_{\beta \in \Gamma} \mathscr{U}_\beta$. For each $U \in \mathscr{V}$, let Δ_U be a largest subset of ω_1 with the following two properties: (a) $|\Delta_U| \leq 2$; (b) for all $\beta \in \Delta_U$, $x_{\xi_\beta \upharpoonright \alpha_0} \in U$. Let $\alpha_1 = \max(\alpha_0, \sup_{U \in \mathscr{V}} \max \Delta_U) + 1$. Continue in the same way with $\alpha_2, \alpha_3, \ldots$, and let $\alpha_\omega = \sup_{n \in \omega} \alpha_1$. We claim that $\alpha_\omega \in C$. Suppose that $N \subseteq X_{\alpha_\omega}$ is clopen, and $(\pi_{\alpha_\omega}^{\omega_1})^{-1}[N] \cap \{x_{\xi_\beta} : \beta < \omega_1\} \neq 0$. Say

$\beta < \omega_1$, $x_{\xi_\beta \restriction \alpha_\omega} \in N$. Write $N = \psi_{\alpha_\omega}[\mathscr{S}a]$ with $a \in A_{\alpha_\omega}$. Say $a \in A_{\alpha_i}$, $i < \omega$. Thus $\psi_{\alpha_\omega}^{-1} x_{\xi_\beta \restriction \alpha_\omega} \in \mathscr{S}a$, so $x_{\xi_\beta \restriction \alpha_i} \in \psi_{\alpha_i}[\mathscr{S}a]$. Then with $\mathscr{V}$ as above, but for α_i rather than α_0, there is a $U \in \mathscr{V}$ such that $x_{\xi_\beta \restriction \alpha_i} \in U$ and $U \subseteq \psi_{\alpha_i}[\mathscr{S}a]$. Hence we get a $\gamma < \alpha_{i+1}$ such that $x_{\xi_\gamma \restriction \alpha_i} \in U$. Thus $\psi_{\alpha_i}^{-1} x_{\xi_\gamma \restriction \alpha_i} \in \mathscr{S}a$, so $\psi_{\alpha_\omega}^{-1} x_{\xi_\gamma \restriction \alpha_\omega} \in \mathscr{S}a$, and hence $N \cap \{x_{\xi_\delta \restriction \alpha_\omega} : \delta < \alpha_\omega\} \neq 0$. The other desired condition is proved similarly. Hence (19) holds.

Now fix $\alpha \in C$ with $\xi \restriction \restriction \alpha = f^\alpha$, where $\langle \xi^\alpha : \alpha < \omega_1 \rangle$ is the special $\Diamond$ sequence. Then

(20) $x_{\xi_\eta \restriction \alpha} \in \overline{\{x_{\xi_\beta^\alpha} : \beta < \alpha\}}$ for all $\eta < \omega_1$.

For, if $x_{\xi_\eta \restriction \alpha} \in N$ with N clopen, then $(\pi_\alpha^{\omega_1})^{-1}[N] \cap \{x_{\xi_\nu} : \nu < \omega_1\} \neq 0$, so $N \cap \{x_{\xi_\beta \restriction \alpha} : \beta < \alpha\} \neq 0$ as desired, since $\alpha \in C$.

(21) For all $\eta < \omega_1$, if $x_{\xi_\eta \restriction \alpha}$ is isolated in $\{x_{\xi_\beta \restriction \alpha} : \beta < \alpha\}$, then $\eta < \alpha$.

For, let N be clopen such that $N \cap \{x_{\xi_\beta \restriction \alpha} : \beta < \alpha\} = \{x_{\xi_\eta \restriction \alpha}\}$. Thus since $\alpha \in C$ we get $|(\pi_\alpha^{\omega_1})^{-1}[N] \cap \{\xi_\beta : \beta < \omega_1\}| = 1$. Say $x_{\xi_\eta \restriction \alpha} = x_{\xi_\gamma \restriction \alpha}$ with $\gamma < \alpha$. Since x_{ξ_η} and x_{ξ_γ} are both members of $(\pi_\alpha^{\omega_1})^{-1}[N] \cap \{\xi_\beta : \beta < \omega_1\}$, it follows that $\eta = \gamma$, as desired.

By (21), $x_{\xi_\alpha \restriction \alpha}$ is an accumulation point of $\{x_{\xi_\beta \restriction \alpha} : \beta < \alpha\} = \{x_{\xi_\beta^\alpha} : \beta < \alpha\}$. Then by the Lemma and (10) we get $x_{\xi_\alpha} \in \overline{\{x_{\xi_\beta \restriction \alpha} : \beta < \alpha\}}$, contradicting left-separatedness.

This finishes Fedorchuk's example.

On the relationship of hd with the other functions, note also that by Theorem 16.1 we have $\pi A \leq \mathrm{hd} A$. πA is strictly less than $\mathrm{hd} A$ in $\mathscr{P}\kappa$, for example. And we have $sA < \mathrm{hd} A$ for A the interval algebra on a Suslin line, and $\mathrm{hd} A < \chi A$ for a Kunen line (Chapter 8).

There is a model of ZFC with a BA A such that $sA, dA < \mathrm{hd} A$ (a remark of I. Juhász in an email message in February, 1995). Namely, take a model with $\mathrm{MA}(\sigma\text{-centered}) + \exists$ a 0-dimensional Susilin line S, let K be a compactification of ω such that $K \backslash \omega = S$, and let $A = \mathrm{clop} K$. (See W. Weiss [84], J. van Mill [84].)

From the result that $\pi A \leq sA \cdot (tA)^+$ it follows that $\mathrm{hd} A \leq sA \cdot (tA)^+$. It is also true that $\mathrm{hd} A \leq \mathrm{Irr} A$. In fact, we have $\mathrm{hd} A = \pi_{\mathrm{H}+} A$, and for any homomorphic image B of A we have $\pi B \leq \mathrm{Irr} B \leq \mathrm{Irr} A$.

If one can construct in ZFC a BA A such that $\mathrm{hL} A < \mathrm{hd} A$, then one can also construct in ZFC a BA B such that $\mathrm{hL} B < dB$ (see problem 53). In fact, A has a homomorphic image such that $\mathrm{hL} A < dB$, and $\mathrm{hL} B \leq \mathrm{hL} A$. Similarly for $\mathrm{hL} < \pi$.

The following problems are open; these are Problems 50 and 51 in Monk [90].

Problem 57. *Can one construct in ZFC a BA A such that $sA < \mathrm{hd} A$?*

Problem 58. *Can one construct in ZFC a BA A such that $\mathrm{hd} A < \chi A$?*

Note that Problem 57 is equivalent to the problem of constructing in ZFC a BA A such that $sA < dA$, and also to the problem of constructing in ZFC a BA A such that $sA < \pi A$; see the argument preceding problem 57. Also note that "yes" for Problem 53 implies "yes" for problem 57.

Problem 58 is equivalent to the problem of constructing in ZFC a BA A such that $hdA < hLA$; see the argument at the end of Chapter 15. And note that "yes" for problem 58 implies "yes" for problem 49.

Bounded versions of hd can be defined as follows. For m a positive integer, a sequence $\langle x_\alpha : \alpha < \kappa \rangle$ of elements of A is said to be m-*left-separated* provided that if $\Gamma \in [\kappa]^m$, $\alpha < \kappa$, and $\alpha < \beta$ for all $\beta \in \Gamma$, then $a_\alpha \cdot \prod_{\beta \in \Gamma} -a_\beta \neq 0$. Then we define

$$hd_m A = \sup\{\kappa : \text{there is an } m\text{-left-separated sequence in } A\}.$$

For this notion, see Rosłanowski, Shelah [94].

17. Incomparability

We begin with one important equivalent definition:

Theorem 17.1. *For any infinite BA A we have* $\mathrm{Inc}\,A = \sup\{|T| : T$ *is a tree included in* $A\}$.

(Note that when we say that T is a tree included in A, we mean merely that T is a subset of A which is a tree under the induced ordering; there is no assumption that incomparable elements (in T) are disjoint (in the dual of A).)

Proof. Since any incomparable set is a tree having only roots, the inequality $\leq$ is clear. To show equality, suppose that κ is regular and A has no incomparable set of size κ; we show that A has no tree of size κ. Suppose T is a tree of size κ. By Theorem 4.25 of Part I of the BA handbook, A has a dense subset D of size $< \kappa$. Now each level of T is an incomparable set, and hence has fewer than κ elements. Hence T has at least κ levels. Let T' be a subset of T of power κ consisting exclusively of elements of successor levels. For each $d \in D$ let

$$M_d = \{t \in T' : \text{if } s \text{ is the immediate predecessor of } t, \text{ then } d \leq t \cdot -s\}.$$

Thus $T' = \bigcup_{d \in D} M_d$, so there is a $d \in D$ such that $|M_d| = \kappa$. But then M_d is incomparable, contradiction: if $y, z \in M_d$ and $y < z$, then $y \leq u$ where u is the immediate predecessor of z, and $d \leq z \cdot -u$, hence $d \cdot y = 0$, contradicting $d \leq y$. $\square$

Note that if $\mathrm{Inc}\,A$ is attained, then it is obviously attained in the tree sense. The converse also holds, as Todorčević pointed out in a letter to the author several years before Monk [90] appeared; this solves Problem 52 in Monk [90]. We give this result here, following the proof in an email message from Shelah of December 1990.

Theorem 17.2. *If A is an infinite BA and there is a tree $T \subseteq A$ with $|T| = \mathrm{Inc}\,A$, then A has an incomparable subset of power $\mathrm{Inc}\,A$.*

Proof. By the proof of Theorem 17.1 we may assume that $\lambda \overset{\text{def}}{=} \mathrm{Inc}\,A$ is singular. Let $\langle \kappa_\alpha : \alpha < \mathrm{cf}\lambda \rangle$ be an increasing sequence of cardinals with supremum λ. Without loss of generality, T has no level of size λ. Now we consider two cases.

    Case 1. For every $\alpha < \mathrm{cf}\lambda$ there is a β such that T has at least κ_α elements of level β. For any ordinal β let $\mathrm{lev}_\beta T$ be the set of elements of T of level β. By an easy construction we obtain a strictly increasing sequence $\langle \beta_\alpha : \alpha < \mathrm{cf}\lambda \rangle$ of ordinals such that $|\mathrm{lev}_{\beta_{\alpha+1}} T| > \max(\kappa_\alpha, |\mathrm{lev}_{\beta_\alpha}|)$ for all $\alpha < \mathrm{cf}\lambda$. For every $\alpha < \mathrm{cf}\lambda$ let S_α be a subset of $\mathrm{lev}_{\beta_{\alpha+1}} T$ of power $(\max(\kappa_\alpha, |\mathrm{lev}_{\beta_\alpha}|))^+$ such that all elements of S_α have the same predecessors at level β_α. Note that if $\alpha < \mathrm{cf}\lambda$ then

$$R_\alpha \overset{\text{def}}{=} \{t \in S_\alpha : t \leq s \text{ for some } s \in S_\gamma \text{ with } \alpha < \gamma < \mathrm{cf}\lambda\}$$

has power at most cfλ. Now the set $\bigcup_{\alpha<\text{cf}\lambda}(S_\alpha\backslash R_\alpha)$ is incomparable of size λ, as desired.

Case 2. Case 1 fails to hold. Then clearly T must have at least λ levels. Hence Depth$A = \lambda$ and c$A = \lambda$, so by the Erdös-Tarski theorem A has a disjoint subset of power λ; it is also an incomparable subset. $\square$

Concerning attainment of Inc, several things are known. Milner and Pouzet [86] proved a general result, of which a special case is that if Inc$A = \lambda$ with cf$\lambda = \omega$, then Inc$A$ is attained. Todorčević has shown that if 2^ω is weakly inaccessible, then there is a BA of size 2^ω with incomparability 2^ω not attained. On the other hand, Theorem 4.25 of the BA handbook shows that if Inc$A$ is a strong limit cardinal, then Inc$A$ is attained. The statement in Monk [90] about attainment due to Shelah has been withdrawn by him. Instead, in Shelah [92a] he proves that Inc$A$ is always attained for singular cardinals. We give this interesting proof here.

Theorem 17.3. Inc *is attained for singular cardinals.*

Proof. Suppose that Inc$B = \lambda$, with λ singular. Now we choose $\langle\lambda_i : i < \text{cf}\lambda\rangle$ and $\langle A_i : i < \text{cf}\lambda\rangle$ so that the following conditions hold:

(1) $\langle\lambda_i : i < \text{cf}\lambda\rangle$ is a strictly increasing sequence of regular cardinals all greater than cfλ and with supremum λ.

(2) $\langle A_i : i < \text{cf}\lambda\rangle$ is a system of incomparable sets in B with $|A_i| = \lambda_i^+$ for all $i < \text{cf}\lambda$.

It is clear that this can be done. In addition, if possible we choose these things so that the following condition holds:

(3) If $i < j < \text{cf}\lambda$, $x \in A_i$, and $y \in A_j$, then $y \not\leq x$.

Now let $A = \bigcup_{i<\text{cf}\lambda} A_i$. The following notation will also be useful. Let $C \subseteq B$. For any $x \in B$, $C \uparrow x = \{y \in C : y \geq x\}$, $C \downarrow x = \{y \in C : y \leq x\}$, and for any cardinal μ, $C \uparrow_\mu = \{x \in C : |C \uparrow x| < \mu\}$ and $C \downarrow_\mu = \{x \in C : |C \downarrow x| < \mu\}$.

Case 1. There is a $\mu < \lambda$ such that $|A \uparrow_\mu| = \lambda$. For each $x \in (A \uparrow_\mu)$ let $fx = (A \uparrow_\mu) \uparrow x$. Thus $f : (A \uparrow_\mu) \rightarrow \mathscr{P}(A \uparrow_\mu)$ and $|fx| < \mu < \lambda$ for all $x \in (A \uparrow_\mu)$. Hence by Hajnal's free set Theorem (see the Handbook, Part III, p. 1231) we get $E \subseteq (A \uparrow_\mu)$ of size λ such that $x \notin fy$ for all distinct $x, y \in E$. So E is incomparable, as desired.

Case 2. There is a $\mu < \lambda$ such that $|A \downarrow_\mu| = \lambda$. Similarly.

Case 3. For every $i < \text{cf}\lambda$ there is an $x_i \in A$ such that $\lambda_i < |A \uparrow x_i| < \lambda$. We now define a function $\mu : \text{cf}\lambda \rightarrow \text{cf}\lambda$ by induction. Having defined μ_j for all $j < i$, choose $\mu_i < \text{cf}\lambda$ so that $\lambda_{\mu_i} > \sum_{j<i}(|A \uparrow x_{\mu_j}| + \lambda_{\mu_j})$. Now let $A_i' = A_{\mu_i}$, $\lambda_i' = \lambda_{\mu_i}$, $x_i' = x_{\mu_i}$ for all $i < \text{cf}\lambda$. Then (1)–(2) hold for A_i' and λ_i', (3) holds for the A_i''s if it held for the A_i's, and

(4) If $j < i < \text{cf}\lambda$, then $\sum_{j<i}|A \uparrow x_j'| < \lambda_i'$.

Now for $i < \mathrm{cf}\lambda$ we have $\left| (A \uparrow x'_i) \backslash \bigcup_{j<i}(A \uparrow x'_j) \right| > \lambda'_i$, $\mathrm{cf}\lambda < \lambda_i$, and $A = \bigcup_{j<\mathrm{cf}\lambda} A_j$, so there is an $\alpha(i) < \mathrm{cf}\lambda$ such that

$$\left| A_{\alpha(i)} \cap (A \uparrow x'_i) \backslash \bigcup_{j<i}(A \uparrow x'_j) \right| > \lambda'_i.$$

Clearly $\alpha(i) \geq \mu_i$. We define $\beta : \mathrm{cf}\lambda \to \mathrm{cf}\lambda$ by induction. If $\beta(j)$ has been defined for all $j < i$, choose $\beta(i) < \mathrm{cf}\lambda$ such that $\sup_{j<i} \max\{\alpha(\beta(j)), \mu_{\beta(j)}\} < \mu_{\beta(i)}$. Now choose

$$A^*_i \subseteq A_{\alpha(\beta(i))} \cap (A \uparrow x'_{\beta(i)}) \backslash \bigcup_{j<i}(A \uparrow x'_{\beta(j)})$$

of size $\lambda'_{\beta(i)}$. Note that $\alpha(\beta(j)) < \mu_{\beta(i)} \leq \alpha(\beta(i))$ for $j < i < \mathrm{cf}\lambda$. Let $\lambda^*_i = \lambda'_{\beta(i)}$. Then (1)–(2) hold for A^*_i and λ^*_i, (3) holds for the A^*_i's if it held for the A_i's (since $A^*_i \subseteq A_{\alpha(\beta(i))}$), and

(5) if $i < j < \mathrm{cf}\lambda$, $x \in A^*_i$, and $y \in A^*_j$, then $x \not\leq y$.

For, otherwise $x'_{\beta(i)} \leq x \leq y \notin (A \uparrow x'_{\beta(i)})$, contradiction. Now let $A^\star_i = \{x : -x \in A^*_i\}$. Then (1)–(3) hold for $A^\star_i$ and λ^*_i. Therefore by the initial choices, (3) itself holds. So, (3) and (5) hold for A^*_i and λ^*_i. It follows that $\bigcup_{i<\mathrm{cf}\lambda} A^*_i$ is an incomparable set of size λ, as desired.

Case 4. For every $i < \mathrm{cf}\lambda$ there is an $x \in A$ such that $\lambda_i < |A \downarrow x| < \lambda$. Similar to Case 3.

Case 5. None of the previous cases. By ¬Case 3, there is an $i(*) < \mathrm{cf}\lambda$ such that for all $x \in A$ we have $\neg(\lambda_{i(*)} < |A \uparrow x| < \lambda)$. By ¬Case 2, $|A \downarrow_{\lambda^+_{i(*)}}| < \lambda$, and by ¬Case 1, $|A \uparrow_{\lambda^+_{i(*)}}| < \lambda$. Choose $x^* \in A \backslash ((A \downarrow_{\lambda^+_{i(*)}}) \cup (A \uparrow_{\lambda^+_{i(*)}}))$. Thus $|A \uparrow x^*| \geq \lambda^+_{i(*)}$, so by the choice of $i(*)$ we have $|A \uparrow x^*| = \lambda$. Also, $|A \downarrow x^*| \geq \lambda^+_{i(*)} > \mathrm{cf}\lambda$, so there is a $j(*) < \mathrm{cf}\lambda$ such that $|(A \downarrow x^*) \cap A_{j(*)}| \geq \mathrm{cf}\lambda$. Choose distinct $y_i \in (A \downarrow x^*) \cap A_{j(*)}$ for $i < \mathrm{cf}\lambda$.

For each $i < \mathrm{cf}\lambda$ let $A'_i = A_i \cap (A \uparrow x^*)$. Thus A'_i is an incomparable set and

$$\left| \bigcup_{i<\mathrm{cf}\lambda} A'_i \right| = \left| (A \uparrow x^*) \cap \bigcup_{i<\mathrm{cf}\lambda} A_i \right| = |(A \uparrow x^*) \cap A| = \lambda.$$

Finally, $\{y_i + x \cdot -x^* : i < \mathrm{cf}\lambda, \ x \neq x^*, \ x \in A'_i\}$ is an incomparable set of size λ, as desired. $\qquad\square$

Now we turn to algebraic operations, as usual. If A is a subalgebra or homomorphic image of B, then $\mathrm{Inc}A \leq \mathrm{Inc}B$. If A is a subalgebra of B, then, easily, $\mathrm{Inc}(A \times B) \geq |A|$; in fact, $\{(a, -a) : a \in A\}$ is an incomparable set in $A \times B$. Hence if A is cardinality-homogeneous and has no incomparable set of size $|A|$, then A is rigid (this follows from some elementary facts concerning automorphisms; see the article in the BA handbook about automorphisms). Thus the incomparability of a product can jump from that in a factor—for example, if A is such that $\mathrm{Inc}A < |A|$, we have $\mathrm{Inc}(A \times A) = |A|$. Finally, $\mathrm{Inc}(A \oplus B) = \max(|A|, |B|)$ if $|A|, |B| \geq 4$, since $A \oplus C \cong A \times A$ if $|C| = 4$.

Ultraproducts: Inc is an ultra-sup function, so Theorems 3.15–3.17 hold, Theorem 3.17 saying that $\mathrm{Inc}\left(\prod_{i \in I} A_i / F\right) \geq \left|\prod_{i \in I} \mathrm{Inc}A_i / F\right|$ for F regular, and Donder's theorem says that under $V = L$ the regularity assumption can be removed. In Shelah [94g] it is shown that $>$ is consistent, but we do not know whether this can be done in ZFC:

Problem 59. *Do there exist in ZFC a system $\langle A_i : i \in I \rangle$ of infinite BAs, I infinite, and a regular ultrafilter F on I such that $\mathrm{Inc}\left(\prod_{i \in I} A_i / F\right) > \left|\prod_{i \in I} \mathrm{Inc}A_i / F\right|$?*

Problem 60. *Is an example with $\mathrm{Inc}\left(\prod_{i \in I} A_i / F\right) < \left|\prod_{i \in I} \mathrm{Inc}A_i / F\right|$ consistent?*

Again this problem may be solved by the methods of Magidor, Shelah [91].

Concerning derived functions of incomparability, we mention only a result of Shelah (email message of December 1990), solving Problem 53 in Monk [90]:

Theorem 17.4. *If $\mathrm{Inc}A = \omega$, then $\mathrm{Card_{H-}}A = \omega$.*

Proof. Suppose that $\mathrm{Inc}A = \omega < \mathrm{Card_{H-}}A$. Without loss of generality, assume that A is a subalgebra of $\mathscr{P}\omega$ containing all of the finite subsets of ω. Hence there is an $a \in A$ such that both a and $\omega \backslash a$ are infinite. Let F be a nonprincipal ultrafilter on $A \restriction a$. Then we can construct $\langle a_\alpha : \alpha < \chi F \rangle$, each $a_\alpha \subseteq a$, such that for each $\alpha < \chi F$, the element a_α is not in the filter generated by $\{a_\beta : \beta < \alpha\}$; in particular, $\beta < \alpha \Rightarrow a_\beta \not\leq a_\alpha$. Similarly we get a nonprincipal ultrafilter G on $A \restriction -a$ and a sequence $\langle b_\alpha : \alpha < \chi G \rangle$ of subelements of $-a$ such that $\beta < \alpha \Rightarrow b_\beta \not\leq b_\alpha$. Say $\chi F \leq \chi G$. Then $\langle a_\alpha + -b_\alpha \cdot -a : \alpha < \chi F \rangle$ is a system of incomparable elements. By 14.5, $\mathrm{a}A \leq \chi F$, so $\mathrm{a}A = \omega$. Hence 14.7 gives a contradiction. $\square$

Now we turn to connections with our other functions. From the Handbook Theorem 4.25 it follows that $\mathrm{d}A \leq \pi A \leq \mathrm{Inc}A$ for any infinite BA A; hence $\mathrm{hd}A \leq \mathrm{Inc}A$, by an easy argument. An example in which they are different is the interval algebra A on the reals. In fact, $\mathrm{hd}A = \omega$ by Theorem 16.1, and an incomparable set of size 2^ω is provided by

$$\{[0, r) \cup [1 + r, 2) : r \in (0, 1)\}.$$

Much effort has been put into constructing BAs A in which $\mathrm{Inc}A < |A|$. An example in ZFC of such an algebra has been given by Shelah (email message, December 1990; see the end of this chapter). Another example is an algebra of Bonnet, Shelah [85]; their algebra is an interval algebra, and has power $\mathrm{cf}(2^\omega)$, so that if $\mathrm{cf}(2^\omega)$ is not limit one gets such an algebra. Rubin's algebra [83] is another example (constructed assuming $\Diamond$). Baumgartner [80] showed that it is consistent to have MA, $2^\omega = \omega_2$, and every uncountable BA has an uncountable incomparable subset.

We give a construction, using $\Diamond$, of a BA A with $\mathrm{Inc}A < |A|$; the example is due to Baumgartner, Komjath [81], and settles another question which is of interest. It depends on some lemmas. For all the lemmas let A be a denumerable atomless subalgebra of $\mathscr{P}\omega$, and let I be a maximal ideal in A. We consider a partial ordering $P = \{(a, b) : a \in I, \ b \in A\backslash I, \ a \subseteq b\}$, ordered by: $(a, b) \preceq (c, d)$ iff $a \supseteq c$ and $b \subseteq d$.

Lemma 17.5. *The following sets are dense in P:*

(i) For each $m \subseteq I$, the set $D_1 m \overset{\text{def}}{=} \{(a, b) \in P : \text{either } \forall c \in m(c \not\subseteq b) \text{ or } \exists c \in m(c \subseteq a)\}$.

(ii) For each $c \in A$, the set $D_2 c \overset{\text{def}}{=} \{(a, b) \in P : \neg(a \subseteq c \subseteq b)\}$.

(iii) For each $c \in I$, the set $D_3 c \overset{\text{def}}{=} \{(a, b) \in P : c \subseteq a \cup (\omega\backslash b)\}$.

(iv) The set $D_4 \overset{\text{def}}{=} \{(a, b) \in P : a \neq 0\}$.

Proof. Suppose $(a, b) \in P$. (i) If $\forall c \in m(c \not\subseteq b)$, then $(a, b) \in D_1 m$, as desired. Otherwise there is a $c \in m$ such that $c \subseteq b$, and then $(a \cup c, b) \in P$, $(a \cup c, b) \preceq (a, b)$, and $(a \cup c, b) \in D_1 m$, as desired.

(ii) If $a \subseteq c \subseteq b$, then there are two cases: Case 1. $c \in I$. Thus $c \subset b$. Choose disjoint non-empty d, e such that $b\backslash c = d \cup e$. Since $1 \notin I$, one of d, e is in I; say $d \in I$. Then $(a \cup d, b) \in P$, $(a \cup d, b) \preceq (a, b)$, and $(a \cup d, b) \in D_2 c$. Case 2. $c \notin I$. We can similarly find $d \subset (c\backslash a)$ such that $d \notin I$; then $(a, a \cup d)$ is the desired element showing that $D_2 c$ is dense.

(iii) The element $(a \cup (c \cap b), b)$ shows that $D_3 c$ is dense.

(iv) If $a \neq 0$, we are through. Otherwise choose non-empty disjoint c, d such that $c \cup d = b$. One of c, d, say c, is in I; then (c, b) is as desired. $\qquad\square$

Lemma 17.6. *Suppose that D is dense in P. Then so are the following sets:*

(i) $SD \overset{\text{def}}{=} \{(a, b) \in P : (\omega\backslash b, \omega\backslash a) \in D\}$.

(ii) For $e, f \in I$, $T(D, e, f) \overset{\text{def}}{=} \{(a, b) \in P : ((a\backslash e) \cup f, (b\backslash e) \cup f) \in D\}$.

Proof. Suppose $(a, b) \in P$. (i) We can choose $(c, d) \preceq (\omega\backslash b, \omega\backslash a)$ so that $(c, d) \in D$. Thus $\omega\backslash b \subseteq c$ and $d \subseteq \omega\backslash a$, so $\omega\backslash c \subseteq b$ and $a \subseteq \omega\backslash d$, which shows that $(\omega\backslash d, \omega\backslash c)$ is a desired element.

(ii) By Lemma 17.5(iii), choose $(a', b') \preceq (a, b)$ such that $e \cup f \subseteq a' \cup (\omega\backslash b')$. Let $e' = a' \cap e$ and $f' = a' \cap f$. By density of D, choose $(x, y) \in D$ such that $(x, y) \preceq ((a'\backslash e) \cup f, (b'\backslash e) \cup f)$. Now let

$$a'' = (x\backslash(e \cup f)) \cup e' \cup f' \text{ and } b'' = (y\backslash(e \cup f)) \cup e' \cup f'.$$

It is easy to check that $(a'', b'') \in T(D, e, f)$ and $(a'', b'') \preceq (a, b)$. $\qquad\square$

Now suppose that M is a countable collection of subsets of I; then we let $\mathcal{D}M$ be the smallest collection of dense sets in P such that

(1) every set $D_1 m, D_2 c, D_3 e, D_4$ is in $\mathcal{D}M$ for $m \in M$, $c \in A$, $e \in I$;

(2) if $D \in \mathcal{D}M$ and $e, f \in I$, then $SD, T(D, e, f) \in \mathcal{D}M$.

A subset $x \subseteq \omega$ is M-$generic$ if for all $D \in \mathcal{D}M$ there is an $(a,b) \in D$ such that $a \subseteq x \subseteq b$. Note that because $D_2 c \in \mathcal{D}M$ for all $c \in A$ we have $x \notin A$ in such a case.

Lemma 17.7. *For every M as above, there is a subset $x \subseteq \omega$ which is M-generic.*

Proof. Let $\langle D_0, D_1 \ldots \rangle$ enumerate all members of $\mathcal{D}M$. Now we define (a_0, b_0), (a_1, b_1), ... by induction: $a_0 = 0$ and $b_0 = \omega$. Having defined (a_i, b_i), choose (a_{i+1}, b_{i+1}) so that $(a_{i+1}, b_{i+1}) \preceq (a_i, b_i)$ and $(a_{i+1}, b_{i+1}) \in D_i$. Let $x = \bigcup_{i < \omega} a_i$. Clearly x is as desired. $\square$

Lemma 17.8. *Let x be M-generic and set $B = \langle A \cup \{x\} \rangle$. Then every element of $B \backslash A$ is M-generic. B is atomless, and $B \supset A$. Moreover, for any $a \in I$ we have $x \cap a \in I$ and $a \backslash x \in I$.*

Proof. First we prove the final statement. In fact, choose $(c, d) \in D_3 a$ so that $c \subseteq x \subseteq d$. Thus $a \subseteq c \cup (\omega \backslash d)$. It follows that $a \cap x \subseteq c \cap a \subseteq a \cap x$, as desired. And $a \backslash x \subseteq a \backslash d \subseteq a \backslash x$, as desired.

Now we claim:

(1) Every element of $B \backslash A$ has one of the two forms $(x \backslash e) \cup f$ or $((\omega \backslash x) \backslash e) \cup f$ for some $e, f \in I$.

In fact, take any element y of $B \backslash A$; we can write it in the form $y = (e \cap x) \cup (f \backslash x)$, where $e, f \in A$. By the above we cannot have $e, f \in I$. Now $-y = ((\omega \backslash e) \cap x) \cup ((\omega \backslash f) \backslash x)$, so by the above we also cannot have $-e, -f \in I$. So we have two cases: Case 1. $e \in I$ and $f \notin I$. Then $y = ((\omega \backslash x) \backslash (\omega \backslash f)) \cup (e \cap x)$, which is in one of the desired forms. Case 2. $e \notin I$ and $f \in I$. Then $y = (x \backslash (\omega \backslash e)) \cup (f \backslash x)$, which again is in one of the desired forms. So (1) holds.

Next

(2) If $e, f \in I$, then $(x \backslash e) \cup f$ is M-generic.

For, given $D \in \mathcal{D}M$, we also have $T(D, e, f) \in \mathcal{D}M$, and hence there is an $(a, b) \in T(D, e, f)$ such that $a \subseteq x \subseteq b$. Then $(a \backslash e) \cup f \subseteq (x \backslash e) \cup f \subseteq (b \backslash e) \cup f$, and $((a \backslash e) \cup f, (b \backslash e) \cup f) \in D$, as desired.

Finally, since $\mathcal{D}M$ is closed under the operation S, it follows easily that $\omega \backslash x$ is M-generic. From (1) and (2) the first conclusion of the lemma now follows.

That B is atomless follows from what has already been shown, plus the fact that $D_4 \in \mathcal{D}M$. And B is a proper extension of A since $D_2 c \in \mathcal{D}M$ for every $c \in A$. $\square$

Lemma 17.9. *Under the hypotheses of Lemma 17.8,*

     (i) for all $a \in I$ and all $b \in B$, if $b \subseteq a$ then $b \in A$.

     (ii) $I' \overset{\text{def}}{=} \langle I \cup \{x\} \rangle^{\text{Id}}$ is a maximal ideal in B.

Proof. (i): By Lemma 17.8, if $b \notin A$ then b is M-generic, so by the last comment of Lemma 17.8, $a \cap b \in A$; so, of course, $b \nsubseteq a$.

(ii): First suppose that I' is not proper; write $\omega = x \cup a$ with $a \in I$. Thus $\omega \backslash a \subseteq x$ so, since $a \backslash x \in A$ by Lemma 17.8, its complement is also in A, and $x = (\omega \backslash a) \cup x \in A$, contradiction. So, I' is a proper ideal.

Since $u \in I'$ or $-u \in I'$ for every $u \in A \cup \{x\}$, and $A \cup \{x\}$ generates B, it follows that I' is maximal.   □

Example 17.10. (The Baumgartner-Komjath algebra.) We construct a BA A such that $\text{Inc}A = \omega = \text{Length}A$, while $\chi A = \omega_1$. $\Diamond$ is assumed.

Let $\langle S_\alpha : \alpha \in \omega_1 \rangle$ be a $\Diamond$-sequence, and let $\langle a_\alpha : \alpha \in \omega_1 \rangle$ be a one-one enumeration of $\mathcal{P}\omega$. For each $\beta < \omega_1$ let $m_\beta = \{a_\alpha : \alpha \in S_\beta\}$.

We define sequences $\langle A_\alpha : \alpha < \omega_1 \rangle$, $\langle I_\alpha : \alpha < \omega_1 \rangle$, $\langle M_\alpha : \alpha < \omega_1 \rangle$ by induction, as follows. Let A_0 be a denumerable atomless subalgebra of $\mathcal{P}\omega$, and let I_0 be a maximal ideal in A_0. If we have defined a denumerable atomless subalgebra A_α of $\mathcal{P}\omega$ and a maximal ideal I_α of A_α, we let $M_\alpha = \{m_\beta : \beta \leq \alpha, \text{ and } m_\beta \subseteq I_\alpha\}$. Let x_α be M_α-generic (with respect to A_α and I_α), and let $A_{\alpha+1} = \langle A_\alpha \cup \{x_\alpha\} \rangle$ and $I_{\alpha+1} = \langle I_\alpha \cup \{x_\alpha\} \rangle^{\text{Id}}$. For α a limit ordinal let $A_\alpha = \bigcup_{\beta < \alpha} A_\beta$ and $I_\alpha = \bigcup_{\beta < \alpha} I_\beta$. Finally, let $A = \bigcup_{\alpha < \omega_1} A_\alpha$ and $I = \bigcup_{\alpha < \omega_1} I_\alpha$.

From the above lemmas it is clear that each A_α is atomless, and hence A is atomless. Furthermore, I is a maximal ideal of A, and $|A \restriction a| = \omega$ for every $a \in I$. Moreover, $|A| = \omega_1$. The filter dual to I has character ω_1, so $\chi A = \omega_1$. In fact, let $F = \{a \in A : -a \in I\}$. Thus F is an ultrafilter on A. Assume that $\chi F = \omega$; say N is a countable generating set for F. Say $N \subseteq A_\alpha$, $\alpha < \omega_1$. Choose $b \in A_{\alpha+1} \backslash A$. Wlog $b \in F$. Then there exist $a_0, \ldots, a_{m-1} \in N$ such that $a_0 \cdot \ldots \cdot a_{m-1} \leq b$. So $-b \leq -a_0 + \cdots + -a_{m-1} \in A_\alpha \cap I$, so by Lemma 17.9, $-b \in A_\alpha$, contradiction.

Suppose that m is an uncountable incomparable set. Now trivially $a \subseteq b$ iff $\omega \backslash b \subseteq \omega \backslash a$, so we may assume that $m \subseteq I$. And wlog $m \cap A_0 = 0$. Let $S = \{\alpha : a_\alpha \in m\}$, and let Z be the set of all α satisfying the following two conditions:

(1) $\{a_\beta : \beta \in S \cap \alpha\} = m \cap A_\alpha$.
(2) For all $b \in A_\alpha \backslash I_\alpha$, if there is a $c \in m$ such that $c \subseteq b$, then there is a $\beta \in S \cap \alpha$ such that $a_\beta \subseteq b$.

Clearly Z is club. Hence by the $\Diamond$ property, choose $\alpha \in Z$ such that $S_\alpha = S \cap \alpha$.

(3) $m_\alpha \subseteq A_\alpha \cap m$.

For, let $x \in m_\alpha$. Say $x = a_\beta$ with $\beta \in S_\alpha = S \cap \alpha$. Since $\alpha \in Z$, we get $x \in m \cap A_\alpha$.

For each $c \in A \backslash A_0$ let ρc be the least β such that $c \in A_{\beta+1} \backslash A_\beta$. Now pick $c \in m \backslash m_\alpha$. Thus $c \notin A_0$. Write $\rho c = \beta$. Now $\beta \geq \alpha$: if $\beta < \alpha$, then $c \in A_\alpha \cap m$, so, since $\alpha \in Z$, we have $c = a_\gamma$ for some $\gamma \in S \cap \alpha = S_\alpha$, so $c \in m_\alpha$, contradiction.

Since c is M_β-generic (with respect to A_β and I_β) and $m_\alpha \in M_\beta$, there is a $(a, b) \in D_1 m_\alpha$ such that $a \subseteq c \subseteq b$. By the definition of $D_1 m_\alpha$ and since c is not comparable with any element of m_α, we must have $\forall c' \in m_\alpha(c' \not\subseteq b)$. Choose b' with $b' \in A_0$ or ($b' \notin A_0$ and $\rho b'$ minimum) such that $c \subseteq b' \notin I_\alpha$ and $\forall c' \in m_\alpha(c' \not\subseteq b')$.

Now $b' \notin A_0$ and $\rho b' \geq \alpha$: suppose not. Then $b' \in A_\alpha \backslash I_\alpha$. Now for any γ, if $\gamma \in S \cap \alpha$ then $\gamma \in S_\alpha$, $a_\gamma \in m_\alpha$, and $a_\gamma \not\subseteq b'$. This contradicts (2) for α.

Say $\rho b' = \gamma \geq \alpha$. Now b' is M_γ-generic (with respect to A_γ and I_γ) and $m_\alpha \in M_\gamma$, so there is a $(a'', b'') \in D_1 m_\alpha$ such that $a'' \subseteq b' \subseteq b''$; note that $a'', b'' \in A_\gamma$. For any $c' \in m_\alpha$ we have $c' \not\subseteq a''$; hence by the definition of $D_1 m_\alpha$ we have $\forall c' \in m_\alpha (c' \not\subseteq b'')$. Since $c \subseteq b'' \notin I_\alpha$ and $b'' \in A_0$ or ($b'' \notin A_0$ and $\rho b'' < \gamma$), this contradicts the minimality of $\rho b'$. Thus we have shown that A has no uncountable incomparable set.

If C is an uncountable chain in A, we may assume that $C \subseteq I$. We define $\langle c_\alpha : \alpha < \omega_1 \rangle$. Suppose $c_\beta \in C$ has been constructed for all $\beta < \alpha$. Say $\{c_\beta : \beta < \alpha\} \subseteq A_\gamma$. Then $\{c : c \in C$ and $c \leq c_\beta$ for some $\beta < \alpha\} \subseteq A_\gamma$ by Lemma 17.9(i). So, we can choose $c_\alpha \in C$ such that $c_\beta < c_\alpha$ for all $\beta < \alpha$. The sequence so constructed shows that $\text{Depth} A = \omega_1$; hence $\text{Inc} A = \omega_1$, contradiction. $\quad\square$

Problem 61. *Can one construct in ZFC a BA A such that $\text{Inc} A < \chi A$?*

This is Problem 54 in Monk [90]. This problem is equivalent to constructing in ZFC a BA A such that $\text{Inc} A < \text{hL} A$; see the argument at the end of Chapter 15. Note that "yes" on problem 61 implies "yes" on both problems 49 and 58.

We should mention in connection with Example 17.10 that Shelah [80], and independently van Wesep, showed that it is consistent to have 2^ω arbitrarily large and to have a BA of size 2^ω whose length and incomparability are countable.

We conclude this chapter with some remarks about incomparability in subalgebras of interval algebras. By Theorem 15.22 of Part I of the BA handbook, if κ is uncountable and regular, and B is a subalgebra of an interval algebra and $|B| = \kappa$, then B has a chain or incomparable subset of size κ. M. Bekkali has shown that it is consistent that this no longer holds for singular cardinals.

An important combinatorial equivalent for incomparability in interval algebras has been established by Shelah. Let μ be an infinite cardinal and let L be a linear order. We say that L is μ-*entangled* if for every $n \in \omega$, every system $\langle t^i_\zeta : i < n, \zeta < \mu \rangle$ of pairwise distinct elements of L with $t^0_\zeta < t^1_\zeta < \cdots < t^{n-1}_\zeta$, and every $w \subseteq n$ there exist $\zeta < \xi < \mu$ such that $\forall i < n(i \in w$ iff $t^i_\zeta < t^i_\xi)$. Shelah [90] showed that the following conditions are equivalent, for μ regular and uncountable:

(1) L is μ-entangled;
(2) If $\langle a_\alpha : \alpha < \mu \rangle$ is a sequence of elements of $\text{Intalg} L$ then there exist $\alpha < \beta < \mu$ such that $a_\alpha \leq a_\beta$;
(3) There is no incomparable subset of $\text{Intalg} I$ with μ elements.

Later Shelah showed in ZFC that for arbitrarily large cardinals λ there is a λ^+-entangled linear order of size λ^+.

18. Hereditary cofinality

Theorem 18.1. *For any infinite BA A, h-cof$A$ is equal to each of:*
$$\sup\{|T| : T \subseteq A,\ T\ \text{well-founded}\};$$
$$\sup\{\kappa : \text{there is an } a \in {}^{\kappa}A \text{ such that for all } \alpha, \beta < \kappa,\ \text{if } \alpha < \beta \text{ then } a_\alpha \not\geq a_\beta\}.$$

Proof. Call these three cardinals κ_0, κ_1. κ_2 respectively. Suppose that $\kappa_1 < \kappa_0$. Let X be a subset of A having no cofinal subset of power $\leq \kappa_1$. We construct elements $\langle x_\alpha : \alpha < \kappa_1^+ \rangle$ by induction: if x_α has been defined for all $\alpha < \beta$, with $\beta < \kappa_1^+$, then $\{x_\alpha : \alpha < \beta\}$ is not cofinal in X, so there is an $x_\beta \in X$ such that $x_\beta \not\leq x_\alpha$ for all $\alpha < \beta$. This finishes the construction. Now $\{x_\alpha : \alpha < \kappa_1^+\}$ is not well-founded, so there exist $\alpha 0, \alpha 1, \ldots < \kappa_1^+$ such that $x_{\alpha 0} > x_{\alpha 1} > \ldots$. Choose $i < j$ such that $\alpha i < \alpha j$. Then $x_{\alpha j} < x_{\alpha i}$ is a contradiction.

Suppose $\kappa_0 < \kappa_1$. Let T be a well-founded subset of A of power κ_0^+. If T has κ_0^+ incomparable elements, this is a contradiction. So T has $\geq \kappa_0^+$ levels. Let T' consist of all elements of T of level $< \kappa_0^+$. Let $X \subseteq T'$ be a cofinal subset of T' of cardinality $\leq \kappa_0$. Then choose $a \in T'$ of level greater than the levels of all members of X; clearly this is impossible.

Thus we have shown that $\kappa_0 = \kappa_1$. Next, we show that $\kappa_2 \leq \kappa_1$. Suppose that κ and a are as in the definition of κ_2; we show that $\{a_\alpha : \alpha < \kappa\}$ is well-founded. Suppose not: say $a_{\alpha_0} > a_{\alpha_1} > \cdots$. Then there exist $m < n$ such that $\alpha_m < \alpha_n$. Since $a_{\alpha_m} > a_{\alpha_n}$, this contradicts the defining property of a.

Finally, suppose that $\kappa_2 < \kappa_1$; we shall get a contradiction. Let T be well-founded, with $|T| > \kappa_2$. Write $T = \{a_\alpha : \alpha < \kappa\}$, with $\kappa > \kappa_2$ and a one-one. Now because $\kappa > \kappa_2$, it follows that for each $\Gamma \in [\kappa]^\kappa$ there are $\alpha, \beta \in \Gamma$ such that $\alpha < \beta$ and $a_\alpha > a_\beta$. Let $\Delta = \{\{\alpha, \beta\} : \alpha < \beta < \kappa \text{ and not}(a_\alpha > a_\beta)\}$. Then from the partition relation $\kappa \to (\kappa, \omega)_2$ we obtain $\alpha_0 < \alpha_1 < \cdots$ in κ such that $a_{\alpha_0} > a_{\alpha_1} > \cdots$, contradicting T well-founded. $\qquad\square$

Concerning ultraproducts, h-cof is a sup-min function, so Theorems 6.1–6.3 hold. h-cof $\left(\prod_{i \in I} A_i/F\right) \geq \left|\prod_{i \in I} \text{h-cof}A_i/F\right|$ under GCH for F regular, by a proof similar to that of Theorem 4.14, and Donder's theorem says that under $V = L$ the regularity assumption can be dropped. The inequality $>$ is consistently possible by Rosłanowski, Shelah [94]. We do not know whether $<$ is consistently possible:

Problem 62. *Is it consistent to have an example with h-cof $\left(\prod_{i \in I} A_i/F\right) < \left|\prod_{i \in I} \text{h-cof}A_i/F\right|$?*

Again, Magidor, Shelah [91] may help for this problem.

Concerning relationships to our other functions, the main facts are that Inc$A \leq$ h-cof$A \leq |A|$ and hL$A \leq$ h-cof$A$. To see that hL$A \leq$ h-cof$A$, suppose that $\langle x_\alpha : \alpha < \kappa \rangle$ is a right-separated sequence of elements of A. Then it is also well-founded. For, suppose that $x_{\alpha_0} > x_{\alpha_1} > \cdots$. Choose $i < j$ with $\alpha_i < \alpha_j$. Then $x_{\alpha_i} > x_{\alpha_j}$, so $x_{\alpha_j} \cdot -x_{\alpha_i}$, contradiction. It is obvious that Inc$A \leq$ h-cof$A \leq |A|$. An example in which hL$A <$ h-cof$A$ is provided by the interval algebra A on the reals. In fact, in Lemma 3.28 we showed that hL$A = \omega$, and in Chapter 17 we showed that

$\mathrm{Inc}A = 2^\omega$, and hence h-cof$A = 2^\omega$. Since $\chi A \leq \mathrm{hL}A \leq$ h-cof$A$, the Baumgartner-Komjath algebra of Chapter 17 provides an example where $\mathrm{Inc}A <$ h-cof$A$.

Problem 63. *Can one construct in ZFC a BA A with the property that $\mathrm{Inc}A <$ h-cof$A$?*

This is Problem 55 in Monk [90]. Note that "yes" on problem 61 implies "yes" on problem 63.

For interval algebras the equality h-cof$A = \mathrm{Inc}A$ holds. This was proved by Shelah [91], as an easy consequence of a result in Shelah [90], namely the equivalences mentioned at the end of the last chapter. In fact, let $A = \mathrm{Intalg}I$, and suppose that $\mathrm{Inc}A <$ h-cof$A$. Let $\mu = (\mathrm{Inc}A)^+$, and by Theorem 18.1 let T be a well-founded subset of A of power μ. Now each level of T is an incomparable set, so there are at least μ levels. Let $\langle a_\alpha : \alpha < \mu \rangle$ be a sequence of elements of T such that a_α has level α for each $\alpha < \mu$. Then by the above result there exist $\alpha < \beta < \mu$ such that $-a_\alpha \leq -a_\beta$. So $a_\beta < a_\alpha$, which is impossible.

To complete the picture, it remains to provide an example in which h-cof is less than cardinality. An example of this in ZFC was given by Shelah [91]; see also Rosłanowski, Shelah [94]. See also the algebra of Bonnet and Shelah [85], where CH is used. We are going to describe a different construction, the algebra of Rubin [83]. It requires $\Diamond$, but it will be used later too. It is relevant to many of our functions and problems.

Example 18.2. Rubin's construction is not direct, but goes by way of more general considerations. Let A be a BA. A *configuration* for A is, for some $n \in \omega$, an $(n+3)$-tuple $\langle a, c_1, c_2, b_1, \ldots, b_n \rangle$ such that $a, b_1, \ldots, b_n$ are pairwise disjoint, each $b_i \neq 0$, $c_1 \subseteq a + \sum_{i=1}^n b_i$, $a + c_1 \leq c_2$, and $(c_2 - c_1) \cdot b_i \neq 0$ for all $i = 1, \ldots, n$. (See Figure 18.3.)

Now we call a subset P of A *nowhere dense for configurations in A*, for brevity nwdc in A, if for every $n \in \omega \backslash 1$ and all disjoint $a, b_1, \ldots, b_n$ with each $b_i \neq 0$, there exist c_1, c_2 such that $\langle a, c_1, c_2, b_1, \ldots, b_n \rangle$ is a configuration and $P \cap (c_1, c_2) = 0$. Rubin's theorem that we are aiming for says that, assuming $\Diamond$, there is an atomless BA A of power ω_1 such that every set which is nwdc in A is countable. Before proceeding to the proof of this theorem, let us check that for such an algebra we have h-cof$A = \omega$. Suppose that P is an uncountable subset of A. Thus P is not nwdc, so we get $n \in \omega \backslash 1$ and disjoint $a, b_1, \ldots, b_n$ with each $b_i \neq 0$ such that

(1) For all c_1, c_2, if $\langle a, c_1, c_2, b_1, \ldots, b_n \rangle$ is a configuration then $P \cap (c_1, c_2) \neq 0$.

Let $c_0 = a + b_1 + \cdots + b_n$. Choose d_0 such that $a \leq d_0 \leq c_0$ and $d_0 \cdot b_i \neq 0 \neq -d_0 \cdot b_i$ for all i; this is possible since A is atomless. Thus $\langle a, d_0, c_0, b_1, \ldots b_n \rangle$ is a configuration, hence choose $c_1 \in P$ with $d_0 < c_1 < c_0$. Then $c_1 \cdot b_i \geq d_0 \cdot b_i \neq 0$ for all i, and $a \leq c_1$. Choose d_1 so that $a \leq d_1 \leq c_1$ and $d_1 \cdot b_i \neq 0 \neq c_1 \cdot -d_1 \cdot b_i$ for all i. then $\langle a, d_1, c_1, b_1, \ldots, b_n \rangle$ is a configuration, so choose $c_2 \in P$ with $d_1 < c_2 < c_1$. Continuing in this fashion, we get elements $c_1, c_2, \ldots$ of P such that $c_1 > c_2 > \cdots$, which means that P is not well-founded. This shows that h-cof$A = \omega$.

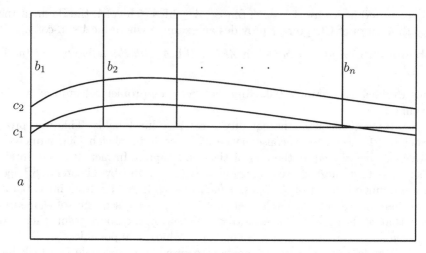

Figure 18.3.

To do the actual construction leading to Rubin's theorem, we need another definition and two lemmas. Let A be a BA and assume that $P \subseteq B \subseteq A$. We say that P is B-*nowhere dense for configurations in* A, for brevity P is B-nwdc in A, if for every $n \in \omega \backslash 1$ and all disjoint $a, b_1, \ldots, b_n \in A$ with each $b_i \neq 0$, there exist $c_1, c_2 \in B$ such that $\langle a, c_1, c_2, b_1, \ldots, b_n \rangle$ is a configuration and $P \cap (c_1, c_2) = 0$. Thus to say that P is nwdc in A is the same as saying that P is A-nwdc in A.

An important tool in the construction is the general notion of the free extension $A(x)$ of a BA A obtained by adjoining an element x (and other elements necessary when it is adjoined); this is the free product of A with a BA with four elements $0, x, -x, 1$. We need only this fact about this procedure:

Lemma 18.4. *Let A be a BA and $A(x)$ the free extension of A by an element x. Suppose that $\langle a_i : i \in I \rangle$ is a system of disjoint elements of A, $\langle b_i : i \in I \rangle$ is another system of elements of A, and $b_i \leq a_i$ for all $i \in I$. Let $I = \langle \{(a_i \cdot x) \triangle b_i : i \in I \rangle^{\mathrm{Id}}$, and let k be the natural homomorphism from $A(x)$ onto $A(x)/I$. Then $k \upharpoonright A$ is one-one.*

Proof. Suppose $kc = 0$, with $c \in A$. Then there exist $i(0), \ldots, i(m) \in I$ such that

$$c \leq (a_{i(0)} \cdot x) \triangle b_{i(0)} + \cdots + (a_{i(m)} \cdot x) \triangle b_{i(m)};$$

letting f be a homomorphism of $A(x)$ into A such that f is the identity on A and $fx = b_{i(0)} + \cdots + b_{i(m)}$, we infer that $c = 0$, as desired. $\qquad\square$

Note that the effect of the ideal I in Lemma 18.4 is to subject x to the condition that $x \cdot a_i = b_i$ for all $i \in \omega$. Now we prove the main lemma:

Lemma 18.5. *Let A be a denumerable atomless BA and for each $i < \omega$ we have $P_i \subseteq B_i \subseteq A$ with P_i is B_i-nwdc for A. Then there is a countable proper extension A' of A such that A is dense in A' and P_i is B_i-nwdc for A' for all $i < \omega$.*

Proof. Let $A(x)$ be a free extension of A by an element x; we shall obtain the desired algebra A' by the procedure of Lemma 18.4; thus we will let $A' = A(x)/I$, with I specified implicitly by defining a_j's and b_j's. Let $\langle s_n : n < \omega \rangle$ be an enumeration of the following set:

$$\{\langle 0, a \rangle : a \in A\} \cup \{\langle 1, a, b, c \rangle : a, b, c \text{ are disjoint elements of } A\} \cup$$
$$\{\langle 2, a, b, c, b_1, \dots, b_k, i \rangle : a, b, c \text{ are disjoint elements of } A, \ k \in \omega \backslash 1$$
$$b_1, \dots, b_k \text{ are disjoint non-zero elements of } A, \text{ and } i < \omega\}.$$

As we shall see, $\langle s_n : n < \omega \rangle$ is a list of things to be done in coming up with the ideal I. We will take care of the objects s_i by induction on i. Suppose that we have already taken care of s_i for $i < n$, having constructed a_j and b_j for this purpose, $j \in J$, so that J is a finite set, $b_j \leq a_j$ for all $j \in J$, the a_j's are pairwise disjoint, and $\sum_{j \in J} a_j < 1$. Let $u = \sum_{j \in J} a_j$ and $v = \sum_{j \in J} b_j$. We want to take care of s_n so that these conditions (called the "list conditions") will still be satisfied. Note that under I, $x \cdot u$ will be equivalent to v, and $-x \cdot u$ will be equivalent to $u \cdot -v$. Now we consider three cases, depending upon the value of the first term of s_n.

Case 1. The first term of s_n is 0; say $s_n = \langle 0, a \rangle$, where $a \in A$. We want to add new elements a_k and b_k to our lists in order to insure that $[x] \neq [a]$ in $A(x)/I$, where in general $[z]$ denotes the equivalence class of $z \in A(x)$ with respect to I. Thus the fact that this case is taken care of for all s_n of this type in our list will insure merely that $A(x)/I$ is a proper extension of A. If $a + u \neq 1$, choose e so that $0 < e < -(a + u)$, and set $a_k = b_k = e$. Then in the end we will have $(e \cdot x) \triangle e \in I$, hence $0 < [e] \leq [x]$, and $[e] \cdot [a] = 0$, so $[x] \neq [a]$. Clearly the list conditions still hold. Now suppose that $a + u = 1$. Thus $-u \leq a$, and $-u \neq 0$. Choose e with $0 < e < -u$. Let $a_k = e$ and $b_k = 0$. Then in the end we will have $e \cdot x \in I$, hence $[e] \cdot [x] = 0$, and $0 < [e] \leq [a]$, so $[a] \neq [x]$. And again the list conditions hold.

Case 2. The first term of s_n is 1; say $s_n = \langle 1, a, b, c \rangle$, where a, b, c are disjoint elements of A. We consider the element $t \stackrel{\text{def}}{=} a + b \cdot x + c \cdot -x$; we want to fix things so that if $[t]$ is non-zero then there will be some element $w \in A$ such that $0 < [w] \leq [t]$. This will insure that A will be dense in $A(x)/I$. Now

$$t = a + b \cdot x \cdot u + b \cdot x \cdot -u + c \cdot -x \cdot u + c \cdot -x \cdot -u,$$

and under I this is equivalent to

$$a + b \cdot v + u \cdot -v \cdot c + b \cdot -u \cdot x + c \cdot -u \cdot -x.$$

Let $a' = a + b \cdot v + u \cdot -v \cdot c$, $b' = b \cdot -u$, $c' = c \cdot -u$; thus a', b', c' are disjoint. If $a' \neq 0$, we don't need to add anything to our lists. Suppose that $b' \neq 0$. Then choose e with $0 < e < b'$, and add a_k, b_k to our lists, where $a_k = b_k = e$; this assures that $[e] \leq [x]$, hence $0 < [e] \leq [t]$; clearly the list conditions hold. If $c' \neq 0$ a similar procedure works. Finally, if $a' = b' = c' = 0$, then $[t] = 0$, and again we do not need to add anything.

Case 3. The first term of s_n is 2; say $s_n = \langle 2, a, b, c, b_1, \dots, b_k, i \rangle$, where a, b, c are disjoint elements of A, $k \in \omega \backslash 1$, $b_1, \dots, b_k$ are disjoint non-zero elements of A, and $i < \omega$. Let t be as in case 2. Case 3 is the crucial case, and here we will do one of three things: (1) make t equivalent to an element of A; (2) make sure that $[t] \cdot [b_j] \neq 0$ for some $j = 1, \ldots, k$; (3) find $c_1, c_2 \in B_i$ with $P_i \cap (c_1, c_2) = 0$ so that $\langle [t], [c_1], [c_2], [b_1], \ldots, [b_k] \rangle$ is a configuration. Thus this step will assure in the end that P_i is B_i-nwdc for A'. In fact, assume that the construction is completed. To show that P_i is B_i-nwdc for A', suppose that $k \in \omega$, $a', b'_1, \ldots, b'_k \in A'$ are disjoint with each $b'_i \neq 0$. Since A is dense in A', choose $b_i \in A$ with $0 < b_i \leq b'_i$ for each i. Write $a' = a + b \cdot [x] + c \cdot [-x]$ with a, b, c pairwise disjoint elements of A. Say $\langle 2, a, b, c, b_1, \ldots, b_k \rangle = s_n$. If (1) was done, the desired conclusion follows since P_i is B_i-nwdc for A. Since a' is disjoint from each b_i, (2) could not have been done. If (3) was done, the desired conclusion is clear.

Let a', b', c' be as in Case 2. If $\sum_{j=1}^{k} b_i \cdot a' \neq 0$, then (2) will automatically hold, and we do not need to add anything to our lists. If there is a j, $1 \leq j \leq k$, such that $b_j \cdot b' \neq 0$, let e be such that $0 < e < b_j \cdot b'$, and adjoin a_l, b_l to our lists, where $a_l = b_l = e$; then we will have $[e] \leq [x]$, and $[b_j] \cdot [t] \neq 0$, which means that (2) holds—and the list conditions are ok. Similarly if $b_j \cdot c' \neq 0$ for some j. If $b' + c' = 0$, then $[t] = [a']$, i.e., (1) holds. Thus we are left with the essential situation: $b' + c' \neq 0$, and $(a' + b' + c') \cdot b_j = 0$ for all $j = 1, \ldots, k$. First of all we use the fact that P_i is B_i-nwdc for A, applied to $a', b' + c', b_1, \ldots, b_k$, to get $c_1, c_2 \in B_i$ such that $P_i \cap (c_1, c_2) = 0$ and $\langle a', c_1, c_2, b' + c', b_1, \ldots, b_k \rangle$ is a configuration. This time we add elements a_l, a_m, b_l, b_m to our lists, where $a_l = c_1 \cdot (b' + c')$, $b_l = c_1 \cdot b'$, $a_m = (b' + c') \cdot -c_2$, and $b_m = c' \cdot -c_2$. Clearly $a_l \cdot a_m = 0$ and both elements are disjoint from previous a_j's. Obviously $b_l \leq a_l$ and $b_m \leq a_m$. Next, since $\langle a', c_1, c_2, b' + c', b_1, \ldots, b + k \rangle$ is a configuration, $c_2 \cdot -c_1 \cdot (b' + c') \neq 0$, and since this element is disjoint from all previous a_j's as well as from a_l and a_m it follows that $u + a_l + a_m < 1$. Thus the list conditions hold. It remains only to show that in the end $\langle [t], [c_1], [c_2], [b_1], \ldots, [b_k] \rangle$ is a configuration. The only things not obvious are that $[c_1] \leq [t + \sum_{j=1}^{k} b_j]$ and $[t] \leq [c_2]$. Since $\langle a', c_1, c_2, b' + c', b_1, \ldots, b_k \rangle$ is a configuration, we have $c_1 \leq a' + b' + c' + b_1 + \cdots + b_k$. Hence to show that $[c_1] \leq [t + \sum_{j=1}^{k} b_j]$, it suffices to prove that $[c_1 \cdot (b' + c')] \leq [t]$, which is done as follows. First note that our added elements a_l, a_m, b_l, b_m assure that $[x \cdot c_1 \cdot (b' + c')] = [c_1 \cdot b']$ and $[x \cdot (b' + c') \cdot -c_2] = [c' \cdot -c_2]$, hence $[c_1 \cdot b'] \leq [x]$ and $[c' \cdot -c_2] \leq [x]$. Now,

$$\begin{aligned}
[t] &\geq [b' \cdot x + c' \cdot -x] \\
&\geq [b' \cdot c_1 \cdot x + c' \cdot -x \cdot c_1] \\
&= [c_1 \cdot b' + c' \cdot c_1 \cdot -x] \\
&= [c_1 \cdot b' + (c' \cdot c_1) \cdot -(c' \cdot c_1 \cdot x)] \\
&\geq [c_1 \cdot b' + (c' \cdot c_1) \cdot -((b' + c') \cdot c_1 \cdot x)] \\
&= [c_1 \cdot b' + (c' \cdot c_1) \cdot -(c_1 \cdot b')] \\
&= [c_1 \cdot b' + c_1 \cdot c'] \\
&= [c_1 \cdot (b' + c')].
\end{aligned}$$

To show that $[t] \leq [c_2]$, it suffices to show that $[t \cdot (b' + c')] \leq [c_2]$, and that is done like this:

$$
\begin{aligned}
[t \cdot (b' + c')] &= [t \cdot (b' + c') \cdot c_2 + t \cdot (b' + c') \cdot -c_2] \\
&\leq [c_2 + t \cdot (b' + c') \cdot -c_2] \\
&= [c_2 + b' \cdot x \cdot (b' + c') \cdot -c_2 + c' \cdot -x \cdot (b' + c') \cdot -c_2] \\
&= [c_2 + b' \cdot c' \cdot -c_2 + c' \cdot -c_2 \cdot -x] \\
&= [c_2].
\end{aligned}
$$

This completes the construction and the proof.   □

Example 18.2 (Conclusion). Recall that we are trying to construct, using $\Diamond$, an atomless BA A of power ω_1 such that every nwdc subset of A is countable. We shall define by induction an increasing sequence $\langle A_\alpha : \alpha < \omega_1, \alpha \text{ limit} \rangle$ of countable BAs, and a sequence $\langle P_\alpha : \alpha < \omega_1, \alpha \text{ limit} \rangle$, such that: the universe of A_α is α; A_ω is atomless and is dense in A_α for all limit $\alpha < \omega_1$; $P_\alpha \subseteq A_\alpha$ for all limit $\alpha < \omega_1$, and P_α is A_α-nwdc for A_β whenever α, β are limit ordinals $< \omega_1$ with $\alpha \leq \beta$.

Let $\langle S_\alpha : \alpha < \omega_1 \rangle$ be a $\Diamond$-sequence. Let A_ω be a denumerable atomless BA. If λ is a limit of limit ordinals, $\lambda < \omega_1$, let $A_\lambda = \bigcup_{\alpha < \lambda} A_\alpha$. If S_λ is nwdc for A_λ, let $P_\lambda = S_\lambda$, and let $P_\lambda = 0$ otherwise. Now suppose that α is a limit ordinal $< \omega_1$, and A_β and P_β have been defined for all limit ordinals $\beta \leq \alpha$. By Lemma 18.5 let $A_{\alpha+\omega}$ be a BA with universe $\alpha + \omega$ such that A_α is dense in $A_{\alpha+\omega}$ and P_β is A_β-nwdc for $A_{\alpha+\omega}$ for all limit $\beta \leq \alpha$. And again choose $P_{\alpha+\omega} = S_{\alpha+\omega}$ if $S_{\alpha+\omega}$ is nwdc for $A_{\alpha+\omega}$, and let it be 0 otherwise. This completes the inductive definition. Let $A = \bigcup \{ A_\alpha : \alpha \text{ limit}, \alpha < \omega_1 \}$.

Clearly A is atomless and of power ω_1. Now suppose, in order to get a contradiction, that P is an uncountable nwdc subset of A. Let

$$
F = \{ \alpha : \alpha < \omega_1, \alpha \text{ limit, and } (A_\alpha, P \cap \alpha) \preceq_{ee} (A, P) \}.
$$

Here $\preceq_{ee}$ means "elementary substructure". Clearly F is club in ω_1. Now by the $\Diamond$-property, the set $S \overset{\text{def}}{=} \{ \alpha < \omega_1 : \alpha \cap P = S_\alpha \}$ is stationary, so we can choose $\alpha \in F \cap S$. Clearly nwdc can be expressed by a set of first-order formulas; so $P \cap \alpha$ is nwdc in A_α. Since $P \cap \alpha = S_\alpha$, the construction then says that $P_\alpha = S_\alpha$. Since P is uncountable, choose $a \in P \backslash P_\alpha$, and then choose $c_1, c_2 \in A_\alpha$ so that $\langle a, c_1, c_2 \rangle$ is a configuration (this means just so that $c_1 \leq a \leq c_2$) and $P_\alpha \cap (c_1, c_2) = 0$; this is possible, since if $a \in A_\beta$ with $\alpha \leq \beta$, then P_α is A_α-nwdc for A_β by the construction. But then we have

$$
\begin{aligned}
(A_\alpha, P_\alpha) &\models \forall x [P(x) \to x \notin (c_1, c_2)]; \\
(A, P) &\models P(a) \wedge x \in (c_1, c_2).
\end{aligned}
$$

This contradicts the fact that $(A_\alpha, P_\alpha) \preceq_{ee} (A, P)$.   □

19. Number of ultrafilters

This cardinal function is rather easy to describe, at least if we do not try to go into the detail that we did for cellularity, for example. If A is a subalgebra or homomorphic image of B, then $|\mathrm{Ult}A| \leq |\mathrm{Ult}B|$. For weak products we have $|\prod_{i\in I}^{\mathrm{w}} A_i| = \max(\omega, \sup_{i\in I}|\mathrm{Ult}A_i|)$. The situation for full products is more complicated:

$$\left|\mathrm{Ult}\left(\prod_{i\in I} A_i\right)\right| \leq 2^{2^{\kappa}},$$

where $\kappa = \sum_{i\in I} \mathrm{d}A_i$. This follows from the following two facts:

$$\prod_{i\in I} A_i \rightarrowtail \prod_{i\in I} \mathscr{P}(\mathrm{d}A_i) \cong \mathscr{P}\left(\overset{\bullet}{\bigcup}_{i\in I} \mathrm{d}A_i\right),$$

where "$\rightarrowtail$" means "is isomorphically embeddable in", and "$\overset{\bullet}{\bigcup}$" means "disjoint union". Next, clearly $|\mathrm{Ult} \oplus_{i\in I} A_i| = \prod_{i\in I} |\mathrm{Ult}A_i|$. We give some observations due to Douglas Peterson concerning ultraproducts and the number of ultrafilters. Clearly $\mathrm{Ult}\left(\prod_{i\in I} A_i/F\right) \geq \left|\prod_{i\in I} \mathrm{Ult}A_i/F\right|$. And if $\mathrm{ess.sup}_{i\in I}^F |A_i| \leq |I|$ and F is regular, then $|\mathrm{Ult}\left(\prod_{i\in I} A_i/F\right)| = 2^{2^{|I|}}$. This follows from one of the results stated for independence, for example.

Concerning relationships to our other functions, we mention only that $|A| \leq |\mathrm{Ult}A|$; and $2^{\mathrm{Ind}A} \leq |\mathrm{Ult}A|$ if $\mathrm{Ind}A$ is attained. This last assumption is needed. For example, if κ is an uncountable strong limit cardinal and A_α is the free BA of size $|\alpha + \omega|$ for each $\alpha < \kappa$, then $\prod_{\alpha<\kappa}^{\mathrm{w}} A_\alpha$ has independence κ and only κ ultrafilters. (These remarks are due to L. Heindorf, and correct a mistake in Monk [90].)

About $|\mathrm{Ult}A|$ for A in special classes of BAs: first recall from Theorem 17.10 of Part I of the BA handbook that $|\mathrm{Ult}A| = |A|$ for A superatomic. If A is not superatomic, then $|\mathrm{Ult}A| \geq 2^{\omega}$, since A has a denumerable atomless subalgebra B, and obviously $|\mathrm{Ult}B| = 2^{\omega}$.

20. Number of automorphisms

This cardinal function is not related very much to the preceding ones. To start with, we state some general facts about the size of automorphism groups in BAs; for proofs or references, see the chapter on automorphism groups in the BA handbook.

1. If A is denumerable, then $|\text{Aut}A| = 2^\omega$.
2. If $0 \neq m \in \omega$ and $\kappa > \omega$, then there is a BA A with $|A| = \kappa$ such that $|\text{Aut}A| = m!$.
3. If $|\text{Aut}A| < \omega$, then $|\text{Aut}A| = m!$ for some $m \in \omega$.
4. If MA and $|\text{Aut}A| = \omega$, then $|A| \geq 2^\omega$.
5. If $2^\omega \leq \kappa$, then there is a BA A such that $|\text{Aut}A| = \omega$ and $|A| = \kappa$.
6. If $\omega < \kappa \leq \lambda$, then there is a BA A with $|A| = \lambda$ and $|\text{Aut}A| = \kappa$.
7. If $\omega \leq \kappa$, then there is a BA A with $|A| = \kappa$ and $|\text{Aut}A| = 2^\kappa$.
8. Any BA can be embedded in a rigid BA.
9. Any BA can be embedded in a homogeneous BA.

Now we discuss algebraic operations on BAs vis-à-vis automorphism groups. If A is a subalgebra or homomorphic image of B, then $|\text{Aut}A|$ can vary in either direction from $|\text{Aut}B|$: embedding a rigid BA A into a homogeneous BA B, we get $|\text{Aut}A| < |\text{Aut}B|$, while if we embed a free BA A in a rigid BA B we get $|\text{Aut}A| > |\text{Aut}B|$; any rigid BA A is the homomorphic image of a free BA B, and then $|\text{Aut}A| < |\text{Aut}B|$; and finally, embed $A \overset{\text{def}}{=} \mathscr{P}\omega$ into a rigid BA B, and then extend the identity on A to a homomorphism from B onto A—this gives $|\text{Aut}A| > |\text{Aut}B|$.

Now we consider products. There are two fundamental, elementary facts here. First, $|A| \leq |\text{Aut}(A \times A)|$ for any BA A. This is easily seen by the following chain of isomorphisms, starting from any element $a \in A$ to produce an automorphism f_a of $A \times A$:

$$A \times A \overset{g}{\cong} (A \restriction a) \times (A \restriction -a) \times (A \restriction a) \times (A \restriction -a)$$
$$\overset{h}{\cong} (A \restriction a) \times (A \restriction -a) \times (A \restriction a) \times (A \restriction -a)$$
$$\overset{g^{-1}}{\cong} A \times A,$$

where g is the natural mapping and h interchanges the first and third factors, leaving the second and fourth fixed. If $a \neq b$, then $f_a \neq f_b$; in fact, say $a \not\leq b$; then $f_a(a,0) = (0,a)$ while $f_b(a,0) = (a \cdot -b, a \cdot b) \neq (0,a)$. This proves that $|A| \leq |\text{Aut}(A \times A)|$. The second fact is that the group $\text{Aut}A \times \text{Aut}B$ embeds isomorphically into $\text{Aut}(A \times B)$; an isomorphism F is defined like this, for any $f \in \text{Aut}A, g \in \text{Aut}B, a \in A, b \in B$: $(F(f,g))(a,b) = (fa, gb)$. Putting these two elementary facts together, we have $|A|, |\text{Aut}A|$ both $\leq |\text{Aut}(A \times A)|$. Shelah in an email message of December 1990 showed that actually equality holds (this solves Problem 56 of Monk [90]):

Theorem 20.1. *If A is an infinite BA, then $|\mathrm{Aut}(A \times A)| = \max(|A|, |\mathrm{Aut}A|)$.*

Proof. First note that $A' \overset{\text{def}}{=} (A \times A) \upharpoonright (1,0)$ and $A'' \overset{\text{def}}{=} (A \times A) \upharpoonright (0,1)$ are both isomorphic to A. Now for any $b \in A \times A$ let $G_b = \{g \in \mathrm{Aut}(A \times A) : g(1,0) = b\}$. Then

(*) For any $b \in A \times A$, $|G_b| \le |\mathrm{Aut}A|^2$.

For, take any $b \in A \times A$ and fix $f \in G_b$ (if $G_b \ne 0$). Note that for any $g \in G_b$, $f^{-1}g(1,0) = (1,0)$; so $(f^{-1} \circ g) \upharpoonright A' \in \mathrm{Aut}A'$, and similarly $(f^{-1} \circ g) \upharpoonright A'' \in \mathrm{Aut}A''$. Now the map

$$g \mapsto ((f^{-1} \circ g) \upharpoonright A', (f^{-1} \circ g) \upharpoonright A'')$$

is clearly one-one, so (*) follows.
    By (*),

$$|\mathrm{Aut}(A \times A)| = \sum_{b \in A \times A} |G_b| \le |A \times A| \cdot |\mathrm{Aut}A|^2,$$

and the theorem follows. $\qquad\qquad\qquad\qquad\qquad\qquad\qquad\qquad\qquad\qquad\qquad\square$

For weak products, we have $\sup_{i \in I} |\mathrm{Aut}A_i| \le |\mathrm{Aut}\left(\prod_{i \in I}^{\mathrm{w}} A_i\right)|$ by the above remarks; a similar statement holds for full products—in fact, the full direct product of groups $\prod_{i \in I} \mathrm{Aut}A_i$ is isomorphically embeddable in $\mathrm{Aut}\left(\prod_{i \in I} A_i\right)$.
    The situation for free products is much like that for products. By Proposition 11.11 of the BA handbook, Part I, every automorphism of A extends to one of $A \oplus B$; so $|\mathrm{Aut}(A \oplus B)| \ge \max(|\mathrm{Aut}A|, |\mathrm{Aut}B|)$. And $|A| \le |\mathrm{Aut}(A \oplus A)|$. In fact, choose $a \in A$ with $0 < a < 1$. Then $|(A \oplus A) \upharpoonright (a \times -a)| = |A|$, $(A \oplus A) \upharpoonright (a \times -a) \cong (A \oplus A) \upharpoonright (-a \times a)$, and

$$A \oplus A \cong [(A \oplus A) \upharpoonright (a \times -a)] \times [(A \oplus A) \upharpoonright (-a \times a)] \times [(A \oplus A) \upharpoonright c]$$

for some c, so our statement follows from the above considerations on products. Actually, Shelah showed that there is an infinite BA A such that $|A|$ and $|\mathrm{Aut}A|$ are both smaller than $|\mathrm{Aut}(A \oplus A)|$ (in an email message of December 1990; S. Koppelberg supplied some details and simplifications in January 1992). This solves Problem 57 in Monk [90]. Namely, we start with an uncountable cardinal κ and a system $\langle B_\alpha : \alpha < \kappa \rangle$ of rigid BAs of size κ such that $B_\alpha \upharpoonright b \not\cong B_\beta \upharpoonright c$ if $\alpha < \beta < \kappa$ and $b \in B_\alpha^+$, $c \in B_\beta^+$; for the existence of such a system see Shelah [83]. Let $A = \prod_{\alpha < \kappa}^{\mathrm{w}} B_\alpha$. Then A is also rigid, as is easy to check. We claim that $A \oplus A$ has 2^κ automorphisms. To see this we use duality. Recall that $\mathrm{Ult}A$ is homeomorphic to the one-point compactification of $\bigcup_{\alpha < \kappa} \mathrm{Ult}B_\alpha$. For each $\Gamma \subseteq \kappa$ we define $f_\Gamma : \mathrm{Ult}A \times \mathrm{Ult}A \to \mathrm{Ult}A \times \mathrm{Ult}A$ by

$$f_\Gamma(F, G) = \begin{cases} (G, F) & \text{if } F, G \in \mathrm{Ult}B_\alpha \text{ for some } \alpha \in \Gamma, \\ (F, G) & \text{otherwise.} \end{cases}$$

Obviously f_γ is one-one and onto, and it is easy to check that it is continuous. Since $f_\Gamma \neq f_\Delta$ for $\Gamma \neq \Delta$, this exhibits 2^κ autohomeomorphisms, as desired.

About the relationship between ultraproducts and automorphisms, it is easy to see that $\mathrm{Aut}\left(\prod_{i \in I} A_i/F\right) \geq \left|\prod_{i \in I} \mathrm{Aut}A_i/F\right|$. If CH holds and A is a rigid BA of power $\aleph_1$, then $^\omega A/F$ has at least $\aleph_1$ automorphisms; this is true because $^\omega A/F$ is ω_1-saturated and of power $\aleph_1$.

As mentioned at the beginning of this section, $|\mathrm{Aut}A|$ is not strongly related to our previous cardinal functions. An example with the property that $|\mathrm{Aut}A| < \mathrm{Depth}A$ is provided by embedding the interval algebra on κ into a rigid BA A. A similar procedure can be applied for independence and π-character, and these three examples show similar things for all of our preceding functions. And recall from the chapter on incomparability that if A is cardinality-homogeneous and has no incomparable subset of size $|A|$, then A is rigid.

Concerning automorphisms of special kinds of BAs, first note that $|\mathrm{Aut}A| = 2^\kappa$ for A the interval algebra on κ. In fact, every automorphism of A is induced by a permutation of κ; so we just need to describe 2^κ permutations of κ that give rise to automorphisms of A. For each $\alpha < \kappa$ we can consider the transposition $(\omega \cdot \alpha + 1, \omega \cdot \alpha + 2)$. For each $\varepsilon \in {}^\kappa 2$ let f_ε be the permutation of κ which, on the interval $[\omega \cdot \alpha, \omega \cdot \alpha + \omega)$, is this transposition if $\varepsilon\alpha = 1$, and is the identity there otherwise. It is easy to see that the function on A induced by f_ε maps into A, and hence is an automorphism, as desired.

If A is infinite and superatomic, then $|\mathrm{Aut}A| \geq 2^\omega$. In fact, we may assume that A is a subalgebra of some power-set algebra $\mathscr{P}\kappa$, and $\{\alpha\} \in A$ for all $\alpha < \kappa$. Let a be a representative of an atom of A at level 1. Suppose that f is a permutation of a such that f^2 is the identity. Extend f to all of κ by letting $f\alpha = \alpha$ if $\alpha \in \kappa\backslash a$. Now we claim that $x \in A$ implies that $f[x] \in A$; this will show that f induces an automorphism of A, hence proving the theorem.

Case 1. $a/I_1 \leq x/I_1$, where I_1 is the ideal of A generated by its atoms. Then $a\backslash x$ is finite. Hence
$$\begin{aligned} f[x] &= f[x \cap a] \cup f[x\backslash a] \\ &= (f[a]\backslash f[a\backslash x]) \cup f[x\backslash a] \\ &= (a\backslash f[a\backslash x]) \cup (x\backslash a) \in A, \end{aligned}$$
since $f[a\backslash x]$ is finite.

Case 2. $a/I_1 \cdot x/I_1 = 0$. Thus $a \cap x$ is finite. Hence
$$f[x] = f[a \cap x] \cup f[x\backslash a] = f[a \cap x] \cup (x\backslash a) \in A,$$
since $f[a \cap x]$ is finite.

21. Number of endomorphisms

The main relationships of $|\mathrm{End}A|$ to our previous functions are the following two easily established facts: $|\mathrm{Ult}A| \leq |\mathrm{End}A|$ and $|\mathrm{Aut}A| \leq |\mathrm{End}A|$. If A is the BA of finite and cofinite subsets of an infinite cardinal κ, then $|\mathrm{Ult}A| = \kappa$ while $|\mathrm{Aut}A| = |\mathrm{End}A| = 2^\kappa$. For an infinite rigid BA A we have $|\mathrm{Aut}A| < |\mathrm{End}A|$. Furthermore, we have:

Theorem 21.1. $|\mathrm{End}A| \leq |\mathrm{Ult}A|^{\mathrm{d}A}$ *for any infinite BA A.*

Proof. Let D be a dense subset of $\mathrm{Ult}A$ of cardinality $\mathrm{d}A$. Then any continuous function from $\mathrm{Ult}A$ into $\mathrm{Ult}A$ is determined by its restriction to D. Hence the theorem follows by duality. $\qquad\square$

It is more interesting to construct a BA A such that $|A| = |\mathrm{Ult}A| = |\mathrm{End}A|$, and we will spend the rest of this chapter discussing this. An easy example of this sort is the interval algebra of the reals, and we first want to generalize the argument for this. (Here we are repeating part of Monk [89].)

Theorem 21.2. *Suppose that L is a complete dense linear ordering of power $\lambda \geq \omega$, and D is a dense subset of L of power κ, where $\lambda^\kappa = \lambda$. Let A be the interval algebra on L. Then $|A| = |\mathrm{End}A| = \lambda$.*

Proof. Recalling the duality for interval algebras from Part I of the BA handbook, we see that $\mathrm{Ult}A$ is a linearly ordered space of size λ with a dense subset (in the topological sense) of power κ. Now apply Theorem 21.1. $\qquad\square$

Corollary 21.3. *If A is the interval algebra on $\mathbb{R}$, then $|A| = |\mathrm{End}A| = 2^\omega$.* $\qquad\square$

Recalling a construction of more general linear orders of the type described in Theorem 21.2 (see Monk [89]), we get

Corollary 21.4. *If μ is an infinite cardinal and $\forall \nu < \mu(\mu^\nu = \mu)$, then there is a BA A such that $|A| = |\mathrm{End}A| = 2^\mu$.* $\qquad\square$

Corollary 21.5. (GCH) *If κ is infinite and regular, then there is a BA A such that $|A| = |\mathrm{End}A| = \kappa^+$.* $\qquad\square$

Corollary 21.6. *Let λ be strong limit, let L consist of all members of $^\lambda 2$ which are not eventually 1, and let A be the interval algebra on L (which is lexicographically ordered). Then $|A| = |\mathrm{End}A| = 2^\lambda$.*

Proof. Let D consist of all members $f \in {}^\lambda 2$ such that there is an α with $f\alpha = 0$ and $f\beta = 1$ for all $\beta > \alpha$. Then D is dense in L and Theorem 21.2 applies. $\qquad\square$

Corollary 21.6 was pointed out by Shelah (answering Problems 58 and 59 in Monk [90].) We mention one more result connecting $|A|$ and $|\mathrm{End}A|$:

Theorem 21.7. *If $|A| = \omega_1$, then $|\mathrm{End}A| \geq 2^\omega$.*

Proof. If A has an atomless subalgebra, then $|\mathrm{End}A| \geq |\mathrm{Ult}A| \geq 2^\omega$. So suppose that A is superatomic. Then there is a homomorphism f from A onto B, the finite-cofinite algebra on ω: if a is an atom of $A/\langle \mathrm{At}A \rangle^{\mathrm{Id}}$, then f can be taken to be the composition of the natural onto mappings

$$A \to A \restriction a \to C \to B,$$

where C is the finite-cofinite algebra on ω or ω_1. There is an isomorphism g of B into A. If X is any subset of ω with $\omega \backslash X$ infinite, then $B/\langle \{i\} : i \in X \rangle^{\mathrm{Id}}$ is isomorphic to B, and so there is an endomorphism k_X of B with kernel $\langle \{i\} : i \in X \rangle^{\mathrm{Id}}$. Clearly the endomorphisms $g \circ k_X \circ f$ of A are distinct for distinct X's. $\square$

Corollary 21.8. $(\omega_1 < 2^\omega)$. *There is no BA A with the property that $|A| = |\mathrm{End}A| = \omega_1$.* $\square$

22. Number of ideals

The main relationships with our earlier functions are: $|\mathrm{Ult}A| \leq |\mathrm{Id}A|$ and $2^{\mathrm{s}A} \leq |\mathrm{Id}A|$; both of these facts are obvious. Also recall the deep Theorem 10.10 from Part I of the BA handbook: if A is an infinite BA, then $|\mathrm{Id}A|^{\omega} = |\mathrm{Id}A|$. This result is due to Shelah [86b]; there he also proves that if κ is a strong limit cardinal of size at most $|A|$, then $|\mathrm{Id}A|^{<\kappa} = |\mathrm{Id}A|$. Note that $|\mathrm{Ult}A| < |\mathrm{Id}A|$ for A the finite-cofinite algebra on an infinite cardinal κ.

Next, we show that $|\mathrm{Id}A| = 2^{\omega}$ for the interval algebra A on the reals; thus A has the property that $|A| = |\mathrm{Ult}A| = |\mathrm{Id}A|$. For each ideal I on A, let $\equiv_I$ be defined as follows: $a \equiv_I b$ iff $a = b$ or else if, say $a < b$, then $[a, b) \in I$. Thus $\equiv_I$ is a convex equivalence relation on $\mathbb{R}$. Now define the function f by setting, for any ideal I,

$$fI = \{(r, s, \varepsilon) : \text{there is an equivalence class } a \text{ under } \equiv_I \text{ such that } |a| > 1$$
$$\text{and } a \text{ has left endpoint } r, \text{ right endpoint } s, \text{ and}$$
$$\varepsilon = 0, 1, 2, 3 \text{ according as } a \text{ is } [r, s], [r, s), (r, s], \text{ or } (r, s)\}.$$

Clearly f is a one-one function; since $fI \in (\mathbb{R} \times \mathbb{R} \times 4)^{\leq \omega}$, it follows that $|\mathrm{Id}\mathbb{R}| = 2^{\omega}$, as desired.

A rigid BA A shows that $|\mathrm{Aut}A| < |\mathrm{Id}A|$ is possible. Koppelberg, Shelah [93] show that if μ is a strong limit cardinal satisfying $\mathrm{cf}(\mu) = \omega$ and $2^{\mu} = \mu^+$, then there is a Boolean algebra B such that $|B| = |\mathrm{End}B| = \mu^+$ and $|\mathrm{Id}B| = 2^{\mu^+}$. This answers Problem 60 of Monk [90]. Also, in an email message of December 1990 Shelah showed that under suitable set-theoretic hypotheses there is a BA A such that $|\mathrm{Id}A| < |\mathrm{Aut}A|$, answering problem 61 from Monk [90]. This result is easy to see from known facts. Namely, let T be a Suslin tree in which each element has infinitely many immediate successors, and with more than ω_1 automorphisms. Assume CH. Then $A \stackrel{\mathrm{def}}{=} \mathrm{Treealg}\,T$ has more than ω_1 automorphisms. By the characterization of the cellularity of tree algebras given in Chapter 3, $cA = \omega$, and since $\mathrm{h}LA = cA$ (see the end of Chapter 15), by the equivalents at the beginning of Chapter 15 every ideal in A is countably generated, and hence A has only ω_1 ideals. The following problem is open.

Problem 64. *Can one construct in ZFC a BA A such that $|\mathrm{Id}A| < |\mathrm{Aut}A|$?*

23. Number of subalgebras

First we note the following simple result:

Proposition 23.1. *If B is a homomorphic image of A, then $|\mathrm{Sub}B| \leq |\mathrm{Sub}A|$.*

Proof. Let f be a homomorphism from A onto B. With each subalgebra C of B associate the subalgebra $f^{-1}[C]$ of A. $\qquad\square$

It is also obvious that if B is a subalgebra of A, then $|\mathrm{Sub}B| \leq |\mathrm{Sub}A|$.

Now we give some results from Shelah [92a]; the main fact is that $|\mathrm{End}A| \leq |\mathrm{Sub}A|$. This answers Problem 63 from Monk [90]. Let $\mathrm{Psub}A$ be the collection of all subsets of A closed under $+$, $\cdot$, and $-$ (as a binary operation—namely, $a - b = a \cdot -b$). The part $|\mathrm{Id}A| \leq |\mathrm{Sub}A|$ in the next theorem is due to James Loats, and can be proved more easily.

Theorem 23.2. $|\mathrm{Psub}A| = |\mathrm{Sub}A|$ *for any infinite BA A. In particular, $|\mathrm{Id}A| \leq |\mathrm{Sub}A|$.*

Proof. First, it is clear that $|A| \leq |\mathrm{Sub}A|$, since $a \mapsto \{0, 1, a, -a\}$ $(a \in A)$ is a 2-to-1 mapping from A into $\mathrm{Sub}A$. Hence for the theorem it suffices to show that the set of infinite members of $\mathrm{Psub}(A)$ has cardinality at most $|\mathrm{Sub}(A)|$. To this end, choose for every infinite $X \in \mathrm{Psub}(A)$ an element a_X of X such that there are elements $u, v \in X$ with $0 < u < a_X < v < 1$; then let $Y[X]$ be the subalgebra generated by $X \upharpoonright a_X$ and let $Z[X]$ be the subalgebra generated by $X \upharpoonright -a_X$. Note that $Y[X]$ consists of all elements of the form $y + -z$ with $y \in X \upharpoonright a_X$, and either $z \in X \upharpoonright a_X$ or $z = 1$, and similarly for $Z[X]$. Now we claim that for $X \neq X'$ we have $Y[X] \neq Y[X']$ or $Z[X] \neq Z[X']$, from which the desired conclusion clearly follows. Suppose that this claim fails; say $X \backslash X' \neq 0$. Let $a = a_X$ and $b = a_{X'}$. We claim next that $-a \leq b$ or $a \leq -b$. For, take any $x \in X \backslash X'$. Then we can write

$$x \cdot a = y + -z, \ y \in X' \upharpoonright b, \text{ and } z \in X' \upharpoonright b \text{ or } z = 1,$$
$$x \cdot -a = u + -v, \ u \in X' \upharpoonright -b, \text{ and } v \in X' \upharpoonright -b \text{ or } v = 1.$$

Since $x \notin X'$, we have $z \neq 1$ or $v \neq 1$; this gives in the first case $-x + -a = z \cdot -y \leq b$, hence $-a \leq b$, and $a \leq -b$ in the second case, proving our latest claim.

Say without loss of generality that $-a \leq b$. Choose $b' \in X'$ such that $b < b' < 1$. So $0 < -b' < -b$. Thus $-b' \in X \upharpoonright -b$, so $-b' = s + -t$, where $s \in X \upharpoonright -a$, and $t \in X \upharpoonright a$ or $t = 1$. If $t = 1$, then $-b' = s \leq -a \leq b$ and $-b' \leq -b$, contradiction. If $t \neq 1$, then $b' = t \cdot -s \leq -a \leq b$, so $-b \leq -b'$, contradiction. $\qquad\square$

Lemma 23.3. $|\mathrm{Aut}A| \leq |\mathrm{Sub}A|$ *for A infinite.*

Proof. The idea is to associate with each automorphism of A a sequence of 8 members of $\mathrm{Psub}A$, in a one-one fashion; clearly this will prove the theorem. Let f be any automorphism. Define

$$J^f = \{x \in A : fy = y \text{ for all } y \leq x\};$$
$$I^f = \{x \in A : x \cdot fx = 0\}.$$

Clearly we have:

(1) If $x \in I^f$ then $fx \in I^f$.
(2) $J^f \cup I^f$ is dense in A.

To prove (2), let $a \in A^+$, and suppose that $a \notin J^f$. Then there is a $b \leq a$ such that $b \neq fb$. If $b \nleq fb$, then $0 \neq b \cdot -fb \leq a$ and $b \cdot -fb \in I^f$. If $fb \nleq b$, then $0 \neq b \cdot -f^{-1}b \leq a$ and $b \cdot -f^{-1}b \in I^f$.

Next, let X be a maximal subset of I^f such that $x \cdot fy = 0$ for all $x, y \in X$. Let $I_1^f = \langle X \rangle^{\mathrm{Id}}$ and $I_2^f = \langle \{fx : x \in X \rangle^{\mathrm{Id}} \}$. Then let

$$I_0^f = \{y \in I^f : fy \in I_1^f \text{ and } y \cdot x = 0 \text{ for all } x \in I_1^f\}.$$

Let $I_3^f = \langle \{fx : x \in I_2^f\} \rangle^{\mathrm{Id}}$. Thus

(3) I_i^f is an ideal $\subseteq I^f$ for $i < 4$.
(4) $I_i^f \cap I_{i+1}^f = \{0\}$ for $i < 3$.
(5) $I_0^f \cup I_1^f \cup I_2^f$ is dense in I^f.

To prove (5), suppose that $x \in (I^f)^+$ but there is no non-zero member of the indicated union which is below it. Since $x \in I^f$, we have $x \cdot fx = 0$. Now since there is no nonzero element of I_1^f below x, we have $x \notin X$, and this gives two cases. *Case 1.* There is a $z \in X$ such that $x \cdot fz \neq 0$. But $x \cdot fz \in I_2^f$, contradiction. *Case 2.* There is a $z \in X$ such that $z \cdot fx \neq 0$. Choose w such that $fw = z \cdot fx$. Since $w \leq x$, we have $w \in (I^f)^+$, and since $fw \leq z$ we have $fw \in I_1^f$. For any $t \in I_1^f$ we have $t \cdot x = 0$ (otherwise $0 \neq t \cdot x \leq x$ and $t \cdot x \in I_1^f$), hence $t \cdot w = 0$. Thus $w \in I_0^f$, contradiction, proving (5).

For $i < 3$ we now define a member C_i^f of Psub$A$: $C_i^f = \{x + fx : x \in I_i^f\}$. Clearly each C_i^f is closed under $+$. For any $x, y \in I_i^f$ we have $x \cdot fy = 0$, and from this it follows that C_i^f is closed under $\cdot$ and $-$.

We have now defined our sequence

$$\langle J^f, I_0^f, I_1^f, I_2^f, I_3^f, C_0^f, C_1^f, C_2^f \rangle$$

of 8 members of Psub$A$. It remains just to show that if two automorphisms f and g give rise to the same sequence $\langle J, I_0, I_1, I_2, I_3, C_0, C_1, C_2 \rangle$, then $f = g$. By (2) and (5) it is enough to show that they agree on $J \cup I_0 \cup I_1 \cup I_2$. Suppose that $y \in I_0$. Thus $y + fy \in C_0$, so there is a $z \in I_0$ such that $y + fy = z + gz$. Now $0 = y \cdot gz$ since $gz \in I_1$, so $y \leq z$. Similarly $z \leq y$, so $y = z$. So $y + fy = y + gy$. But $y \in I^f$, so $y \cdot fy = 0$. Similarly $y \cdot gy = 0$, so $fy = gy$. The cases of I_1 and I_2 are similar. $\square$

Theorem 23.4. $|\mathrm{End}A| \leq |\mathrm{Sub}A|$ *for any infinite BA A.*

Proof. Let $\mu = |\mathrm{Sub}A|$, and suppose that $\mu < |\mathrm{End}A|$. With each $f \in \mathrm{End}A$ associate the pair $(\mathrm{Kernel}f, \mathrm{Range}f)$. The number of such pairs is at most $|\mathrm{Id}A| \times$

$|\text{Sub}A|$, which by 23.2 is μ. Thus there is a set E of μ^+ endomorphisms with the same kernel I and range R. For each $f \in E$ we define a mapping g_f from A/I onto R by $g_f(x/I) = fx$; clearly g_f is well-defined and is an isomorphism from A/I onto R. Fix $h \in E$. Then $\{g_f \circ g_h^{-1} : f \in E\}$ is a set of μ^+ different automorphisms of R. Thus by 23.3,

$$\mu < |\text{Aut}R| \le |\text{Sub}R| \le |\text{Sub}A|,$$

contradiction.   □

Another result in Shelah [92a] is that $|\text{Aut}A|^\omega \le |\text{Sub}A|$; we shall not give the proof.

For any BA A, let $\text{Pend}A = \{f : f$ is a homomorphism from a subalgebra of A onto another subalgebra of $A\}$. Such homomorphisms are called *partial endomorphisms* of A.

Theorem 23.5. $|\text{Pend}A| = |\text{Sub}A|$ *for any infinite BA* A.

Proof. $\ge$ is clear, since with each subalgebra B one can associate the identity mapping on B, a partial endomorphism of A. For $\le$ we proceed as in the proof of Theorem 23.4: this time we associate with each partial endomorphism a triple consisting of its domain, kernel, and range; otherwise the details are similar.   □

Theorem 23.6. *If* $A \times B$ *is infinite, then* $|\text{Sub}(A \times B)| = \max(|\text{Sub}A|, |\text{Sub}B|)$.

Proof. We prove the equivalent statement that if A is infinite and $a \in A$, then $|\text{Sub}A| = \max(|\text{Sub}(A \restriction a)|, |\text{Sub}(A \restriction -a)|)$. The inequality $\ge$ is obvious. Let $\mu = \max(|\text{Sub}(A \restriction a)|, |\text{Sub}(A \restriction -a)|)$, and suppose that $\mu < |\text{Sub}A|$. Now we associate with each subalgebra B of A five objects:

$$C_0^B = \{x \cdot a : x \in B\}, \text{ a subalgebra of } A \restriction a;$$
$$C_1^B = \{x \cdot -a : x \in B\}, \text{ a subalgebra of } A \restriction -a;$$
$$I_0^B = \{x \in B : x \le a\}, \text{ an ideal of } C_0^B;$$
$$I_1^B = \{x \in B : x \le -a\}, \text{ an ideal of } C_1^B;$$

and g_B, the isomorphism from C_0^B/I_0^B onto C_1^B/I_1^B such that $g_B((x \cdot a)/I_0^B) = (x \cdot -a)/I_1^B$ for all $x \in B$. Now B can be reconstructed from these five objects:

$$B = \langle I_0^B \cup I_1^B \cup \{x + y : x \in C_0^B, y \in C_1^B, \text{ and } g_B(x/I_0^B) = y/I_1^B\}\rangle$$

In fact, the direction $\subseteq$ is clear. For $\supseteq$ it suffices to show that if $x \in C_0^B$, $y \in C_1^B$, and $g_B(x/I_0^B) = y/I_1^B$ then $x + y \in B$. Say $x = x' \cdot a$ and $y = y' \cdot -a$, with $x', y' \in B$. Now $(x' \cdot -a)/I_1^B = (y' \cdot -a)/I_1^B$, so $(x' \cdot -a)\triangle(y' \cdot -a) \in B$. Now

$$(x + y)\triangle(x' \cdot -a)\triangle(y' \cdot -a) = (x' \cdot a)\triangle(y' \cdot -a)\triangle(x' \cdot -a)\triangle(y' \cdot -a)$$
$$= (x' \cdot a)\triangle(x' \cdot -a) = x' \in B,$$

so $x + y \in B$.

Now there is a set X of μ^+ subalgebras of A such that $C_i^B = C_i^D$ and $I_i^B = I_i^D$ for all $B, D \in X$ and all $i < 2$. Fix $B \in X$. Now $\{g_D^{-1} \circ g_B : D \in X\}$ is a set of automorphisms of C_0^B/I_0^B, and by 23.1 and 23.3,

$$|\text{Aut}(C_0^B/I_0)| \leq |\text{Sub}(C_0^B/I_0)| \leq |\text{Sub}C_0^B| \leq \mu,$$

so there are distinct $D, E \in X$ for which $g_D = g_E$. This contradicts the noted fact about reconstruction.   □

Note that $2^{\text{Irr}A} \leq |\text{Sub}A|$ if $\text{Irr}A$ is attained. An example A for which $|\text{Id}A| < |\text{Sub}A|$ is provided by the interval algebra A on the reals. We noted in the last chapter that $|\text{Id}A| = 2^\omega$. Since $\text{Irr}A = 2^\omega$ attained, we have $|\text{Sub}A| = 2^{2^\omega}$.

The above theorems imply that $|\text{Sub}A|$ is our biggest cardinal function. Its size is, of course, always at most $2^{|A|}$. For most algebras, this value is actually attained. It is also quite interesting to construct a BA A in which $|\text{Sub}A|$ is as small as possible. The only algebra we know of where this is the case is Rubin's algebra A from Chapter 18. We now go through the proof that $|\text{Sub}A| = \omega_1$; thus $|A| = |\text{Ult}A| = |\text{Id}A| = |\text{Sub}A|$. We call an element $a \in A$ *countable* provided that $A \restriction a$ is countable.

Lemma 23.7. *A has only countably many countable elements.*

Proof. Let P be the set of all countable elements of A, and suppose that P is uncountable. Then it is easy to construct $\langle a_\alpha : \alpha < \omega_1 \rangle \in {}^{\omega_1}P$ such that if $\alpha < \beta < \omega_1$ then $a_\beta \not\leq a_\alpha$. Since h-cof$(A) = \omega$, the set $P' \overset{\text{def}}{=} \{a_\alpha : \alpha < \omega_1\}$ is not well-founded; say $a_{\alpha(0)} > a_{\alpha(1)} > \cdots$. Choose $i, j \in \omega$ such that $i < j$ and $\alpha(i) < \alpha(j)$. Then $a_{\alpha(i)} > a_{\alpha(j)}$ contradicts the choice of the a_β's.   □

Note that the collection of countable elements of A forms an ideal, which we denote by $C(A)$. To proceed further, we have to go back to the main property of Rubin's algebra. For this purpose we introduce the following notation. A *preconfiguration* in a BA A is a sequence $\langle a, b_1, \ldots, b_n \rangle$ of pairwise disjoint elements of A with each $b_i \neq 0$, $n > 0$. Given such a preconfiguration, a subset P of A is *dense at* $\langle a, b_1, \ldots, b_n \rangle$ provided that for all c_1, c_2, if $\langle a, c_1, c_2, b_1, \ldots, b_n \rangle$ is a configuration then $P \cap (c_1, c_2) \neq 0$. Thus the main property of Rubin's BA A is that if P is an uncountable subset of A then there is a preconfiguration $\langle a, b_1, \ldots, b_n \rangle$ of A such that P is dense at $\langle a, b_1, \ldots, b_n \rangle$. For both of the next two lemmas we advise the reader to draw a diagram along the lines of the one in Chapter 18 to see what is going on.

Lemma 23.8. *Assume that $P \subseteq A$, P is uncountable, $\langle a, b_1, \ldots, b_n \rangle$ is a preconfiguration of A, P is dense at $\langle a, b_1, \ldots, b_n \rangle$, $a \leq b \leq a + \sum_{i=1}^n b_i$, $b \cdot b_i \neq 0$ for $i = 1, \ldots, n$, and $b \cdot -a \notin C(A)$. Then $P \cap [a, b]$ is uncountable.*

Proof. Suppose that $P \cap [a, b]$ is countable, and let Q be the closure of $P \cap [a, b]$ under $+$; so Q is countable also. Say $b \cdot b_i \notin C(A)$. Pick any $c \leq b \cdot b_i$, $c \neq 0$, such that $c' \overset{\text{def}}{=} a + c + \sum_{j \neq i} b_j \notin Q$. Then pick c_1, c_2 so that $a + c_i < c'$ and

$\langle a, c_i, c', b_1, \ldots, b_n \rangle$ is a configuration for $i = 1, 2$, and $c_1 + c_2 = c'$. Then pick $d_1, d_2 \in P$ so that $c_i < d_i < c'$ for $i = 1, 2$. But $d_1, d_2 \in [a, b]$ and $d_1 + d_2 = c'$, so $c' \in Q$, contradiction. $\qquad \square$

Lemma 23.9. *Every subalgebra of A is the union of countably many closed intervals.*

Proof. Suppose that B is a subalgebra of A which is not the union of countably many closed intervals. Let $\langle [x_\alpha, y_\alpha] : \alpha < \omega_1 \rangle$ enumerate all of the closed intervals contained in B. Now B contains ω_1 elements pairwise inequivalent with respect to $C(A)$; hence it is easy to construct a sequence $\langle z_\alpha : \alpha < \omega_1 \rangle \in {}^{\omega_1}B$ with the following two properties:

(1) $z_\alpha \notin \bigcup_{\beta < \alpha} [x_\beta, y_\beta]$ for each $\alpha < \omega_1$;
(2) $z_\alpha \triangle z_\beta \notin C(A)$ for distinct $\alpha, \beta < \omega_1$.

Let $D = \{ z_\alpha : \alpha < \omega_1 \}$. Since D is somewhere dense, let $\langle a, b_1, \ldots, b_n \rangle$ be a preconfiguration of A such that D is dense at $\langle a, b_1, \ldots, b_n \rangle$. Choose any b such that $a \leq b \leq a + \sum_{i=1}^{n} b_i$ and $b \cdot b_i \neq 0 \neq b_i \cdot -b$ for all $i = 1, \ldots, n$. We show that $[a, b] \subseteq B$. Take any $d \in [a, b]$. Choose e_1, e_2 with the following properties: $d = e_1 \cdot e_2$; $a \leq e_i \leq a + \sum_{j=1}^{n} b_j$; $e_i \cdot b_j \neq 0$ for $i = 1, 2$, $j = 1, \ldots, n$. Thus $\langle a, d, e_i, b_1, \ldots, b_n \rangle$ is a configuration, so we can choose $f_i \in D \cap (d, e_i)$ for $i = 1, 2$. Then $f_1 \cdot f_2 = d$, and so $d \in B$ (since $D \subseteq B$). Thus, indeed, $[a, b] \subseteq B$. By an easy argument, $[a, b] \cap D$ has at least two elements. Since distinct elements of D are inequivalent mod $C(A)$, it follows that $b \cdot -a \notin C(A)$. Hence by Lemma 23.8, $D \cap [a, b]$ is uncountable. But this clearly contradicts the construction of D. $\qquad \square$

With these lemmas available we now prove that $|\mathrm{Sub}A| = \omega_1$. We claim that each subalgebra B of A is generated by an ideal along with a countable set. In fact, write $B = \bigcup_{i < \omega} [a_i, b_i]$. Let I be the ideal generated by $\{ b_i \cdot -a_i : i < \omega \}$. Then clearly B is generated by $I \cup \{ b_i : i < \omega \}$, as required. Now every ideal is countably generated. This follows from the fact that $\mathrm{hL}A \leq \mathrm{h\text{-}cof}A = \omega$, proved in Chapter 18, and one of the equivalents of hL given in Chapter 15. This being the case, it follows that there are exactly ω_1 ideals in A, since $\omega_1^\omega = \omega_1$ by virtue of CH (which follows from $\diamondsuit$, which we are assuming). Now $|\mathrm{Sub}A| = \omega_1$ is clear, again using CH.

In Cummings, Shelah [95] it is shown that it is consistent (relative to a large cardinal assumption) that every infinite BA A has $2^{|A|}$ subalgebras. This answers Problem 62 in Monk [90]. It is possible to have $|\mathrm{Aut}A| < |\mathrm{Sub}A|$: take a rigid BA. One can even have $|\mathrm{End}A| < |\mathrm{Sub}A|$, for example in the interval algebra on the reals.

24. Other cardinal functions

There are many other cardinal functions besides the 21 that we have discussed in the preceding chapters. In this chapter we give a list of some natural ones; some of these have been explicitly mentioned earlier. We also mention some facts and problems about them, without trying to be exhaustive. In particular, many of the problems may be easy, so we do not list them among our formal problems.

Functions mentioned in the previous text

1. c_{mm}. See the text following 3.26 for the definition and properties of this function. It is clear that $c_{mm}A \leq cA$ for any infinite BA A.

2. The following function is related to c_{mm}; see the text following 3.26:

$$\mathfrak{p}A = \min\{|Y| : Y \subseteq A, \sum Y = 1,$$
$$\text{and } \sum Y' \neq 1 \text{ for every finite subset } Y' \text{ of } Y\}.$$

3. $_d\mathrm{Depth}_{S-}$. See the text following 4.23. Obviously $_d\mathrm{Depth}_{S-}A \leq \mathrm{Depth}A$.

4. tow. See the text following 4.23. Clearly $\mathrm{tow}A \leq \mathrm{Depth}A$; tow is a variant of Depth.

5. d_n. See the discussion following 5.14. It is noted in 5.15 that $d_nA \leq dA$ for any infinite A. These functions are finite variants of d.

6. π_{S+}. See the discussion following 6.10. It is proved there that $\pi A \leq \pi_{S+}A \leq \mathrm{hd}A$ for any infinite BA A, with strict inequality possible in both cases.

7. Length_{H+}. See the discussion after 7.7. We have $tA \leq \mathrm{Length}_{H+}A$, with $<$ possible.

8. Length_{H-}. See the discussion after 7.9; 7.8 and 7.9 are also relevant.

9. Length_{h-}. It is possible that this function always has value ω (Problem 23).

10. Length_{h+}. See the end of Chapter 7. This function is related to Depth_{h+} and Length_{H+}.

11. $_d\mathrm{Length}_{S-}$. This is related to the above function $_d\mathrm{Depth}_{S-}$. See the end of Chapter 7.

12. Irr_{mn}. These are finite versions of irredundance, defined at the end of Chapter 8.

13. Card_{H-}. See Chapter 9 for information.

14. Card_{h-}. See Chapter 9. It is possible to have $\mathrm{Card}_{h-}A > |\mathrm{Ult}A|$.

15. Ind_{H-}. See Chapter 10, discussion around Problem 29.

16. Ind_{h+}. See Chapter 10, discussion around Problem 30.

17. Ind_{h-}. See Chapter 10, discussion around Problem 31.

18. $_d\mathrm{Ind}_{S-}$. See Chapter 10.

19. Ind_n. See the end of Chapter 10.

20. $\pi\chi_{S+}$. See the discussion preceding 11.9.

21. $_d\pi\chi_{S-}$. See the discrussion preceding 11.9.

22. $\pi\chi_{\inf}$. See 11.9.

23. wd. This is the function *weak density*, related to $\pi\chi_{\inf}$; see the discussion following 11.9.

24. hwd. This is *hereditary weak density*. See the discussion following Problem 40 in Chapter 11.

25. t_{H-}. See the discussion after Problem 44.

26. $_d t_{S-}$. See the discussion after Problem 44.

27. t_{mn}. This is a finite version of tightness; see the text following the proof of 12.12.

28. t_m. Another finite version of tightness; see above.

29. ut_{mn}. Another finite version of tightness; see above.

30. ut_m. Another finite version of tightness; see above.

31. s_{H-}. See the discussion before 13.7.

32. $_d s_{S-}$. See the discussion before 13.7.

33. dd. This function is related to s; see 13.7.

34. s_m. This is a finite version of spread; see the end of Chapter 13.

35. χ_{H-}. See 14.4–14.7.

36. a. This is the altitude function, related to χ_{H-}; see 14.4–14.7.

37. $\chi_{\inf}$. See 14.3.

38. χ_{S+}. See the discussion before 14.3.

39. $(\chi_{\text{npinf}})_{H-}$. See 14.5.

40. hL_m. This is a finite version of hL; see the text following Problem 53.

41. hd_m. This is a finite version of hd; see the end of Chapter 16.

Some additional natural functions

42. **The tree algebra number.** For any BA A, the *tree algebra number* of A, denoted by ta$A$, is the supremum of cardinalities of subalgebras of A isomorphic to tree algebras. This number is clearly greater or equal to cellularity. It is dominated by d, since if A is isomorphic to a tree algebra and A is a subalgebra of B, then $|A| = dA \le dB$. Note that an infinite free algebra always has tree algebra number $\aleph_0$. So ta$B$ can be much smaller than dB. If $cB < \text{ta}B$, then there is a generalized Suslin tree of size $(cB)^+$.

43. **The pseudo-tree algebra number.** For any BA A, the *pseudo-tree algebra number* of A, denoted by pta$A$, is the supremum of cardinalities of subalgebras of A isomorphic to pseudo-tree algebras. Clearly Length$A \le$ pta$A$, ta$A \le$ pta$A$, and

pta$A \leq$ Irr$A$. Any infinite free algebra has pseudo-tree number $\aleph_0$. In a large free algebra we thus have pta$A = \aleph_0$ while dA is large. So the place of pta with respect to the standard functions is clear.

44. The semigroup algebra number. For any BA A, the *semigroup algebra number* of A, denoted by sa$A$, is the supremum of cardinalities of subalgebras of A which are semigroup algebras. We have Ind$A \leq$ sa$A$ and pta$A \leq$ sa$A$. It is not clear whether sa$A \leq$ Irr$A$.

45. The tail algebra number. For any BA A, the *tail algebra number* of A, denoted by tla$A$, is the supremum of cardinalities of subalgebras of A isomorphic to tail algebras. Clearly sa$A \leq$ tla$A$. It may be that tla$A = |A|$ for every infinite BA A.

46. Disjunctiveness. The *disjunctiveness* of A, dj$A$, is the supremum of cardinalities of disjunctive subsets of A. Clearly tla$A \leq$ dj$A$ and s$A \leq$ dj$A$. It may be that dj$A = |A|$ for every infinite BA A.

47. Minimality. The *minimality* of A, mA, is the supremum of cardinalities of minimally generated subalgebras of A. Clearly pta$A \leq$ mA. For every infinite free BA A we have m$A = \aleph_0$.

48. Initial chain algebra number. This number, denoted by ic$A$, is the supremum of cardinalities of subalgebras of A isomorphic to the initial chain algebra on some tree. If A is free, then ic$A = \omega$.

49. Initial chain algebra number for pseudo trees. Similarly, for pseudo-trees; denoted by icp$A$. Thus ic$A \leq$ icp$A$. If A is free, then icp$A = \omega$.

50. Superatomic number. This is the supremum of cardinalities of superatomic subalgebras of a BA; denoted by spa. If A is free, then spa$A = \omega$. We have ic$A \leq$ spa$A$.

51. Pseudoaltitude. The *pseudo-altitude* of a BA A is $\min\{\chi_{\inf}B : B$ is an infinite homomorphic image of $A\}$ and is denoted by pa$A$. See cofinality, below.

52. Cofinality. The *cofinality* of a BA A, denoted by cf$A$, is the smallest infinite cardinal κ such that A can be written as the union of an increasing chain of type κ of subalgebras of A. This notion and the previous two are discussed in van Douwen [89]. We have a $\leq$ pa $\leq$ cf $\leq$ Card$_{H-} \leq 2^\omega$. It is open to show in ZFC that cf $\leq \omega_1$.

53. The order-ideal number. This is the supremum of the cardinality of a system of ideals ordered by inclusion; we denote it by oi$A$. Clearly hL $\leq$ oi and hd $\leq$ oi. It is possible to have $|A| <$ oi$A$; this is true for $A = $ Intalg$\mathbb{Q}$, for example. It is not clear whether oi $<$ Card is possible.

Dimensions of Boolean algebras

Heindorf [91] introduces an interesting notion of *dimension* of Boolean algebras, which gives rise to several cardinal functions. Let $\mathscr{A}$ be a non-empty class of non-trivial BAs. Since every BA is embeddable in a product of two-element BAs, every BA is embeddable in a product of members of $\mathscr{A}$. Hence the following definition makes sense: the $\mathscr{A}$-*dimension* of a BA A is the smallest cardinal number κ such that A can be embedded in a product of κ members of $\mathscr{A}$ (not necessarily distinct members). This $\mathscr{A}$-dimension is denoted by $\mathscr{A}$-dim$A$. It is natural to consider this notion for natural classes $\mathscr{A}$. Various one-element classes $\mathscr{A}$ are natural, of course. At the opposite extreme we can consider the notion for our natural proper classes—the class of all free algebras, of all superatomic algebras, etc. In Heindorf [91] the three cases (1) all free algebras, (2) all superatomic algebras, (3) all interval algebras, are investigated. Some cases are trivial; for example, with $\mathscr{A}$ the class of all complete BAs, the $\mathscr{A}$-dimension of any BA is 1, since any BA can be embedded in a complete BA. Another obvious remark is that each dimension function is dominated by the function d. We list some dimension functions which may be interesting:

54. int-dim, where $\mathscr{A}$ is the class of all interval algebras.

55. sa-dim, where $\mathscr{A}$ is the class of all superatomic algebras.

56. free-dim, where $\mathscr{A}$ is the class of all free algebras.

57. tree-dim, where $\mathscr{A}$ is the class of all tree algebras.

58. ptree-dim, where $\mathscr{A}$ is the class of all pseudo-tree algebras.

59. sg-dim, where $\mathscr{A}$ is the class of all semigroup algebras.

60. mg-dim, where $\mathscr{A}$ is the class of all minimally generated algebras.

61. ic-dim, where $\mathscr{A}$ is the class of all initial chain algebras.

62. dj-dim, where $\mathscr{A}$ is the class of all disjunctively generated algebras.

63. finco-dim, where $\mathscr{A}$ is the class of all finite-cofinite algebras on some infinite set.

64. $\{$Finco$\omega\}$-dim.

65. $\{$Intalg$\mathbb{R}\}$-dim.

25. Diagrams
General case

l = difference can be large; s = "small" difference

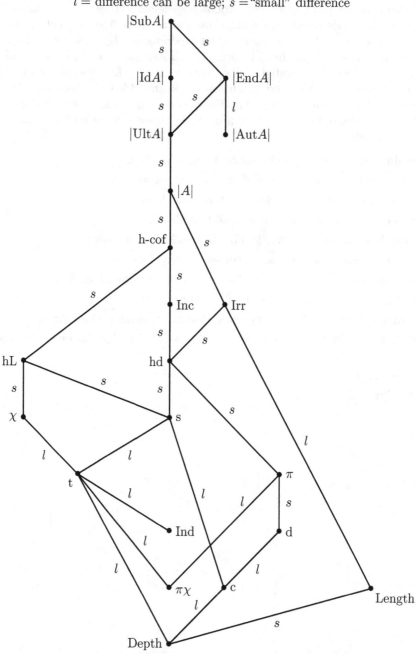

In this chapter we give several diagrams for the relationships between the main 21 functions that we have considered. Thus this chapter summarizes the main text. But it also turns out that we have some new things to say upon considering these relationships thoroughly.

For each of the diagrams, we need to do the following: (1) For each edge, indicate where the relation is proved, and give an example where the functions involved are different. Also, if the difference is indicated as "small", indicate where that is stated in the text, while if the difference is "large", indicate an example. Recall that a difference is "small" if there is some limitation on the difference. It is "large" if for every infinite cardinal κ, there is an example where the difference is at least κ. (2) Show that there are not any relations except those indicated in the diagrams. It suffices to do this just for crucial places in the diagram. For example, if in the diagram for the general case we give an algebra A in which $\text{Length}A < \pi\chi A$, this will also be an example for $\text{Length}A < \text{h-cof}A$ and $\text{Depth}A < \pi\chi A$. As will be seen, we have not been completely successful in either of these two tasks; there are several open problems left.

The main diagram, edges and "large" and "small" indications.

25.1. Depth $\leq$ Length. This relation is obvious from the definitions. The difference is small by the Erdös, Rado theorem; see the end of Chapter 7.

25.2. Depth $\leq$ c. Again, this is obvious from the definitions. The difference is large in the finite-cofinite algebra on an infinite cardinal κ.

25.3. Depth $\leq$ t. This is proved in Chapter 4. The difference is big in a free algebra.

25.4. $\pi\chi \leq$ t. See Chapter 11. The difference can be large in an interval algebra.

25.5. $\pi\chi \leq \pi$. Obvious from the definitions. The difference is large in a finite-cofinite algebra; see Chapter 11.

25.6. c $\leq$ s. This is a consequence of a theorem in Chapter 3. The difference is large in free algebras.

25.7. c $\leq$ d. See Chapter 5. The difference is large in free algebras.

25.8. Length $\leq$ Irr. Obvious from the definitions. The difference is large in free algebras.

25.9. Ind $\leq$ t. See the beginning of Chapter 12. The difference can be large in some interval algebras, since Depth $\leq$ t.

25.10. d $\leq \pi$. Obvious from the topological versions of these functions. The difference is small. See the end of Chapter 6 for an example where they differ.

25.11. t $\leq \chi$. Obvious from the definitions. The difference is big in a finite-cofinite algebra on a large cardinal κ.

25.12. t $\leq$ s. See Theorem 5.11. The difference is big in a finite-cofinite algebra on a cardinal κ.

25.13. $\pi \leq$ hd. See Theorem 6.10. The difference is small, since $\pi A \leq \text{hd}A \leq |A| \leq 2^{\pi A}$. The functions differ in $\mathcal{P}\kappa$, for example.

25.14. hd $\leq$ Irr. See the end of Chapter 16. They differ in the interval algebra on the reals.

25.15. $\chi \leq$ hL. See Chapter 15. The difference is small, since $|\text{Ult}A| \leq 2^{\chi A}$; they differ on the Aleksandroff duplicate of a free algebra; see Chapter 14.

25.16. s $\leq$ hL. Obvious from the definitions. The difference is small, since $|A| \leq 2^{sA}$ for any BA A by Chapter 13. They differ in a Kunen line (constructed under CH; see the end of Chapter 15). Whether there is an example in ZFC is an open question; see the end of chapter 15, and Problem 49.

25.17. s $\leq$ hd. Obvious from the definitions. The difference is small (see above). They differ on the interval algebra of a Suslin line. Whether there is an example in ZFC is open (Problem 57).

25.18. Irr $\leq$ Card. Obvious from the definitions. The difference is small; from Theorem 4.23 of Part I of the Boolean algebra handbook it follows that $|A| \leq 2^{\text{Irr}A}$. A compact Kunen line (constructed under CH) gives a BA in which they are different (see Chapter 8). It is open to give an example in ZFC. (Problem 28)

25.19. hL $\leq$ h-cof. See Chapter 18. The difference is small, since $\chi A \leq$ hLA, and so hL$A \leq$ h-cof$A \leq |A| \leq 2^{\chi A} \leq 2^{\text{hL}A}$, using Chapter 14. They differ on the interval algebra on the reals; see the beginning of Chapter 18.

25.20. hd $\leq$ Inc. This is an easy consequence of Theorem 4.25 of the BA handbook, part I. The difference is small, since s $\leq$ hd, Inc $\leq$ Card, and $|A| \leq 2^{sA}$ for any BA. They differ on Intalg $\mathbb{R}$.

25.21. Inc $\leq$ h-cof. Obvious from Theorem 18.1. The difference is small since by the above $|A| \leq 2^{\text{Inc}A}$. They differ on the Baumgartner, Komjath algebra (see the beginning of Chapter 18); this was constructed using $\Diamond$, and it remains a problem to get an example with weaker assumptions. (Problem 63)

25.22. h-cof $\leq$ Card. Obvious from the definitions. The difference is small (see above). An example where they differ can be found in Rosłanowski, Shelah [94].

25.23. Card $\leq |\text{Ult}|$. This is well-known; see the Handbook Part I, Theorem 5.31. The difference is, of course, small. They differ in an infinite free algebra.

25.24. $|\text{Ult}| \leq |\text{End}|$. This is obvious. The difference is small. They differ for the finite-cofinite algebra on an infinite cardinal.

25.25. $|\text{Aut}| \leq |\text{End}|$. Also obvious. The difference is large, as shown by a rigid BA.

25.26. $|\text{Ult}| \leq |\text{Id}|$. Again obvious. The difference is small. They differ on the finite-cofinite algebra on an infinite cardinal.

25.27. $|\text{Id}| \leq |\text{Sub}|$. See Chapter 23. The difference is small. They differ for the interval algebra on the reals.

25.28. $|\text{End}| \leq |\text{Sub}|$. See Chapter 23. The difference is small. They differ for the interval algebra on the reals.

The main diagram: no other relationships. Keep in mind that we only treat "crucial" relations; other possibilities are supposed to follow from these.

25.29. Length $< \pi\chi$: an uncountable free algebra: see Chapter 11.

25.30. Length $<$ c: the finite-cofinite algebra on an uncountable cardinal.

25.31. Length $<$ Ind: an uncountable free algebra.

25.32. $\pi\chi <$ Depth: see the example in Chapter 11.

25.33. d $< \pi\chi$: some free algebras.

25.34. Ind $<$ Depth: the interval algebra on an uncountable cardinal.

25.35. Ind $< \pi\chi$: true in the interval algebra on an uncountable cardinal; see Chapter 11.

25.36. $\pi <$ Ind: $\mathcal{P}\kappa$. The difference is small.

25.37. $\pi\chi <$ Ind: the difference can be large; see chapter 11.

25.38. $\chi <$ c: the Aleksandroff duplicate of an infinite free algebra; Chapter 14.

25.39. hL $<$ Length: Intalg $\mathbb{R}$.

25.40. hL $<$ d: The interval algebra of a complete Suslin line. It is not known if this is possible in ZFC. (Problem 54)

25.41. hL $<$ Inc: The interval algebra on the reals; see Chapter 17.

25.42. Inc $< \chi$: The Baumgartner-Komjath algebra, constructed under $\diamondsuit$; see Chapter 17. It is not known if this is possible under weaker hypotheses. Weaker problems are hd $< \chi$? and s $< \chi$?. See Problems 49, 59, 62.

25.43. Inc $<$ Length: Constructed by Shelah in ZFC using entangled linear orders.

25.44. Irr $< \chi$: The compact Kunen line, constructed using CH. No example is known in ZFC.

Problem 65. *Can one construct in ZFC a BA A such that* $\mathrm{Irr}A < \chi A$?

This problem is equivalent to the problem of constructing in ZFC a BA A such that $\mathrm{Irr}A < \mathrm{hL}A$; see the argument at the end of Chapter 15.

25.45. h-cof $<$ Length: Constructed by Shelah in ZFC using entangled linear orders; see Shelah [91].

25.46. $|\mathrm{Aut}| <$ Depth: embed a large interval algebra in a rigid algebra.

25.47. $|\mathrm{Aut}| < \pi\chi$: a rigid complete BA of large cellularity gives an example.

25.48. $|\mathrm{Aut}| <$ Ind: embed a large free algebra in a rigid algebra.

25.49. $|\mathrm{Id}| < |\mathrm{Aut}|$: see Chapter 22; possible under some set-theoretic assumptions. No example is known in ZFC (Problem 64).

25.50. $|\mathrm{Ult}| < |\mathrm{Aut}|$: the finite-cofinite algebra on an infinite cardinal.

25.51. $|\mathrm{End}| < |\mathrm{Id}|$: See below in the treatment of superatomic algebras.

Other possibilities follow from the above crucial relations, but in the case of examples mentioned above that involve additional axioms of set theory there are some additional problems; see arguments at the end of Chapters 15 and 16. And some more problems arise:

25.52. Irr $<$ Inc using a Kunen line; see the examples chapter. No example is known in ZFC:

Problem 66. *Is there an example in ZFC of a BA A such that* $\mathrm{Irr}A < \mathrm{Inc}A$?

25.53. Irr $<$ hL by the example for Irr $< \chi$, but no example in ZFC is known. This is actually equivalent to Problem 65. In fact, if A is such that $\mathrm{Irr}A < \mathrm{hL}A$, then there is an ideal I in A not generated by fewer than $(\mathrm{Irr}A)^+$ elements, and then with $B \overset{\mathrm{def}}{=} I \cup -I$ we have $\mathrm{Irr}B < \chi B$.

25.54. Irr < h-cof by the example for Irr < χ, but no example in ZFC is known:

Problem 67. *Is there an example in ZFC of a BA A such that* IrrA < h-cofA?

Note that "yes" on either of problems 65 or 66 implies "yes" on problem 67.

25.55. Inc < hL by the example for Inc < χ, but no example in ZFC is known. This is equivalent to Problem 61 by a familiar argument.

25.56. |Id| < |End| by the example for |Id| < |Aut|, but this problem is open:

Problem 68. *Is there an example in ZFC of a BA A such that* |IdA| < |EndA|?

The interval algebra diagram: the edges, indicated equalities, and the "large" and "small" indications. See below.

25.57. Ind = ω: this is one of the main results about interval algebras; see Part I of the BA handbook.

25.58. $\omega \leq \pi\chi$, difference possibly large. See the description of $\pi\chi$ for interval algebras in Chapter 11.

25.59. Depth = t = χ: see Chapter 14.

25.60. $\pi\chi \leq$ Depth, difference possibly large. See Chapter 11.

25.61. c=s=hL: see Chapter 15.

25.62. Depth $\leq$ c, with the difference small. The difference is small since $|A| \leq 2^{\mathrm{Depth}A}$ for an interval algebra A; *this implies smallness for the next few that we consider also.* For an example where they differ, see Chapter 4.

25.63. d = π = hd: obvious from the retractiveness of interval algebras.

25.64. c $\leq$ d. They differ in the interval algebra of a Suslin line; see Chapter 5. In fact, for any infinite cardinal κ the following two conditions are equivalent:

(1) there is an interval algebra A such that $\kappa = cA < dA$;

(2) there is a κ^+-Suslin line (or tree).

Thus the problem of getting an example in ZFC of a BA A such that $cA < dA$ is equivalent to the set-theoretical question of proving in ZFC that there is for some infinite κ a κ^+-Suslin tree.

25.65. d $\leq$ Inc. They differ in the interval algebra on the reals.

25.66. Inc = h-cof. See Chapter 18.

25.67. h-cof $\leq$ Card. Shelah [91] constructed an example where they differ in ZFC.

25.68. Card $\leq$ |Ult|. They differ for the interval algebra on the rationals.

25.69. |Ult| $\leq$ |Id|. They differ on the interval algebra on κ.

25.70. |Aut| $\leq$ |End|, the difference large: take an infinite rigid interval algebra.

25.71. |Id| $\leq$ |End|: follows from retractiveness. A Suslin line with more than ω_1 automorphisms gives an example where they are different, assuming CH. And if an example of an interval algebra A such that $|IdA| < |EndA|$ can be given in ZFC, then assuming GCH + (there is no uncountable inaccessible) one can show that for some infinite κ there is a κ^+-Suslin tree. For, suppose that A is an interval algebra such that $|IdA| < |EndA|$, GCH holds, and there are no uncountable inaccessibles. Thus $|A| = |UltA| = |IdA|$ and $|EndA| = |A|^+$. If $cA = |A|$, then cA is attained (since there are no uncountable inaccessibles), and hence $|A| < |\mathrm{Id}A|$, contradiction. Thus $cA < |A|$, and $|A| = |cA|^+$. If $dA < |A|$, then $|\mathrm{End}A| \le |\mathrm{Ult}A|^{dA} = |A|^{dA} = |A|$, contradiction. So $cA < dA$, and A must be the interval algebra on a $(cA)^+$-Suslin tree. So, the problem of existence of such interval algebras A in ZFC is stronger than the above set-theoretical question:

Interval algebras
$l =$ difference can be large; $s =$ "small" difference

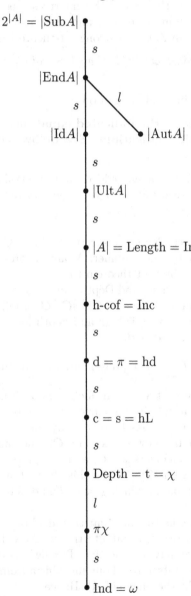

Problem 69. *Can one construct in ZFC an interval algebra A such that $|\mathrm{Id}A| <$ $|\mathrm{End}A|$?*

25.72. $|\mathrm{End}| \leq |\mathrm{Sub}|$. They differ on the interval algebra on the reals.

The interval algebra diagram: no other relationships.

25.73. $|\mathrm{Aut}| <$ Ind: an infinite rigid interval algebra.
25.74. $|\mathrm{Id}| < |\mathrm{Aut}|$: as in the case of Suslin trees (see Chapter 22), one can construct a Suslin line with more than ω_1 automorphisms. Again, the question of existence of such algebras in ZFC is a strong set-theoretical hypothesis:

Problem 70. *Can one construct in ZFC an interval algebra A such that $|\mathrm{Id}A| <$ $|\mathrm{Aut}A|$?*

25.75. $|\mathrm{Ult}| < |\mathrm{Aut}|$: the interval algebra on κ.

The tree algebra diagram: the indicated equalities and inequalities, and the "large" and "small" indications. See below. Let T be a tree, and let $A = \mathrm{Treealg}\, T$.

25.76. $\mathrm{Ind}A = \omega$, since A can be embedded in an interval algebra.
25.77. The difference between Ind and $\pi\chi$ can be arbitrarily large by the description of $\pi\chi$ for tree algebras.
25.78. $\pi\chi \leq \mathrm{t}$ in general.
25.79. Depth $= \mathrm{t}$, since tree algebras are retractive; see Chapter 4.
25.80. Length $=$ Depth by the Brenner, Monk theorem (Handbook, p. 269). $\mathrm{Depth}(A) = \mathrm{Depth}(T) + \omega$ by that theorem too.
25.81. The difference between $\pi\chi$ and Depth can be arbitrarily large; Chapter 11.
25.82. $\chi A = \sup\{|\text{set of immed. succ. of } C|, \mathrm{cf}C : C \text{ an initial chain}\}$; see Chapter 14. Finco κ is an example where χ is high and depth low.
25.83. In Chapter 3 we showed that

$$\mathrm{c}A = \max\{|\{t \in T : t \text{ has finitely many immed. succ.}\}|, \mathrm{Inc}(T)\}.$$

From this it is easy to see that $\chi \leq \mathrm{c}$. In fact, suppose that C is an initial chain of T whose order type is an infinite regular cardinal. Obviously |set of immediate succ. of $C| \leq \mathrm{c}$. If $|\{x \in C : x \text{ has finitely many immediate successors}\}| = |C|$, clearly $|C| \leq \mathrm{c}$. If this set has power less than $|C|$, one can choose an element to the side of x for $|C|$ many elements $x \in C$, and this gives an incomparable subset of T of size $|C|$, and so again $|C| \leq \mathrm{c}$. So this shows that $\chi \leq \mathrm{c}$. An Aronszajn non-Suslin tree is an example in which $\chi < \mathrm{c}$. The difference has to be small by the general diagram.
25.84. s$=$c by Chapter 3 plus the fact that tree algebras are retractive.
25.85. To see that s$=$hL, take any infinite tree T. Now $\mathrm{Treealg}\, T$ embeds in an interval algebra A, and we may assume that $\mathrm{Treealg}\, T$ is dense in A (extend the identity from $\mathrm{Treealg}\, T$ onto itself to a homomorphism from A into the completion of $\mathrm{Treealg}\, T$, and then take the image of A). Hence

Tree algebras

l = difference can be large; s = "small" difference

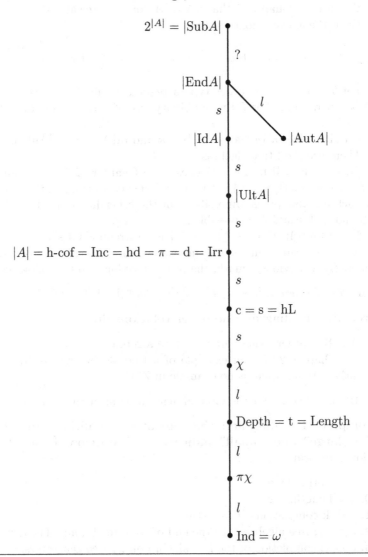

$$\mathrm{hL}A \geq \mathrm{hL}(\mathrm{Treealg}\,T) \geq \mathrm{c}(\mathrm{Treealg}\,T) = \mathrm{c}A = \mathrm{hL}A.$$

25.86. Suppose that $\mathrm{c}A < |A|$. Then clearly T is a tree such that $\mathrm{Inc}(T) < |T|$ and $|T|$ has height $|T|$ but no chains of length $|T|$. Moreover, since $\mathrm{c}A < |A|$, the cardinal $|A|$ must be a successor. Thus T is a generalized Suslin tree on a successor cardinal.

25.87. d=hd, since tree algebras are retractive, d$B \leq$ dA for B a subalgebra of A, and by Theorem 14.1.

25.88. Recall from Chapter 5 that $\pi A = |A|$ for A a tree algebra.

25.89. From the above it follows that

$$|A| = \text{h-cof} = \text{Inc} = \text{hd} = \pi = \text{d} = \text{Irr}.$$

25.90. For T a chain with order type a regular cardinal we have $|A| = |\text{Ult}A|$, since A is superatomic. For T the full binary tree of height ω we have $|A| = \omega$ and $|\text{Ult}A| = 2^\omega$.

25.91. For $A = \text{Finco}\,\kappa$ we have $|\text{Ult}A| = \kappa$ and $|\text{Id}A| = 2^\kappa = |\text{Aut}A|$.

25.92. There are rigid tree algebras.

25.93. $|\text{Id}A| \leq |\text{End}A|$ by retractiveness. See Chapter 22 for an example where they differ (consistently). If c$A = |A|$ and cellularity is attained, then $|\text{Id}A| = 2^{|A|}$, so that such an example is impossible. On the other hand, if c$A < |A|$, then see 25.86. So no such example is possible in ZFC alone.

25.94. $2^{|A|} = |\text{Sub}A|$ since $\{T \uparrow t : t \in T\}$ is an irredundant set.

25.95. $|\text{End}A| \leq |\text{Sub}A|$ in general; no example of a tree algebra is known where they differ. By a previous remark, the problem reduces to the following question.

Problem 71. *Is there a tree algebra A such that $|\text{End}A| < 2^{|A|}$?*

The tree algebra diagram: no other relationships.

25.96. $|\text{Aut}A| <$ others: there are rigid tree algebras.

25.97. See Chapter 22 for an example of a tree algebra where $|\text{Id}A| < |\text{Aut}A|$ (consistently). Again, there is no example in ZFC.

25.98. $|\text{Ult}A| < |\text{Aut}A|$ for the interval algebra on an infinite cardinal.

The complete BA diagram: the indicated equalities and inequalities and the "large" and "small" indications. See below. In fact, the "small" indications are clear.

25.99. c $=$ Depth: obvious.

25.100. c $<$ Length: $\mathcal{P}\omega$.

25.101. c $<$ d: completions of free algebras.

25.102. c $< \pi\chi$: free algebras; see the end of Chapter 11, and Theorem 11.6.

25.103. d $< \pi$: completion of the free algebra on ω_1 free generators.

25.104. $\pi\chi < \pi$: under GCH these are equal, by an argument of Bozeman. It is not known if this is true in ZFC. See Problem 40.

25.105. Length $<$ Card: a large ccc algebra.

25.106. $\pi <$ Card: $\mathcal{P}\kappa$.

25.107. $|A| = \text{Ind} = \text{t} = \text{s} = \chi = \text{hL} = \text{Irr} = \text{s} = \text{hd} = \text{Inc} = \text{h-cof}$: these equalities all follow from $|\text{Ind}| = \text{Card}$ (attained), which is a consequence of the Balcar, Franěk theorem.

25.108. $|\mathrm{Ult}A| = |\mathrm{Id}A| = |\mathrm{Sub}A| = |\mathrm{End}A| = 2^{|A|}$: again true since any infinite complete BA A has an independent subset of size $|A|$.

Complete BAs
l = difference can be small; s = "small" difference

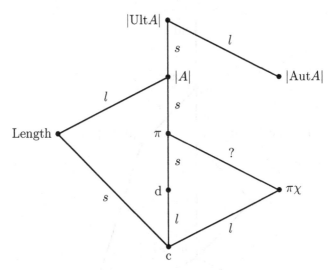

c=Depth
$|A| = \mathrm{Ind} = \mathrm{t} = \mathrm{s} = \chi = \mathrm{hL} = \mathrm{Irr} = \mathrm{s} = \mathrm{hd} = \mathrm{Inc} = \mathrm{h\text{-}cof}$
$|\mathrm{Ult}A| = |\mathrm{Id}A| = |\mathrm{Sub}A| = |\mathrm{End}A| = 2^{|A|}$

25.109. $|\mathrm{Aut}| < |\mathrm{Ult}|$: a rigid complete algebra.

The complete BA diagram: no other relations.

25.110. $\pi < \mathrm{Length}$: $\mathscr{P}\omega$.

25.111. $\mathrm{Length} < \pi\chi$: the completion of a free algebra.

25.112. $\mathrm{Length} < \mathrm{d}$: the completion of a free algebra.

25.113. Under GCH we have $\mathrm{d} \le \pi\chi$, by the result of Bozeman and Chapter 11; it is not known whether this holds in ZFC. This is equivalent to Problem 40, by Chapter 11.

25.114. $|\mathrm{Aut}| < \mathrm{c}$: embed $\mathcal{P}\kappa$ in a rigid BA.

25.115. $\mathrm{Card} < |\mathrm{Aut}|$: the completion of the free BA of size 2^{ω}.

Diagram for superatomic BAs: the indicated relations, and the "large" and "small" indications. See below.

25.116. $\mathrm{Ind} = \omega$ since every superatomic BA has countable independence.

25.117. $\mathrm{Ind} < \mathrm{Depth}$: Any interval algebra on an uncountable cardinal provides an example; the difference can be arbitrarily large.

25.118. $\mathrm{Depth} = \mathrm{Length}$ by Rosenstein [82] Corollary 5.29, p. 88.

Superatomic BAs

l = difference can be small; s ="small" difference

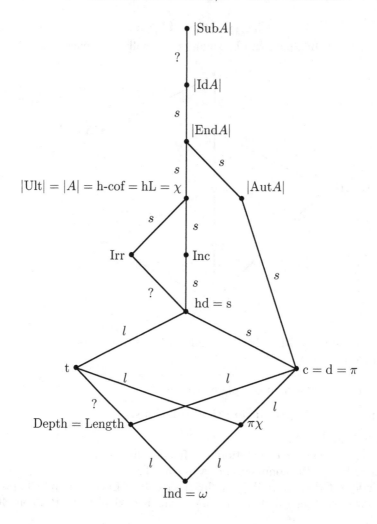

25.119. Ind $< \pi\chi$: the interval algebra on a cardinal provides an example with the difference arbitrarily large.

25.120. Depth $<$ t. There is an even stronger example, with c $<$ t. Let $\langle a_\alpha : \alpha < \omega_1 \rangle$ be a system of subsets of ω such that for $\alpha < \beta < \omega_1$ we have $a_\alpha \backslash a_\beta$ finite and $a_\beta \backslash a_\alpha$ infinite. Let A be the subalgebra of $\mathscr{P}\omega$ generated by the singletons together with the a_α's. Clearly A is as desired. On the other hand, in Dow, Monk [94] it is shown that if $\kappa \to (\kappa)_2^{<\omega}$, then any superatomic BA with tightness κ^+ also has depth at least κ. This shows that the difference between tightness and depth cannot be arbitrarily large. But we do not know how big the gap can be;

recall Problem 44.

25.121. $\pi\chi < t$: see Chapter 11; the difference can be large.

25.122. Obviously $c = d = \pi = $ number of atoms.

25.123. Depth $< c$: they differ in a finite-cofinite algebra, where the difference can be large.

25.124. $\pi\chi < c$: see Chapter 11; the difference can be large.

25.125. hd $= s$: see the characterizations of hd and s.

25.126. $t < s$, and the difference can be arbitrarily large: a finite-cofinite algebra.

25.127. $c < s$: Take a family $\mathscr{A}$ of 2^ω almost disjoint subsets of ω, and consider the BA generated by $\mathscr{A} \cup \{\{i\} : i \in \omega\}$.

25.128. $s < $ Inc. Under $\diamondsuit$, Shelah constructed a thin-tall BA A with countable spread. Then $A \times A$ has countable spread too, while its incomparability is ω_1. The example of Bonnet, Rubin [92] can also be used for this purpose. We do not know whether there is an example in ZFC:

Problem 72. *Is there an example in ZFC of a superatomic BA A such that $sA < \mathrm{Inc}A$?*

25.129. We do not know of an example of a superatomic BA in which s is less than Irr:

Problem 73. *Is there a superatomic BA A such that $sA < \mathrm{Irr}A$?*

25.130. $|A| = $ h-cof $= $ hL $= \chi$: $\chi = $ Card by Chapter 14, and the other equalities follow.

25.131. In Bonnet, Rubin [92] a superatomic algebra of power ω_1 is constructed using $\diamondsuit$ in which Inc and Irr are countable.

Problem 74. *Can one construct in ZFC a superatomic algebra A with the property that $\mathrm{Inc}A < |A|$?*

Problem 75. *Can one construct in ZFC a superatomic algebra A with the property that $\mathrm{Irr}A < |A|$?*

25.132. $|A| = |\mathrm{Ult}A|$: see the Handbook.

25.133. $|A| < |\mathrm{End}A|$ in a finite-cofinite algebra.

25.134. M. Rubin constructed under $\diamondsuit$ a BA A such that $|\mathrm{Aut}A| < |A|$ (unpublished, December 1992) in particular, $|\mathrm{Aut}A| < |\mathrm{End}A|$. We do not know whether this can be done in ZFC:

Problem 76. *Can one construct in ZFC a superatomic BA A such that $|\mathrm{Aut}A| < |\mathrm{End}A|$?*

25.135. $c < |\mathrm{Aut}|$. The BA of finite and cofinite subsets of κ gives an example where they differ.

25.136. $|\mathrm{End}A| \leq |\mathrm{Id}A|$ for A superatomic. For, let κ be the number of atoms of A. Then $|\mathrm{Ult}A| = |A| \leq 2^\kappa \leq |\mathrm{Id}A|$, and hence by Chapter 21, $|\mathrm{End}A| \leq |\mathrm{Ult}A|^{\mathrm{d}A} \leq 2^\kappa \leq |\mathrm{Id}A|$. In the example of 25.127 we have $|\mathrm{End}A| = 2^\omega$ and $|\mathrm{Id}A| = 2^{2^\omega}$.

25.137. We do not have an example where $|\mathrm{Id}A| < |\mathrm{Sub}A|$:

Problem 77. *Can one have* $|\mathrm{Id}A| < |\mathrm{Sub}A|$ *in a superatomic BA?*

Superatomic BAs, no additional relationships:

25.138. $\pi\chi <$ Depth: see Chapter 11.
25.139. Depth $< \pi\chi$: Dow, Monk [94] constructed an example.
25.140. t $<$ c: the finite-cofinite algebra on κ.
25.141. We do not have any example of a superatomic BA A with the property that $\mathrm{Inc}A < \mathrm{Irr}A$:

Problem 78. *Is there, under any set-theoretic assumptions, a superatomic BA A such that* $\mathrm{Inc}A < \mathrm{Irr}A$?

25.142. We also do not have an example of a superatomic BA A with the property that $\mathrm{Irr}A < \mathrm{Inc}A$:

Problem 79. *Is there, under any set-theoretic assumptions, a superatomic BA A such that* $\mathrm{Irr}A < \mathrm{Inc}A$?

25.143. Card $< |\mathrm{Aut}|$: a finite-cofinite algebra.
25.144. $|\mathrm{Aut}|$ small relative to "lower" functions. Recall 25.134. But we have the following problem:

Problem 80. *Is there in ZFC a superatomic BA A with* $|\mathrm{Aut}A| < |A|$?

Now the algebra of Rubin in 25.134 has $\aleph_1$ atoms, and also $\aleph_1$ automorphisms. Also recall from Chapter 20 that any infinite superatomic BA has at least 2^ω automorphisms. Also, it is clear that any atomic BA A has at least as many automorphisms as atoms, since every finite permutation of the atoms extends to an automorphism of the algebra. The strongest remaining problem is as follows.

Problem 81. *Under any set-theoretic assumptions, is there a superatomic BA A such that* $|\mathrm{Aut}A| < \mathrm{t}A$?

25.145. The relationship between Card and $|\mathrm{Aut}|$ is not completely clear, but we indicate some other facts; see 25.143. Recall from Chapter 20 that an infinite superatomic BA has at least 2^ω automorphisms. The initial chain algebra A on $^{\leq\omega}2$ is such that $|A| = |\mathrm{Aut}A| = 2^\omega$.

Now assume that $2^\omega = \omega_2$, $2^{\omega_1} = \omega_3$, and $2^{\omega_2} = \omega_4$. Let T be the tree $^{\leq\omega}\omega_1$, and let $A = \mathrm{Init}\,T$, the initial chain algebra on T. Note that $|A| = \omega_2$. For each permutation φ of ω_1 there is an automorphism φ' of A such that $\varphi'(T \downarrow t) = T \downarrow (\varphi \circ t)$ for every $t \in T$. If $\varphi \neq \psi$, then $\varphi' \neq \psi'$. Thus this gives $2^{\omega_1} = \omega_3$ automorphisms; since A has only ω_1 atoms, it follows that $|\mathrm{Aut}A| = \omega_3$. Note that $2^{|A|} = \omega_4$. Thus $|A| < |\mathrm{Aut}A| < 2^{|A|}$.
25.146. $|\mathrm{Aut}| < |\mathrm{Id}|$: the algebra of 25.145.

Atomic BAs

$l =$ difference can be large; $s =$ small difference

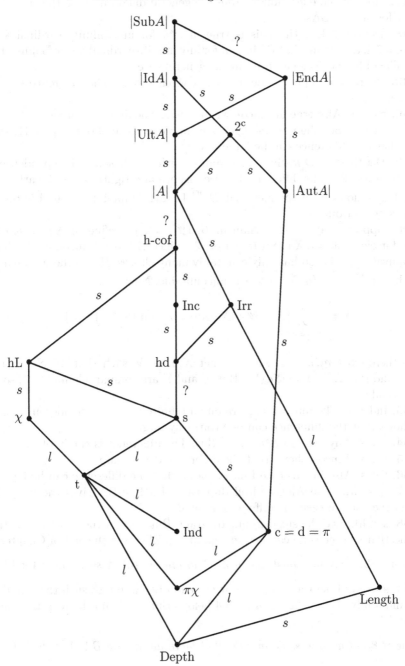

Atomic diagram, edges and "large" and "small" indications.

25.147. Most of the edges follow from the general diagram. Note that $c = d = \pi$ is clear for atomic BAs.

25.148. Depth < Length. This is true in $\mathscr{P}\kappa$ for any infinite cardinal κ; see Theorem 7.4, and recall that $\mathrm{Ded}\kappa > \kappa$ for any infinite cardinal κ (see Baumgartner [76]). The difference is small even in the general case.

25.149. Depth < c. This is true in $\mathrm{Finco}\kappa$ when $\kappa > \omega$. The difference here can be large.

25.150. $\pi\chi$ < c. Also true in $\mathrm{Finco}\kappa$ when $\kappa > \omega$. The difference here can be large.

25.151. $\pi\chi$ < t. See the interval algebra example at the end of Chapter 11, which shows that the difference can be large.

25.152. Depth < t. $\mathscr{P}\kappa$ gives an example with depth κ and independence 2^κ, hence tightness 2^κ. The difference between Depth and tightness can be arbitrarily large. This follows from the fact that $B \stackrel{\mathrm{def}}{=} \mathrm{Length}(\mathrm{Dup}(A)) = \omega$ if A is the free BA on κ generators.

In fact, suppose that $\mathscr{X}$ is a chain in B, $|\mathscr{X}| = \omega_1$. Define $(a, X) \equiv (b, Y)$ iff $a = b$, for elements (a, X) and (b, Y) of $\mathscr{X}$. Since A has no uncountable chains, this equivalence relation has only countably many classes. Hence there is an $a \in A$ such that $\mathscr{Y} \stackrel{\mathrm{def}}{=} \{X : (a, X) \in \mathscr{X}\}$ is uncountable. Now

$$\mathscr{Y} = \bigcup_{m,n \in \omega} \{X \in \mathscr{Y} : |X \backslash Sa| = m \text{ and } |Sa \backslash X| = n\}.$$

Hence there exist $m, n \in \omega$ with distinct $X, Y \in \mathscr{Y}$ such that $|X \backslash Sa| = m = |Y \backslash Sa|$ and $|Sa \backslash X| = n = |Sa \backslash Y|$. But X and Y are comparable under $\subseteq$, so this is impossible.

25.153. Ind < t. The interval algebra on an infinite cardinal provides an example and shows that the difference can be arbitrarily large.

25.154. c < s. Any algebra $\mathscr{P}\kappa$ shows this. The difference is small, of course.

25.155. t < χ. $\mathrm{Finco}\kappa$ shows that the difference can be large.

25.156. t < s. Also in this case $\mathrm{Finco}\kappa$ shows that the difference can be large.

25.157. χ < hL. The Aleksandroff duplicate of a free BA gives the inequality. Even in the general case the difference is small.

25.158. s < hL. See the superatomic diagram. But even in the general case there is a question whether one can get an example in ZFC; see the end of Chapter 15.

Problem 82. *Can one construct in ZFC an atomic BA A such that $sA < hLA$?*

25.159. s < hd. Take the algebra A at the end of Chapter 16 such that $sA, dA < hdA$, and apply the argument in 25.164 below. But we do not know if this can be done in ZFC:

Problem 83. *Can one show in ZFC that there an atomic BA A such that $sA < hdA$?*

25.160. Length < Irr. The finite-cofinite algebra on an infinite cardinal furnishes an example.

25.161. hd < Inc. Take the interval $[0,1)$ and replace each rational r by two elements r_0, r_1 with $r_0 < r_1$ and no elements between them, thereby forming a linear ordering L. Then Intalg$L$ is the desired example.

25.162. hd < Irr. The previous example works.

25.163. hL < h-cof. That example works here too.

25.164. Inc < h-cof. We prove, in ZFC, that if A is a BA such that Inc$A$ < h-cof$A$, then there is an atomic BA B such that Inc$B$ < h-cof$B$. This argument, and variants of it below, are due to M. Rubin. Without loss of generality A is a subalgebra of $\mathscr{P}\kappa$, where $\kappa = $ d$A \leq $ Inc$A$. Let $\lambda = $ Inc$A$. Let $B = \langle A \cup \{\{\alpha\} : \alpha < \kappa\}\rangle^{\mathscr{P}\kappa}$. We claim that Inc$B = \lambda$. For, suppose that $X \in [B]^{\lambda^+}$. For all $x \in X$ there are $F_x, G_x \in [\kappa]^{<\omega}$ and $a_x \in A$ such that $x = (a_x \backslash F_x) \cup G_x$, $F_x \subseteq a_x$, and $G_x \cap a_x = 0$. Then there exist $H, K \in [\kappa]^{<\omega}$ and $Y \in [X]^{\lambda^+}$ such that $F_x = H$ and $G_x = K$ for all $x \in Y$. $\{a_x : x \in Y\}$ is not incomparable, so there exist distinct a_{x_0}, a_{x_1} such that $x_0, x_1 \in Y$ and $a_{x_0} < a_{x_1}$. Then $x_0 < x_1$, as desired.

It follows that, assuming $\Diamond$, there is an atomic BA B such that Inc$B$ < h-cof$B$. The problem of finding an example in ZFC is equivalent to the problem for arbitrary BAs; see Problem 55.

25.165. Irr < $|A|$. The situation is very similar here, but the above argument of Rubin has to be supplemented. We show in ZFC that if Irr$A$ < $|A|$, then there is an atomic BA B such that Irr$B$ < $|B|$. Again, since $\kappa \stackrel{\text{def}}{=} $ d$A \leq $ Irr$A$, we may assume that A is a subalgebra of $\mathscr{P}\kappa$, and we let $B = \langle A \cup \{\{\alpha\} : \alpha < \kappa\}\rangle^{\mathscr{P}\kappa}$. Let $\lambda = $ Irr$A$, and suppose that $D \in [B]^{\lambda^+}$. As in 25.164 we may assume that there exist $H, K \in [\kappa]^{<\omega}$ such that for all $x \in D$ there is an $a_x \in A$ such that $H \subseteq a_x$, $K \cap a_x = 0$, and $x = (a_x \backslash H) \cup K$. For each $y \in A$, let $fy = y \backslash H$. Then f is a homomorphism of A into $B \restriction (\kappa \backslash H)$. For each $b \in B$ let $gb = (b \backslash H, b \cap K)$. So g is a homomorphism from B into $(B \restriction (\kappa \backslash H)) \times \mathscr{P}H$. If $x \in D$, then $gx = (a_x \backslash H, K)$. Hence $g \restriction \langle D\rangle^B$ is a homomorphism of $\langle D\rangle^B$ into $f[A] \times \mathscr{P}K$. Now Irr$(f[A]) \leq $ Irr$A = \lambda$, so we contradict Corollary 8.2 by showing that $g \restriction \langle D\rangle^B$ is one-one. Suppose that

$$gx_0 \cap \ldots \cap gx_{m-1} \cap -gy_0 \ldots \cap -gy_{n-1} = 0$$

with $x_0, \ldots, x_{m-1}, y_0, \ldots, y_{n-1} \in D$.

Case 1. $n = 0$. Then $a_{x_0} \cap \ldots \cap a_{x_{m-1}} \subseteq H$, and $K = 0$. It follows that $x_0 \cap \ldots \cap x_{m-1} = 0$.

Case 2. $n > 0$. Note that $-gz = (-a - H, -a \cap H)$ for all $z \in B$. It follows that $-y_0 \cap \ldots \cap -y_{n-1} \subseteq H$; since $H \subseteq y_i$ for each i, we must have $-y_0 \cap \ldots \cap -y_{n-1} = 0$, as desired.

It follows from Chapter 8 that under CH there is an atomic BA B such that Irr$B$ < $|B|$. The problem of existence of such an algebra in ZFC is equivalent to the problem for arbitrary BAs; see Problem 25.

25.166. h-cof $< |A|$. The situation is like that for Inc $<$ h-cof. By essentially the same argument, one can show in ZFC that if A is a BA such that h-cof$A < |A|$, then there is an atomic BA of this sort. As mentioned in Chapter 18, Shelah has constructed a BA A of this sort.

25.167. $|A| < |\text{Ult}A|$. $\mathscr{P}\kappa$ is an example.

25.168. $|\text{Ult}A| < |\text{Id}A|$. Finco$\kappa$ furnishes an example.

25.169. $|\text{Ult}A| < |\text{End}A|$. Again Finco$\kappa$ furnishes an example.

25.170. $c < |\text{Aut}|$. This is clear since any finite permutation of the atoms extends to an automorphism of the algebra. $<$ holds for $\mathscr{P}\kappa$, for example.

25.171. $|\text{Aut}A| < 2^c$. $\leq$ true since any automorphism is induced by a permutation of the atoms. An atomic BA of size 2^ω with countable automorphism group is an example where $<$ holds.

25.172. $|\text{Aut}A| < |\text{End}A|$. The same example works.

25.173. $2^c \leq |\text{Id}|$. Every subset of the set of atoms determines the ideal generated by those atoms. In $\mathscr{P}\kappa$ the difference is strict.

25.174. $|\text{Id}A| < |\text{Sub}A|$. As in the example with hd $<$ Irr.

25.175. $|\text{End}A| < |\text{Sub}A|$. Let L be the linear order obtained from $\mathbb{R}$ by replacing each rational by two adjacent points. Then $A = \text{Intalg}\, L$ is as desired.

Atomic diagram, no other relations

25.176. Length $< \pi\chi$. An example of Dow, Monk [94] works here: length ω, $\pi\chi$ ω_1. To get the difference arbitrarily large, one can work as follows. Let κ be a regular infinite cardinal. Let $\langle x_\alpha : \alpha < \kappa \rangle$ be a system of independent elements of $\mathscr{P}\kappa$ such that all the elementary products

$$\bigcap_{\alpha \in \Gamma} x_\alpha \cap \bigcap_{\alpha \in \Delta} (\kappa \backslash x_\beta)$$

have size κ (Γ and Δ finite disjoint subsets of κ). For each $\alpha < \kappa$ let $y_\alpha = x_\alpha \backslash \alpha$. Then let A be the subalgebra of $\mathscr{P}\kappa$ generated by

$$\{\{\alpha\} : \alpha < \kappa\} \cup \{y_\alpha : \alpha < \kappa\}.$$

First we check that $\pi\chi A = \kappa$. For, let F be the ultrafilter on A such that all cofinite subsets of κ are in F, and also each $y_\alpha \in F$. Suppose D is dense in F, and $|D| < \kappa$. We may assume that D is a collection of singletons, say $D = \{\{\alpha\} : \alpha \in \Gamma\}$, where $\Gamma \in [\kappa]^{<\kappa}$. Choose $\alpha < \kappa$ such that all members of Γ are less than α. Then there is no member of D below y_α, contradiction. Next, Length$A = \omega$. To see this, first note that A/fin is isomorphic to the free BA on κ generators. Now if L is an uncountable chain in A, define an equivalence relation $\equiv$ on L by setting $a \equiv b$ iff $a \triangle b$ is finite. Clearly each equivalence class is countable. Hence there are uncountably many equivalence classes. This means that we get an uncountable chain in A/fin, contradiction.

25.177. $|\text{Ult}| < |\text{Aut}|$: the finite-cofinite algebra on an infinite cardinal.

25.178. Length < Ind: see 25.152.

25.179. $\pi\chi$ < Depth: see Chapter 11.

25.180. Ind < Depth: Intalgκ.

25.181. $|\mathrm{Aut}|$ < Ind. An example is given in McKenzie, Monk [75].

25.182. c < Ind: $\mathscr{P}\kappa$. One can even get the difference arbitrarily large. Recall from Chapter 11 that there is an atomless BA A with $\pi\chi A = \omega$ and the independence of A any prescribed value κ; A also has size κ. We may assume that A is actually a subalgebra of $\mathscr{P}\kappa$. Let $B = \langle A \cup \{\{\alpha\} : \alpha < \kappa\}$. Now let F be any nonprincipal ultrafilter on B. Let $D \subseteq A^+$ be a countable set dense in $F \cap A$. Since A is atomless, each $d \in D$ is an infinite subset of κ. For each $d \in D$ let Γ_d be a countably infinite subset of d. Let $E = \{\{\alpha\} : \alpha \in \Gamma_d \text{ for some } d \in D\}$. So E is a countable subset of B. We claim that it is dense in F. For, take any element $x \in F$. We can write $x = (a \backslash M) \cup N$, where $a \in A$ and M and N are finite subsets of κ. Clearly $a \in D$, so there is a $d \in D$ such that $d \le a$. Then choose $\alpha \in \Gamma_d \backslash M$. Thus $\{\alpha\} \in E$ and $\{\alpha\} \subseteq x$, as desired.

25.183. Ind < $\pi\chi$: Intalgκ.

25.184. h-cof < Length. The situation is as for 25.166; atomic examples exist in ZFC.

25.185. Inc < Length. The situation is as for 25.164; examples exist in ZFC.

25.186. hd < Length: see 25.161.

25.187. hL < Length: see 25.161.

25.188. χ < c: the Aleksandroff duplicate of a free BA gives an example; see Chapter 14.

25.189. Irr < χ. See 25.165; the situation here is similar.

25.190. Inc < χ. See 25.164; examples exist in ZFC.

25.191. $|\mathrm{End}|$ < $|\mathrm{Id}|$. See 25.136 for a superatomic example.

Atomless algebras. The diagram here is the same as the general case, but to check this we have to give some new examples in some cases. One new example is the weak power of a denumerable atomless BA, used in many of the cases in the general diagram in which a finite-cofinite algebra was used. Where an interval algebra of a discrete linear order L was used, one can use instead the linear order obtained from L by replacing each point by a copy of the rationals. In several examples, the Aleksandroff duplicate was used. Like in the examples just mentioned, here also one can replace atoms by the denumerable atomless BA. We go through the details of this in one case:

25.192. $\chi \le$ hL. Let B be the Aleksandroff duplicate of a free algebra on κ free generators. To replace atoms by the denumerable atomless BA, we use set products; see Chapter 1. We may assume that B is a field of subsets of some set I containing all singletons. For each $i \in I$ suppose that A_i is a denumerable atomless field of subsets of a set J_i, where the J_i's are pairwise disjoint. The algebra we want is $C \stackrel{\text{def}}{=} \prod_{i \in I}^B A_i$. Recall that each element of C can be written uniquely in the form $h(b, F, a)$, where $b \in B$, F is a finite subset of I disjoint from b, and a is a member of $\prod_{i \in F}(A_i \backslash \{0,1\})$. We want first to describe all the ultrafilters of C. They are of two types:

Type 1. Let H be a nonprincipal ultrafilter on B. Then $G \stackrel{\text{def}}{=} \{h(b,F,a) : \text{conditions}$ as above, and $b \in H\}$ is an ultrafilter on C. This follows easily from the following easy facts, where we assume that $h(b,F,a)$ and $h(b',F',a')$ both satisfy the above conditions:

(1) $h(b,F,a) \subseteq h(b',F',a')$ iff $b \subseteq b'$ and for all $i \in F$, either $i \in b'$ or else $i \in F'$ and $a_i \subseteq a_i'$.

(2) $h(b,F,a) \cap h(b',F',a') = h(b \cap b', G, a'')$, where

$$G = (F \cap b') \cup (F' \cap b) \cup \{i \in F \cap F' : 0 \neq a_i \cap a_i'\}$$

and for all $i \in G$,

$$a_i'' = \begin{cases} a_i & \text{if } i \in F \cap b', \\ a_i' & \text{if } i \in F' \cap b, \\ a_i \cap a_i' & \text{otherwise.} \end{cases}$$

(3) $h(b,F,a) \cup h(b',F',a') = h(c,G,a'')$, where

$$c = b \cup b' \cup \{i \in F \cap F' : a_i \cup a_i' = J_i\},$$

$$G = (F \backslash (b' \cup F')) \cup (F' \backslash (b \cup F)) \cup \{i \in F \cap F' : a_i \cup a_i' \neq J_i\},$$

and, for any $i \in G$, a_i'' takes on the obvious value.

(4) $K \backslash h(b,F,a) = h(I \backslash (b \cup F), F, c)$, where for any $i \in F$, $c_i = J_i \backslash a_i$.

Type 2. For any $i \in I$ and any ultrafilter H on A_i, the set $G \stackrel{\text{def}}{=} \{h(b,F,a) : \text{conditions as above, and either } i \in b \text{ or else } i \in F \text{ and } a_i \in H\}$.

Now let L be any ultrafilter on C; we claim that L is of type 1 or of type 2. *Case 1.* For every finite subset M of I, the element $h(I \backslash M, 0, 0)$ is in L. Let $H = \{b \in B : h(b,0,0) \in L\}$. It is easy to check that H is a nonprincipal ultrafilter on B, and that L is obtained from H as indicated in the type 1 description. *Case 2.* There is a finite subset M of I such that $h(M,0,0)$ is in L. Then there is an $i \in I$ such that $h(\{i\},0,0) \in L$. Let $H = \{x \in A_i : h(0,\{i\},a) \in L \text{ with } a_i = x\} \cup \{J_i\}$. Then it is easy to check that H is an ultrafilter on A_i and L is obtained from H as in the type 2 description.

This completes the description of the ultrafilters on C.

If G is of type 1, obtained from H, then $\chi G = \chi H$. Similarly for type 2. It follows that $\chi C = \max\{\kappa, 2^\omega\}$, while clearly $cC = 2^\kappa$. So for $\kappa \geq 2^\omega$ we have $\chi C < \text{hL}C$.

Semigroup algebras. Here again the main diagram applies, except that there are a number of open problems where we used special algebras in the discussion above.

25.193. $\pi < \mathrm{hd}$. One can take $\langle\{\{i\} : i < \omega\} \cup \{a_\alpha : \alpha < 2^\omega\}\rangle^{\mathscr{P}\omega}$, where $\langle a_\alpha : \alpha < 2^\omega\rangle$ is an independent family of subsets of ω.

25.194. $\mathrm{s} < \mathrm{hL}$. In the general diagram we used a Kunen line for this, assuming CH. We do not know if a Kunen line is a semigroup algebra.

Problem 84. *Is the Kunen line a semigroup algebra?*

Problem 85. *Is there a semigroup algebra A such that $\mathrm{s}A < \mathrm{hL}A$?*

25.195. $\mathrm{Irr} < \mathrm{Card}$. In the general case there were a number of examples of this; we presented three: the Kunen line, a Todorčević algebra, and Rubin's algebra. We do not know whether the Kunen line or the Todorčević algebra are semigroup algebras; see Problem 84 and the following problem.

Problem 86. *Is the Todorčević algebra of Chapter 8 a semigroup algebra?*

We now show that Rubin's algebra is not a semigroup algebra. In fact, take the notation of 18.2, and suppose that H is a subset showing that A is a semigroup algebra. Let $P = H\backslash\{0, 1\}$. Then $|P| = \omega_1$, hence it is not nwdc, and so we get $n \in \omega\backslash 1$ and disjoint $a, b_1, \ldots, b_n$ with each $b_i \neq 0$ such that

(1) For all c_1, c_2, if $\langle a, c_1, c_2, b_1, \ldots, b_n\rangle$ is a configuration it follows that $P \cap (c_1, c_2) \neq 0$.

Now as in 18.2 we can get an element $c_1 \in P$ such that $a \leq c_1 \leq a + b_1 + \cdots + b_n$ and $c_1 \cdot b_i \neq 0 \neq b_i \cdot -c_1$ for all $i = 1, \ldots, n$. Then it is easy to apply (1) to get two more elements c_2, c_3 of P such that each one is properly less than c_1, while $c_1 = c_2 + c_3$. This contradicts the disjunctiveness of P.

Problem 87. *Is there a semigroup algebra A such that $\mathrm{Irr}A < |A|$?*

25.196. $\mathrm{Inc} < \mathrm{h\text{-}cof}$. Recall that the Baumgartner-Komjath algebra works for this in the general case.

Problem 88. *Is the Baumgartner-Komjath algebra a semigroup algebra?*

Problem 89. *Is there a semigroup algebra A such that $\mathrm{Inc}A < \mathrm{h\text{-}cof}A$?*

25.197. $\mathrm{h\text{-}cof} < \mathrm{Card}$. See Chapters 17 and 18 for an example.

25.198. $\pi < \mathrm{Ind}$. See 25.193.
25.199. $\pi\chi < \mathrm{Ind}$. See 25.193
25.200. $\mathrm{Inc} < \chi$. In the general case the Baumgartner-Komjath algebra works. See problem 88.

Problem 90. *Is there a semigroup algebra A such that $\mathrm{Inc}A < \mathrm{Irr}A$?*

25.201. $\mathrm{Irr} < \chi$. See above.

Problem 91. *Is there a semigroup algebra A such that $\mathrm{Irr}A < \chi A$?*

25.202. $|\mathrm{Aut}| < \pi\chi$. In the general case we used a rigid complete BA, but we have no example which is a semigroup algebra.

Problem 92. *Is there a semigroup algebra A such that $|\text{Aut}A| < \pi\chi A$?*

25.203. $|\text{Aut}| < \text{Ind}$. We do not have an example.

Problem 93. *Is there a semigroup algebra A such that $|\text{Aut}A| < \text{Ind}A$?*

25.204. Finally, we mention two problems concerning the number of ideals.

Problem 94. *Is there a semigroup algebra A such that $|\text{Id}A| < |\text{Aut}A|$?*

Problem 95. *Is there a semigroup algebra A such that $|\text{End}A| < |\text{Id}A|$?*

Pseudo-tree algebras. For the diagram, see below. The edges and counterexamples follow by looking at the interval algebra and tree algebra descriptions.

Minimally generated BAs. See below for the diagram. The edges follow from those for interval algebras and superatomic algebras; we just make two comments.

25.205. c < d. For interval algebras a Suslin line gives an example. The difference is small, and the existence of an example with < is connected to the generalized Suslin problem.

25.206. c < s. For superatomic algebras there is an example, but the difference is not large. In fact, the difference is small for minimally generated algebras. For if A has spread at least $(2^{cA})^+$, then it has at least that size, and so by Theorem 10.1 of the BA Handbook, A has an independent subset of that size; so A cannot be minimally generated.

Pseudo-tree algebras

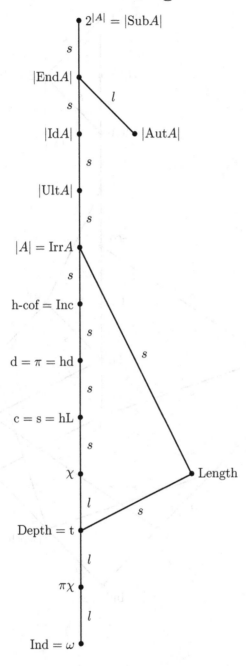

Minimally generated BAs

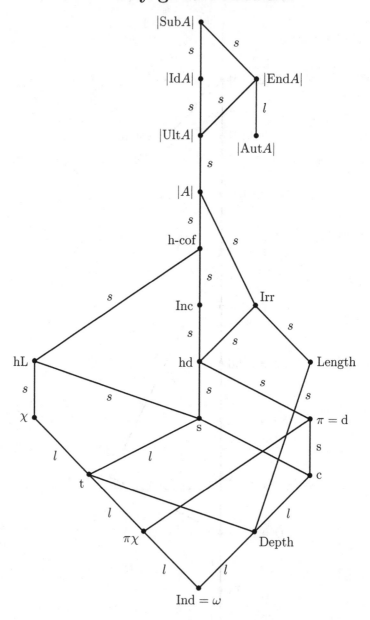

26. Examples

We determine our cardinal functions on the following examples, as much as possible; see also the following table:

1. The finite-cofinite algebra on κ.
2. The free algebra on κ free generators.
3. The interval algebra on the reals.
4. $\mathscr{P}\kappa$.
5. The interval algebra on κ.
6. $\mathscr{P}\omega/\mathrm{Fin}$.
7. The Aleksandroff duplicate of a free algebra.
8. The completion of a free algebra.
9. The countable-cocountable algebra on ω_1.
10. A compact Kunen line.
11. The Baumgartner-Komjath algebra.
12. The Rubin algebra.

We do not have to consider all of our 21 functions for each of them, since usually the determination of some key functions says what the rest are; see the diagrams.

1. The finite-cofinite algebra on κ.

1. $cA = \kappa$.
2. $tA = \omega$. See the beginning of Chapter 12.
3. $|\mathrm{Ult}A| = \kappa$.
4. $|\mathrm{Aut}A| = 2^\kappa$.

2. The free BA on κ free generators.

1. $\mathrm{Length}A = cA = \omega$. See Handbook, Part I, Corollaries 9.17 and 9.18.
2. $dA =$ the smallest cardinal λ such that $\kappa \leq 2^\lambda$; see Corollary 5.7.
3. $\pi\chi A = \kappa$: see Theorem 11.6.
4. $\mathrm{Ind}A = \kappa$.
5. $|\mathrm{Ult}A| = 2^\kappa$.
6. $|\mathrm{Aut}A| = 2^\kappa$.

3. The interval algebra on the reals.

1. $\pi A = \omega$.
2. $\mathrm{Inc}A = 2^\omega$. For example, $\{[r, r+1) : r \in \mathbb{R}\}$ is incomparable.
3. $|\mathrm{End}A| = 2^\omega$; Corollary 21.3.
4. $|\mathrm{Aut}A| = 2^\omega$; this is clear by the above, since it is easy to exhibit 2^ω automorphisms.

4. $\mathscr{P}\kappa$.

1. $cA = \kappa$.
2. $\pi A = \kappa$.
3. $\pi\chi A = \kappa$. An easy argument gives this.
4. $\mathrm{Length}A = \mathrm{Ded}\kappa$. See Chapter 7.
5. $|\mathrm{Ult}A| = 2^{2^{\kappa}}$.
6. $|\mathrm{Aut}A| = 2^{\kappa}$.

5. The interval algebra on κ.

1. $\pi\chi A = \kappa$. See the end of Chapter 11. $\pi\chi A$ is attained if κ is regular, otherwise not.
2. $|\mathrm{Ult}A| = \kappa$. See Theorem 17.10 of Part I of the BA handbook.
3. $|\mathrm{Aut}A| = 2^{\kappa}$. See the end of Chapter 20.

6. $\mathscr{P}\omega/\mathrm{Fin}$

1. $\mathrm{Depth}A \geq \omega_1$. It is consistent that it is ω_1 and $\neg$CH holds. Under MA, it is 2^{ω}. See the end of Chapter 4.
2. $\mathrm{Length}A = 2^{\omega}$.
3. $cA = 2^{\omega}$.
4. $\pi\chi A \geq \mathrm{cf}2^{\omega}$. $\mathrm{Con}(2^{\omega} = \aleph_{\omega_1} + \pi\chi A = \omega_1)$; see van Mill [84], p. 558.
5. $\mathrm{Ind}A = 2^{\omega}$.
6. $|\mathrm{Ult}A| = 2^{2^{\omega}}$.
7. $|\mathrm{Aut}A|$ can consistently be 2^{ω} or $2^{2^{\omega}}$; see van Mill [84], p. 537.

7. The Aleksandroff duplicate of a free BA.

We use notation as in Chapter 14. Thus B is a free BA of size κ and $\mathrm{Dup}B$ is its Aleksandroff duplicate.
1. $c(\mathrm{Dup}B) = 2^{\kappa}$.
2. $\chi\mathrm{Dup}B = \kappa$. See Chapter 14.
3. $\mathrm{Length}(\mathrm{Dup}B) = \omega$. In fact, suppose that $Y \subseteq \mathrm{Dup}B$, Y a chain, $|Y| = \omega_1$. Define $(a, X) \equiv (b, Z)$ iff $(a, X), (b, Z) \in Y$ and $a = b$. Then since B has no uncountable chains, there are only countably many $\equiv$-classes. So there is a class, say K, which has ω_1 elements; say that a is the first member of each ordered pair in K. Then $M \stackrel{\mathrm{def}}{=} \{X : (a, X) \in K\}$ is of size ω_1, is a chain under inclusion, and if $X, Z \in M$ with $X \subseteq Z$, then $Z\backslash X$ is finite. Clearly this is impossible.
4. $\mathrm{Ind}(\mathrm{Dup}B) = \kappa$.
5. $\pi\chi\mathrm{Dup}B = \omega$. (This corrects a mistake in Monk [90].) For, let H be a non-principal ultrafilter on $\mathrm{Ult}\mathrm{Dup}B$. By the description of ultrafilters on $\mathrm{Dup}B$ in Chapter 14, there is an ultrafilter G on B such that

$$H = \{(a, X) : a \in G,\ X \subseteq \mathrm{Ult}B,\ \mathcal{S}a\triangle X \text{ is finite}\}.$$

Let $\langle x_\alpha : \alpha < \kappa\rangle$ be the system of free generators of B (without repetitions). Choose $\varepsilon \in {}^{\kappa}2$ such that $x_\alpha^{\varepsilon\alpha} \in G$ for all $\alpha < \kappa$. For each $n < \omega$ let F_n be an ultrafilter on B such that $x_\alpha^{\varepsilon\alpha} \in F_n$ for all $\alpha \neq n$ and $x_n^{1-\varepsilon n} \in F_n$. We claim that $\{(0, \{F_n\}) : n < \omega\}$ is dense in H. For, let $y \in H$. Without loss of generality y has the form

$$(x_{\alpha_1}^{\varepsilon\alpha_1} \cdot \ldots \cdot x_{\alpha_m}^{\varepsilon\alpha_m}, X), \quad \mathcal{S}(x_{\alpha_1}^{\varepsilon\alpha_1} \cdot \ldots \cdot x_{\alpha_m}^{\varepsilon\alpha_m}) \triangle X \text{ finite.}$$

Choose $n < \omega$ so that $n \neq \alpha_1, \ldots, \alpha_m$ and $F_n \notin \mathcal{S}(x_{\alpha_1}^{\varepsilon\alpha_1} \cdot \ldots \cdot x_{\alpha_m}^{\varepsilon\alpha_m}) \triangle X)$. Since $F_n \in \mathcal{S}(x_{\alpha_1}^{\varepsilon\alpha_1} \cdot \ldots \cdot x_{\alpha_m}^{\varepsilon\alpha_m})$, it follows that $F_n \in X$, as desired.

6. $|\mathrm{Ult}(\mathrm{Dup}(B))| = 2^\kappa$. See the description of ultrafilters in Chapter 14.

7. $|\mathrm{Id}(\mathrm{Dup}(B))| = 2^{2^\kappa}$.

8. In an email message of January 1992, Sabine Koppelberg shows that $\mathrm{Aut}(\mathrm{Dup}A)$ has exactly 2^κ elements, answering Problem 64 in Monk [90]. She also showed that $|\mathrm{End}(\mathrm{Dup}\,A)| = 2^{2^\kappa}$. We give her proofs here.

(a) For any BA A, let

$$\mathrm{Dup}'A = \langle \mathcal{S}[A] \cup \{\{F\} : F \in \mathrm{Ult}A\}\rangle^{\mathscr{P}\mathrm{Ult}\,A}.$$

For A atomless, $\mathrm{Dup}\,A$ is isomorphic to $\mathrm{Dup}'A$; an isomorphism is given by $f(a, X) = X$ for all $(a, X) \in \mathrm{Dup}\,A$, as is easily checked.

(b) Any automorphism of A induces an automorphism of $\mathrm{Dup}\,A$. Namely, if f is an automorphism of A, define $f^+(a, X) = (fa, \{f[F] : F \in X\})$; it is easy to check that f^+ is an automorphism of $\mathrm{Dup}\,A$. Clearly $f^+ \neq g^+$ for distinct f, g.

(c) In $\mathrm{Dup}'A$, $\{F\} = \bigcap_{a\in F} \mathcal{S}a$. Hence if f and g are automorphisms of $\mathrm{Dup}'A$ and $f \upharpoonright \mathcal{S}[A] = g \upharpoonright \mathcal{S}[A]$, then $f = g$.

(d) From (a)–(c) it follows that $|\mathrm{Aut}(\mathrm{Dup}\,A)| = 2^\kappa$ for A free on κ free generators.

(e) Suppose that A is atomless. Let f be a homomorphism from $\mathcal{S}[A]$ into $\mathrm{Dup}'A$ and g a homomorphism from $\mathrm{Finco}\,(\mathrm{Ult}A)$ into $\mathrm{Dup}'A$. Then $f \cup g$ extends to an endomorphism of $\mathrm{Dup}'A$ iff the following condition holds:

(*) If $a \in A$, M is a finite subset of $\mathrm{Ult}A$, and $M \subseteq \mathcal{S}a$, then $gM \subseteq f(\mathcal{S}a)$.

In fact, $\Rightarrow$ is obvious. For $\Leftarrow$, to apply Sikorski's criterion suppose that $\mathcal{S}a \cap N = 0$, where $a \in A$ and $N \in \mathrm{Finco}\,(\mathrm{Ult}A)$. If N is cofinite, then $\mathcal{S}a$ is finite, hence $a = 0$, so $f\mathcal{S}a \cap gN = 0$. If N is finite, then $f\mathcal{S}a \cap gN = 0$ by (*).

(f) Suppose that π is a one-one mapping of $\mathrm{Ult}A$ into $\mathrm{Ult}A$. Then we can define an endomorphism π^+ of $\mathrm{Finco}\,(\mathrm{Ult}A)$ by setting $\pi^+\mathscr{F} = \{\pi F : F \in \mathscr{F}\}$ for $\mathscr{F}$ a finite subset of $\mathrm{Ult}A$, and $\pi^+\mathscr{F} = \mathrm{Ult}A\backslash\pi^+(\mathrm{Ult}A\backslash\mathscr{F})$ for $\mathscr{F}$ a cofinite subset of $\mathrm{Ult}A$. Then for f a homomorphism from $\mathcal{S}[A]$ into $\mathrm{Dup}'A$ the criterion (*) for $f \cup \pi^+$ to extend to an endomorphism of $\mathrm{Dup}'A$ becomes

(**) If $a \in A$ and $F \in \mathcal{S}a$, then $\pi F \in f(\mathcal{S}a)$.

(g) Now suppose that A is free on κ free generators. We show that $\mathrm{Dup}\,A$ has 2^{2^κ} endomorphisms. Now A is isomorphic to $A \oplus A$; let h be an isomorphism of $A \oplus A$ onto A. Let k be the embedding of A onto the first factor of $A \oplus A$, and l the embedding onto the second factor. Set $f = h \circ k$. Thus f is a one-one endomorphism of A, so the dual mapping f^{-1} from $\mathrm{Ult}A$ to $\mathrm{Ult}A$ is onto. Now

($\star$) For each $G \in \mathrm{Ult}A$, the set $\{F : f^{-1}[F] = G\}$ has at least two elements.

In fact, we claim that for any non-zero element b of A the set $f[G] \cup \{hlb\}$ has the finite intersection property. If not, there is an element $a \in G$ such that $fa \cdot hlb = 0$; since $f = h \circ k$, it follows that $ka \cdot lb = 0$, contradiction. This claim being true, if we take $b \in A \backslash \{0, 1\}$, we can get ultrafilters F, F' in A such that $f[G] \cup \{hlb\} \subseteq F$ and $f[G] \cup \{hl(-b)\} \subseteq F'$. Then $F \neq F'$ and $f^{-1}[F] = G = f^{-1}[F']$, as desired in ($\star$).

Now take a function π from $\mathrm{Ult}A$ into $\mathrm{Ult}A$ such that $\pi G \in \{F : f^{-1}[F] = G\}$ for all $G \in \mathrm{Ult}A$. Let $f'[Sa] = S(fa)$ for all $a \in A$. Then f' is a homomorphism from $S[A]$ into $\mathrm{Dup}'A$. If $G \in Sa$, then $\pi G \in \{F : f^{-1}[F] = G\}$, so $f^{-1}[\pi G] = G$, $a \in f^{-1}[\pi G]$, $fa \in \pi G$, and $\pi G \in S(fa) = f'[Sa]$. This means that ($**$) holds for π and f', and so $f' \cup \pi^+$ extends to an endomorphism of $\mathrm{Dup}'A$. Clearly $\pi^+ \neq \sigma^+$ for distinct π, σ, so we have exhibited 2^{2^κ} endomorphisms of $\mathrm{Dup}'A$, as desired.

8. The completion of a free algebra.

Let B be a free algebra of size κ, A its completion.
1. $cA = \omega$.
2. $\mathrm{Length}A = 2^\omega$. In fact, $\geq$ is clear. Suppose that L is a chain of size $(2^\omega)^+$. Using a well-ordering of L and the partition relation $(2^\omega)^+ \rightarrow (\omega_1)^2_\omega$, we get an uncountable well-ordered chain in A, contradiction.
3. dA is the least cardinal λ such that $\kappa \leq 2^\lambda$. For, this is true for B itself by Chapter 5, and an application of Sikorski's extension theorem shows that it is true of A.
4. $\pi\chi A = \kappa$. For, $\leq$ is clear. Suppose that F is an ultrafilter on A and D is a π-base for F. Without loss of generality $D \subseteq B$. Then D is dense in $F \cap B$, so $|D| \geq \kappa$.
5. $\pi A = \kappa$ by the same argument.
6. $|A| = \kappa^\omega$.
7. $|\mathrm{Ult}A| = 2^{\kappa^\omega}$.
8. $|\mathrm{Aut}A| = 2^\kappa$. In fact, any automorphism of A is uniquely determined by its restriction to B, and there are only 2^κ mappings of B into A. On the other hand, there are at least 2^κ automorphisms of A.

9. The countable-cocountable algebra on ω_1.

1. $\mathrm{Depth}A = \omega_1$, by an easy argument.
2. $\mathrm{Length}A = 2^\omega$.
3. $\pi A = \omega_1$.
4. $\mathrm{Ind}A = 2^\omega$.
5. $\pi\chi A = \omega_1$: let F be the ultrafilter of cocountable sets. Suppose that D is dense in F, with $|D| \leq \omega$. Without loss of generality the members of D are singletons. But then $\omega_1 \backslash \bigcup D \in F$, contradiction.
6. $|A| = 2^\omega$.
7. $|\mathrm{Ult}A| = 2^{2^\omega}$.

8. $|\text{Aut}A| = 2^{\omega_1}$.

10. A compact Kunen line.

Recall that this Boolean algebra was constructed using CH.
1. $\text{Irr}A = \omega$. See Chapter 8.
2. $\chi A = \omega_1$. This was proved in the discussion following Theorem 14.3.
3. $|A| = |\text{Ult}A| = \omega_1$. Clear from the construction.
4. $|\text{End}A| = \omega_1$ by 3 and Theorem 21.1.
5. Although we have not been able to determine $\text{Inc}A$, the algebra $A \times A$ has incomparability ω_1, and still has the other important properties of A: $|A \times A| = |\text{Ult}(A \times A)| = \omega_1$, $\chi(A \times A) = \omega_1$, and $\text{Irr}(A \times A) = \omega$. The set $\{(a, -a) : a \in A\}$ is incomparable, showing that $\text{Inc}(A \times A) = \omega_1$. Obviously $|A \times A| = |\text{Ult}(A \times A)| = \omega_1$ and $\chi(A \times A) = \omega_1$. To see that $\text{Irr}(A \times A) = \omega$, note from the Handbook volume 1, example 11.6, that $(A \times A) \oplus (A \times A) \cong (A \oplus A)^4$, and then apply Heindorf's theorem in Chapter 8.

Problem 96. *Determine* $\text{Inc}A$, $|\text{Aut}A|$, $|\text{Id}A|$, *and* $|\text{Sub}A|$ *for the compact Kunen line of Chapter 8.*

This is part of problem 65 in the Monk [90].

11. The Baumgartner, Komjath algebra.

Recall that this BA was constructed using $\diamondsuit$. See Chapter 17.
1. $\text{Inc}A = \omega$.
2. $\text{Length}A = \omega$.
3. $\chi A = \omega_1$.
4. $|\text{Ult}A| = \omega_1$. To see this, first note that each ultrafilter on A is determined by the membership of the elements x_α or their complements. Hence this equality follows from the following fact:

(*) If F and G are ultrafilters on A, $x_\alpha \in F \cap G$, and $F \cap A_{\alpha+1} = G \cap A_{\alpha+1}$, then $F = G$.

In fact, suppose that $\beta \in (\alpha, \omega_1)$. If $x_\beta \in F$, then $x_\alpha \cap x_\beta \in F$; but by construction $x_\alpha \cap x_\beta \in A_{\alpha+1}$, so $x_\alpha \cap x_\beta \in G$ and so $x_\beta \in G$. The same argument works if $\omega \backslash x_\beta \in F$, so $F \subseteq G$ and hence $F = G$.
5. $|\text{End}A| = \omega_1$ by 4 and Theorem 21.1.
6. Since A has a nonzero element a such that $A \restriction a$ is countable, $A \restriction a$ has ω_1 automorphisms, and so the same is true of A itself.

Problem 97. *Determine* $\text{Irr}A$, $|\text{Id}A|$, *and* $|\text{Sub}A|$ *for the Baumgartner, Komjath algebra.*

This is part of problem 66 in the Monk [90].

12. The Rubin algebra.

This algebra was also constructed using $\diamondsuit$.

1. h-cof$A = \omega$; see Chapter 18.
2. Irr$A = \omega$; see Rubin [83].
3. $|A| = \omega_1$. This is clear from the construction in Chapter 18.
4. $|\text{Sub}A| = \omega_1$. See Chapter 23.

We do not know about $|\text{Aut}A|$, although the construction can be changed to make A rigid; see Shelah [91] for more details.

Table of examples

Example	Depth	$\pi\chi$	c	Length	Ind	d	t	π	χ	s	Irr
Fincoκ	ω	ω	κ	ω	ω	κ	ω	κ	κ	κ	κ
Frκ	ω	κ	ω	ω	κ	(1)	κ	κ	κ	κ	κ
Intalg$\mathbb{R}$	ω	ω	ω	2^ω	ω	ω	ω	ω	ω	ω	2^ω
$\mathscr{P}\kappa$	κ	κ	κ	Dedκ	2^κ	κ	2^κ	κ	2^κ	2^κ	2^κ
Intalgκ	κ	κ	κ	κ	ω	κ	κ	κ	κ	κ	κ
$\mathscr{P}\omega/\text{fin}$	(2)	(3)	2^ω	2^ω	2^ω	2^ω	2^ω	2^ω	2^ω	2^ω	2^ω
Dup	ω	ω	2^κ	ω	κ	2^κ	κ	2^κ	κ	2^κ	2^κ
$\overline{\text{Fr}\kappa}$	ω	κ	ω	2^ω	κ^ω	(1)	κ^ω	κ	κ^ω	κ^ω	κ^ω
Cblcoω_1	ω_1	ω_1	ω_1	2^ω	2^ω	ω_1	2^ω	ω_1	2^ω	2^ω	2^ω
CKL	ω	ω	ω	ω	ω	ω	ω	ω	ω_1	ω	ω
BK	ω	ω	ω	ω	ω	ω	ω	ω	ω_1	ω	?
Rubin	ω	ω	ω	ω	ω	ω	ω	ω	ω	ω	ω

Notes:
Dup is the Aleksandroff duplicate of the free BA of size κ.
CKL is the compact Kunen line constructed in chapter 8.
BK is the Baumgartner, Komjath algebra constructed in chapter 17.
Rubin is the algebra constructed in chapter 18.

(1) The least λ such that $\kappa \leq 2^\lambda$.
(2) The depth is $\geq \omega_1$; various possibilities are consistent.
(3) $\geq \text{cf}2^\omega$.

(*Table continued on the next page*)

Table 277

Example	hL	hd	Inc	h-cof	Card	$\|\mathrm{Ult}\|$	$\|\mathrm{Aut}\|$	$\|\mathrm{Id}\|$	$\|\mathrm{End}\|$	$\|\mathrm{Sub}\|$
Fincoκ	κ	κ	κ	κ	κ	κ	2^κ	2^κ	2^κ	2^κ
Frκ	κ	κ	κ	κ	κ	2^κ	2^κ	2^κ	2^κ	2^κ
Intalg$\mathbb{R}$	ω	ω	2^ω	2^ω	2^ω	2^ω	2^ω	2^ω	2^ω	2^{2^ω}
$\mathscr{P}\kappa$	2^κ	2^κ	2^κ	2^κ	2^κ	2^{2^κ}	2^κ	2^{2^κ}	2^{2^κ}	2^{2^κ}
Intalgκ	κ	κ	κ	κ	κ	κ	2^κ	2^κ	2^κ	2^κ
$\mathscr{P}\omega/\mathrm{fin}$	2^ω	2^ω	2^ω	2^ω	2^ω	2^{2^ω}	(1)	2^{2^ω}	2^{2^ω}	2^{2^ω}
Dup	2^κ	2^κ	2^κ	2^κ	2^κ	2^κ	2^κ	2^{2^κ}	2^{2^κ}	2^{2^κ}
$\overline{\mathrm{Fr}\kappa}$	κ^ω	κ^ω	κ^ω	κ^ω	κ^ω	2^{κ^ω}	2^κ	2^{κ^ω}	2^{κ^ω}	2^{κ^ω}
Cblcoω_1	2^ω	2^ω	2^ω	2^ω	2^ω	2^{2^ω}	2^{ω_1}	2^{2^ω}	2^{2^ω}	2^{2^ω}
CKL	ω_1	ω	?	ω_1	ω_1	ω_1	?	?	ω_1	?
BK	ω_1	ω	ω	ω_1	ω_1	ω_1	ω_1	?	ω_1	?
Rubin	ω	ω	ω	ω	ω_1	ω_1	?	ω_1	ω_1	ω_1

Notes:

Dup is the Aleksandroff duplicate of the free BA of size κ.

CKL is the compact Kunen line constructed in chapter 8.

BK is the Baumgartner, Komjath algebra constructed in chapter 17.

Rubin is the algebra constructed in chapter 18.

(1) Consistently 2^ω or 2^{2^ω}.

References

Arhangelskiĭ, A. [78]. *The structure and classification of topological spaces and cardinal invariants.* Russian Mathematical Surveys 33, no. 6, 33–96.

Balcar, B.; Simon, P. [91] *On minimal π-character of points in extremally disconnected compact spaces.* Topol. Appl. 41, no. 1–3, 133–145.

Balcar, B.; Simon, P. [92] *Reaping number and π-character of Boolean algebras.* Discrete Math. 108, no. 1–3, 5–12.

Baumgartner, J. [76]. *Almost disjoint sets, the dense set problem and the partition calculus,* Ann. Math. Logic 9, 401–439.

Baumgartner, J. [80]. *Chains and antichains in $\mathcal{P}\omega$.* J. Symb. Logic 45, 85–92.

Baumgartner, J.; Komjath, P. [81]. *Boolean algebras in which every chain and antichain is countable.* Fund. Math. 111, 125–133.

Baumgartner, J.; Taylor, A.; Wagon, S. [82]. **Structural properties of ideals.** Dissert. Math. 197, 95pp.

Bekkali, M. [91] **Topics in set theory.** Springer-Verlag Lecture Notes in Mathematics 1476, 120pp.

Bekkali, M. [92] *Length in free product.* Abstracts Amer. Math. Soc. 13, no. 3, 336, 92T-06-67.

Bell, M.; Ginsburg, J.; Todorčević, S. [82]. *Countable spread of* $\exp Y$ *and* λY. Topol. Appl. 14, 1–12.

Bonnet, R.; Rubin, M. [92]. *A thin tall Boolean algebra which is isomorphic to each of its uncountable subalgebras.* Preprint.

Bonnet, R.; Shelah, S. [85]. *Narrow Boolean algebras.* Annals Pure Appl. Logic 28,1–12. Publication 210 of Shelah.

Bozeman, K. [91]. *On the relationship between density and weak density in Boolean algebras.* Proc. Amer. Math. Soc. 112, no. 4, 1137–1141.

Brenner, G. [82]. **Tree algebras.** Ph. D. thesis, University of Colorado.

Burris, S. [75]. *Boolean powers.* Alg. Univ. 5, no. 3, 341–360.

Burris, S.; Sankappanavar, H. [81]. **A course in universal algebra.** Springer-Verlag, xvi+276pp.

Chang, C. C.; Keisler, H. J. [73]. **Model Theory.** North-Holland, 550pp.

Comfort, W. W. [71]. *A survey of cardinal invariants.* Gen. Topol. Appl. 1, 163–199.

Comfort, W. W.; Negrepontis, S. [74]. **The theory of ultrafilters.** Springer-Verlag, x+482 pp.

Comfort, W. W.; Negrepontis, S. [82]. **Chain conditions in topology.** Cambridge University Press, xi+300pp.

Cramer, T. [74] *Extensions of free Boolean algebras.* J. London Math. Soc. 8, 226–230.

Cummings, J.; Shelah, S. [95] *A model in which every Boolean algebra has many subalgebras.* Preprint. Publ. 530.

Day, G. W. [67] *Superatomic Boolean algebras.* Pacific J. Math. 23, no. 2, 479–489.

Donder, H. [88] *Regularity of ultrafilters and the core model.* Israel J. Math. 63, 289–322.

van Douwen, E. K. [84] *The integers and topology.* In **Handbook of set-theoretic topology**, North-Holland, 111–167.

van Douwen, E. K. [89] *Cardinal functions on Boolean spaces.* In **Handbook of Boolean algebras**, North-Holland, 417–467.

Dow, A.; Monk, J. D. [94] *Depth, π-character, and tightness for superatomic Boolean algebras.* Preprint.

Dwinger, Ph. [82] *Completeness of Boolean powers of Boolean algebras.* Universal algebra, Colloq. Math. Soc. János Bolyai 29, 209–217.

Engelking, R. [77] **General topology.** Polish Scientific Publishers, Monografie Matematyczne, v. 60, 626pp.

Erdös, P.; Hajnal, A.; Máté, A., Rado, R. [84]. **Combinatorial set theory: partition relations for cardinals.** Adadémiai Kiadó, 347pp.

Fedorchuk, V. [75] *On the cardinality of hereditarily separable compact Hausdorff spaces.* (Russian) Dokl. Akad. Nauk SSSR 222, no. 2, 651–655. English translation: Sov. Math. Dokl. 16, 651–655.

Foreman, M.; Laver, R. [88] *Some downwards transfer properties for $\aleph_2$.* Adv. in Math. 67, no. 2, 230–238.

Gardner, R.; Pfeffer, W. [84] *Borel measures.* **Handbook of set-theoretic topology**, 961–1044, North-Holland.

Grätzer, G.; Lakser, H. [69] *Chain conditions in the distributive free product of lattices.* Trans. Amer. Math. Soc. 144, 301–312.

Hajnal, A.; Juhász, I. [71] *A consequence of Martin's axiom.* Indag. Math. 33, 457–463.

Hechler, S. [72] *Short complete nested sequences in $\beta N \backslash N$ and small almost-disjoint families.* General Topol. Appl. 2, 139–149.

Heindorf, L. [87] *A decidability proof for the theory of countable Boolean algebras in the language with quantification over ideals.* 5th Easter Conference on Model Theory, Sem. Berichte 93, Humboldt Univ., Sekt. Math., 34–45.

Heindorf, L. [89a] *A note on irredundant sets.* Alg. Univ. 26, 216–221.

Heindorf, L. [89b] *Boolean semigroup rings and exponentials of compact zero-dimensional spaces.* Fund. Math. 135, no. 1, 37–47.

Heindorf, L. [90] *Moderate families in Boolean algebras.* Annals Pure Appl. Logic 57, 217–250.

Heindorf, L. [91] *Dimensions of Boolean algebras.* Preprint.

Heindorf, L. [92] *An embedding of free Boolean algebras.* Preprint.

Hodel, R. [84]. *Cardinal functions I.* In **Handbook of set-theoretic topology**, North-Holland, 1–61.

Jech, T. [78] **Set theory.** Academic Press, xi+621pp.

Jech, T. [86] **Multiple forcing.** Cambridge University Press, viii+136pp.

Juhász, I. [71]. **Cardinal functions in topology.** Math. Centre Tracts 34, Amsterdam, 150pp. (This book contains some material not found in its revised version, which follows:)

Juhász, I. [80]. **Cardinal functions in topology – ten years later.** Math. Centre Tracts 123, Amsterdam, 160pp.

Juhász, I. [84]. *Cardinal functions II.* In **Handbook of set-theoretic topology**, North-Holland, 63–109.

Juhász, I. [93] *On the weight-spectrum of a compact space.* Preprint.

Juhász, I., Kunen, K, Rudin, M. E. [76] *Two more hereditarily separable non-Lindelöf spaces*, Can. J. Math 28, 998–1005.

Juhász, I.; Szentmiklóssy, Z. [92] *Convergent free sequences in compact spaces.* Proc. Amer. Math. Soc. 116, no. 4, 1153–1160.

Just, W. [88] *Remark on the altitude of Boolean algebras.* Alg. Univ. 25, 283–289.

Just, W.; Koszmider, P. [91] *Remarks on cofinalities and homomorphism types of Boolean algebras.* Alg. Univ. 28, no. 1, 138–149.

Just, W.; Weese, M. [91] *On independent subsets of Boolean algebras.* Alg. Univ. 30, no. 4, 521–525.

Keisler, H. J.; Prikry, K. [74] *A result concerning cardinalities of ultraproducts.* J. Symb. Logic 39, no. 1, 43–48.

Koppelberg, S. [77] *Boolean algebras as unions of chains of subalgebras*, Alg. Univ. 7, 195–203.

Koppelberg, S. [89a] **General theory of Boolean algebras**, Part I of **Handbook of Boolean algebras**. North-Holland, 312pp.

Koppelberg, S. [89b] *Minimally generated Boolean algebras.* Order, 5, 393–406.

Koppelberg, S. [92] Handwritten notes.

Koppelberg, S.; Monk, J. D. [92] *Pseudo-trees and Boolean algebras.* Order 8, 359–374.

Koppelberg, S.; Shelah, S. [93] *Densities of ultraproducts of Boolean algebras.* Preprint. Publ. 415.

Kunen, K. [75] *Seminar notes: large compact S-spaces.* Unpublished handwritten notes.

Kunen, K. [78] *Saturated ideals.* J. Symb. Logic 43, 65–76.

Kunen, K. [80] **Set Theory**, North-Holland, xvi+313pp.

Kuratowski, K. [58] **Topologie**, vol. 1, fourth edition, 494pp.

Kurepa, G. [35] *Ensembles linéaires et une classe de tableaux ramifiés (tableaux ramifiés de M. Aronszajn)* Putl. Math. Univ. Belgrade 6, 129–160.

Kurepa, G. [57] *Partitive sets and ordered chains.* "Rad" de l'Acad. Yougoslave 302, 197–235.

Kurepa, G. [62] *The Cartesian multiplication and the cellularity number.* Publ. Inst. Math. (Beograd) (N.S.) 2 (16), 121–139.

Magidor, M.; Shelah, S. [91] *On the length of ultraproducts of Boolean algebras.*

Malyhin, V. [72] *On tightness and Souslin number in* $\exp X$ *and in a product of spaces*, Sov. Math. Dok. 13, 496–499.

Marjanović, M. M. [72] *Exponentially complete spaces III.* Publ. Inst. Math. (Belgrade) 14 (28), 97–109.

McKenzie, R.; Monk, J. D. [75] *On automorphism groups of Boolean algebras.* Colloq. Math. Soc. J. Bolyai, 951–988.

McKenzie, R.; Monk, J. D. [82] *Chains in Boolean algebras.* Ann. Math. Logic 22, 137–175.

van Mill, J. [84] *An introduction to* $\beta\omega$. In **Handbook of set-theoretic topology**, North-Holland, 503–567.

Milner, E.; Pouzet, M. [86] *On the width of ordered sets and Boolean algebras.* Algebra Universalis 23, 242–253.

Monk, J. D. [83] *Independence in Boolean algebras.* Per. Math. Hung. 14, 269–308.

Monk, J. D. [84]. *Cardinal functions on Boolean algebras.* In Orders: Descriptions and Roles, Annals of Discrete Mathematics 23, 9–37.

Monk, J. D. [89a] *Endomorphisms of Boolean algebras.* In **Handbook of Boolean algebras**, North-Holland, 491–516.

Monk, J. D. [89b] *Appendix on set theory.* In **Handbook of Boolean algebras**, North-Holland, 1213–1233.

Monk, J. D. [90] **Cardinal functions on Boolean algebras.** 152pp. Birkhäuser Verlag.

Nyikos, P. [90] *Dichotomies in compact spaces.* Preprint.

Parovichenko, I. I. *The branching hypothesis and the correlation between local weight and power to topological spaces.* (Russian), Dokl. Akad. Nauk SSSR 174, no. 1; English translation: Soviet Math. Dokl. 8, no. 3, 589–591.

Purisch, S. [94] *Solution of problem H10.* Topology Proceedings 17, 412–413.

Peterson, D. [93] **Cardinal functions on ultraproducts, and reaping numbers.** Ph. D. Thesis, Univ. of Colo.

Peterson, D. [95] *Cardinal functions on ultraproducts of Boolean algebras.* Preprint.

Peterson, D. [95] *Reaping numbers and operations on Boolean algebras.*

Quackenbush, R. W. [72] *Free products of bounded distributive lattices.* Alg. Univ. 2/3, 793–794.

Rosenstein, J. [82] **Linear Orderings**, Acad. Press, xvi + 487pp.

Rosłanowski, A.; Shelah, S. [94] *F-99: Notes on cardinal invariants and ultraproducts of Boolean algebras.* Preprint. Publ. 534.

Rubin, M. [83] *A Boolean algebra with few subalgebras, interval Boolean algebras, and retractiveness.* Trans. Amer. Math. Soc. 278, 65–89.

Shapiro, L. [76a] *The space of closed subsets of $D^{\aleph_2}$ is not a dyadic bicompact.* (Russian) Dokl. Akad. Nauk SSSR 228, no. 6; English translation: Soviet Math. Dokl. 17, no. 3, 937–941.

Shapiro, L. [76b] *On spaces of closed subsets of bicompacts.* (Russian) Dokl. Akad. Nauk SSSR 231, no. 2; English translation: Soviet Math. Dokl. 17, no. 6, 1567–1571.

Shelah, S. [79] *Boolean algebras with few endomorphisms.* Proc. Amer. Math. Soc. 74, 135–142. Publ. 89.

Shelah, S. [80] *Remarks on Boolean algebras.* Alg. Univ. 11, 77–89. Publ. 92.

Shelah, S. [83] *Constructions of many complicated uncountable structures and Boolean algebras.* Israel J. Math. 45, 100–146. Publ. 136.

Shelah, S. [86b] *Remarks on the number of ideals of Boolean algebras and open sets of a topology.* In **Around classification theory of models**, Lecture Notes in Math. 1182, 151–187. Publ. 233.

Shelah, S. [87] Handwritten notes.

Shelah, S. [88a] *Successors of singulars, cofinalities of reduced products of cardinals and productivity of chain conditions.* Israel J. Math. 62, no. 2, 213–256. Publ. 282.

Shelah, S. [88b] *On successors of singulars.* Abstracts Amer. Math. Soc. 9, no. 6, p. 500. (no. 88T-03-242)

Shelah, S. [88e] *Was Sierpinski right? I* Isr. J. Math. 62, 355–380. Publ. 276.

Shelah, S. [89] *Notes in set theory.* Abstracts Amer. Math. Soc. 10, no. 4, 89T-03-125, p. 302.

Shelah, S. [90] *Products of regular cardinals and cardinal invariants of products of Boolean algebras.* Israel J. Math. 70, no. 2, 129–187. Publ. 345.

Shelah, S. [91a] e-mail messages of December 1990 and various dates in 1991.

Shelah, S. [91b] *Strong negative partition relations below the continuum.* Acta Math. Hung. 58, no. 1–2, 95–100. Publ. 327.

Shelah, S. [92a] *Factor = quotient, uncountable Boolean algebras, number of endomorphism and width.* Math. Japonica 37, no. 1, 1–19. Publ. 397.

Shelah, S. [92b] *On Monk's questions.* Preprint. Publ. 479.

Shelah, S. [94a] $\aleph_{\omega+1}$ *has a Jónsson algebra.* In **Cardinal Arithmetic**, Chapter 2, Oxford University Press. Publ. 355.

Shelah, S. [94b] *There are Jónsson algebras in many inaccessible cardinals.* In **Cardinal Arithmetic**, Chapter 3, Oxford University Press. Publ. 365.

Shelah, S. [94c] *Cellularity of free products of Boolean algebras (or topologies).* Preprint. Publ. 575

Shelah, S. [94d] *The number of independent elements in the product of interval Boolean algebras.* Mathematica Japon. 39, 1–5. Publ. 503.

Shelah, S. [94e] **Cardinal arithmetic.** Oxford Univ. Press, 481pp.

Shelah, S. [94f] *Further on coloring.* In preparation. Publ. 535.

Shelah, S. [94g] *σ-entangled linear orders and narrowness of products of Boolean algebras.* Preprint. Publ. 462.

Shelah, S. [95a] *The pcf theorem revisited.* Preprint. Publ. 506.

Shelah, S. [95b] *Colouring and non-productivity of $\aleph_2$-c.c.* Publ. 572. Preprint.

Shelah, S.; Soukup, L. [89] *Some remarks on a question of J. D. Monk.* Preprint. Publ. 376.

Sirota, S. *Spectral representation of spaces of closed subsets of bicompacta.* (Russian) Dokl. Akad. Nauk SSSR 181, no. 5; English translation: Soviet Math. Dokl. 9, no. 4, 997–1000.

Solovay, R.; Tennenbaum, S. [71] *Iterated Cohen extensions and Souslin's problem.* Ann. of Math. 94, 201–245.

Takahashi, M. [88] *Completeness of Boolean powers of Boolean algebras.* J. Math. Soc. Japan 40, no. 3, 445–456.

Todorčević, S. [83] *Forcing positive partition relations,* Trans. Amer. Math. Soc. 280, 703–720.

Todorčević, S. [85] *Remarks on chain conditions in products.* Compos. Math. 55, no. 3, 295–302.

Todorčević, S. [86] *Remarks on cellularity in products.* Compos. Math. 57, 357–372.

Todorčević, S. [87a] *On the cellularity of Boolean algebras.* Handwritten notes.

Todorčević. S. [87b] *Partitioning pairs of countable ordinals.* Acta math. 159, 261–294.

Todorčević, S. [89] **Partition problems in general topology.** Contemporary Mathematics, v. 84, Amer. Math. Soc., xii+116pp.

Todorčević, S. [90a] *Free sequences.* Topol. Appl. 35, 235–238.

Todorčević, S. [90b] *Irredundant sets in Boolean algebras.* Preprint.

Weese, M. [80] *A new product for Boolean algebras and a conjecture of Feiner.* Wiss. Z. Humboldt-Univ. Berlin Math.-Natur. Reihe 29, 441–443.

Weiss, W. [84] *Versions of Martin's axiom.* In **Handbook of General Topology,** 827–886, North-Holland.

Index of problems

Problem 1. *Is it true that for every singular cardinal κ there exist BAs A and B such that $cA = \kappa$, $cf\kappa \leq cB < \kappa$, and $c(A \oplus B) > \kappa$?* Page 46.

Problem 2. *Is it consistent that there is a an infinite set I, a system $\langle A_i : i \in I \rangle$ of infinite BAs, and an ultrafilter F on I such that $c\left(\prod_{i \in I} A_i/F\right) < \left|\prod_{i \in I} cA_i/F\right|$?* Page 62.

Problem 3. *Give a purely cardinal number characterization of c_{Sr}.* Page 75.

Problem 4. *Is there a BA A with $c_{Sr}A = \{(\omega, \omega), (\omega, \omega_1), (\omega_1, \omega_1), (\omega, \omega_2), (\omega_2, \omega_2)\}$? Equivalently, is there a BA A such that $|A| = \omega_2 = cA$, A has a ccc subalgebra of power ω_2, and every subalgebra of A of size ω_2 either has cellularity ω or ω_2?* Page 77.

Problem 5. *Can one construct in ZFC BAs with c_{Sr} equal to the following relations?*
(i) $\{(\omega, \omega), (\omega, \omega_1), (\omega_1, \omega_1), (\omega_1, \omega_2)\}$.
(ii) $\{(\omega, \omega), (\omega, \omega_1), (\omega_1, \omega_1), (\omega_1, \omega_2), (\omega_2, \omega_2)\}$. Page 77.

Problem 6. *Describe in cardinal number terms the relation c_{Hr}.* Page 79.

Problem 7. *Can one prove in ZFC that BAs with the following relations c_{Hr} exist?*
(i) $\{(\omega, \omega), (\omega, \omega_1), (\omega_1, \omega_1), (\omega_1, \omega_2)\}$.
(ii) $\{(\omega, \omega), (\omega, \omega_1), (\omega_1, \omega_1), (\omega_1, \omega_2), (\omega_2, \omega_2)\}$. Page 79.

Problem 8. *Is it consistent that BAs with the following relations c_{Hr} exist?*
(i) $\{(\omega, \omega_1), (\omega_1, \omega_1), (\omega_2, \omega_2)\}$.
(ii) $\{(\omega, \omega_1), (\omega_1, \omega_1), (\omega_1, \omega_2)\}$.
(iii) $\{(\omega, \omega_1), (\omega_1, \omega_1), (\omega_1, \omega_2), (\omega_2, \omega_2)\}$.
(iv) $\{(\omega, \omega_1), (\omega_1, \omega_1), (\omega, \omega_2), (\omega_2, \omega_2)\}$. Page 79.

Problem 9. *Describe cellularity for pseudo-tree algebras.* Page 85.

Problem 10. *Is it true that for every infinite BA A there is a cardinal κ such that if B and C are extensions of A with depth at least κ then $\mathrm{Depth}(B \oplus_A C) = \max(\mathrm{Depth}B, \mathrm{Depth}C)$?* Page 90.

Problem 11. *Is it true that for every infinite BA A there exist extensions B and C of A and an infinite cardinal κ such that B and C have no chains of order type κ but $B \oplus_A C$ does?* Page 90.

Problem 12. *Is an example with $\mathrm{Depth}\left(\prod_{i \in I} A_i/F\right) > \left|\prod_{i \in I} \mathrm{Depth}A_i/F\right|$ possible in ZFC?* Page 92.

Problem 13. *Is $tB \in \mathrm{Depth}_{Hs}B$ for every infinite BA B?* Page 102.

Problem 14. *Are there an infinite cardinal κ and a BA A such that $(\kappa, (2^\kappa)^+) \in \mathrm{Depth}_{Sr}A$, while $(\omega, (2^\kappa)^+) \notin \mathrm{Depth}_{Sr}A$?* Page 102.

Problem 15. *Characterize the relation* $\mathrm{Depth}_{\mathrm{Sr}}$. *Page 102.*

Problem 16. *Are there an infinite cardinal* κ *and a BA* A *such that* $(\kappa, (2^\kappa)^+) \in \mathrm{Depth}_{\mathrm{Hr}} A$, *while* $(\omega, (2^\kappa)^+) \notin \mathrm{Depth}_{\mathrm{Hr}} A$? *Page 103.*

Problem 17. *Characterize the relation* $\mathrm{Depth}_{\mathrm{Hr}}$. *Page 103.*

Problem 18. *Is it true that* $[\omega, \mathrm{hd} A) \subseteq \mathrm{d}_{\mathrm{Hs}} A$ *for every infinite BA* A? *Page 112.*

Problem 19. *Completely describe* d_{Hs}. *Page 112.*

Problem 20. *Can one find in ZFC a BA* A *such that* $\pi_{\mathrm{S}+} A$ *is not attained?* *Page 122.*

Problem 21. *Is it true that for every infinite BA* A *we have*

$$\pi_{\mathrm{Hs}} A = \begin{cases} [\omega, \mathrm{hd} A], & \text{if } \mathrm{hd} A \text{ is attained,} \\ [\omega, \mathrm{hd} A), & \text{otherwise?} \end{cases}$$

Page 123.

Problem 22. *Can one prove in ZFC that there exist a system* $\langle A_i : i \in I \rangle$ *of infinite BAs,* I *infinite, and an ultrafilter* F *on* I *such that* $\mathrm{Length}\left(\prod_{i \in I} A_i / F\right) > \left|\prod_{i \in I} \mathrm{Length} A_i / F\right|$? *Page 127.*

Problem 23. *Is always* $\mathrm{Length}_{\mathrm{h}-} A = \omega$? *Page 132.*

Problem 24. *Is* $\mathrm{Irr}(A \times B) = \max\{\mathrm{Irr} A, \mathrm{Irr} B\}$? *Page 133. Page 133.*

Problem 25. *Is there an example of a system* $\langle A_i : i \in I \rangle$ *of infinite BAs, with* I *infinite, and a uniform ultrafilter* F *on* I *such that* $\mathrm{Irr}\left(\prod_{i \in I} A_i / F\right) < \left|\prod_{i \in I} \mathrm{Irr} A_i / F\right|$? *Page 134.*

Problem 26. *Is there an example of a system* $\langle A_i : i \in I \rangle$ *of infinite BAs, with* I *infinite, and a uniform ultrafilter* F *on* I *such that* $\mathrm{Irr}\left(\prod_{i \in I} A_i / F\right) > \left|\prod_{i \in I} \mathrm{Irr} A_i / F\right|$? *Page 134.*

Problem 27. *Is it true that* $\mathrm{Irr} A = \mathrm{s}(A \oplus A)$ *for every infinite BA* A? *Page 138.*

Problem 28. *Can one construct in ZFC a BA* A *such that* $\mathrm{Irr} A < |A|$? *Page 144.*

Problem 29. *Can one construct in ZFC a BA* A *with the property that* $\mathrm{Ind}_{\mathrm{H}-} A < \mathrm{Card}_{\mathrm{H}-} A$? *Page 151.*

Problem 30. *Is* $\mathrm{Ind}_{\mathrm{h}+} A = \mathrm{Card}_{\mathrm{h}+} A$ *for every infinite BA* A? *Page 151.*

Problem 31. *Is* $\mathrm{Ind}_{\mathrm{h}-} A = \mathrm{Ind}_{\mathrm{H}-} A$ *for every infinite BA* A? *Page 151.*

Problem 32. *Assume that* $\rho < \nu < \kappa \leq 2^\rho < \lambda \leq 2^\nu$ *with* κ *and* λ *regular. Is there a* κ-*cc BA* A *of power* λ *with no independent subset of power* λ? *Page 151.*

Problem 33. *Can one prove the following in ZFC? Suppose that* $\mathrm{cf}\mu < \kappa < \mu < \lambda \leq \mu^{\mathrm{cf}\mu} = \mu^{<\kappa}$ *and* $\forall \rho < \mu(\rho^{<\kappa} < \mu)$. *Then there is a BA of power* λ *satisfying the* κ-*cc with no independent subset of power* λ. *Page 152.*

Problem 34. *Suppose that κ is uncountable and weakly inaccessible, $2^\nu < \lambda$ for all $\nu < \kappa$, $2^{<\kappa} = \lambda$, and λ is singular. Is there a κ-cc BA of power λ with no independent subset of power λ?* Page 152.

Problem 35. *For all $n \in \omega$ let A_n be the free BA on $\beth_n$ free generators. Does $\prod_{n \in \omega} A_n$ have free caliber $\beth_\omega^+$?* Page 152.

Problem 36. *Let A be free on a set of size $\beth_{\omega+1}$. Is $\beth_{\omega+1} \in \overline{\text{Freecal}A}$?* Page 152.

Problem 37. *Is there for every μ a complete BA A of power 2^μ such that $\text{Freecal}A = 0$?* Page 152.

Problem 38. *If K is a set of regular cardinals with $\mu = \min K$ and $\psi = \sup K$, and if K satisfies $(1)(i)$–(iii), is there a BA A such that K is the set of regular members of $\text{Freecal}A$?* Page 153.

Problem 39. *Can one prove the following in ZFC? For every $m \in \omega$ with $m \geq 2$ there is an interval algebra having a subset P of size ω_1 such that for all $Q \in [P]^{\omega_1}$, Q has m pairwise comparable elements and also m independent elements.* Page 153.

Problem 40. *Can one show in ZFC that $\pi A = \pi\chi A$ for A complete?* Page 161.

Problem 41. *Does attainment of tightness imply attainment in the free sequence sense?* Page 167.

Problem 42. *Does attainment of tightness imply attainment in the $\pi\chi_{H+}$ sense?* Page 167.

Problem 43. *Does attainment of tightness in the $\pi\chi_{H+}$ sense imply attainment in the sense of the definition?* Page 167.

Problem 44. *Is the following true? Let κ and λ be infinite cardinals, with λ regular. Then the following conditions are equivalent:*
(i) $\text{cf}\kappa = \lambda$.
(ii) There is a strictly increasing sequence $\langle A_\alpha : \alpha < \lambda \rangle$ of Boolean algebras each having no ultrafilter with tightness κ such that $\bigcup_{\alpha<\lambda} A_\alpha$ has an ultrafilter with tightness κ. page 171.

Problem 45. *Is there a superatomic BA A such that $\text{t}A = (2^\omega)^+$ and $\text{Depth}A = \omega$?* Page 174.

Problem 46. *Can one construct an example with $\text{s}\left(\prod_{i \in I} A_i/F\right) > \left|\prod_{i \in I} \text{s}A_i/F\right|$ in ZFC?* Page 176.

Problem 47. *Is an example with $\text{s}\left(\prod_{i \in I} A_i/F\right) < \left|\prod_{i \in I} \text{s}A_i/F\right|$ consistent?* Page 176.

Problem 48. *Is it consistent that there exist a system $\langle A_i : i \in I \rangle$ of infinite BAs with I infinite, and an ultrafilter F such that $\chi\left(\prod_{i \in I} A_i/F\right) < \left|\prod_{i \in I} \chi A_i/F\right|$?* Page 183.

Problem 49. *Can one construct in ZFC a BA A such that $\text{s}A < \chi A$?* Page 186.

Problem 50. *Describe the implications between attainment of* hL *as defined, in the ideal-generation sense, and in the right-separated sense.* Page 191.

Problem 51. *Can one construct in ZFC an example with* hL $\left(\prod_{i \in I} A_i/F\right) > \left|\prod_{i \in I} \mathrm{hL}A_i/F\right|$? Page 192. We do not know whether $<$ is possible:

Problem 52. *Can one have* hL $\left(\prod_{i \in I} A_i/F\right) < \left|\prod_{i \in I} \mathrm{hL}A_i/F\right|$ *for some system of BAs (consistently)?* Page 192.

Problem 53. *Is there an example in ZFC of a BA* A *such that* hL$A < $ dA? Page 194.

Problem 54. *Describe completely the attainment relations for the equivalent definitions of* hd. Page 197.

Problem 55. *Can one get an example with* hd $\left(\prod_{i \in I} A_i/F\right) > \left|\prod_{i \in I} \mathrm{hd}A_i/F\right|$ *in ZFC?* Page 197.

Problem 56. *Is an example with* hd $\left(\prod_{i \in I} A_i/F\right) < \left|\prod_{i \in I} \mathrm{hd}A_i/F\right|$ *consistent?* Page 197.

Problem 57. *Can one construct in ZFC a BA* A *such that* s$A < $ hd$A$? Page 216.

Problem 58. *Can one construct in ZFC a BA* A *such that* hd$A < \chi A$? Page 216.

Problem 59. *Do there exist in ZFC a system* $\langle A_i : i \in I \rangle$ *of infinite BAs, I infinite, and a regular ultrafilter F on I such that* Inc $\left(\prod_{i \in I} A_i/F\right) > \left|\prod_{i \in I} \mathrm{Inc}A_i/F\right|$? Page 221.

Problem 60. *Is an example with* Inc $\left(\prod_{i \in I} A_i/F\right) < \left|\prod_{i \in I} \mathrm{Inc}A_i/F\right|$ *consistent?* Page 221.

Problem 61. *Can one construct in ZFC a BA* A *such that* Inc$A < \chi A$? Page 225.

Problem 62. *Is it consistent to have an example with* h-cof $\left(\prod_{i \in I} A_i/F\right) < \left|\prod_{i \in I} \text{h-cof}A_i/F\right|$? Page 226.

Problem 63. *Can one construct in ZFC a BA* A *with the property that* Inc$A < $ h-cof$A$? Page 227.

Problem 64. *Can one construct in ZFC a BA* A *such that* $|\mathrm{Id}A| < |\mathrm{Aut}A|$? Page 238.

Problem 65. *Can one construct in ZFC a BA* A *such that* Irr$A < \chi A$? Page 251.

Problem 66. *Is there an example in ZFC of a BA* A *such that* Irr$A < $ Inc$A$? Page 251.

Problem 67. *Is there an example in ZFC of a BA* A *such that* Irr$A < $ h-cof$A$? Page 252.

Problem 68. *Is there an example in ZFC of a BA* A *such that* $|\mathrm{Id}A| < |\mathrm{End}A|$? page 252.

Problem 69. *Can one construct in ZFC an interval algebra A such that $|\mathrm{Id}A| < |\mathrm{End}A|$? Page 254.*

Problem 70. *Can one construct in ZFC an interval algebra A such that $|\mathrm{Id}A| < |\mathrm{Aut}A|$? Page 254.*

Problem 71. *Is there a tree algebra A such that $|\mathrm{End}A| < 2^{|A|}$? Page 256.*

Problem 72. *Is there an example in ZFC of a superatomic BA A such that $\mathrm{s}A < \mathrm{Inc}A$? Page 259.*

Problem 73. *Is there a superatomic BA A such that $\mathrm{s}A < \mathrm{Irr}A$? Page 259.*

Problem 74. *Can one construct in ZFC a superatomic algebra A with the property that $\mathrm{Inc}A < |A|$? Page 259.*

Problem 75. *Can one construct in ZFC a superatomic algebra A with the property that $\mathrm{Irr}A < |A|$? Page 259.*

Problem 76. *Can one construct in ZFC a superatomic BA A such that $|\mathrm{Aut}A| < |\mathrm{End}A|$? Page 259.*

Problem 77. *Can one have $|\mathrm{Id}A| < |\mathrm{Sub}A|$ in a superatomic BA? Page 260.*

Problem 78. *Is there, under any set-theoretic assumptions, a superatomic BA A such that $\mathrm{Inc}A < \mathrm{Irr}A$? Page 260.*

Problem 79. *Is there, under any set-theoretic assumptions, a superatomic BA A such that $\mathrm{Irr}A < \mathrm{Inc}A$? Page 260.*

Problem 80. *Is there in ZFC a superatomic BA A with $|\mathrm{Aut}A| < |A|$? Page 260.*

Problem 81. *Under any set-theoretic assumptions, is there a superatomic BA A such that $|\mathrm{Aut}A| < \mathrm{t}A$? Page 260.*

Problem 82. *Can one construct in ZFC an atomic BA A such that $\mathrm{s}A < \mathrm{hL}A$? Page 262.*

Problem 83. *Can one show in ZFC that there an atomic BA A such that $\mathrm{s}A < \mathrm{hd}A$? Page 262.*

Problem 84. *Is the Kunen line a semigroup algebra? Page 267.*

Problem 85. *Is there a semigroup algebra A such that $\mathrm{s}A < \mathrm{hL}A$? Page 267.*

Problem 86. *Is the Todorčević algebra of Chapter 8 a semigroup algebra? Page 267.*

Problem 87. *Is there a semigroup algebra A such that $\mathrm{Irr}A < |A|$? Page 267.*

Problem 88. *Is the Baumgartner-Komjath algebra a semigroup algebra? Page 267.*

Problem 89. *Is there a semigroup algebra A such that $\mathrm{Inc}A < \mathrm{h\text{-}cof}A$? Page 267.*

Problem 90. *Is there a semigroup algebra A such that* $\mathrm{Inc}A < \mathrm{Irr}A$*?* Page 267.

Problem 91. *Is there a semigroup algebra A such that* $\mathrm{Irr}A < \chi A$*?* Page 267.

Problem 92. *Is there a semigroup algebra A such that* $|\mathrm{Aut}A| < \pi\chi A$*?* Page 268.

Problem 93. *Is there a semigroup algebra A such that* $|\mathrm{Aut}A| < \mathrm{Ind}A$*?* Page 268.

Problem 94. *Is there a semigroup algebra A such that* $|\mathrm{Id}A| < |\mathrm{Aut}A|$*?* Page 268.

Problem 95. *Is there a semigroup algebra A such that* $|\mathrm{End}A| < |\mathrm{Id}A|$*?* Page 268.

Problem 96. *Determine* $\mathrm{Inc}A$, $|\mathrm{Aut}A|$, $|\mathrm{Id}A|$, *and* $|\mathrm{Sub}A|$ *for the compact Kunen line of Chapter 8.* Page 275.

Problem 97. *Determine* $\mathrm{Irr}A$, $|\mathrm{Id}A|$, *and* $|\mathrm{Sub}A|$ *for the Baumgartner, Komjath algebra.* Page 275.

Index of symbols

Index of names and words